W9-ARW-754

Process Quality Control

Other McGraw-Hill Quality Control Books of Interest

Process Quality Control

Troubleshooting and Interpretation of Data

Ellis R. Ott
Late Professor Emeritus
The State University of New Jersey, Rutgers

Edward G. Schilling
Paul A. Miller Distinguished Professor
Rochester Institute of Technology

Second Edition

McGraw-Hill Publishing Company

New York St. Louis San Francisco Auckland Bogotá
Caracas Hamburg Lisbon London Madrid Mexico
Milan Montreal New Delhi Oklahoma City
Paris San Juan São Paulo Singapore
Sydney Tokyo Toronto

Library of Congress Cataloging-in-Publication Data

Ott, Ellis R. (Ellis Raymond), date.
Process quality control : troubleshooting and interpretation of data / Ellis R. Ott, Edward G. Schilling. — 2nd ed.
p. cm.
ISBN 0-07-047924-0
1. Quality control—Statistical methods. 2. Process control—Statistical methods. I. Schilling, Edward G., date.
II. Title.
TS156.086 1990
670.42′7—dc20 89-28280
CIP

1 2 3 4 5 6 7 8 9 0 DOC/DOC 9 5 4 3 2 1 0

ISBN 0-07-047924-0

The sponsoring editor for this book was Robert W. Hauserman, the editing supervisor was Gretlyn Cline, the designer was Naomi Auerbach, and the production supervisor was Dianne L. Walber. It was set in Century Schoolbook. It was composed by the McGraw-Hill Publishing Company Professional and Reference Division composition unit.

Printed and bound by R. R. Donnelley & Sons Company.

To Virginia and Jean, with love and appreciation

Contents

Case Histories

Preface

Mistakes are a fact of life. It is the response to error that counts.

NIKKI GIOVANNI

I am now sure that as Rachmaninoff toiled on his "Rhapsody on a Theme of Paganini," he felt the intensity of the intellect and the heartbeat of that composer many times over, hoping that the genius of the original composition would be reborn in a contemporary mold. There is no pretense here of the mastery of Rachmaninoff, only the hope that Ott's genius might sparkle the more in a new setting to further enlighten those who would interpret data, aware of the concepts of statistical control, for troubleshooting and process improvement. In preparation of the second edition, I have walked again and again down a familiar road in the company of an old friend. The path may have become a little different and the footing not quite the same, but the road is still there for you also. Come walk with us through the real world of statistics and enjoy the majesty of its methodology and the sure insight it provides into our world and ourselves.

Process control is based on a philosophy of continual improvement. Ellis Ott believed that. In completing the first edition he recognized the need for future revision. Accordingly, the second edition builds upon his work and the experience of those who have used the text over the intervening years. The format has been modified to provide self-contained sections on interpretation of data, process control, and troubleshooting, with examples to facilitate use in in-plant training as well as a college text. Problems and case studies have been extended so that every chapter is covered. Many advances in analysis of means, such as exact limits, extension to multilevel factorial experiments, including crossed and nested designs, and the like, have been made and have been included here. Chapters were added dealing specifically with the philosophy and methods of process control and basic principles of experiment design. These comprise a variety of additional control charts (median, midrange, s, geometric moving average, CUSUM, acceptance control charts, etc.) and notions of design. Also very basic methods of data analysis have been included such as stem-and-leaf, box-plots, and recognition of the importance of the use of s in recent years. These and other additions, changes, and corrections have been made necessary by the development of the field over the years.

I wish to thank those who helped so much in this revision. Many of the contributors cited by Ellis in the first edition were of tremendous help here. Others, like Stan Marash of Stat-a-matrix, Larry Rabinowitz of William and Mary, Maria Ramos-Nenno of Gordon S. Black Corporation, John Sheesley of Air Products Corporation, Dan Smialek and Thomas Witt of the Rochester Institute of Technology, Dan Sommers of General Electric, Lowell Tomlinson of Western Electric, Dick Zwickl of Western Electric, and a host of those who must remain unmentioned, gave so freely of their time and talent. Special mention must be made, however, of the efforts of Roger Berger of Iowa State University who provided most of the new examples to be found in this text, together with invaluable editorial comment. Some of these individuals knew Ellis Ott, others did not. Ellis would have thought of them all as his own.

Special thanks also to my wife, Jean, whose love for words is eclipsed only by her love for children. Without her help in composition, editing, and critical review, this revision would never have been made. My deep gratitude also to Professor Richard N. Schmidt and the late Professor Sigmund Zobel of SUNY Buffalo who kindled my interest in statistics, and to the late Carl Mentch who taught me how to be a successful process control engineer and saw to it that I became one. All three were students of Dr. Ott. Ellis drew strength not only from those who bear academic degrees signed in his name, but even more from that host of those unnamed whose belief in statistics and in themselves was made stronger, even a little bit, by his counsel.

Who is this book for? Not for the close-minded who would use the data as a means to an end, but for the open-minded whose end is what the data means. For statistics as a science has its ultimate meaning only insofar as it is developed and used in the search for truth. The romance of statistics is the dream of reality.

Dr. Ott loved cats. He loved statistical cats even more. So, paraphrasing a line from the musical *Cats*, this book is for STASTICAL CATS and not for MYSTICAL CATS, and for all those who know in their hearts that God did not make all cats the same and want to delight in the difference.

Rochester, New York *Edward G. Schilling*

Preface to the First Edition

Double, double, toil and trouble;
Fire burn and cauldron bubble.
SHAKESPEARE, *Macbeth*

Every manufacturing operation has problems. Some are painfully obvious; others require ingenuity and hard work to identify. Every production engineer and supervisor has to be a troubleshooter. Anyone hoping to become a production engineer or supervisor must learn about process improvement and troubleshooting that uses systematic data obtained from the process. There are two approaches to reducing trouble: learn to prevent it and learn to cure it when it develops. This book has something to say on each approach.

Some methods are presented for identifying opportunities for process improvements and for locating important differences. The knowledge that such differences do exist and the pinpointing of them are vital information. Experience has shown that those familiar with the process can then find ways to make improvements and corrections, if they can first be convinced that the differences actually do exist. This is one reason for the emphasis on graphical presentations.

Methods, examples, and broader case histories will relate to a wide variety of experiences. The case histories are real experiences, typical of ever-recurring problems. Many persons find it meaningful to read about an experience relating specifically to their own problems. However, the readers are encouraged to give even more attention to case histories in those industries not even apparently relevant to their own. It is surprising to find how often the case histories bring up ideas and suggestions applicable to their very own problems.

Some readers will want to browse through the first five introductory chapters. The material will be familiar, though presenting some new ideas. Many readers will want to study them more carefully. They present some basic methods of obtaining and diagnosing data. The book also presents some useful statistical methods and concepts having important practical applications. Derivations of a mathematical nature are included occasionally when they seem to support applications.

Many books emphasize control charts and acceptance sampling plans. Such books are important, and their readers will find broad new avenues in this book extending them into important areas of investigation. Some books emphasize management aspects. Many emphasize experimental design and analysis, especially on technical investigation in agriculture, the biological sciences, and chemical engineering. Some elements of these different areas have been included and related here, but emphasis in this book is on combining a number of approaches to troubleshooting and process improvement in manufacturing, including basic methods and principles of statistical quality control and experimental designs.

Some books emphasize the testing of hypotheses, but usually the problem confronting a production problem solver is entirely different: discovering hypotheses that warrant testing (see Chap. 9, especially). Troubleshooting cannot be entirely formalized, and there is no substitute for being inquisitive and exercising ingenuity. Many problems can be resolved without recourse to formal data collection. The ideas presented here are intended to support and extend the science of processes, not to replace them; the support and cooperation of technical and production personnel are to be cultivated by troubleshooters.

The business of providing product to the consumer requires many major functions. Production itself has major subdivisions: design and specify, purchase and acquire, manufacture, package, inspect and assure quality. Each of these critical functions observes Murphy's first law: *If anything can go wrong, it will.* Although troubleshooting and process improvement projects are as old as civilization itself, today they are still usually left to individuals who are relatively untrained in the area and who regard it as a chore. It is our intent here to gather together and organize some procedures rather universally applicable to troubleshooting. This procedure usually employs data collection and/or logic in addition to the process science and know-how.

A fortunate series of events led me from a university life into industrial quality control early in World War II, then back into a progressive state university with permission to build a graduate program in applied statistics. Hundreds of students there, many from neighboring industries, delighted their professor with choice sets of data—data from real, in-plant problems and operations. Many of these students became colleagues and personal friends. All of them contributed and exchanged experiences and ideas on industrial problems. William C. Frey, Edwin S. Shecter, Carl Mentch, Edward G. Schilling, Frank W. Wehrfritz are among the "old-timers." Living nearby were many pioneers in quality control, all of whom were so very generous in giving advice in many ways: Walter A. Shewhart, Harold F. Dodge,

Paul S. Olmstead, Enoch B. Ferrell shared their ideas, their time, and their support. It was a rich heritage. References are made to numerous articles from publications of the American Society for Quality Control, especially from *Industrial Quality Control* and the *Journal of Quality Technology*. It pleases me to acknowledge their permission to use various excerpts from them. The patient, and delightful technical and editorial assistance of Allegra Rodgers, a neighboring Texan and former "student," has been invaluable. And if Virginia, an amazingly versatile wife, had not given such steady support throughout the project, completion of this manuscript would have been impossible or long delayed—in spite of her exceptionally able typing. To all these, and to so many others unnamed here, who have contributed so many ideas and given so much support, it is a pleasure to give thanks.

It will please me if this book can pass along some of the heritage which came my way by contributing ideas and principles which will lead you to new levels of competence.

Lake Marble Falls, Texas *Ellis R. Ott*

An adequate science of control for management should take into account the fact that measurements of phenomena in both social and natural science for the most part obey neither deterministic nor statistical laws, until assignable causes of variability have been found and removed.

W. A. SHEWHART
"Statistical Quality Control,"
Trans. ASME, *Ten Year Management Report, May 1942*

Part 1

Basics of Interpretation of Data

Chapter

1

Variables Data: An Introduction

1.1 Introduction: An Experience with Data

Look around you and you will see variation. Often we ignore it, sometimes we curse it, always we live with it in an imperfect world. Ellis Ott's first industrial experience with variation pertained to the thickness of mica pieces being supplied by a vendor. The pieces had a design pattern of punched holes. They were carefully arranged to hold various grids, plates, and other radio tube components in their proper places. But the failure of mica pieces to conform to thickness specifications for radio tubes presented a problem. In particular, there were too many thin pieces that were found in many different types of mica. The vendor was aware of the problem but was quite sure that there was nothing he could do to resolve it. "The world supply of first-grade mica was cut off by World War II, and only an inferior quality is obtainable. My workers split the mica blocks quite properly to the specified dimensions.[1] Because of the poor mica quality," he insisted, "these pieces subsequently split in handling producing two thin pieces." He was sorry, but there was nothing he could do about it!

Now there are some general principles to be recognized by troubleshooters:

RULE 1 Don't expect many people to advance the idea that the problem is their own fault. Rather it is the fault of raw materials and components, a wornout machine, or something else beyond their own control. *"It's not my fault!"*

RULE 2 Get some *data* on the problem; do not spend too much time in initial planning. (An exception is when data collection requires a long time or is very expensive; very careful planning is then important.)

RULE 3 *Always graph* your data in some simple way—*always.* In this case,

[1]Purchase specifications were 8.5 to 15 thousandths (0.0085 to 0.015 in) with an industry-accepted allowance of 5 percent over and 5 percent under these dimensions. Very thin pieces did not give enough support to the assemblage. Very thick pieces were difficult to assemble. (It should be noted that specifications which designate such an allowance outside the stated "specifications" are quite unusual.)

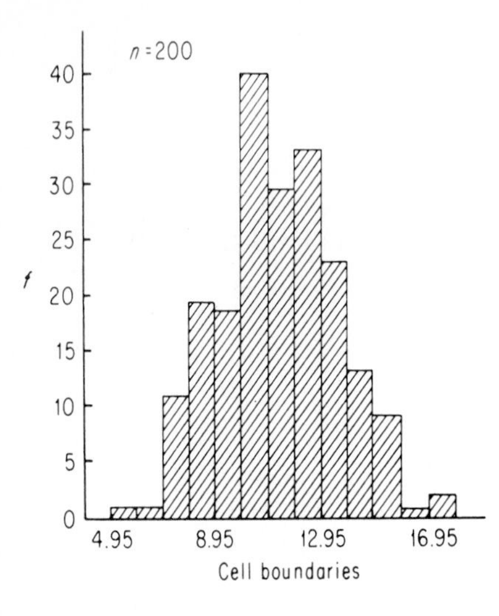

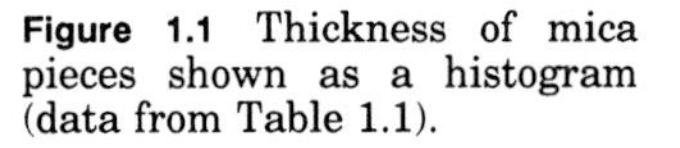
Figure 1.1 Thickness of mica pieces shown as a histogram (data from Table 1.1).

the engineer took mica samples from a recent shipment and measured the thickness of 200 pieces.[2] The resulting data shown in Table 1.1 are presented in Fig. 1.1 as a histogram.

Discussion: Figure 1.1 shows some important things.

1. A substantial number[3] of mica pieces are too thin and some are too thick when compared to the upper and lower specifications limits of 8.5 and 15.

2. The center of the two specifications is 0.5(8.5 + 15) = 11.75 thousandths; the peak of the thickness distribution is to the left of 11.75 at 10.25 thousandths. If the splitting blades were adjusted to increase the thickness by about 0.5 thousandth, the peak would be moved near the center of the specifications; the number of thin pieces would be reduced slightly more than the number of thick pieces would be increased. The adjusted process would produce fewer nonconforming pieces but would still produce more outside the specifications than a 5 percent allowable deviation on each side.

3. It is conceivable that a few of the mica pieces had split during handling, as the vendor believed. However, if more than an occasional one were splitting, a bimodal[4] pattern with two humps would be formed. Consequently, it is neither logical nor productive to attribute the problem of thin micas to the splitting process.

What might the vendor investigate to reduce the variability in the splitting process?

---

[2]Experience with problem solving suggests that it is best to ask for only about 50 measurements—certainly not more than 100. An exception would arise when an overwhelming mass of data is required, usually for psychological reasons.

[3]A count shows that 24 of the 200 pieces are under 8.5 thousandths of an inch and 7 are over 15 thousandths.

[4]See Fig. 1.4.

Answer: There was more than one operator hand-splitting this particular type of mica piece. Differences between these operators are almost certainly contributing to variation in the process.

Also, differences in thickness from an individual operator would be expected to develop over a period of a few hours because *changes* in knife sharpness and *operator fatigue* could produce important variations.

At the vendor's suggestion, quality control charts were instituted on individual operators. These charts helped reduce variability in the process and produce micas conforming to specifications.

In the larger study of the mica thickness problem, samples were examined from several different mica types; many pieces were found to be too thin and relatively few pieces too thick with each type. What then?

Economic factors often exert an influence on manufacturing processes, either consciously or unconsciously. In this splitting operation, the vendor bought the mica by the pound but sold it by the piece. One can imagine a possible reluctance to direct the mica-splitting process to produce any greater thickness than absolutely necessary.

A more formal discussion of data display will be presented in following sections. Mechanics of grouping the data in Table 1.1 and of constructing Fig. 1.1 will also be explained in Secs. 1.3 and 1.4.

1.2 Variability

In every manufacturing operation there is variability. The variability becomes evident whenever a quality characteristic of the product is measured. There are two basically different reasons for variability, and it is very important to distinguish between them.

Variability inherent in the process

It is important to learn how much of the product variability is actually inherent in the process. Is the variation a result of *random* effects of components and raw materials? Is it from small mechanical linkage variations in a machine producing random variation in the product? Is it from slight variations in an operator's performance? Many factors influence a process and each contributes to the inherent variation affecting the resulting product. These are sometimes called common causes. There is also variation in test equipment and test procedures—whether used to measure a physical dimension, an electronic or a chemical characteristic, or any other characteristic. This inherent variation in testing is a factor contributing to the observed measurement of product characteristics—sometimes an important factor.[5]

There *is* variation in a process even when all adjustable factors known to affect the process have been set and held constant during its operations.

Also there is a *pattern* to the inherent variation of a specific stable process, and there are different basic characteristic patterns of data from different pro-

[5]See Case History 2.4.

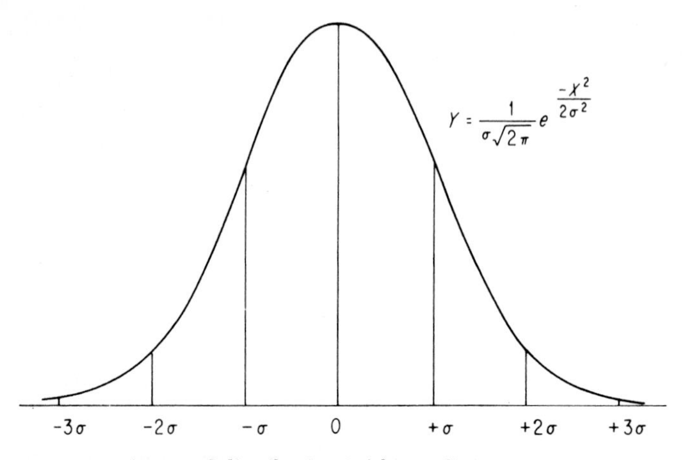

Figure 1.2 A normal distribution (with $\mu = 0$).

cesses. However, the most frequent and useful one is called the *normal distribution;* its idealized mathematical form is shown in Fig. 1.2; it is discussed further in Sec. 1.8.

The mica thickness data in Fig. 1.1 have a general resemblance to the normal distribution of Fig. 1.2. A majority of observations are clustered around a central value, there are tails on each end, and it is relatively symmetrical about a vertical line drawn at about 11.2 thousandths.

There are other basic patterns of variability; they are referred to as *nonnormal* distributions. The *log normal* is fairly common when making acoustical measurements and certain measurements of electronic products. If the logarithms of the measurements are plotted, the resulting pattern is a normal distribution—hence its name. A sketch of a log-normal pattern is shown in Fig. 1.3; it has a longer tail to the right.

Based on industrial experience, the log-normal distribution does not exist as frequently as some analysts believe. Many apparent basically log-normal distributions of data are *not* the consequence of a *stable log-normal* process but of *two basically normal distributions* with a large percentage produced at one level. The net result can produce a bimodal distribution as in Fig. 1.4, which presents a false appearance of being inherently log normal.

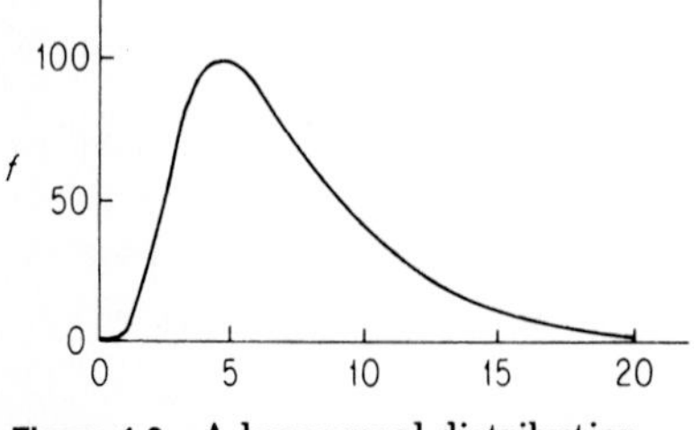

Figure 1.3 A log-normal distribution.

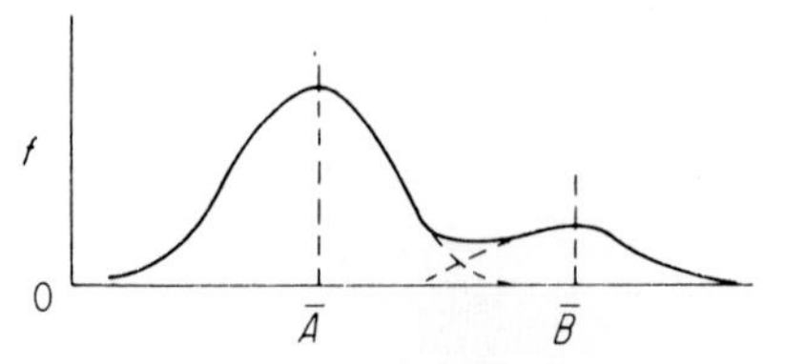

Figure 1.4 A bimodal distribution composed of two normal distributions.

The production troubleshooter needs help in identifying the *nature* of the causes producing variation. If different operators or machines are performing the same operation, it is important to learn whether some are performing better or worse than others. Specific differences, when discovered, will often lead to ideas for improvement when those performing differently—either better or worse—are compared. Some causes may be common to all machines and operators. In reducing process variability caused by them, the troubleshooter may need to identify a variety of small improvements which can be extended over all machines and operators.

Variability from assignable causes

There are other important causes of variability which Dr. Walter Shewhart called *assignable* causes.[6] These are sometimes called special causes.

This second type of variation often contributes in large part to the overall variability of the process. Evidence of this type of variation offers important opportunities for improving the uniformity of product. The process average may change gradually as a result of gradual changes in temperature, tool wear, or operator fatigue. Or the process may be unnecessarily variable because two operators or machines are performing at different averages. Variability resulting from two or more processes operating at different levels, or of a single source operating with an unstable average, are typical of production processes. *They are the rule,* not the exception to the rule.

This second type of variability must be studied by various techniques of data analysis which will be presented in this book with the aim of separating it from the first. Then after responsible factors have been identified and corrected, a continuing control[7] of the process will be needed.

1.3 Organizing Data

Certain concepts and methods about the analysis and interpretation of data appear to be simple. Yet they are not easily acquired and assimilated. The discussion of methods extends over the first five chapters. Some patience will be needed. After some concepts and methodologies have been considered, the weaving of them together with actual experiences (case histories) will begin to make sense.

The data presented in Table 1.1 are measurements of small pieces of mica delivered as one shipment. These readings were made with a dial indicator gauge on a sample of $n = 200$ pieces of one mica type. (Table 1.1 is the data source for Fig. 1.1.)

The observations in Table 1.1 could also be displayed individually along a horizontal scale. Such a display would show extremes and any indication of clustering. When n is large, we often decide to present the data in the con-

[6]Walter A. Shewhart, *Economic Control of Quality of Manufactured Product,* D. Van Nostrand Company, Inc., New York, 1931.

[7]See Chap. 2.

TABLE 1.1 Mica Thickness, Thousandths of an Inch

8.0	12.5	12.5	14.0	13.5	12.0	14.0	12.0	10.0	14.5
10.0	10.5	8.0	15.0	9.0	13.0	11.0	10.0	14.0	11.0
12.0	10.5	13.5	11.5	12.0	15.5	14.0	7.5	11.5	11.0
12.0	12.5	15.5	13.5	12.5	17.0	8.0	11.0	11.5	17.0
11.5	9.0	9.5	11.5	12.5	14.0	11.5	13.0	13.0	15.0
8.0	13.0	15.0	9.5	12.5	15.0	13.5	12.0	11.0	11.0
11.5	11.5	10.0	12.5	9.0	13.0	11.5	16.0	10.5	9.0
9.5	14.5	10.0	5.0	13.5	7.5	11.0	9.0	10.5	14.0
9.5	13.5	9.0	8.0	12.5	12.0	9.5	10.0	7.5	10.5
10.5	12.5	14.5	13.0	12.5	12.0	13.0	8.5	10.5	10.5
13.0	10.0	11.0	8.5	10.5	7.0	10.0	12.0	12.0	10.5
13.5	10.5	10.5	7.5	8.0	12.5	10.5	14.5	12.0	8.0
11.0	8.0	11.5	10.0	8.5	10.5	12.0	10.5	11.0	10.5
14.5	13.0	8.5	11.0	13.5	8.5	11.0	11.0	10.0	12.5
12.0	7.0	8.0	13.5	13.0	6.0	10.0	10.0	12.0	14.5
13.0	8.0	10.0	9.0	13.0	15.0	10.0	13.5	11.5	7.5
11.0	7.0	7.5	15.5	13.0	15.5	11.5	10.5	9.5	9.5
10.5	7.0	10.0	12.5	9.5	10.0	10.0	12.0	8.5	10.0
9.5	9.5	12.5	7.0	9.5	12.0	10.0	10.0	8.5	12.0
11.5	11.5	8.0	10.5	14.5	8.5	10.0	12.5	12.5	11.0

SOURCE: Lewis M. Reagan, Ellis R. Ott, and Daniel T. Sigley, *College Algebra*, rev. ed., chap. 18, Holt Rinehart and Company, Inc., New York, 1940. (Reprinted by permission of Holt, Rinehart & Winston, Inc.)

densed form of a *grouped frequency distribution* (Table 1.2) or *histogram.* A histogram is a picture of the distribution in the form of a bar chart with the bars juxtaposed.

Table 1.2 was prepared[8] by selecting *cell boundaries* to form equal intervals of width $m = 1$ called *cells.* A tally mark was then entered in the proper cell corresponding to each measurement in the table. The number of measurements which fall in a particular cell is called the *frequency* (f_i) for that ith cell; also f_i/n is called the *relative frequency,* and $100\ f_i/n$ is the *percent frequency.* An immediate observation from Table 1.2 is that mica pieces vary from a thickness of about 5 thousandths at one extreme to 17 at the other. Also the frequency of measurements is greatest near the center, 10 or 11 thousandths, and tails off to low frequencies on each end.

1.4 Grouping Data When *n* Is Large

Cells—how many and how wide?

Table 1.1 has been presented in Table 1.2 as a *tally sheet.* Many times there are advantages in recording the data *initially* on a blank tally sheet instead of

[8]A discussion of the grouping procedure is given in Sec. 1.4.

TABLE 1.2 Data: Mica Thickness as a Tally Sheet

Data of Table 1.1 grouped into cells whose boundaries and midpoints are shown in columns at the left

Cell boundaries	Cell midpoint	Tally	Observed frequency	Percent frequency
4.75				
	5.25	/	1	0.5
5.75				
	6.25	/	1	0.5
6.75				
	7.25	𝍸 𝍸 /	11	5.5
7.75				
	8.25	𝍸 𝍸 𝍸 ////	19	9.5
8.75				
	9.25	𝍸 𝍸 𝍸 ///	18	9.0
9.75				
	10.25	𝍸 𝍸 𝍸 𝍸 𝍸 𝍸 𝍸 𝍸	40	20.0
10.75				
	11.25	𝍸 𝍸 𝍸 𝍸 𝍸 ////	29	14.5
11.75				
	12.25	𝍸 𝍸 𝍸 𝍸 𝍸 𝍸 ///	33	16.5
12.75				
	13.25	𝍸 𝍸 𝍸 𝍸 ///	23	11.5
13.75				
	14.25	𝍸 𝍸 ///	13	6.5
14.75				
	15.25	𝍸 ////	9	4.5
15.75				
	16.25	/	1	0.5
16.75				
	17.25	//	2	1.0
17.75				
			$n = 200$	100%

listing the numbers as in Table 1.1 In preparing a frequency distribution, it is usually best:

1. To make the *cell intervals* equal, of width m.

2. To choose the *cell boundaries* halfway between two possible observations. This simplifies classification. For example, in Table 1.1 observations were recorded to the nearest half (0.5); cell boundaries were chosen beginning with 4.75, i.e., halfway between 4.5 and 5.0.

3. The *number of cells* in grouping data from a very large sample should be between 13 and 20. Sturges rule of thumb for cell size gives the following relationship between the number of cells c and sample size n:

$$c = 1 + 3.3 \log_{10} n$$

and since $2^{3.3} \cong 10$, the relationship simplifies[9] to a useful rule of thumb

$$2n = 2^c$$

$$n = 2^{c-1}$$

This leads to the following rough starting points for the number of cells in a frequency distribution using Sturges rule.

Sample size n	Number of cells c	Sample size n	Number of cells c
6–11	4	377–756	10
12–23	5	757–1519	11
24–46	6	1520–3053	12
47–93	7	3054–6135	13
94–187	8	6136–12,328	14
188–376	9	12,329–24,770	15

Note that a frequency distribution is a form of data display. The values given are rough starting points only and thereafter the number of cells should be adjusted to make the distribution reveal as much about the data as possible. The number of cells c is directly related to the cell width: $c = \Delta/m$. In Table 1.1, a large[10] observation is 17.0; a small one is 6.0. Their difference Δ is read "delta."

$$\Delta = 17.0 - 6.0 = 11.0$$

Now if the cell width m is chosen to be $m = 1$, we expect at least $\Delta/m = 11$ cells; if chosen to be $m = 0.5$, we expect the data to extend over at least

[9]Converting Sturges rule to base 2 we obtain

$$c = 1 + 3.3 \log_{10} n$$
$$c = 1 + 3.3 \log_2 n \log_{10} 2$$
$$c = 1 + 3.3(0.301) \log_2 n$$
$$c = 1 + 0.993 \log_2 n$$
$$c - 1 = \log_2 n$$
$$n = 2^{c-1}$$
$$2n = 2^c$$

[10]Whether these are actually the very largest or smallest is not critical. Using $n = 2$, about 6 cells would be required.

$11/0.5 = 22$ cells. The tally (Table 1.2) was prepared with $m = 1$, resulting in 13 cells. (See Exercise 1.*a* for $m = 0.5$.)

4. *Choose cell boundaries which will simplify the tallying.* The choice of 4.75, 5.75, 6.75, and so forth, as cell boundaries when grouping the data of Table 1.1 results in classifying all numbers beginning with a 5 into the same cell; its midpoint is 5.25, halfway between 5.0 and 5.5.

Similarly, all readings beginning with a 6 are grouped into the cell whose midpoint is 6.25, and so forth. This makes tallying quite simple.

5. The *midpoint of a cell* is the average of its two boundaries. Midpoints begin with

$$0.5(4.75 + 5.75) = 5.25$$

then increase successively by the cell width m.

6. A frequency distribution representing a set of data makes it possible to compute two different numbers (see Table 1.3) that give objective and useful information about the *location* and *spread* of the distribution. These computed numbers are especially important in data analysis. The number[11] $\bar{X}$ is an *estimate* of the central location of the process, μ. It is the arithmetic average, or mean, of the observations. Estimates of the process mean are symbolized by $\hat{\mu}$. Thus, when the sample mean is used as an estimate, $\hat{\mu} = \bar{X}$. The number $\hat{\sigma}$ is an *estimate* of the process variation σ; some interpretations of $\hat{\sigma}$ will be discussed in Sec. 1.8. Both $\bar{X}$ and $\hat{\sigma}$ can be computed directly from the values themselves or after they have been organized as a frequency distribution; the computations are shown in Table 1.3.

1.5 The Arithmetic Average or Mean—Central Value

There are $n = 200$ measurements of mica thickness recorded in Table 1.1. The *arithmetic average* of this sample could be found by adding the 200 numbers and then dividing by 200. We do this when n is small and sometimes when using a calculator or computer. The average $\bar{X}$ obtained in this way is 11.1525.

More generally, let the n measurements be

$$X_1, X_2, X_3, \ldots, X_n \tag{1.1}$$

A shorthand notation is commonly used to represent sums of numbers. The capital Greek letter *sigma*, written Σ, indicates summation. Then the average

[11] Read "X bar" for the symbol $\bar{X}$; and read "sigma hat" for $\hat{\sigma}$.

TABLE 1.3 Mica Thickness

Computation of $\overline{X}$ and $\hat{\sigma}$; data of Table 1.1

Cell boundaries	Cell midpts	Tally	f_i	d_i	$f_i d_i$	$f_i d_i^2$	Accumulated tally $\sum f_i$	$\sum$ %
	5.25	/	1	−5	−5	25	1	0.5
5.75								
	6.25	/	1	−4	−4	16	2	1.0
6.75								
	7.25	𝍸 𝍸 /	11	−3	−33	99	13	6.5
7.75								
	8.25	𝍸 𝍸 𝍸 ////	19	−2	−38	76	32	16.0
8.75								
	9.25	𝍸 𝍸 𝍸 ///	18	−1	−18	18	50	25.0
9.75								
	$A =$ 10.25	𝍸 𝍸 𝍸 𝍸 𝍸 𝍸 𝍸 𝍸	40	0	0	0	90	45.0
10.75								
	11.25	𝍸 𝍸 𝍸 𝍸 𝍸 ////	29	+1	29	29	119	59.5
11.75								
	12.25	𝍸 𝍸 𝍸 𝍸 𝍸 𝍸 ///	33	+2	66	132	152	76.0
12.75								
	13.25	𝍸 𝍸 𝍸 𝍸 ///	23	+3	69	207	175	87.5
13.75								
	14.25	𝍸 𝍸 ///	13	+4	52	208	188	94.0
14.75								
	15.25	𝍸 ////	9	+5	45	225	197	98.5
15.75								
	16.25	/	1	+6	6	36	198	99.0
16.75								
	17.25	//	2	+7	14	98	200	100.0
17.75								
		$n =$	200		+183	1169		

$$\sum f_i d_i = 183 \qquad \sum f_i d_i^2 = 1169$$

$$\overline{X} = A + \sum \frac{f_i d_i}{n} = 10.25 + (1)(0.915) = 11.165 \tag{1.3}$$

$$\hat{\sigma} = m \sqrt{\frac{n\left(\sum f_i d_i^2\right) - \left(\sum f_i d_i\right)^2}{n(n-1)}} \tag{1.4b}$$

$$= (1) \sqrt{\frac{200(1169) - (183)^2}{200(199)}} = 2.244$$

$\overline{X}$ of the n numbers in Eq. (1.1) is written symbolically[12] as

$$\overline{X} = \frac{\sum_{i=1}^{n} X_i}{n} \tag{1.2}$$

Median and midrange

A *second* measure of the center of a distribution is the *median,* $\tilde{X}$. When there is an odd number of observations, the middle one is called the median. When there is an even number, the median is defined to be halfway between the two central values, i.e., their arithmetic average. In brief, half of the observations are greater than the median and half are smaller. For the mica data, the median is 11.0. A *third* measure is the midrange which is halfway between the extreme observations, i.e., their mean. For the mica data the midrange is $(17 + 5)/2 = 11$. The midrange, for example, is used to represent average daily temperature.

1.6 Measures of Variation

Computing a standard deviation

Figure 1.5 shows two distributions having the same average. It is very clear that the average alone does not represent a distribution adequately. The distribution in (2) spreads out more than the one in (1). Thus some measure is needed to describe the *spread* or *variability* of a frequency distribution. The variability of the distribution in (2) appears to be about twice that in (1).

A useful *measure* of variability is called the *standard deviation* $\hat{\sigma}$. We shall present the calculation and then discuss some uses and interpretations of $\hat{\sigma}$. This is the small Greek letter sigma with a "hat" to indicate that it is an estimate of the *unknown measure of population variability.* The symbol $\hat{\sigma}$ is used as an omnibus symbol to represent *any* estimate of the unknown process parameter σ, just as $\hat{\mu}$ represents *any* estimate of μ. Their specific meaning is taken in context.

---

[12]The expression

$$\sum_{i=1}^{n} X_i$$

is read "the summation of X_i from $i = 1$ to n." The letter i (or j or whatever letter is used) is called the *index of summation.* The numbers written above and below, or following Σ indicate that the index i is to be given successively each integral value from 1 to n, inclusively.

The expression

$$\sum_{i=1}^{n} X_i^2$$

represents the sum

$$X_1^2 + X_2^2 + X_3^2 + \cdots + X_{10}^2$$

and sometimes the index of summation is omitted when *all* the observations are used. Thus,

$$\sum (X - \overline{X})^2 = (X_1 - \overline{X})^2 + (X_2 - \overline{X})^2 + (X_3 - \overline{X})^2 + \cdots + (X_n - \overline{X})^2$$

Both these expressions are used frequently in data analysis.

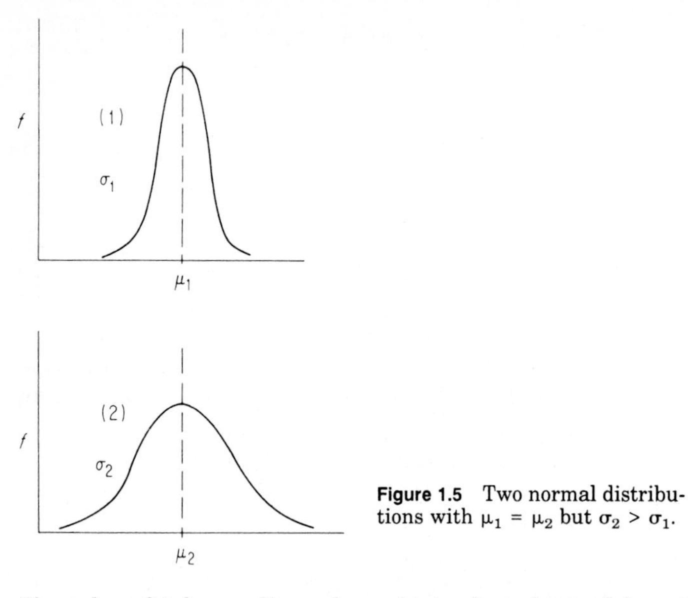

Figure 1.5 Two normal distributions with $\mu_1 = \mu_2$ but $\sigma_2 > \sigma_1$.

The value of $\hat{\sigma}$ for smaller values of n is often obtained from the formula

$$\hat{\sigma} = s = \sqrt{\frac{\Sigma(X - \overline{X})2}{n - 1}} = \sqrt{\frac{n\Sigma X^2 - (\Sigma X)^2}{n(n - 1)}} \tag{1.4a}$$

The first formula is the definition of s, while the algebraically equivalent second formula is often easier to compute. To illustrate the calculation required, consider the first five observations of mica thickness in the first column of Table 1.1, i.e., 8.0, 10.0, 12.0, 12.0, 11.5. Their mean is $\overline{X} = 53.5/5 = 10.7$. We have

	X	X^2	$(X - \overline{X})$	$(X - \overline{X})^2$
	8.0	64	− 2.7	7.29
	10.0	100	− 0.7	0.49
	12.0	144	1.3	1.69
	12.0	144	1.3	1.69
	11.5	132.25	0.8	0.64
Total	53.5	584.25	0.0	11.80

$$\hat{\sigma} = s = \sqrt{\frac{11.8}{4}} = \sqrt{\frac{5(584.25) - (53.5)^2}{5(4)}} = 1.7176$$

For the complete sample of 200 measurements of mica thickness we find

$$\Sigma X = 2230.5$$

$$\Sigma X^2 = 25{,}881.75$$

so, using all the observations

$$\overline{X} = 2230.5/200 = 11.1525$$

and

$$s = \sqrt{\frac{n\Sigma X_i^2 - (\Sigma X_i)^2}{n(n-1)}} = \sqrt{\frac{200(25881.75) - (2230.5)^2}{200(199)}}$$
$$= \sqrt{5.055772613} = 2.248505 \approx 2.249$$

Occasionally, data are available only in grouped form, or it is necessary to group for other purposes. When this happens a simple computational procedure can be used to obtain the arithmetic average and the standard deviation of the data.

1. Begin by selecting an arbitrary cell midpoint A as the origin; the computation is simplified when A is chosen near the center of the distribution. In Table 1.3, we have chosen $A = 10.25$.

2. Fill in the d_i column beginning with a zero opposite the chosen A cell; this is the origin. The d_i indicate the *deviation in cells* from A. Number increasing cell midpoints consecutively with +1, +2, +3, etc., and number decreasing cell midpoints with −1, −2, −3, etc.

3. Complete the f_id_i column by multiplying corresponding f_i and d_i.

4. Add the numbers in the f_id_i column to obtain $\Sigma f_id_i = 183$, and divide by $n = 200$ to obtain

$$\frac{\Sigma f_i d_i}{n} = 0.915$$

5. Return to the original scale by computing

$$\overline{X} = A + \frac{m\Sigma f_i d_i}{n} = 10.25 + (1)(0.915) = 11.165 \tag{1.3}$$

This value 11.165 has been obtained after assigning all measurements within a cell to have the value of the midpoint of that cell. The result compares closely with the arithmetic average 11.1525 computed by adding all individual measurements and dividing by 200.

6. The computation of $\hat{\sigma} = s$ in Table 1.3 requires one more column than for $\overline{X}$; it is labeled $f_id_i^2$. It is obtained by multiplying each number in the d_i column by the corresponding number in the f_id_i column. Then compute

$$\Sigma f_i d_i^2 = 1169$$

and obtain $\hat{\sigma}$ from Eq. (1.4*b*):

$$\hat{\sigma} = s = m\sqrt{\frac{n\Sigma f_i d_i^2 - (\Sigma f_i d_i)^2}{n(n-1)}} \tag{1.4b}$$

$$= (1)\sqrt{\frac{200(1169) - (183)^2}{200(199)}}$$

$$= (1)\sqrt{5.032940} = 2.244$$

These computation procedures are a great simplification over other procedures when n is large. This result also compares closely with the sample standard deviation $s = 2.249$ obtained from all the individual measurements. When n is small, see Chap. 4. Some interpretations and uses of $\hat{\sigma}$ will be given in subsequent sections.

The estimate s has certain desirable properties, for instance, the sample variance s^2 is an unbiased estimate of the true but unknown process variance σ^2. There are, of course, other possible estimates of the standard deviation, each with its own properties. Differences in the behavior of these estimates give important clues in troubleshooting. For example,[13] there are the maximum likelihood estimate, the range estimate, the mean deviation from the median, the best linear estimate, and others. For sample size 5, these are

Maximum likelihood estimate:

$$\hat{\sigma} = \sqrt{\frac{n-1}{n}}\, s = 0.89s \quad \text{(100\% efficiency)}$$

Range estimate:

$$\hat{\sigma} = 0.43(X_{(5)} - X_{(1)}) \qquad \text{(95.5\% efficiency)}$$

Mean deviation from median:

$$\hat{\sigma} = 0.3016(X_{(5)} + X_{(4)} - X_2 - X_1) \qquad \text{(99\% efficiency)}$$

Best linear estimate:

$$\hat{\sigma} = 0.3724(\mathrm{X}_{(5)} - \mathrm{X}_{(1)}) + 0.1532(\mathrm{X}_4 - \mathrm{X}_2) \qquad \text{(98.8\% efficiency)}$$

where $X_{(i)}$ indicates the ith *ordered* observation. For the first sample of 5 from the mica data, ordered as 8, 10, 11.5, 12, 12, these are

Maximum likelihood estimate:

$$\hat{\sigma} = \sqrt{\frac{4}{5}}(1.7176) = 1.5363$$

Range estimate:

$$\hat{\sigma} = 0.43(12 - 8) = 1.72$$

Mean deviation from median:

$$\hat{\sigma} = 0.3016(12 + 12 - 10 - 8) = 1.8096$$

Best linear estimate:

$$\hat{\sigma} = 0.3724(12 - 8) + 0.1532(12 - 10) = 1.796$$

It should not be surprising that we come out with different values. We are estimating and hence there is no "correct" answer. We don't know what the

[13]For other examples of simple estimates, see W. J. Dixon and F. J. Massey, *Introduction to Statistical Analysis*, McGraw-Hill Book Company, New York, 1969.

true population σ is; if we did there would be no need to calculate an estimate. Since each of these methods has different properties, any (or all) of them may be appropriate depending on the circumstances.

The desirable properties of s, or more properly its square s^2, has led to the popularity of that estimate in characterizing sample variation. It can also be calculated rather simply from a frequency distribution.

Note: The procedures of Table 1.3 not only provide values of $\overline{X}$ and $\hat{\sigma}$ but also provide a display of the data. If the histogram approximated by the tally shows a definite bimodal shape, for example, any interpretations of either $\overline{X}$ or $\hat{\sigma}$ must be made carefully.

One advantage of plotting a histogram is to check on whether the data appear to come from a single source or from perhaps two or more sources having different averages. As stated previously, the mica manufacturer believed that many of the mica pieces were splitting during handling. Then a definite bimodal shape should be expected in Table 1.3. The data *do not* lend support to this belief.

The mica-splitting data (Tables 1.1, 1.2, 1.3) were obtained by measuring the thickness of mica pieces at the incoming inspection department. The data almost surely came from a process representing production over some time period of different workers on different knife splitters; it just is not reasonable to expect all conditions to be the same. The data represent *what was actually* shipped to us—not necessarily what the production process was capable of producing.

Some coding of data (optional)[14]

The computations of $\overline{X}$ and $\hat{\sigma}$ in Table 1.3 have used some important properties and methods of coding (transforming) data. These are used in change of scale. Consider again the n measurements in Eq. (1.1).

$$X_1, X_2, X_3, \ldots, X_n \tag{1.1}$$

- What happens to their average $\overline{X}$ and standard deviation $\hat{\sigma}$ if we translate the origin by *adding* a constant c to each?

$$X_1 + c, X_2 + c, X_3 + c, \ldots, X_n + c \tag{1.5}$$

1. The *average* of this new set of numbers will be the original average increased by c:

$$\text{New average} = \frac{\Sigma(X_i + c)}{n} = \frac{\Sigma X_i + nc}{n} = \overline{X} + c$$

2. The *standard deviation* of this new set of numbers in Eq. (1.5) is not changed by a translation of origin; their standard deviation is still $\hat{\sigma}$.

[14]This procedure may be omitted. Simple algebra is sufficient to prove the following relations pertaining to *standard deviations;* simple but tedious. The proofs are omitted.

What happens to the average $\overline{X}$ and standard deviation $\hat{\sigma}$ if we multiply each number in Eq. (1.1) by a constant c?

$$cX_1, cX_2, cX_3, \ldots, cX_n \tag{1.6}$$

1. The average of these numbers will be the original average multiplied by c:

$$\text{New average} = \frac{\Sigma cX_i}{n} = \frac{c\Sigma X_i}{n} = c\overline{X}$$

2. The standard deviation of these numbers will be the original multiplied by c; that is,

$$new\ \hat{\sigma} = c\hat{\sigma}$$

1.7 Plotting on Normal Probability Paper

This graphical method of presenting data is often helpful in checking on the stability of the source producing the data. The *accumulated percents* of mica-thickness data are shown in Table 1.3, right-hand column. These have been plotted on normal probability paper in Fig. 1.6. There are 13 cells in Table 1.3; a convenient scale has been chosen on the base line (Fig. 1.6) to accommodate the 13 cells. The upper cell boundaries have been printed on the base scale; the chart shows the accumulated percent frequencies to the upper cell boundaries.

Normal probability paper is scaled in such a way that a truly *normal curve* will be represented by a straight line. A line can be drawn using a clear plastic ruler to approximate the points; it is not unusual for one or two points on each end to deviate slightly, as in Fig. 1.6, even if the source of the data is essentially a normal curve. The data line up rather surprisingly well.

The *median* and the *standard deviation* of the data can be estimated from the straight-line graph on normal probability paper.

The median is simply the 50 percent point. A perpendicular line has been dropped from the intersection of the plotted line and the 50 percent horizontal line. This cuts the base line at the median. Its estimate is

$$10.75 + 0.4 = 11.15$$

This is in close agreement with the computed $\overline{X} = 11.1525$ from all the measurements.

Estimating the standard deviation involves more arithmetic. One method is to determine where horizontal lines corresponding to the accumulated

16% and 84%

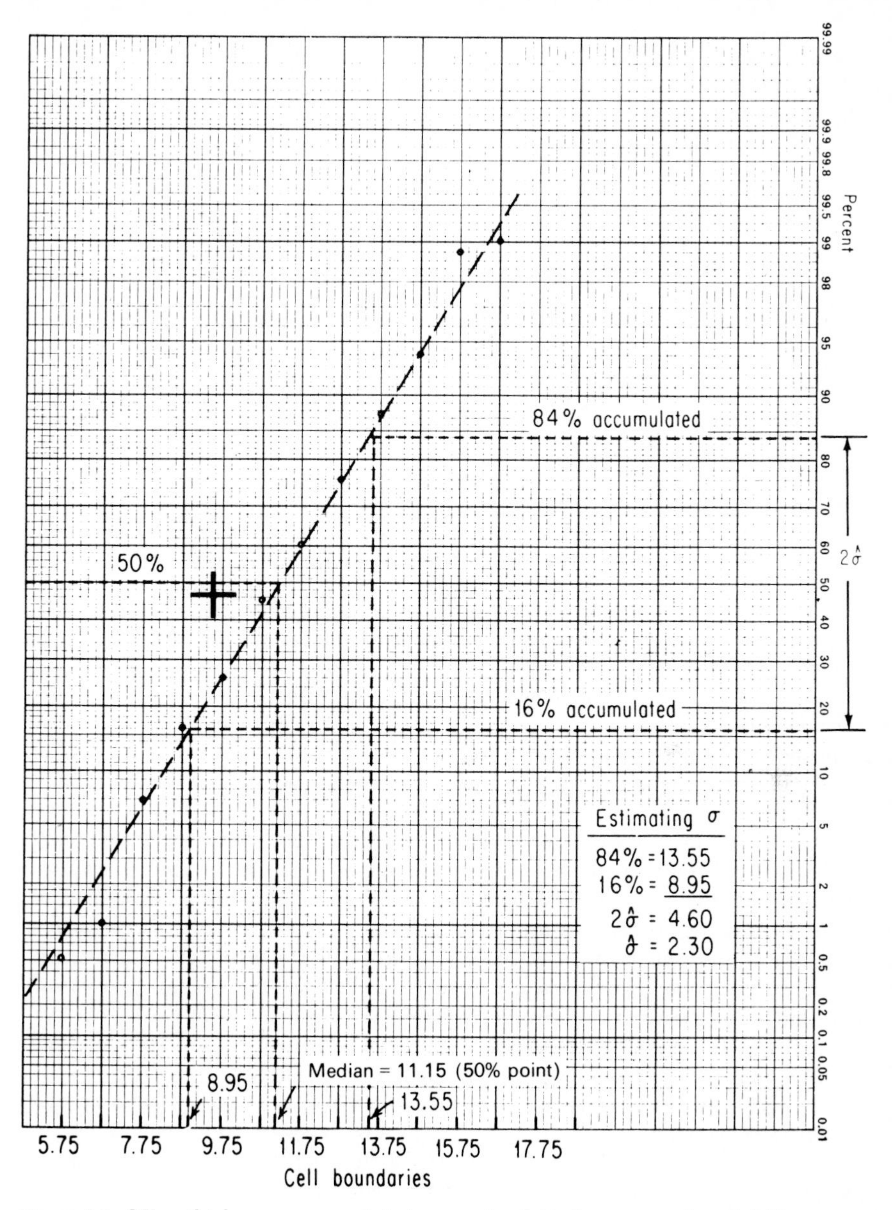

Figure 1.6 Mica thickness; accumulated percents plotted on normal probability paper (data from Table 1.3).

cut the line. These numbers correspond to areas under the normal curve to the left of ordinates drawn at $\overline{X} - \hat{\sigma}$ and $\overline{X} + \hat{\sigma}$, that is, they differ by an estimated $2\hat{\sigma}$. See Eq. (1.10) of the following Sec. 1.8.

In Fig. 1.6, corresponding vertical lines appear to cut the base line at

$$\begin{aligned}
&\text{84\% point:} \quad 13.75 - 0.2 \cong 13.55 \\
&\text{16\% point:} \quad 8.75 + 0.2 \cong 8.95 \\
&2\hat{\sigma} \cong 4.60 \\
&\hat{\sigma} \cong 2.30
\end{aligned}$$
[15]

This agrees reasonably well with the previously computed value

$$\hat{\sigma} = s = 2.249$$

from all 200 measurements.

Interpretation: The only possible evidence of mica pieces splitting into two during handling is the pair of points at the lower left end of the line. But splitting is certainly not a major factor—rather, the process average should be increased by about 0.6 thousandths (11.75 − 11.15 = 0.6) since the center of specifications is at

$$\tfrac{1}{2}(8.5 + 15.0) = 11.75 \text{ thousandths}$$

Example 1.1 Depth-of-cut data are shown as a frequency distribution in Table 1.4 with accumulated frequencies on the right (data from Table 1.8).

These accumulated frequencies have been plotted on normal probability paper in Fig. 1.7. The points give evidence of fitting *two* line segments; a single line does not fit them well. There is a *run* of length 5 below the initial line. Although it is possible mechanically to compute an $\bar{X}$ and a $\hat{\sigma}$, we should be hesitant to do so. These data represent *two* different processes.

This set of data is discussed again in Chap. 2, Case History 2.1.

1.8 Predictions Regarding Sampling Variation: The Normal Curve

This topic is of primary importance in process maintenance and improvement.

Consider pieces of mica being split by one operator. The operator produces many thousands a day. We can imagine that this process will continue for many months or years. The number produced is large—so large that we can consider it to be an infinite universe. In production operations, we are concerned not only with the mica pieces which are actually produced *and* examined but with those which were produced and *not* examined. We are also concerned with those which are yet to be produced. We want to make inferences about them. This is possible provided the process is stable.

We can think of the process as operating at some *fixed stable level* and with some *fixed stable standard deviation.* We refer to these two concepts by the Greek letters μ and σ,[16] respectively. The actual values of μ and σ can never be learned in practice; they are abstract concepts. Yet they can be estimated

---

[15]The symbol $\cong$ means "approximately equal to."

[16]μ is pronounced "mew." It designates an *assumed true, but usually unknown, process average.* When n items of a random sample from a process are measured/tested, their average $\bar{X}$ designates an *estimate* of μ: $\bar{X} = \hat{\mu}$.

Another important concept is that of a *desired or specified average:* It is commonly designated by the symbol $\bar{X}'$ (read "X bar prime").

The symbol σ' (read "sigma prime") is sometimes used to designate a *desired or specified measure of process variability.*

TABLE 1.4 Data: Depth of Cut
Data from Table 1.8 displayed on a tally sheet

Cell boundaries	Cell interval	Tally	f	Σf	$\Sigma\%$
	1610–11	//	2	125	100%
1609.5					
	1608–09		0	123	98.4
1607.5					
	1606–07	////	4	123	98.4
1605.5					
	1604–05	//	2	119	95.2
1603.5					
	1602–03	𝍸 //	7	117	93.6
1601.5					
	1600–01	𝍸 𝍸 /	11	110	88.0
1599.5					
	1598–99	𝍸 𝍸 𝍸 /	16	99	79.2
1597.5					
	1596–97	𝍸 𝍸 𝍸 𝍸 𝍸 ////	29	83	66.4
1595.5					
	1594–95	𝍸 𝍸 𝍸 𝍸 𝍸 ///	28	54	43.2
1593.5					
	1592–93	𝍸 𝍸 𝍸 /	16	26	20.8
1591.5					
	1590–91	𝍸	5	10	8.0
1589.5					
	1588–89	///	3	5	4.0
1587.5					
	1586–87		0	2	1.6
1585.5					
	1584–85	/	1	2	1.6
1583.5					
	1582–83	/	1	1	0.8

as closely as we please by computing $\overline{X}$ and $\hat{\sigma}$ from large enough samples. How large the sample must be is answered in the following discussions (Sec. 1.11).

If we have two operators splitting micas, it is not unusual to find differences in their output either in average thickness or in variability of product. If they are found to have equal averages and standard deviations, then they can be considered to be a single population source.

It is important to know how much variation can be predicted in a succession of samples from a stable[17] process. What can be predicted about a second sample of 200 mica pieces which we might have obtained from the shipment which

[17]Unstable processes are unpredictable. Few processes are stable for very long periods of time whether in a laboratory or in production.

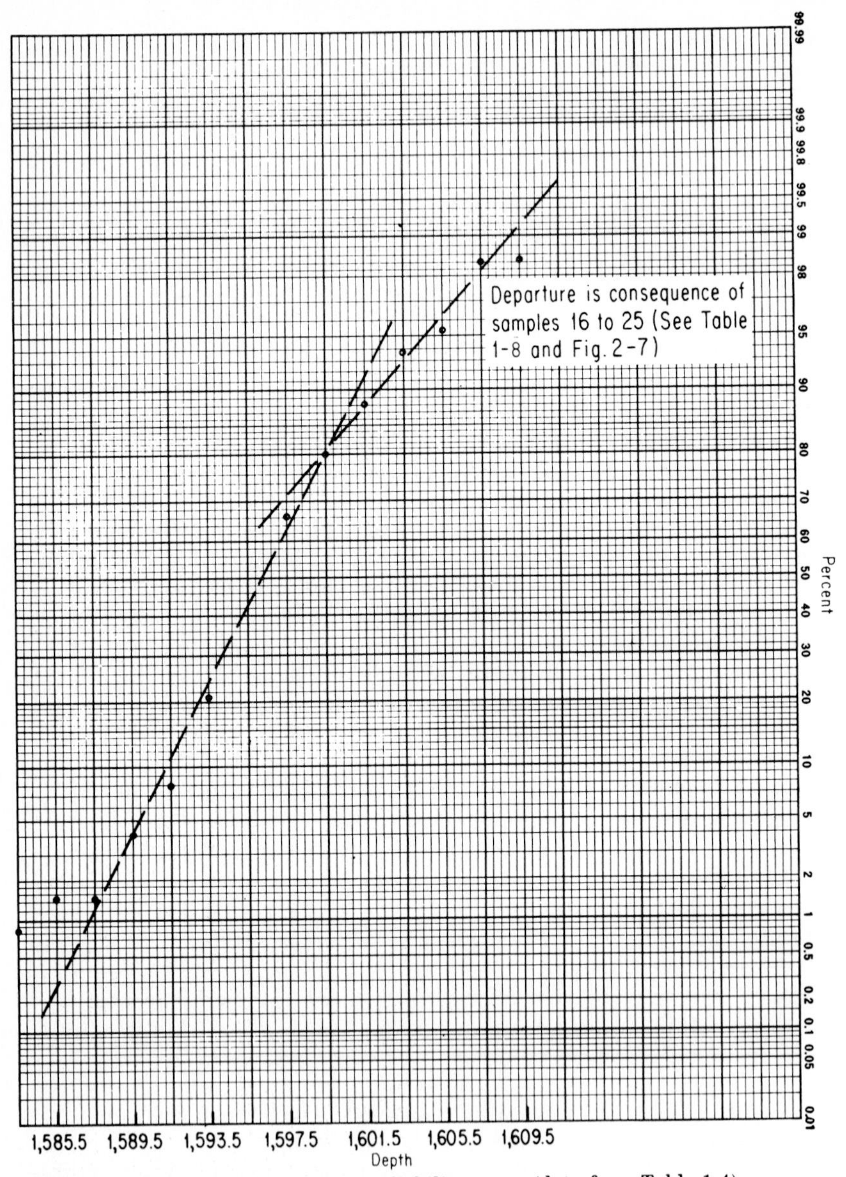

Figure 1.7 Depth of cut on normal probability paper (data from Table 1.4).

provided the data in Table 1.1? It would be surprising if the newly computed $\overline{X}$ were exactly 11.1525 thousandths as it was for the first sample; it would be equally surprising if the computed $\hat{\sigma}$ were exactly 2.249 again. The following two theorems relate to the amount of variation expected among sample averages $\overline{X}_i$ and standard deviations $\hat{\sigma}_i$ of samples of n each drawn from a stable process (or which might be so drawn).

The k averages

$$\overline{X}_1, \overline{X}_2, \overline{X}_3, \ldots, \overline{X}_k$$

of the k samples will vary and will themselves form a frequency distribution. The sample averages will vary "considerably less" than individuals vary.

THEOREM 1: *The standard deviation $\hat{\sigma}_{\overline{X}}$ of averages of samples of size n drawn from a process will be related to the standard deviation of individual observations by the relation:*

$$\hat{\sigma}_{\overline{X}} = \hat{\sigma}/\sqrt{n} \tag{1.7}$$

This theorem says that averages of $n = 4$, for example, are predicted to vary half as much as individuals and that averages of $n = 100$ are predicted to vary one-tenth as much as individuals.

From each of the k samples, we can also compute a standard deviation

$$\hat{\sigma}_1, \hat{\sigma}_2, \hat{\sigma}_3, \ldots, \hat{\sigma}_k$$

These also form a distribution. What can be predicted about the variation among these standard deviations computed from samples of size n?

THEOREM 2: *The standard deviation of sample standard deviations will be related to the standard deviation of individual measurements ($\hat{\sigma}$) by the relation:*

$$\hat{\sigma}_{\hat{\sigma}} \cong \frac{\hat{\sigma}}{\sqrt{2n}} \tag{1.8}$$

These two theorems are important in the study of industrial processes. The basic theorems about the predicted variation in $\sigma_{\overline{X}}$ and $\sigma_{\hat{\sigma}}$ relate to idealized mathematical distributions. In applying them to real data, we must obtain estimates $\hat{\sigma}_x$ and $\hat{\sigma}_{\hat{\sigma}}$, these estimates are given in Eqs. (1.7) and (1.8).

Distributions of sample averages from parent universes (populations) of different shapes are similar.

Consider averages of samples of size n drawn from a parent population or process. It had been known that sample averages were essentially *normally* distributed:

1. When n was "large," certainly when n approached infinity.
2. Usually regardless of the shape of the parent population, normally distributed or not.

In the late 1920s, Dr. Walter A. Shewhart conducted some basic and industrially important chip drawings. Numbers were written on small metal-rimmed tags, placed in a brown kitchen bowl, and experimental drawings (with replacement of chips) made from it. Among other things he wanted to see if there were predictable patterns (shapes) to distributions of averages of size n drawn from some simple populations. He recognized the portent of using *small* samples for industrial applications provided more was known about small samplings from a stable universe, such as drawing numbered chips from a bowl. Three different

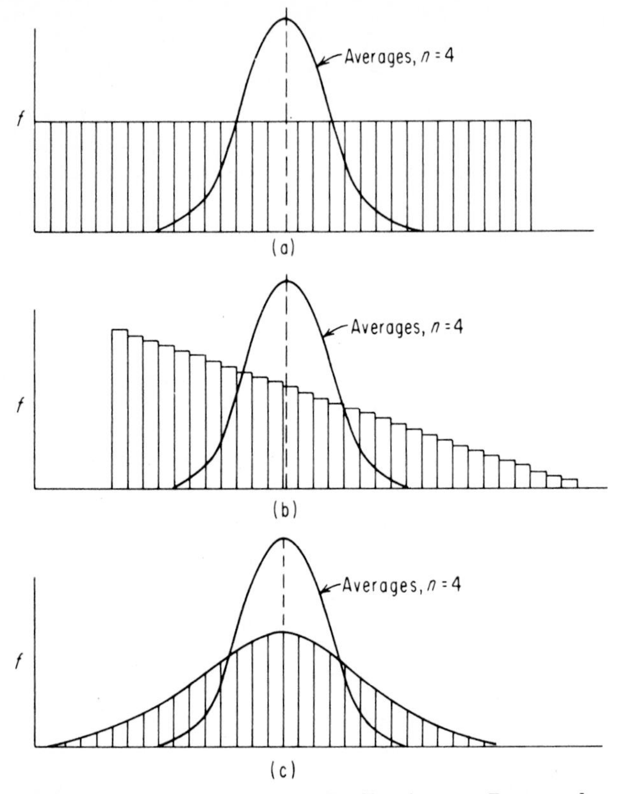

Figure 1.8 Distributions sampled by Shewhart (*a*) Rectangular parent population; (*b*) right-triangular parent population; (*c*) normal parent population.

sets of chips were used: one represented a rectangular universe; another, a right-triangular distribution universe; and the third, a normal distribution. In each experiment, many sample averages were obtained using $n = 3$, 4, and 5. One important consequence is given in Theorem 3.

THEOREM 3: *Even with samples as small as $n = 4$, the distribution of averages of random samples drawn from almost any shaped parent population*[18] *will be essentially normal.*

Figure 1.8 portrays the relationship of the distribution of sample averages to their parent universes even for samples as small as $n = 4$. For sample sizes larger than $n = 4$, the shape of the curve of averages also tends to normality. Averages (with n as small as 4) from these different parent populations tend to be normally distributed (Theorem 3).

The normal curve is symmetrical and bell-shaped; it has an equation whose idealized form is

[18]With a finite variance.

$$Y = \frac{1}{\sigma\sqrt{2\pi}} e^{[-(X-\mu)^2/2\sigma^2]} \tag{1.9}$$

The term *normal* is a technical term; it is not synonymous with *usual* nor the opposite of abnormal.

Areas under the normal curve (Fig. 1.2) can be calculated, but the calculations are tedious. Values are given in Appendix Table A.1. However, there are a few important area relationships which are used so frequently that they should be memorized: the following are obtained from Table A.1:

Between	Percent of area under normal curve
$\mu - 3\sigma$ and $\mu + 3\sigma$	99.73 ≅ 99.7, that is, "almost all"
$\mu - 2\sigma$ and $\mu + 2\sigma$	95.44 ≅ 95
$\mu - \sigma$ and $\mu + \sigma$	68.26 ≅ 68

(1.10)

In practice, of course, we do not know either μ or σ; they are replaced by their estimates $\bar{X}$ and $\hat{\sigma}$, computed from a representative sample of the population.

In other words, about 95 percent of all production from a well-controlled (stable) process can be expected to lie within a range of $\pm 2\sigma$ about the process average, and almost all—99.7 percent—within a range of $\pm 3\sigma$ about the average.

Example 1.2. Two Applications

1. *Within what region can we predict that mica thickness will vary in the shipment from which the sample of Table 1.1 came?*

To obtain the answer, we assume a stable process producing normally distributed thicknesses.

Answer A From the 200 observations we computed $\bar{X} = 11.152$ and $\hat{\sigma} = 2.249$. From relation in Eq. (1.10), we expect almost all (about 99.7 percent) to be between

$$\bar{X} + 3\hat{\sigma} = 11.152 + 3(2.249) = 17.90 \text{ thousandths}$$

and

$$\bar{X} - 3\hat{\sigma} = 11.152 - 3(2.249) = 4.40 \text{ thousandths}$$

Also from relation in Eq. (1.10), we expect about 95 percent to be between

$$\bar{X} + 2\hat{\sigma} = 15.65 \quad \text{and} \quad \bar{X} - 2\hat{\sigma} = 6.65$$

Answer B In Table 1.1, we find one thinnest piece to be 5.0; also, two thickest ones to be 17.0. This is in agreement with the $\pm 3\sigma$ prediction of Answer A.

2. *What percent of nonconforming mica pieces do we expect to find in the entire shipment of which the data in Table 1.1 comprise a sample?*

Answer The specifications on the mica thickness were 8.5 to 15.0 thousandths of inches as shown in Fig. 1.9. We can compute the distance from $\bar{X} = 11.152$ to each of the specifications expressed in standard deviations

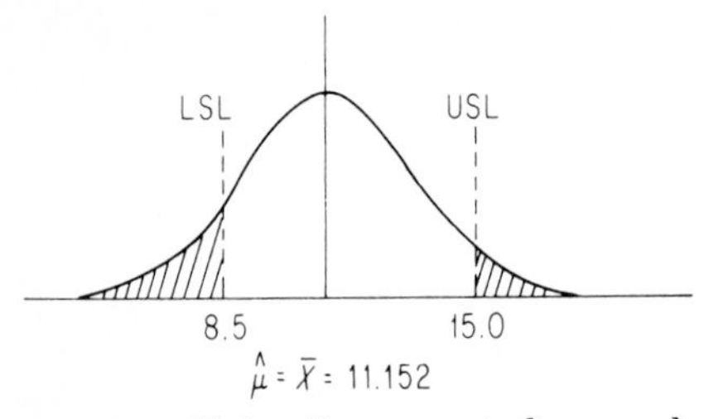

Figure 1.9 Estimating percent of a normal curve outside given specifications (related to data of Table 1.1).

$$Z_1 = \frac{\overline{X} - \text{LSL}}{\hat{\sigma}} = \frac{11.152 - 8.5}{2.249} = 1.18 \tag{1.11a}$$

From Appendix Table A.1, we find the corresponding percent below 8.5 (that is, below $\overline{X} - 1.18\hat{\sigma}$) to be 11.9 percent. (The actual count from Table 1.1 is 24, that is, 12 percent.) Also,

$$Z_2 = \frac{\text{USL} - \overline{X}}{\hat{\sigma}} = 1.71 \tag{1.11b}$$

Again from Table A.1, we find the expected percent above 15 (that is, above $\overline{X} + 1.71\hat{\sigma}$) to be about 4.4 percent. (The actual count is 7, that is, 3.5 percent.)

Discussion: There are different possible explanations for the excessive variability of the mica-splitting operation: variation in the mica hardness, variation in each of the operators who split the blocks of mica using a small bench knife, and any variations between operators.

An important method of future process surveillance was recommended—a Shewhart control chart which is discussed in Chap. 2.

Example 1.3. Using the Mica-Thickness Data The average and standard deviation were computed from the sample of 200 measurements to be

$$\overline{X} = 11.152 \text{ thousandths} \quad \text{and} \quad \hat{\sigma} = 2.249 \text{ thousandths}$$

- Then from *Theorem* 2 for a series of averages of samples of $n = 200$ from this same process, assumed stable,

$$\hat{\sigma}_{\overline{X}} = \frac{\hat{\sigma}}{\sqrt{200}} = \frac{2.249}{14.14} = 0.159$$

An estimate of the variation of averages to be expected in random, representative samples of $n = 200$ from a process with $\hat{\sigma} = 2.249$ and average $\overline{X} = 11.152''$ is then[19]

$$\overline{X} \pm 2\hat{\sigma}_{\overline{X}} = 11.152 \pm 2(0.159) = 11.152 \pm 0.318 \text{ thousandths} \quad \text{(with about 95\% confidence)}$$

$$\overline{X} \pm 3\hat{\sigma}_{\overline{X}} = 11.152 \pm 0.477 \text{ thousandths} \quad \text{(with about 99.7\% confidence)}$$

[19]See Eq. (1.10) and Theorem 1.

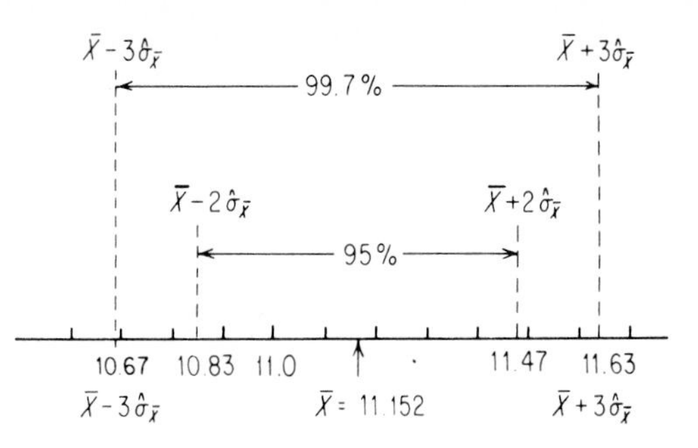

Figure 1.10 Estimating confidence intervals of unknown process average.

Also from *Theorem* 1, we can estimate the location of the assumed *true* but *unknown average* μ of the mica-splitting process. This converse use of Theorem 1 is applicable when n is as large as 30 or more. A modification, not discussed in this text, is required for smaller sample sizes. For $n = 200$

$$\bar{X} - 2\hat{\sigma}_{\bar{x}} = 11.152 - 0.318 \cong 10.83$$

and

$$\bar{X} + 2\hat{\sigma}_{\bar{x}} = 11.152 + 0.318 \cong 11.47$$

that is, $10.83 < \mu < 11.47$ thousandths (with about 95 percent confidence). Also, we can estimate the location of the unknown average μ to be between

$$\bar{X} - 3\hat{\sigma}_{\bar{x}} = 11.152 - 0.477 \cong 10.67$$

and

$$\bar{X} + 3\hat{\sigma}_{\bar{x}} = 11.152 + 0.477 \cong 11.63$$

that is, $10.67 < \mu < 11.63$ with 99.7 percent confidence.

In Fig. 1.10, we see the increase in interval required to change the confidence in our estimate from 95.5 to 99.7 percent.

1.9 Series of Small Samples from a Production Process

The amount of variation expected in $\bar{X}$ and $\hat{\sigma}$ from a succession of samples from an industrial process can be predicted only when the process average and variability are stable.[20] In the following discussion we assume that the process is stable and make predictions about the expected variation in samples obtained randomly from it.

1.10 Change in Sample Size: Predictions about $\bar{X}$ and $\hat{\sigma}$

We might have taken a smaller sample from the mica shipment. For example, the measurements in the top five rows constitute a sample of $n = 50$ from the

[20]Actually, of course, the lack of basic stability in a process average is the usual situation; it is the major reason for troubleshooting and process-improvement studies. Methods of using a succession of small samples in studying lack of stability in a process will be considered in Chap. 2 and subsequently.

mica-splitting process which produced the shipment. We expect large random samples to provide more accurate estimates of the true process average and standard deviation than smaller samples. Smaller samples, however, often provide answers which are entirely adequate. The two theorems of this chapter give useful information pertaining to sample size.

The computed values $\bar{X}$ and $\hat{\sigma}$ from all 200 values were

$$\bar{X} = 11.152 \qquad \text{and} \qquad \hat{\sigma} = 2.249 \qquad \text{with } n = 200$$

(These values are the ones computed from the sample of 200; we shall now assume them to be the true values of an assumed stable process average and standard deviation—a very risky assumption in practice.)

The values, $\bar{X}$ and $\hat{\sigma}$, of the top 50 measurements from Table 1.1 are computed below from the original observations.

$$\Sigma X = 604.5 \qquad \Sigma X^2 = 7552.25 \qquad n = 50$$

$$\bar{X} = \frac{\Sigma X_i}{n} = \frac{604.5}{50} = 12.09$$

$$\hat{\sigma} = s = \sqrt{\frac{n\Sigma X^2 - (\Sigma X)^2}{n(n-1)}} = \sqrt{\frac{50(7552.25) - (604.5)^2}{50(49)}} = 2.231$$

First, what can we predict about $\bar{X}$ and $\hat{\sigma}$ of samples of size $n = 50$ drawn from a population with

$$\mu = 11.152 \qquad \text{and} \qquad \sigma = 2.249$$

Variation of $\bar{X}$

From Eq. (1.10) and Theorem 1 the expected average of a sample of $n = 50$, assuming stability, is in the region

$$\mu \pm 2\sigma_{\bar{x}} = 11.152 \pm 2\left(\frac{2.249}{\sqrt{50}}\right) \qquad (95\% \text{ confidence}) = 11.152 \pm 0.636$$

$$\text{that is} \quad 11.152 - 0.636 < \bar{X} < 11.152 + 0.636$$

$$\text{and} \quad 10.516 < \bar{X} < 11.788 \qquad (95\% \text{ confidence})$$

The sample average of the first 50 observations from Table 1.1, $\bar{X} = 12.09$, falls outside this interval. We conclude the data of Table 1.1 are not a single homogeneous universe. (See Sec. 2.5.)

Variation of $\hat{\sigma}$

The expected value of $\hat{\sigma}$ for samples of $n = 50$, assuming stability, is in the region

$$\sigma \pm 2\sigma_{\sigma} = 2.249 \pm 2\left(\frac{2.249}{\sqrt{100}}\right) \qquad (95\% \text{ confidence}) = 2.249 \pm 2(0.225)$$

$$= 2.249 \pm 0.450$$

$$\text{Therefore,} \quad 1.799 < \hat{\sigma} < 2.699$$

The sample standard deviation of the first 50 observations from Table 1.1, $\hat{\sigma} = 2.231$, falls within this 95 percent confidence interval.

1.11 How Large a Sample Is Needed to Estimate a Process Average?

There are many things to consider when answering this question. In fact, the question itself requires modification before answering. Is the test destructive, nondestructive, or semidestructive? How expensive is it to obtain and test a sample of n units? How close an answer is needed? How much variation among measurements is expected? What level of confidence is adequate? All these questions must be considered.

Discussion: We begin by returning to the discussion of variation expected in averages of random samples of n items about the true but unknown process average μ. The expected variation of sample averages about μ[21] is

$$\pm 2\frac{\sigma}{\sqrt{n}} \qquad \text{(confidence about 95\%)}$$

and

$$\pm 3\frac{\sigma}{\sqrt{n}} \qquad \text{(confidence 99.7\%)}$$

Now let the allowable deviation (error) in estimating μ be $\pm\Delta$ (read "delta"); also let an estimate or guess of σ be $\hat{\sigma}$; then

$$\Delta \cong \frac{2\hat{\sigma}}{\sqrt{n}} \quad \text{and} \quad n \cong \left(\frac{2\hat{\sigma}}{\Delta}\right)^2 \qquad \text{(about 95\% confidence)} \qquad (1.12a)$$

also

$$\Delta \cong \frac{3\hat{\sigma}}{\sqrt{n}} \quad \text{and} \quad n \cong \left(\frac{3\hat{\sigma}}{\Delta}\right)^2 \qquad \text{(99.7\% confidence)} \qquad (1.12b)$$

Confidence levels other than the two shown in Eq. (1.12) can be used by referring to Table A.1. When our estimate or guess of a required sample size n is even as small as $n = 4$, then Theorem 3 applies.

Example 1.4 The mica manufacturer wants to estimate the true process average of one of his operators (data of Table 1.1). How large a random sample will he need?

- In this simple nondestructive testing situation, cost associated with *sample* size selection and test are of little concern.
- What is a reasonable choice of Δ? Since specifications are 8.5 to 15 thousandths, an allowance of $\pm\Delta = \pm 0.001$ seems reasonable to use in estimating a sample size.
- What is an estimate of σ? No information is available here for any one operator; we do have an estimate of overall variation, $\hat{\sigma} = 2.244$ from Table 1.3. This estimate probably includes variation resulting from several operators and thus is larger than for any one. However, the best available estimate is from Eq. (1.4*b*): $\hat{\sigma} = 2.249$. Then from Eq. (1.12*a*) $n = (4.498/1)^2 \cong 20$, (about 95% confidence).

[21]See Sec. 1.8, Theorems 1 and 3; also Eq. (1.10).

Decision: A sample size of $n = 20$ to 25 should be adequate to approximate the process average μ. However, a somewhat larger sample might be selected since it would cost but little more and might be accepted more readily by other persons associated with the project.

1.12 Sampling and a Second Method of Computing $\hat{\sigma}$

The method of this section is basic to many procedures for studying production processes.

The data in Table 1.1 represent a sample of 200 thickness measurements from pieces of mica delivered in one shipment. We have also considered a smaller sample from the shipment and used it to make inferences about the average and variability of the entire shipment.

There are definite advantages in subdividing sample data already in hand into smaller samples, such as breaking the mica sample of $n = 200$ into $k = 40$ subsamples or groups of size $n_g = 5$. Table 1.5 shows the data of Table 1.1 displayed in 40 sets of five each. The decision to choose five vertically aligned samples is an arbitrary one; there is no known physical significance to the order of manufacture in this set. Where there is a known order—either of manufacture or measurement—such an order should be preserved in representing the data, as in Fig. 1.11.

We have computed two numbers from each of these 40 subsamples: the average $\bar{X}$ and range, R, are shown directly below each sample. The range of a sample is simply:

$$R = \text{the largest observation minus the smallest}$$

The range is a measure of the variation within each small sample; the *average* of the ranges is designated by a bar over the R, that is, $\bar{R}$, and one reads "R bar." There is an amazingly simple and useful relationship (theorem)[22] between the *average range,* $\bar{R}$, and the *standard deviation* σ of the process of which these $k = 40$ groups of $n_g = 5$ are subsamples. The theorem is very important in industrial applications.

THEOREM 4: *Consider k small random samples (k > 20, usually) of size n_g drawn from a normally distributed stable process. Compute the ranges for the k samples and their average $\bar{R}$. Then the standard deviation (σ) of the stable process is estimated by*

$$\hat{\sigma} = \bar{R}/d_2 \qquad (1.13)$$

[22] Acheson J. Duncan, "The Use of Ranges in Comparing Variabilities," *Ind. Qual. Control,* vol. 11, no. 5, February, 1955, pp. 18, 19, and 22. E. S. Pearson, "A Further Note on the Distribution of Range in Samples from a Normal Population," *Biom.,* vol, 24, 1932, p. 404.

TABLE 1.5 Mica Thickness Data in Subgroups of n_g = 5 with Their Averages and Ranges

Data of Table 1.1

	8.0	12.5	12.5	14.0	13.5	12.0	14.0	12.0	10.0	14.5
	10.0	10.5	8.0	15.0	9.0	13.0	11.0	10.0	14.0	11.0
	12.0	10.5	13.5	11.5	12.0	15.5	14.0	7.5	11.5	11.0
	12.0	12.5	15.5	13.5	12.5	17.0	8.0	11.0	11.5	17.0
	11.5	9.0	9.5	11.5	12.5	14.0	11.5	13.0	13.0	15.0
$\bar{X}$:	10.7	11.0	11.9	13.1	11.9	14.3	11.7	10.7	12.0	13.7
R:	4.0	3.5	7.5	3.5	4.5	5.0	6.0	5.5	4.0	6.0
	8.0	13.0	15.0	9.5	12.5	15.0	13.5	12.0	11.0	11.0
	11.5	11.5	10.0	12.5	9.0	13.0	11.5	16.0	10.5	9.0
	9.5	14.5	10.0	5.0	13.5	7.5	11.0	9.0	10.5	14.0
	9.5	13.5	9.0	8.0	12.5	12.0	9.5	10.0	7.5	10.5
	10.5	12.5	14.5	13.0	12.5	12.0	13.0	8.5	10.5	10.5
$\bar{X}$:	9.8	13.0	11.7	9.6	12.0	11.9	11.7	11.1	10.0	11.0
R:	3.5	3.0	6.0	8.0	4.5	7.5	4.0	7.5	3.5	5.0
	13.0	10.0	11.0	8.5	10.5	7.0	10.0	12.0	12.0	10.5
	13.5	10.5	10.5	7.5	8.0	12.5	10.5	14.5	12.0	8.0
	11.0	8.0	11.5	10.0	8.5	10.5	12.0	10.5	11.0	10.5
	14.5	13.0	8.5	11.0	13.5	8.5	11.0	11.0	10.0	12.5
	12.0	7.0	8.0	13.5	13.0	6.0	10.0	10.0	12.0	14.5
$\bar{X}$:	12.8	9.7	9.9	10.1	10.7	8.9	10.7	11.6	11.4	11.2
R:	3.5	6.0	3.5	6.0	5.5	6.5	2.0	4.5	2.0	6.5
	13.0	8.0	10.0	9.0	13.0	15.0	10.0	13.5	11.5	7.5
	11.0	7.0	7.5	15.5	13.0	15.5	11.5	10.5	9.5	9.5
	10.5	7.0	10.0	12.5	9.5	10.0	10.0	12.0	8.5	10.0
	9.5	9.5	12.5	7.0	9.5	12.0	10.0	10.0	8.5	12.0
	11.5	11.5	8.0	10.5	14.5	8.5	10.0	12.5	12.5	11.0
$\bar{X}$:	11.1	8.6	9.6	10.9	11.9	12.2	10.3	11.7	10.1	10.0
R:	3.5	4.5	5.0	8.5	5.0	7.0	1.5	3.5	4.0	4.5

where d_2 is a constant depending upon the subsample size, n_g. Some frequently used values of d_2 are given in Table 1.6.

In other words, an estimate of the standard deviation of the process can be obtained either from direct calculation, the grouped method of Table 1.3 or from $\bar{R}$ in Eq. (1.13). For additional discussion, see Secs. 2.5 and 4.4.

Example 1.5. Data of Table 1.5 The ranges (n_g = 5) have been plotted in Fig. 1.11*b;* the average of the 40 ranges is $\bar{R}$ = 4.875. Then from Eq. (1.13) and Table 1.6

$$\hat{\sigma} = (4.875)/2.33 = 2.092$$

This estimate of σ is somewhat smaller than the value 2.249 obtained by direct computation. There are several possible reasons why the two estimates of σ are not exactly equal:

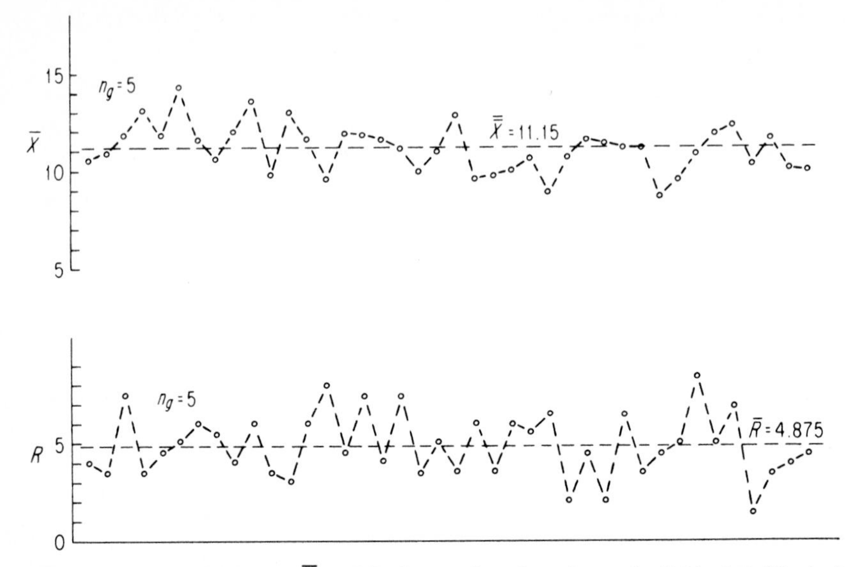

Figure 1.11 Mica thickness, $\bar{X}$ and R charts; data in order as in Table 1.5. (Control limits on this data are shown in Fig. 2.5.)

TABLE 1.6 Values of the Constant d_2

See also Table A.4

n_g	d_2
2	1.13
3	1.69
4	2.06
5	2.33
6	2.53

1. Theorem 1 is based on the concept of a process whose average is stable; this is a condition seldom justified in real life. Almost every process, even those which are stable for most practical purposes, shows gradual trends and abrupt shifts in average when analyzed carefully by control-chart methods.[23]

2. The difference is simply due to sampling error in the way the estimates are computed. Actually, the difference between the first and second estimates is small when compared to repeat sampling variation from the same stable process.

[23]NOTE: The methods of Chap. 2 consider practical methods of examining data from a process for excessive variation in its average and variability.

1.13 Some Important Remarks about the Two Estimates

Variation can be measured in terms of overall variation in *all* the numbers taken together, $\hat{\sigma}_1$, or in terms of an estimate of the variation *within subgroups* of the data $\hat{\sigma}_2$. When the process producing the data is stable $\hat{\sigma}_1 \cong \hat{\sigma}_2$. However, when the process is unstable we may find evidence of lack of stability in observing that $\sigma_1 > \sigma_2$. Thus,

$$\hat{\sigma}_1 = s = \sqrt{\frac{\Sigma(X - \overline{X})^2}{n - 1}} \qquad \hat{\sigma}_2 = \overline{R}/d_2$$

Figure 1.12 portrays a situation typified by machining a hole in the end of a shaft. The shifting average of the process is represented in the figure by a succession of small curves at 8 A.M., 8:30 A.M., and so forth. The short-term variation of the process is considered to be unchanging. The shift in average may be steady or irregular. Consider successive small samples, say of $n = 5$, taken from the process at 30-min intervals beginning at 8 A.M.

The variability, σ, of the machining process *over a short time interval* is measured by

$$\hat{\sigma}_2 = \overline{R}/d_2 = \overline{R}/2.33$$

This is a measure of the inherent capability *provided it were operating at a stable average;* this stability is possible only if ways can be found to remove those factors causing evident changes in the process average. (These include such possible factors as tool wear or slippage of chuck fastenings.)

The shaded area to the right of Fig. 1.12 represents the accumulated mea-

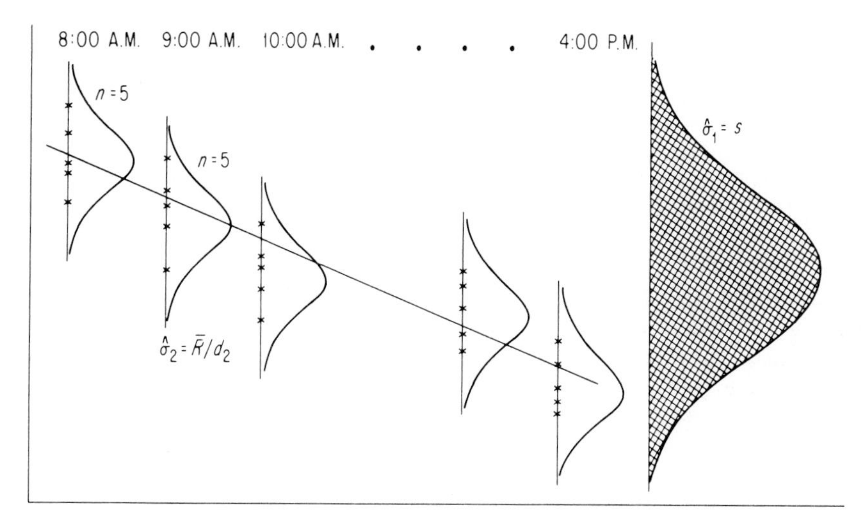

Figure 1.12 Schematic of a tobogganing production process.

surements of individuals from samples of five obtained successively beginning at 8 A.M. The variability of these accumulated sample measurements is *not* measured by $\hat{\sigma}_2 = \bar{R}/d_2$, but from the overall standard deviation

$$\hat{\sigma}_1 = s = \sqrt{\frac{\Sigma(X - \bar{X})^2}{n - 1}}$$

This latter is an estimate of the variation in the accumulated *total* production from 8 A.M. to 4:30 P.M.

In Fig. 1.12, the value of $\hat{\sigma}_1$ appears to be about twice $\hat{\sigma}_2$, since the spread of the accumulated shaded area is about twice that of each smaller distribution.

A gradually diminishing average diameter having a normal distribution (with spread $6\hat{\sigma}_2 = 6\bar{R}/d_2$) at any given time is shown in Fig. 1.12. The product accumulated over a period of time will be much more widely spread, often appearing to be almost normal, and with spread

$$6\hat{\sigma}_1 = 6s = 6\sqrt{\frac{\Sigma(X - \bar{X})^2}{n - 1}}$$

Evidently $6\hat{\sigma}_1$ will be substantially larger than $6\hat{\sigma}_2$.

A comparison of the two estimates from the same set of data, $\hat{\sigma}_1 = s$ and $\hat{\sigma}_2 = \bar{R}/d_2$, is frequently helpful in troubleshooting. If they differ substantially, the process average is suspected of instability.[24] The second, $\hat{\sigma}_2$, estimates the *within subgroup variability* of individuals. The first, $\hat{\sigma}_1$, estimates the variability of individuals produced over the period of sampling. If the process is stable, we expect the two estimates to be in fairly close agreement.

The following pertinent discussion is from the *ASTM Manual on Presentation of Data:*[25]

> *Breaking up data into rational subgroups.* One of the essential features of the method...is classifying the total set of test observations under consideration into subgroups or samples, within which variations may be considered to be due to nonassignable chance causes only, but between which there may be differences due to assignable causes whose presence is suspected or considered possible.
>
> This part of the problem is obviously not statistical in character but depends on technical knowledge and familiarity with the conditions under which the material sampled was produced and the conditions under which the data were taken.

The production person has a problem in deciding what constitutes a reasonable procedure for obtaining rational subgroups from a process. Experience and knowledge of the process will suggest certain possible sources which should be kept separate: Product from different machines, operators, or shifts; from different heads or positions on the same machine; from different molds or

---

[24]More discussion of testing for stability is given in Sec. 2.5.

[25]American Society for Testing and Materials, *ASTM Manual on Presentation of Data,* ASTM, Philadelphia, Pa., 1937, p. 48.

8.	96
9.	86796
10.	7701779310
11.	099779710642197
12.	0082
13.	170
14.	3

Figure 1.13 Stem-and-leaf of mica data means.

cavities in the same mold; from different time periods. Such problems will be considered throughout this book.

1.14 Stem-and-Leaf

When a relatively small amount of data is collected, it is sometimes desirable to order it in a stem-and-leaf pattern to observe the shape of the distribution and to facilitate further analysis. This technique was developed by John Tukey[26] and is particularly useful in troubleshooting with few observations. A stem-and-leaf diagram is constructed as follows:

1. Find the extremes of the data, drop the rightmost digit, and form a vertical column of the consecutive values between these extremes. This column is called the stem.
2. Go through the data and record the rightmost digit of each across from the appropriate value on the stem to fill out the number.
3. If an ordered stem-and-leaf diagram is desired, place the leaves in numerical order.

Consider the means of the subgroups in Table 1.5 as follows:

10.7	11.0	11.9	13.1	11.9	14.3	11.7	10.7	12.0	13.7
9.8	13.0	11.7	9.6	12.0	11.9	11.7	11.1	10.0	11.0
12.8	9.7	9.9	10.1	10.7	8.9	10.7	11.6	11.4	11.2
11.1	8.6	9.6	10.9	11.9	12.2	10.3	11.7	10.1	10.0

The resulting stem-and-leaf is shown in Fig. 1.13. Notice that it has the normal shape and is much tighter in spread than the frequency distribution of the individual observations shown in Table 1.3 as predicted by Theorem 1 and Theorem 3.

An ordered stem-and-leaf diagram of the means from the mica data is shown in Fig. 1.14.

The median, quartiles, and extremes are easily obtained from this type of plot. Note that there are $n = 40$ observations. The slash (/) shows the position of the quartiles. Thus we must count *through* $n/2 = 20$ observations to get to the me-

[26]Tukey, John W., *Exploratory Data Analysis,* Addison-Wesley Publishing Company, Inc., Reading, Mass., 1977.

8.	69
9.	66789
10.	001/1377779
11.	001/1246777799/99
12.	0028
13.	017
14	3

Figure 1.14 Ordered stem-and-leaf of mica data.

dian. The median is the $(n + 1)/2 = 20.5$th observation and is taken halfway between 11.1 and 11.1, which is, of course, 11.1. Similarly, the lower quartile (1/4 through the ordered data) is the $(n + 1)/4 = 10.25$th observation. Thus, it is one quarter of the way between 10.1 and 10.1, and hence is 10.1. The third quartile (3/4 through the data) is the $(3/4)(n + 1) = 30.75$th observation and hence is 3/4 of the way between 11.9 and 11.9, and since $3/4(0) = 0$, it is 11.9. [Note that the first decile (1/10 through the ordered data) is the $(n + 1)/10 = 4.1$th observation and is 1/10 of the distance between the 4th and 5th observation, so it is 9.61.]

1.15 Box-Plots

Another innovation by Tukey[27] is particularly useful in comparing distributions. It is known as the box-plot. To set up a box-plot

1. Order the data
2. Find the lowest and highest values, $X_{(1)}$ and $X_{(n)}$
3. Find the median, $\tilde{X}$
4. Obtain the first and third quartiles, Q_1 and Q_3
5. Obtain the mean (optional) $\bar{X}$

The form of the box-plot is then as shown in Fig. 1.15. It depicts some essential measures of the distribution in a way that allows comparison of various distributions. For example, if we wish to compare the distribution of the 40 means from Table 1.5 with the distribution of the first 40 individual values, the sample size chosen is in order to keep the number of observations equal since the appearance of the box plot is sample-size dependent. An ordered stem-and-leaf for the 40 individual measurements is shown in Fig. 1.16. A comparison of these distributions is shown in Fig. 1.17.

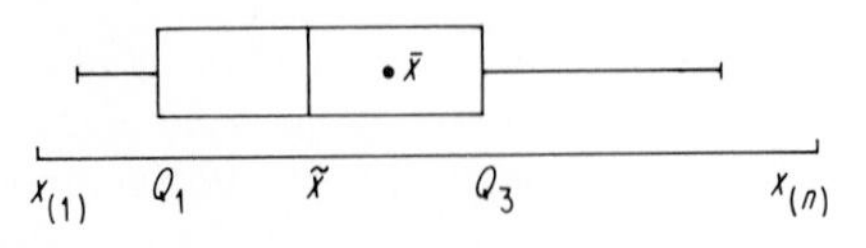

Figure 1.15 Form of box-plot.

[27]Tukey, loc. cit.

7.	5	$X_{(1)}$	=	7.5
8.	000	X_{40}	=	17.0
9.	005	$\tilde{X}$	=	12.0
10.	005/5	Q_1	=	10.5
11.	005555	Q_3	=	13.5
12.	000/0055555	$\overline{X}$	=	11.9
13.	005/55			
14.	0000			
15.	055			
16.				
17.	0			

Figure 1.16 Ordered stem-and-leaf of 40 individual mica measurements.

As predicted by Theorem 1, the distribution of the means is tighter than that of the individuals and we also see reasonable symmetry in the distribution of means as predicted by Theorem 3.

It may also be informative to form the box-plot of the entire distribution of mica thicknesses as shown in Table 1.3. Here it is desirable to estimate the median and quartiles from the frequency distribution itself. This is done by locating the class in which the quartile is to be found and applying the following formulas:

$$Q_1 = L + \left(\frac{n/4 - c}{f}\right)m = 8.75 + \left(\frac{200/4 - 32}{18}\right)1 = 9.75$$

$$\tilde{x} = Q_2 = L + \left(\frac{n/2 - c}{f}\right)m = 10.75 + \left(\frac{200/2 - 90}{29}\right)1 = 11.09$$

$$Q_3 = L = \left(\frac{3n/4 - c}{f}\right)m = 11.75 + \left(\frac{600/4 - 119}{33}\right)1 = 12.69$$

where L = lower class boundary of class containing quartile
n = total frequencies
c = cumulative frequencies up to the quartile class
f = frequency of class containing quartile
m = class width

We then have $X_{(1)} = 5.0$, $X_{(200)} = 17.0$, $\tilde{X} = 11.09$, $Q_1 = 9.75$, $Q_3 = 12.69$, $\overline{X} = 11.16$ and our box-plot appears as in Fig. 1.18. A comparison of Fig. 1.18

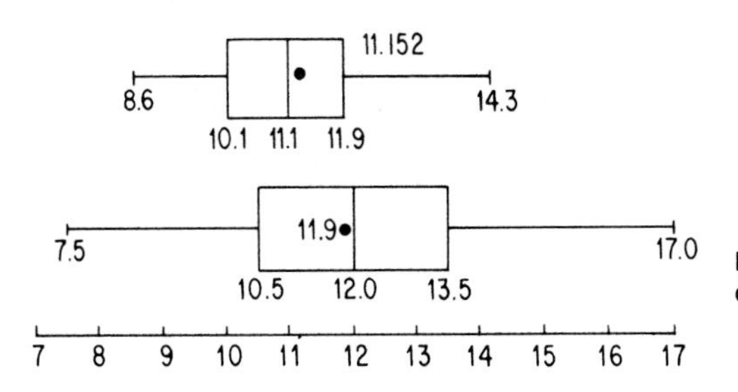

Figure 1.17 Box-plot of mica individuals and means.

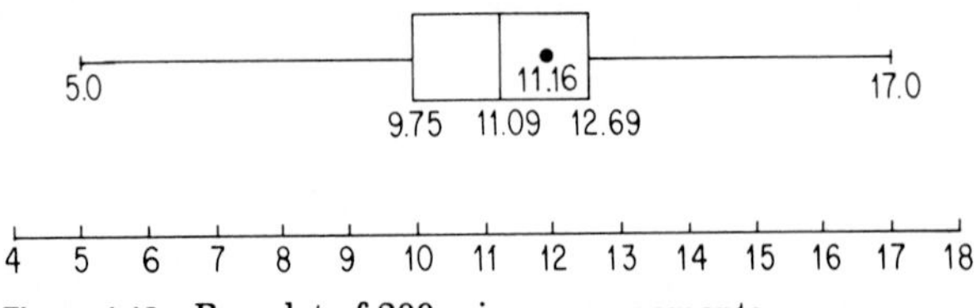

Figure 1.18 Box-plot of 200 mica measurements.

with Fig. 1.17 will show how increasing the sample size of the frequency distribution of individuals changes the box-plot by widening the extremes.

Finally, it should be noted that, for a normal distribution, the semi-interquartile range, that is $(Q_3 - Q_1)/2$, is equal to $2/3\sigma$. It is possible, then, to get an approximate feel for the size of the standard deviation by visualizing 1.5 times the average distance from the median to the quartiles when the box-plot appears reasonably symmetric.

The stem-and-leaf and the box-plot are primarily tools of communication. They help visualize the distribution and are used as vehicles of comparison. Used in conjunction with other methods they can be important vehicles for visualization and understanding. The approach taken here differs slightly from the formal approach to box-plot development as proposed by Tukey.[28] Rather, it is based on a more elementary approach attributed to Chatfield and discussed at some length by Heyes.[29]

1.16 A Note on Notation

In this chapter and in the remainder of this book, we have used the following notation:

$$n = \text{sample size}$$

$$k = \text{number of subgroups or number of points plotted}$$

$$n_g = \text{subgroup size}$$

so

$$n = kn_g$$

and when

$$k = 1$$

$$n = n_g$$

Also, in later chapters, and particularly with regard to design of experiments, we will use

[28]Tukey, John W., *Exploratory Data Analysis,* Addison-Wesley Publishing Company, Inc., Reading, Mass., 1977.

[29]Heyes, Gerald B., "The Box-Plot," *Quality Progress.* December 1985, pp. 13–17.

p = number of factors in an experiment

r = number of replicate observations per cell

Note that when the number of observations per subgroup is constant, the treatment total or the means of c cells is calculated from

$$n_g = cr$$

observations.

The reader is cautioned that the literature of industrial statistics incorporates a variety of notations, so that some sources use the symbol n to represent both sample size and subsample sizes and tables are indexed accordingly.

CASE HISTORY 1.1

Solder Joints

A solder joint is a simple thing. In a hearing aid, there are some 85. Many of our everyday items, a small transistor radio, a telephone switchboard, a kitchen toaster, all are dependent on solder joints.

When we asked Fritz, the head of a department which assembles hearing-aid chassis, how many defective solder joints he had, the reply was, "Well, we don't really know, but not too many." Having no basis for even a wild guess, we suggested one in a hundred. "Well, maybe," said Fritz. So we talked to the quality control supervisor.

How does one proceed to improve soldering? There are many answers: "better" soldering irons or "better" solder, a quality motivation program or improved instructions to the foreman and operators. We began a small study by recording the number of defects found on a sample of just ten hearing aids per day. We recorded the location of defects by making tally marks on a blown-up diagram of the circuitry. After a few days, it was evident that there were six or seven positions, of the possible 87, responsible for the great majority of defects.

Initial data showed about one defect per 100 solder joints—such as cold solder joints, open joints, shorting contacts. Spacings were very close (not like a big telephone switchboard or a guided missile, critics argued), and some thought that 1:100 was as good as could be expected. Besides, solder joints were inspected 100 percent, so why the concern?

Data was taken and analyzed. Reductions in defects came quickly. One wire at a soldering position was given a pretinning. Another was given a sleeve to eliminate possible shorting. Specific instructions to individual operators on their soldering techniques also helped. A control chart was posted and was a surprisingly good motivational factor.

In three months, the continuing samples of 10 units per day showed a reduction of soldering defects to about 1:10,000. More important (and surprising to many) was the marked improvement in quality of the completed hearing aids.

Once again, quality can only be manufactured into the product—not inspected into it.

Improvements in the hearing-aid assembly required a detailed analysis of individual operator performance (the individual is, indeed, important, and each may require specific help). Group motivation can be helpful, too, in some situations.

Are you investigating the few positions in your operations which account for most of the defects? Are you then establishing continuing control charts, graphical reports, and other aspects of a quality feedback system which will help maintain improvements and point to the beginning of later problems which will surely develop? Getting a quality system organized is not easy.

What is the quality problem? What is a good way to attack the problem? Production will have one answer, design may have another, purchasing and testing another, and so on. But you can be sure of one thing: everyone will tell you in some indirect way, "It isn't my fault!" To anticipate this is not cynicism, it is merely a recognition of human nature, shared by all of us.

Almost everyone would like a magic wand, an overall panacea, applying with equal effectiveness to all machines, operators, and situations, thereby eliminating the need for us to give attention to piece-by-piece operation. There are different types of magic wands:

"Give us better soldering irons, or better solder, or better components, or better raw materials and equipment. This will solve the problem!" Of course, one or more such changes may be helpful, but they will not excuse us from the responsibility of working with specific details of processes within our own control to obtain optimum performance.

"Give us operators who care," also known as "if we could only find a way to interest operators on the line in their assignments and get them to pay attention to instructions!" Of course. But in the hearing-aid experience, operators primarily needed to be well-instructed (and reinstructed) in the details of their operations. It was the system of taking samples of 10 units per day that provided clues as to which operators (or machines) needed specific types of instructions.

Each of these improvements can be helpful. Indeed, they were helpful in one way or another in the improvement of hearing-aid defects from a rate of 1:100 units to 1:10,000 units. The critical decision, however, was the one to keep records on individual solder positions in such a way that individual trouble points could be pinpointed and kept under surveillance.

1.17 Summary

An orderly collection and plotting of a moderate data sample will frequently suggest trouble, and if not solutions, then sources of trouble which warrant further investigation. This chapter considers a case history of 200 samples taken from a mass repetitive process.

From this sample, methods of calculating various statistical estimates are set forth: the standard deviation, average (mean), and range. In a discussion of basic probability distributions a foundation is laid for comparing what is expected to happen with what is actually happening.

This chapter introduces concepts about samples from a process or other population source. It presents methods of estimating the central tendency of the process μ and its inherent process capability σ. The importance of an estimate computed from a large sample is compared with that computed from a set of k small samples from the same process. These concepts are basic in troubleshooting. They are also basic to the methods of Chap. 2.

1.18 Practice Exercises

The following exercises are suggestions which may be used in association with the indicated sets of data. Working with sets of data is helpful in understanding the basic concepts which have been discussed. If you have real sets of data from your own experience, however, you are urged to use these same methods with them.

1. *a*. Make a tally sheet for the data in Table 1.1 using cell width $m = 0.5$ thousandth (inch).
   b. From the tally choose an A, then compute $\overline{X}$ and $\hat{\sigma}$; make a casual comparison with the previous results using $m = 1$ in Table 1.3. (They should agree closely but not exactly.)
2. The top half of the data in Table 1.1 is also a sample, $n = 100$, of the mica-splitting process.
   a. Compute $\overline{X}$ and $\hat{\sigma}$ from Eqs. (1.3) and (1.4*b*) for this top half.
   b. Also, compute $\hat{\sigma} = \overline{R}/d_2$ from Eq. (1.13) using $k = 20$ vertical sets of size $n_g = 5$.
   c. Compare $\hat{\sigma}$ obtained in (b) with that obtained in Ex. 1.*b* and Table 1.3.
   d. Also compute $\hat{\sigma}$ from Eq. (1.13) using horizontal subgroups, $n_g = 5$.
3. The first four columns of Table 1.1 comprise a sample, $n = 80$, of the mica-splitting process.
   a. Compute $\overline{X}$ and $\hat{\sigma}$ from Eqs. (1.2) and (1.4*a*).
   b. Compute $\hat{\sigma} = \overline{R}/d_2$ from Eq. (1.13) using $k = 20$ sets of $n_g = 4$, grouped horizontally. Are the results of (a) and (b) similar?
4. Prepare a frequency distribution for the "depth-of-cut" data of Table 1.8. (We suggest that you make your tally marks in one color for the first 16 rows of samples and in a contrasting color for the last 9 rows. Then note the contrast in location of the two sets.)
   a. Compute $\overline{X}$ and $\hat{\sigma}$ from Eqs. (1.3) and (1.4*b*).
   b. Compute $\hat{\sigma}_2$ from Eq. (1.13) using ranges from the rows of samples, $n_g = 5$.
   c. Compute $\hat{\sigma}_1$ from the histogram of the first 16 rows or all 25. *Note:* See Case History 2.1 for some discussion.
5. Prepare suitable frequency distributions of the sets of data referred to in the exercises below. Find $\overline{X}$ and $\hat{\sigma}$ for each set. (Also, compare $\hat{\sigma}$ obtained from the frequency distribution with $\hat{\sigma} = \overline{R}/d_2$ from a suitable range chart, using any grouping you choose or may be assigned.)
   a. The 77 measurements in Table 1.7 on an electrical characteristic.
   b. The 125 measurements in Table 1.8, depth of cut.
   c. Consider again the process which produced the data in Table 1.1. If we assume that the average of the process could be increased to be at the center of the specifications, what percent would be expected to be under the LSL and what percent over the USL? Assume no change in σ.
6. If the specifications are as listed below, find the expected percentages nonconforming produced by the process which produced the corresponding samples:
   a. In Table 1.7, below LSL = 14.5 dB and above 17 dB.
   b. In Table 1.8, below LSL = 0.159 in and above USL = 0.160 in
7. Prepare a graph on normal-probability paper for all the data of Table 1.7.
   a. Is there seeming evidence of more than one principal parent universe?
   b. Estimate σ from the normal-probability graph. Compare it with $\hat{\sigma} = \overline{R}/d_2$. Do they disagree "substantially," or are they "in the same ball park?"
8. A design engineer made measurements of a particular electrical character-

TABLE 1.7 Electrical Characteristics (in Decibels) of Final Assemblies from 11 Strips of Ceramic: Case History 15.1

	1	2	3	4	5	6	7	8	9	10	11
	16.5	15.7	17.3	16.9	15.5	13.5	16.5	16.5	14.5	16.9	16.5
	17.2	17.6	15.8	15.8	16.6	13.5	14.3	16.9	14.9	16.5	16.7
	16.6	16.3	16.8	16.9	15.9	16.0	16.9	16.8	15.6	17.1	16.3
	15.0	14.6	17.2	16.8	16.5	15.9	14.6	16.1	16.8	15.8	14.0
	14.4	14.9	16.2	16.6	16.1	13.7	17.5	16.9	12.9	15.7	14.9
	16.5	15.2	16.9	16.0	16.2	15.2	15.5	15.0	16.6	13.0	15.6
	15.5	16.1	14.9	16.6	15.7	15.9	16.1	16.1	10.9	15.0	16.8
$\bar{X}=$	16.0	15.8	16.4	16.5	16.1	15.0	15.9	16.3	14.6	15.7	15.8
$R=$	2.8	3.0	2.4	1.1	1.1	2.5	3.2	1.9	5.9	4.1	2.8

SOURCE: Ellis R. Ott, Variables Control Charts in Production Research, *Ind. Qual. Control*, vol. 6, no. 3, p. 30, 1949. (Reprinted by permission of the editor.)

TABLE 1.8 Air-Receiver Magnetic Assembly: Case History 2.1

Measurements (depth of cut) in inches on each of five items in a sample taken at 15-min intervals during production

Sample no.			$(n_g = 5)$		
1	.1600	.1595	.1596	.1597	.1597
2	.1597	.1595	.1595	.1595	.1600
3	.1592	.1597	.1597	.1595	.1602
4	.1595	.1597	.1592	.1592	.1591
5	.1596	.1593	.1596	.1595	.1594
6	.1598	.1605	.1602	.1593	.1595
7	.1597	.1602	.1595	.1590	.1597
8	.1592	.1596	.1596	.1600	.1599
9	.1594	.1597	.1593	.1599	.1595
10	.1595	.1602	.1595	.1589	.1595
11	.1594	.1583	.1596	.1598	.1598
12	.1595	.1597	.1600	.1593	.1594
13	.1597	.1595	.1593	.1594	.1592
14	.1593	.1597	.1599	.1585	.1595
15	.1597	.1591	.1588	.1606	.1591
16	.1591	.1594	.1589	.1596	.1597
17	.1592	.1600	.1598	.1598	.1597
18	.1600	.1605	.1599	.1603	.1593
19	.1599	.1601	.1597	.1596	.1593
20	.1595	.1595	.1606	.1606	.1598
21	.1599	.1597	.1599	.1595	.1610
22	.1596	.1611	.1595	.1597	.1595
23	.1598	.1602	.1594	.1600	.1597
24	.1593	.1606	.1603	.1599	.1600
25	.1593	.1598	.1597	.1601	.1601

istic on an initial sample of hearing aids. Find $\overline{X}$ and s for these eight measurements: 1.71, 2.20, 1.58, 1.69, 2.00, 1.59, 1.52, 2.52.

a. Estimate the range within which 99.7 percent of the product will fall.

b. Give a 95 percent confidence interval for the true mean and standard deviation of the process from which this sample was taken.

c. Are these data normal? Check with a probability plot.

Chapter

2

Ideas from Time Sequences of Observations

2.1 Introduction

A gradual change in a critical adjustment or condition in a process is expected to produce a gradual change in the data pattern. An abrupt change in the process is expected to produce an abrupt change in the data pattern. We need ways of identifying the *presence* and *nature* of these patterns. The fairly standard practice of examining any regular data reports simply by looking at them is grossly inadequate. Such reports are far more valuable when analyzed by methods discussed in the following sections.

There is no single way for a medical doctor to diagnose the ailment of a patient. Consideration is given to information from a thermometer, a stethoscope, pulse rates, chemical and biological analysis, x-rays, and many other tests.

Neither is there just one way to obtain or diagnose data from the operation of a process. Simple processes are often adjusted without reference to any data. But data from even the simplest process will provide unsuspected information on its behavior. In order to benefit from data coming either regularly or in a special study from a temperamental process, it is important to follow one important and basic rule:

Plot the data[1] *in a time sequence.*

Different general methods are employed to diagnose the behavior of time-sequence data after plotting. Two important ones will be discussed in this chapter.

1. Use of certain *run* criteria (Sec. 2.4).
2. Control charts with control limits and various other criteria (Sec. 2.5) signaling the presence of assignable causes.

[1] It is standard practice to scan a data report, then file it, and forget it. We suggest instead that you plot important data and usually dispose of the report. Or, perhaps record the data initially in graphical form.

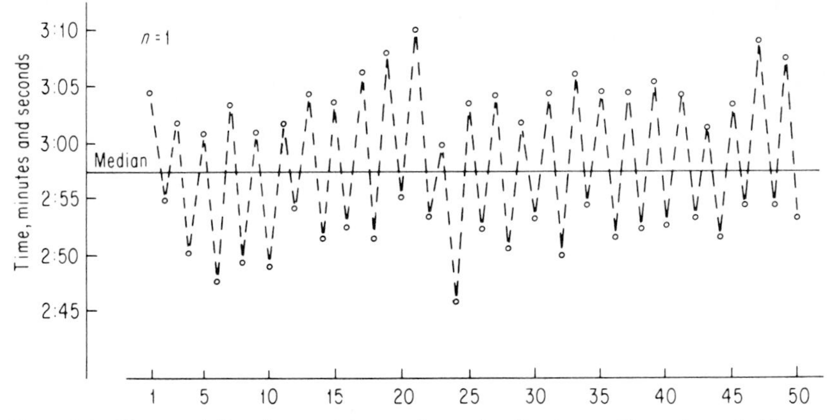

Figure 2.1 Measured time for sand to run through a 3-min egg timer (recorded in order of observation). (*Courtesy of Mrs. Elaine Amoresano Rose.*)

Example 2.1. A Look At Some Data In a graduate course, primarily of students in statistics but including graduate students from the natural and social sciences, Dr. Ott's first assignment for each student had been to "obtain a time sequence of k subgroups from some process," asking them if possible, to "choose a process considered to be stable." A young lady[2] elected to complete her assignment by measuring times for sand to run through a 3-min egg timer in successive tests. The time was measured by a stop watch. Data are shown in Fig. 2.1. Does this set of data appear to represent a stable (random) process? Also, is it a "3-min egg timer"?

Some Casual Observations

- The median of the 50 observations is about 2½ sec below 3 min. This is a slight bias (inaccuracy) but should not affect the taste quality of boiled eggs.
- Almost no point is "close" to the median! Half of the points lie about 10 sec above the median and the other half 5 to 10 sec below.
- The points are alternately high and low—a perfect "sawtooth" pattern *indicating two causes* operating alternately to produce the pattern. This agrees with the preceding observation.
- There appears to be a steady increase from the eighth to the twentieth point on each side of the egg timer.
- Beginning with the twenty-third observation, there is an abrupt drop, both on the "slow half" and the "fast half." After the drop, the process operates near the initial level.

Discussion—Egg Timer Data An egg timer is surely a simple machine; one would hardly predict nonrandomness in successive trials with it. However, once the peculiar patterns are seen, what are possible explanations?

- *The sawtooth:* The egg timer had two halves. Elaine recognized this as an obvious "probable" explanation for the sawtooth pattern; she then made a few more measurements to identify the "fast" and "slow" sides of the timer.

[2]Mrs. Elaine Amoresano Rose, former graduate student in Applied and Mathematical Statistics, Rutgers Statistics Center.

- *The abrupt shift downward (twenty-third point).* There are three possibilities:

  1. *The egg timer;* the sand may be affected by *humidity* and *temperature.* Elaine said she took a break after the twenty-third experiment. Perhaps she laid the timer in the sun or on a warm stove. A change in heat or humidity of the sand may be the explanation for the drop at the twenty-fourth experiment.

  2. *The stop watch* used in timing. There is no obvious reason to think that its performance might have produced the sawtooth pattern, but thought should be given to the possibility. It does seem possible that a change in the temperature of the watch might have occurred during the break. Or, was the watch possibly inaccurate?

  3. *The operator observer.* Was there an unconscious systematic parallax effect introduced by the operator herself? Or some other operator effect?

  Thus when studying *any process* to determine the cause for peculiarities in the data, one must consider in general: (1) the manufacturing process, (2) the measuring process, and (3) the way the data are taken and recorded.

Summary Regarding Egg Timer Data in Fig. 2.1 Figure 2.1 shows the presence of two types of nonrandomness, neither of which was foreseen by the experimenter. Nonrandomness *will almost always* occur; such occurrence is *the rule and not the exception.*

An egg timer is a very simple system in comparison with the real-life scientific systems we must learn to diagnose and operate. The unsuspected behaviors of large-scale scientific systems are much more complicated; yet they can be investigated in the same manner as the egg timer. Data from the process showed differences: between the two sides of the egg timer, *and* a change with time. Sometimes the causes of unusual data patterns can be identified easily. It is logical here to surmise that a slight difference exists in the shape of the two sides, Fig. 2.2 (*a*) and (*b*); often the identification is more elusive. Knowing that nonrandomness exists is most important information to the experimenter. However, this knowledge must be supported by follow-up investigations to *identify the causes* of any nonrandomness which are of practical interest.

Note: This set of data, Fig. 2.1, will be discussed further in Sec. 2.4. The very powerful graphical display and the "look test" provide the important information. But more formal methods are usually needed, and two important

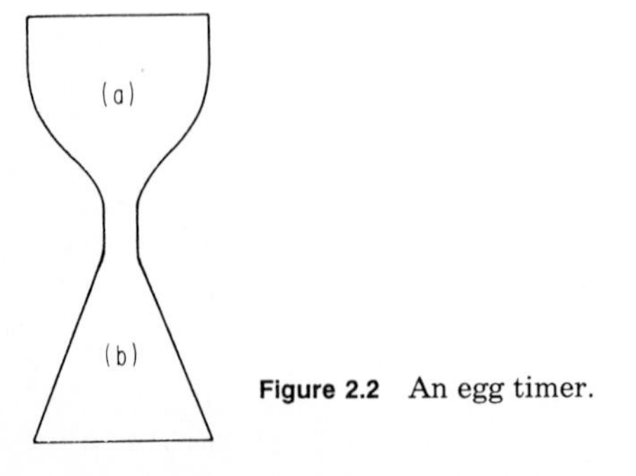

Figure 2.2 An egg timer.

direct methods of analysis are discussed later in the chapter. The methods are *run analysis* and *control charts*. We shall first consider two general approaches to process troubleshooting.

2.2 Data from a Scientific or Production Process

It is standard practice to study a scientific process by changing different variables suspected of contributing to the variation of the process. The resulting data are then analyzed in some fashion to determine whether the changes made in these variables have had an effect which appears significant either scientifically or economically.

A less-utilized but important method is to *hold constant all variables* which are suspected of contributing to variations of the process, and then decide whether the *resulting pattern* of observations actually represents a stable, uniform process—i.e., whether the process is "well-behaved" or whether there is evidence of previously unknown nonstability (statistical nonrandomness). Different patterns indicate different causes of nonrandomness and often suggest the type of factors which have influenced the behavior of the process—even though neither their existence nor identity may have been suspected. This unsuspected nonrandom behavior occurs frequently, and recognition of its existence may prompt studies to identify the unknown factors, that is, *lead to scientific discovery.*

Whenever a sequence of observations in order of time can be obtained from a process, an analysis of its data pattern can provide important clues regarding variables or factors which may be affecting the behavior of the process.

2.3 Signals and Risks

Dr. Paul Olmstead[3] remarks:

> To the extent that we as engineers have been able to associate physical data with assignable causes, these causes may be classified by the types of physical data that they produce, namely:
>
> 1. Gross error or blunder (shift in an individual)
> 2. Shift in average or level
> 3. Shift in spread or variability
> 4. Gradual change in average or level (trend)
> 5. A regular pattern of change in level (cycle)

When combinations of two or more assignable causes occur frequently in a process, they will then produce combinations of data patterns. Learning that something in the system is affecting it, either to its advantage or disadvantage, is important information. The province and ability of specialists—engi-

[3]Paul S. Olmstead, "How to Detect the Type of an Assignable Cause," *Ind. Qual. Control,* vol. 9, no. 3, p. 32 and vol. 9, no. 4, p. 22. (Reprinted by permission of the author and editor.)

neers, scientists, production experts—are to find compensating corrections and adjustments. Their know-how results both from formal training and practical experience. However, experience has shown that they welcome suggestions. Certain patterns of data have origins which can be associated with causes; origins are suggested at times by data patterns.

When analyzing process data, we need criteria which will signal the presence of important process changes of behavior but which will not signal the presence of rather minor process changes. Or, when we are studying the effects of different conditions in a research and development study, we want criteria which will identify those different conditions (factors) which may contribute substantially either to potential improvements or to difficulties. If we tried to establish signals which never were in error when indicating the presence of important changes, then those signals would sometimes *fail to signal* the presence of important conditions. It is *not possible to have perfection.*

The facts are that we must take risks in any scientific study just as in all other aspects of life. There are two kinds of risks: (1) that we shall institute investigations which are unwarranted either economically or scientifically; this is called the alpha risk (α risk), the *sin of commission,* and (2) that we shall miss important opportunities to investigate; this is called the beta risk (β risk), the *sin of omission.*

We aspire to sets of decision criteria which will provide a reasonable compromise between the α and β risks. A reduction in the α risk will increase the β risk unless compensations of some kind are provided. The risk situation is directly analogous to the person contemplating the acceptance of a new position, of beginning a new business, of hiring a new employee, or of buying a stock on the Stock Exchange.

What risks are proper? There is no single answer, of course. When a process is stable, we want our system of signals to indicate stability; when there is *enough* change to be of possible scientific interest, the signaling system should *usually* indicate the presence of assignable causes. Statisticians usually establish unduly low risks for α, often $\alpha = 0.05$ or $\alpha = 0.01$. They are reluctant to advise engineering, production, or other scientists to investigate conditions unless they are almost certain of identifying an important factor or condition. However, the scientist is the one who has to decide the approximate level of compromise between "looking unnecessarily for the presence of assignable causes" and "missing opportunities for important improvements." A scientist in research will often want to pursue possibilities corresponding to appreciably *larger* values of α and *lower* values of β, especially in exploratory studies; and may later want to specify *smaller* values of α when publishing the results of important research. Values of $\alpha = 0.10$ and even larger are often sensible to accept when making a decision whether to investigate possible process improvement. In diagnosis for troubleshooting, we expect to make some unnecessary investigations. It is prudent to investigate many times knowing that we may fail to identify a cause in the process. Perhaps $\alpha = 0.10$ or $\alpha = 0.25$ is economically practical. Not even the best professional baseball player bats as high as 0.500 or 0.600.

The relationship of these risks to some procedures of data analysis will be discussed in the following sections. Some methods of *runs* are considered first. Then methods of *control charts* are considered in the following chapter.

Some signals to observe

When it has been decided to study a process by obtaining data from its performance, then the data should be plotted in some appropriate form and in the order it is being gathered. Every set of k subgroups offers two opportunities:

1. To *test the hypothesis* that the data represent *random variation* from stable sources. Was the source of data apparently stable or is there evidence of nonrandomness?
2. To infer the *nature* of the source(s) responsible for any nonrandomness (from the data pattern), that is, to infer previously unsuspected hypotheses.

This is the essence of the scientific method. Two major types of criteria are discussed in Secs. 2.4 and 2.5.

2.4 Run Criteria

Introduction

When someone repeatedly tosses a coin and produces a run of six heads in succession, we realize that something is unusual. It *could* happen; the probability is $(0.5)^6 = 0.015$. We would then usually ask to see both sides of the coin because these would be very unlikely runs from an ordinary coin having a head and a tail. When we then question the coin's integrity, our *risk* of being unreasonably suspicious is $\alpha = 0.015$.

A *median* line is one with half of the points above and half below. (Probability that a single observation falls above is $P_a = 0.5$ and that it falls below is $P_b = 0.5$, when k is even). The use of runs is formalized in the following sections; it is a most useful procedure to suggest clues from an analysis of ordered data from a process.

When exactly three consecutive points are above the median, this is a *run above the median* of length 3. In Fig. 2.3 consecutive runs *above and below the median* are of length 3, 1, 1, 1, 2, and 4. The total number of runs is $N_R = 6$. We usually count the number directly from the figure once the median line

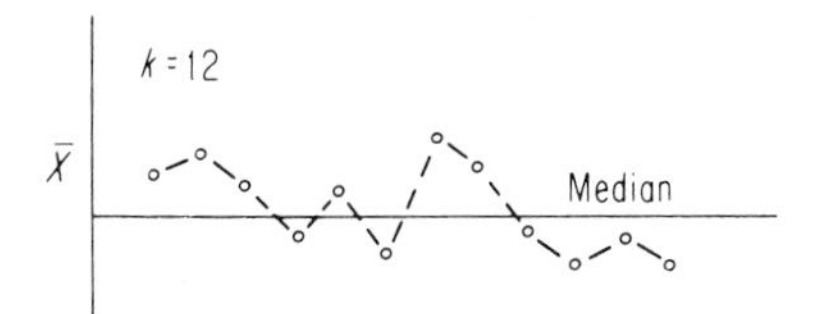

Figure 2.3 Twelve averages showing six runs above and below the median.

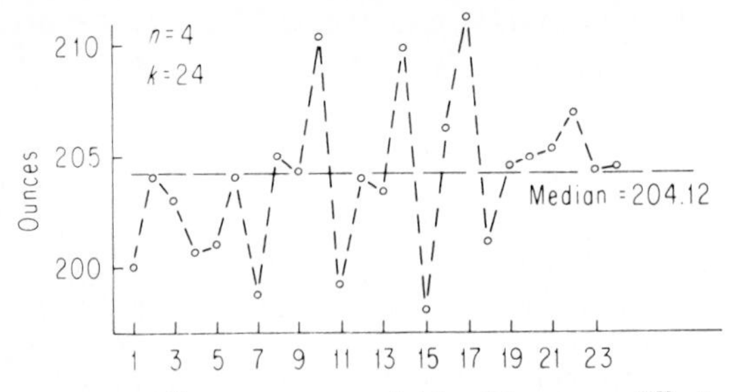

Figure 2.4 Gross average weights of ice-cream fill at 10-min intervals (data from Table 2.5). (*Courtesy of David Lipman.*)

has been drawn. Locating the median when k is larger can be expedited by adjusting a clear, plastic ruler or the edge of a card.

Example 2.2 The data plotted in Fig. 2.4 represent $k = 24$ averages (gross weights of ice cream) in order of production. Data represent $\bar{X}$ values, each computed from $n_g = 4$ observations in Table 2.5.

The median is halfway between 204.00 and 204.25. The *total number of runs above and below the median* is $N_R = 8$. The runs are of length 7, 3, 3, 1, 1, 2, 1, and 6, respectively.

Before presenting a formal direct run analysis, let us consider the first run; it is of length 7. It is just as improbable for a stable process to produce the first seven observations on the same side of the median as for a coin to flip seven heads (tails) at the start of a demonstration. *We do not believe that the process represented in Fig. 2.4 was stable* over the first 4-hr period of manufacture. This is not surprising; it is *unlikely that any industrial process* will function at an entirely stable level over a 4-hr period. Whether the magnitude of the changes here is economically important is a different matter. Now let us consider a direct analysis using *runs.*

Some interpretations of runs

Too many runs above and below the median indicate the following possible engineering reasons:

1. Samples being drawn alternately from two different populations (sources), resulting in a "sawtooth" effect. These occur fairly frequently in *portions* of a set of data. Their explanation is usually found to be two different sources—analysts, machines, raw materials—which enter the process alternately or nearly alternately.
2. Three or four different sources which enter the process in a cyclical manner.

Too few runs are quite common. Their explanations include:

1. A general shift in the process average
2. An abrupt shift in the process average
3. A slow cyclical change in averages

Sources of variation in a process can usually be determined (identified) by the engineer, chemist, or production supervisor once aware that they exist.

Formal criteria: total number of runs about the median

The total number of runs in the data of Fig. 2.4 is $N_R = 8$. How many such runs are expected in a set of k random points? In answering the question, it is sufficient to consider even numbers only, that is, numbers of the form $k = 2m$. The *average expected number* is

$$\overline{N}_R = \frac{k+2}{2} = m + 1 \tag{2.1}$$

and the *standard deviation* of the sampling distribution of N_R is

$$\sigma = \sqrt{\frac{m(m-1)}{2m-1}} = \sqrt{\frac{\frac{k}{2}\left(\frac{k}{2}-1\right)}{k-1}} = \sqrt{\frac{k(k-2)}{4(k-1)}} = \frac{1}{2}\sqrt{\frac{k(k-2)}{k-1}} \cong \frac{\sqrt{k}}{2} \tag{2.2}$$

Equation (2.1) is used so frequently that it helps to remember it. An aid to memory is the following:

1. The *minimum* possible number of runs is 2.
2. The *maximum* possible number is k.
3. The *expected* number is the average of these two. This is an aid to memory, not a proof.[4]

Of course, the number of runs actually observed will often be more or less than $\overline{N}_R = m + 1$. By how much? The answer is obtained easily from Table A.2; it lists one-tailed significantly small critical values and significantly large critical values corresponding to risks $\alpha = 0.05$ and 0.01.

Example 2.3. Use of Table A.2 In Fig. 2.4 with $k = 24$, the expected number of runs is $0.5(24 + 2) = 13$. The number we observe is only 8. From Table A.2, small critical values of N_R corresponding to $k = 24$ are seen to be 8 and 7 for $\alpha = 0.05$ and 0.01, respectively.

Thus the count of 8 runs is less than expected with a 5 percent risk, but not significantly less with a 1 percent risk. This evidence of a nonstable (nonrandom) process behavior agrees with the presence of a run of length 7 below the median. We would ordinarily investigate the process expecting to identify sources of assignable causes.

Expected number of runs of exactly length *s* (optional)

A set of data may display the expected total number of runs but may have an unusual distribution of long and short runs. Table A.3 has two columns. The first

[4] Churchill Eisenhart and Freda S. Swed, "Tables for Testing Randomness of Grouping in a Sequence of Alternatives," *Ann. Math. Stat.*, vol 14, 1943, pp. 66–87.

one lists the expected number of runs $\overline{N}_{R,s}$ of exactly length s; the second lists the expected number of runs, $\overline{N}_{R \geq s}$, that is, of length greater than or equal to s.

From the second column, for example, it can be seen that when $k = 2^6 = 64$, only one run of *length 6 or longer* is expected. When $k = 2^5 = 32$, only one-half a run of length 6 or longer is expected (i.e., a run of length 6 or longer is expected about half the time).

Example 2.4 Consider again the ice-cream fill data of Fig. 2.4, $k = 24$. The number of runs of exactly length $s = 1$, for example, is 6.8 as shown in Table 2.1. The number expected from the approximation, Table A.3, is $24/4 = 6$; the number actually in the data is 3. Other comparisons of the number expected with the number observed are given in this table. It indicates two long runs: the first is a run of 7 below average; then a run of 6 high weights at the end. This pattern suggests an increase in filling weight during the study; it is not clear whether the increase was gradual or abrupt.

A χ^2 test might be performed to check the significance of the lengths of runs exhibited in Table 2.1. The formula for χ^2 is

$$\chi^2 = \Sigma \frac{(\text{observed} - \text{expected})^2}{\text{expected}}$$

and critical values of χ^2 are tabulated elsewhere in most statistics texts. For length of run 1, ≥ 2, we obtain

$$\chi^2 = \frac{(3 - 6.8)^2}{6.8} + \frac{(5 - 6.2)^2}{6.2} = 2.12 + 0.23 = 2.35$$

The critical value at the $\alpha = 0.05$ level is $\chi_{0.05}{}^2 = 3.84$ so we are unable to assert that nonrandomness exists from this test.

Example 2.5 Consider the sawtooth pattern of the egg timer, Fig. 2.1. From Table A.2, the critical values for the total expected numbers of runs N_R above and below the median for $k = 50$ are

$$17 \leq N_R \leq 34$$

TABLE 2.1 A comparison of the expected Number of Runs* and the Observed Number

Data of Fig. 2.4, where $k = 24$

s	Expected number of runs of exactly length s	Number observed	Expected number of runs of length $\geq s$	Number observed
1	6.8	3	13.0	8
2	3.4	1	6.2	5
3	1.6	2	2.8	4
4	0.7	0	1.2	2
5	0.3	0	0.5	2
6	0.1	1	0.2	2
7	0.1	1	0.1	1

* *Note:* Values for expected number of runs in this table have been computed from the exact formulae in Table A-3. This is not usually advisable since the computation is laborious and the approximate values are sufficiently close for most practical purposes.

The observed number is 50. This is much larger than the critical value of 34 corresponding to $\alpha = 0.01$. We conclude (again) that the data are not random. The pattern produced by the egg timer is a perfect sawtooth indicating two alternating sources. The sources are evidently the two sides of the egg timer.

The longest run-up or run-down

In Fig. 2.4. there are four *increases* in $\overline{X}$ beginning with sample 18 and concluding with 22. This four-stage increase is preceded and followed by a decrease; this is said to be a *run-up of length exactly 4.* In counting runs up and down we count the intervals between the points, and not the points themselves.

It is easy to recognize a long run-up or run-down once the data have been plotted. A long run-up or run-down is typical of a substantial shift in the process average. Expected values of both long and short extreme lengths in a random display of k observations are sometimes of value when analyzing a set of data. Here k is the number of points in the sequence being analyzed. A few one-tailed critical values have been given in Table 2.2.

It is easy to see and remember that a run-up of length 6 or 7 is quite unusual for sets of data even as large as $k = 200$. Even a run of 5 may warrant investigating.

Summary—run analysis

Important criteria which indicate the presence of assignable causes by unusual runs in a set of k subgroups of n_g each have been described. They are applicable either to sets of data representing k individual observations ($n = 1$) or to k averages ($n_g > 1$); or, to k percents defective found by inspecting k lots of an item.

TABLE 2.2 Critical Extreme Length of a Run-Up or a Run-Down in a Random Set of k Observations (One-Tail)

	$\alpha = .01$		$\alpha = .05$	
k	Small critical value	Large critical value	Small critical value	Large critical value
10	0	6	1	5
20	1	6	1	5
40	1	7	1	6
60	1	7	2	6
100	2	7	2	6
200	2	7	2	7

SOURCE: These tabular values were sent by Paul S. Olmstead based on his article: Distribution of Sample Arrangements for Runs-Up and Runs-Down, *Ann. Math. Stat.*, vol. 17, pp. 24–33, March 1946. (They are reproduced by permission of the author.)

The ultimate importance of any criterion in analyzing data is its usefulness in identifying factors which are important to the behavior of the process. The application of runs ranks high in this respect.

RUN CRITERIA

1. *Total number* of runs N_R about the median:[5] Count the runs.
   a. Expected number is $\overline{N}_R = k + 2/2$.
   b. The fewest and largest number of expected runs are given in Table A.2 for certain risks, α.
2. A run *above* or *below the median* of length greater than six is evidence of an assignable cause warranting investigation, even when k is as large as 200. Count the points.
3. The *distribution* of runs of length s about the median (See Table A.3). A set of data may display about the expected total number of runs yet have too many or too few short runs (or long ones). Count the points.
4. A *long run-up or run-down* usually indicates a gradual shift in the process average. A run-up or run-down of length five or six is usually longer than expected. Count the lines between the points. Note that the run length and the number of runs are quite different.

2.5 Shewhart Control Charts for Variables

Introduction

The Shewhart control chart is a well-known, powerful method of checking on the stability of a process. It was conceived as a device to help production in its routine hour-by-hour adjustments; its value in this regard is unequaled. It is applicable to quality characteristics, either of a variable or attribute type. The control chart provides a graphical time sequence of data from the process itself. This then permits the application of run analyses to study the historical behavior patterns of a process. Further, the control chart provides *additional* signals to the current behavior of the process; the upper and lower control limits (UCL and LCL) are limits to the maximum expected variation of the process. The mechanics of preparing variable and attribute control charts will be presented. Their application to troubleshooting is a second reason for their importance.

[5]Sometimes we apply the criteria of runs to the *average* line instead of the median; we do this as tentative criteria.

Mechanics of preparing control charts (variables)

The control chart is a method of studying a process from a sequence of small random samples from the process. The basic idea of the procedure is to collect small samples of size n_g (usually at regular time intervals) from the process being studied. Samples of size $n_g = 4$ or 5 are usually best. It will sometimes be expedient to use $n_g = 1$, 2, or 3; sample sizes larger than 6 or 7 are not recommended. A quality characteristic of each unit of the sample is then measured, and the measurements are (usually) recorded *but are always charted.*

The importance of *rational subgroups* must be emphasized when specifying the source of the $n_g = 4$ or 5 items in a sample. Since our aim is to actually locate the trouble, as well as to determine whether or not it exists, we must break down the data in a logical fashion. "The man who is successful in dividing his data initially into *rational* subgroups based upon rational hypotheses is therefore inherently better off in the long run than the one who is not thus successful."[6]

In starting a control chart, it is necessary to collect some data to provide preliminary information determining central lines on average $\overline{X}$ and ranges R. It is usually recommended that $k = 20$ to $k = 25$ subgroups of n_g each be obtained, but $k < 20$ may be used initially to avoid delay. (Modifications may be made to adjust for unequal subgroup sizes.) The formal routine of preparing the control chart once the k data subgroups have been obtained are as follows:

STEP 1: Compute the average $\overline{X}$ and the range R of each sample. Plot the k points on the $\overline{X}$ chart and R chart being sure to preserve the order in which they were produced. (It is very important to write the sample size on every chart and in a regular place, usually in the upper left-hand side as in Fig. 2.5*a*, 2.5*b*.)

STEP 2: Compute the two averages, $\overline{\overline{X}}$ and $\overline{R}$; draw them as lines.

STEP 3: Compute the following 3-sigma control limits for the R chart and draw them as lines

$$\text{UCL}(R) = D_4\overline{R}$$

$$\text{LCL}(R) = D_3\overline{R}$$

Observe whether any ranges fall above $D_4\overline{R}$ or below $D_3\overline{R}$. If not, accept the concept (tentatively) that the variation of the process is homogeneous, and proceed to Step 4.[7]

---

[6]Walter A. Shewhart, *Economic Control of Quality of Manufactured Product,* D. Van Nostrand Company, Inc., New York, 1931, p. 299.

[7]When the R chart has a single outage, we sometimes do two things: (*a*) check the sample for a maverick, and (*b*) exclude the outage subgroup and recompute $\overline{R}$. Usually this recomputing is not worth the effort.

When the R chart has several outages, the variability of the process is unstable, and it will not be reasonable to compute a $\hat{\sigma}$. The process needs attention.

Other examples treating an R chart with outages are discussed in other sections of this book.

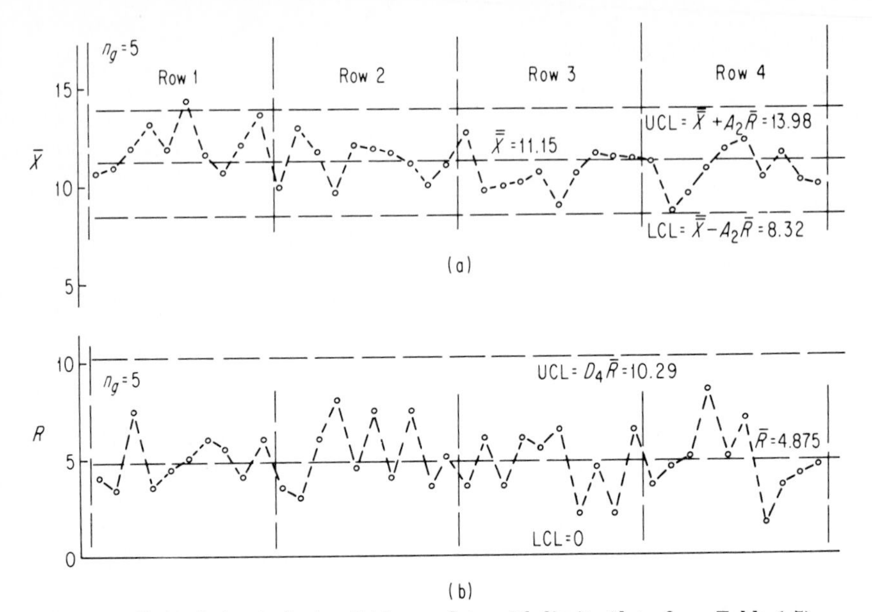

Figure 2.5 Control chart of mica thickness data with limits (data from Table 1.5).

Note: The distribution of R is not symmetrical—it is "skewed" with a tail for larger values. Although we want values of $(\bar{R} + 3\hat{\sigma}_R)$, values of $3\hat{\sigma}_R$ are not obtained simply. The easiest calculation is to use the D_4 factors given in Table 2.3, where $D_4\bar{R} = \bar{R} + 3\hat{\sigma}_R$.

STEP 4: Compute $A_2\bar{R}$, and obtain 3-sigma control limits on $\bar{X}$:

TABLE 2.3 Factors to Use with $\bar{X}$, R Control Charts for Variables

Choose n_g to be less than seven when feasible; these factors assume sampling from a normal universe; see also Table A.4

n_g	D_3	D_4	A_2	d_2
2	0	3.27	1.88	1.13
3	0	2.57	1.02	1.69
4	0	2.28	0.73	2.06
5	0	2.11	0.58	2.33
6	0	2.00	0.48	2.53
7	0.08	1.92	0.42	2.70
8	0.14	1.86	0.37	2.85
9	0.18	1.82	0.34	2.97
10	0.22	1.78	0.31	3.08

$$\text{UCL}(\overline{X}): \overline{\overline{X}} + A_2\overline{R} = \overline{\overline{X}} + 3\hat{\sigma}_{\bar{x}}$$

$$\text{LCL}(\overline{X}): \overline{\overline{X}} - A_2\overline{R} = \overline{\overline{X}} - 3\hat{\sigma}_{\bar{x}}$$

where $\hat{\sigma}_{\bar{x}} = \hat{\sigma}/\sqrt{n_g}$ and $\hat{\sigma} = \overline{R}/d_2$.

STEP 5: Draw dotted lines corresponding to UCL($\overline{X}$) and LCL($\overline{X}$).

STEP 6: Consider whether there is evidence of assignable causes (see following discussion). If any point falls outside UCL and LCL, we call this an "outage" which indicates the existence of an *assignable cause.*

Some discussion

The recommendation to use 3-sigma control limits was made by Dr. Shewhart after extensive study of data from production processes. It was found that almost every set of production data having as many as 25 or 30 subsets will show outages. Further, the nature of the assignable causes signaled by the outages using 3-sigma limits was usually important and identifiable by process personnel.

Upper and lower 3-sigma control limits on $\overline{X}$ are lines to judge "excessive" variation of *averages* of samples of size n_g

$$\overline{\overline{X}} \pm 3\hat{\sigma}_{\bar{x}} = \overline{\overline{X}} \pm \frac{3\hat{\sigma}}{\sqrt{n_g}} = \overline{\overline{X}} \pm \frac{3\overline{R}}{d_2\sqrt{n_g}}$$

However, computation is simplified by using the A_2 factor from Table 2.3

$$\text{UCL}(\overline{X}) = \overline{\overline{X}} + A_2\overline{R}$$

where $A_2 = \dfrac{3}{d_2\sqrt{n_g}}$.

It was also found from experience that it was practical to investigate production sources signaled by certain *run criteria* in data. These run criteria are recommended as adjuncts to outages.

The choice of a reasonable or *rational* subgroup is important but not always easy to make in practice. Items produced on the same machine, at about the same time, and with the same operator will often be a sensible choice—but not always. A machine may have only one head or several; a mold may have one cavity or several. A decision will have to be made whether to limit the sample to just one head or cavity or allow all heads or cavities to be included. Initially, the decision may be to include several heads or cavities and then change to individual heads if large differences are found.

Sample sizes of 4 or 5 are usually best. They are large enough to signal important changes in a process; they are not large enough usually to signal smaller less important changes. Some discussion of the sensitivity of sample size is given in connection with operating-characteristic curves.

Example of control-chart limits, mica-thickness data

In Fig. 1.11, charts of $\overline{X}$ and R points, $n_g = 5$, were made for the mica-thickness data of Table 1.5. We assumed there that the range chart represented a stable process; under that assumption, we computed

$$\hat{\sigma} = \overline{R}/d_2 = 2.09$$

We may now use the procedure outlined above to compute 3-sigma control limits for each chart and to consider different criteria to check on stability of the process.

Control-chart limits

STEP 1: See Table 1.5: An $\overline{X}$ and R have been computed for each subgroup.

STEP 2: $\overline{\overline{X}} = 11.15$ and $\overline{R} = 4.875$

STEP 3: $\text{UCL}(R) = D_4\overline{R} = (2.11)(4.875) = 10.29$
$\text{LCL}(R) = D_3\overline{R} = 0$
All points fall below UCL(R); see Fig. 2.5*b*.

STEP 4: $\overline{\overline{X}} + A_2\overline{R} = 11.15 + (0.58)(4.875) = 11.15 + 2.83 = 13.98$
$\overline{\overline{X}} - A_2\overline{R} = 11.15 - 2.83 = 8.32$

STEP 5: See Fig. 2.5*a*, control limits are plotted.

STEP 6: Based on criteria below.

Discussion: *R chart (Fig. 2.5b)*
There is no point above UCL(R), nor any above (or close to) $\overline{R} + 3\sigma_R$. Neither is there a long run on either side of $\overline{R}$; there is one run of length 4 below and one of length 4 above. Runs of length 4 are expected.

Conclusion: The chart suggests no unreasonable process variability; all points are below the upper control limit.

Discussion: $\overline{X}$ chart (Fig. 2.5a)
Point 6 ($\overline{X} = 14.3$) is above UCL($\overline{X}$); the outage indicates a difference in the process average. Also, of the eight points 3 to 10, inclusive, there are seven points above $\overline{\overline{X}}$. This run criterion suggests that the average of the first group of about 10 points is somewhat higher than the average of the entire set of data. The difference is not large; but it does indicate that something in the manufacturing or measuring process was "not quite" stable.

This set of data was chosen initially in order to discuss a process which was much more stable than ordinary. It is almost impossible to find $k = 20$ or more data subsets from an industrial process without an indication of instability. In process improvement and troubleshooting, these bits of evidence can be important.

Summary: some criteria for statistical control (stability)

Routine production: criteria for action. On the production floor, definite and uncomplicated signals and procedures to be used by production personnel work best. Recommended control-chart criteria to use as evidence of assignable causes requiring process adjustment or possible investigations are:

1. One point outside lines at

$$\bar{\bar{X}} \pm A_2\bar{R} = \bar{\bar{X}} \pm 3\hat{\sigma}_{\bar{x}} \qquad (\alpha \cong 3/1000)$$

A process shift of as much as 1σ is not immediately detected by a point falling outside 3-sigma limits; the probability is about 1/6 for $k = 4$ (see Fig. 2.8).

2. Two consecutive points (on the same side) outside either

$$\bar{\bar{X}} + 2\hat{\sigma}_{\bar{x}} \qquad \text{or} \qquad \bar{\bar{X}} - 2\hat{\sigma}_{\bar{x}} \qquad (\alpha \cong 1/800)$$

3. A run of seven consecutive points above (or below) the process average or median $(\alpha \cong 1/64)$

The first criterion is the one in ordinary usage; these last two should be used when it is important not to miss shifts in the average and there is someone to supervise the process adjustments.

Process improvement and troubleshooting. Since we are now anxious to investigate opportunities to learn more about the process or to adjust it, it is sensible to accept greater risks of making investigations or adjustments which may be futile perhaps as often as 10 percent of the time (allow a risk of $\alpha = 0.10$). Besides the three criteria just listed above, some or all of the following may be practical for you:

1. One point outside $\bar{\bar{X}} \pm 2\hat{\sigma}_{\bar{x}}$ $(\alpha \cong 0.05)$

2. A run of the last five points (consecutive) on the same side of $\bar{\bar{X}}$.[8] $(\alpha \cong 0.06)$

Note: The risk associated with any run of the last k points, $k > 5$, is less than for $k = 5$.

3. Six of the last seven points (6/7) on the same side $(\alpha \cong 0.10)$

Note: Risk for k out of the last $(k + 1)$, $k > 7$, is less than 0.10.

4. Eight of the last ten points (8/10) on the same side $(\alpha \cong 0.10)$

5. The last three points outside $\bar{\bar{X}} \pm \hat{\sigma}_{\bar{x}}$ (on the same side) $(\alpha \cong 0.01)$

[8]Probabilities associated with runs here and below are based on runs about the median of the data but are only slightly different when applied to runs about their mean (arithmetic average).

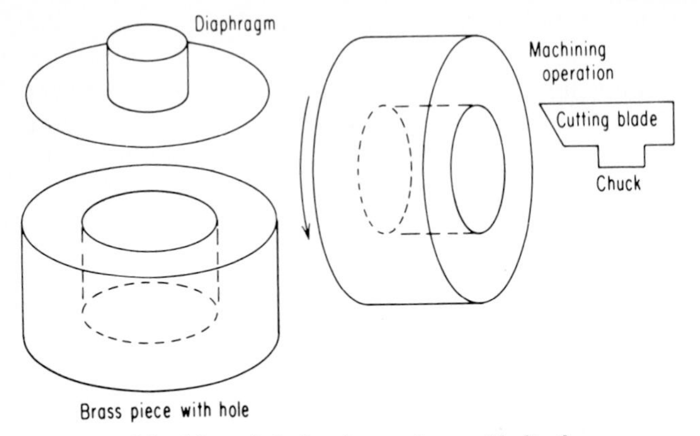

Figure 2.6 Matching a hole in a brass piece, with diaphragm assembly.

6. The last two points outside $\bar{\bar{X}} \pm 1.5\hat{\sigma}_{\bar{X}}$ (on the same side) $(\alpha \cong 0.01)$

There are different types of assignable causes, and some will be signaled by one of these criteria sooner than by another. Any signal which results in process improvement is helpful. It provides signals to those with the process know-how to investigate and gradually allows some of the art of manufacturing to be replaced by science.

CASE HISTORY 2.1

Depth of Cut

Ellis Ott recounts a typical troubleshooting adventure on the manufacturing floor.

> While walking through a department in a hearing-aid plant rather early one morning, I stopped to watch a small assembly operation. A worker was performing a series of operations; a diaphragm would be picked from a pile, placed as a cover on a small brass piece (see Fig. 2.6), and then the assembly placed in an electronic meter where a reading was observed. If the reading were within certain limits, the assembly would be sent on to the next stage in production. If not, the diaphragm was removed, another tried, and the testing repeated. After five or six such trials,[9] satisfactory mates were usually found.
>
> I had a discussion with the engineer. Why was this selective assembly necessary? The explanation was that the lathe being used to cut the hole was too old to provide the necessary precision—there was too much variation in the depth of cut. "However," the engineer said, "management has now become convinced that a new lathe is a necessity, and one is on order." This was not an entirely satisfying explanation. "Could we get 20 or 25 sets of measurements at 15-min intervals as a special project today?" I asked. "Well, yes." Plans were

[9]This is a fairly typical selective-assembly operation. They are characteristically expensive, although they may be a necessary temporary evil: (1) they are expensive in operator-assembly and test time, (2) there always comes a day when acceptable mating parts are impossible to find, but assembly "must be continued," (3) it serves as an excuse for delaying corrective action on the process producing the components.

made to have one inspector work this into the day's assignments. I returned to look at the data about 4 P.M.

The individual measurements which had been collected were given in Table 1.8; they are repeated, with the additional $\overline{X}$ and R columns, in Table 2.4. (A histogram displays them in Table 1.4; also see the plot on normal probability paper in Fig. 1.7.)

The steps below relate to the previous numbering in preparing a control chart:

1. The averages and ranges have been plotted in Fig. 2.7.
2. The average $\overline{\overline{X}} = 159.67$ mils and $\overline{R} = 0.98$ mils have been computed and lines drawn in Fig. 2.7.
3. $\text{UCL}(R) = D_4\overline{R} = (2.11)(0.98) = 2.07$.

TABLE 2.4 Data: Air-Receiver Magnetic Assembly (Depth of Cut in Mils)
Taken at 15-min intervals in order of production

						$\overline{X}$	Range R
1	160.0	159.5	159.6	159.7	159.7	159.7	0.5
2	159.7	159.5	159.5	159.5	160.0	159.6	0.5
3	159.2	159.7	159.7	159.5	160.2	159.7	1.0
4	159.5	159.7	159.2	159.2	159.1	159.3	0.6
5	159.6	159.3	159.6	159.5	159.4	159.5	0.3
6	159.8	160.5	160.2	159.3	159.5	159.9	1.2
7	159.7	160.2	159.5	159.0	159.7	159.6	1.2
8	159.2	159.6	159.6	160.0	159.9	159.7	0.8
9	159.4	159.7	159.3	159.9	159.5	159.6	0.6
10	159.5	160.2	159.5	158.9	159.5	159.5	1.3
11	159.4	158.3	159.6	159.8	159.8	159.4	1.5
12	159.5	159.7	160.0	159.3	159.4	159.6	0.7
13	159.7	159.5	159.3	159.4	159.2	159.4	0.5
14	159.3	159.7	159.9	158.5	159.5	159.4	1.4
15	159.7	159.1	158.8	160.6	159.1	159.5	1.8
16	159.1	159.4	158.9	159.6	159.7	159.5	0.8
17	159.2	160.0	159.8	159.8	159.7	159.7	0.8
18	160.0	160.5	159.9	160.3	159.3	160.0	1.2
19	159.9	160.1	159.7	159.6	159.3	159.7	0.8
20	159.5	159.5	160.6	160.6	159.8	159.9	1.1
21	159.9	159.7	159.9	159.5	161.0	160.0	1.5
22	159.6	161.1	159.5	159.7	159.5	159.9	1.6
23	159.8	160.2	159.4	160.0	159.7	159.8	0.8
24	159.3	160.6	160.3	159.9	160.0	160.0	1.3
25	159.3	159.8	159.7	160.1	160.1	159.8	0.8
						$\overline{\overline{X}} = 159.67$	$\overline{R} = 0.98$

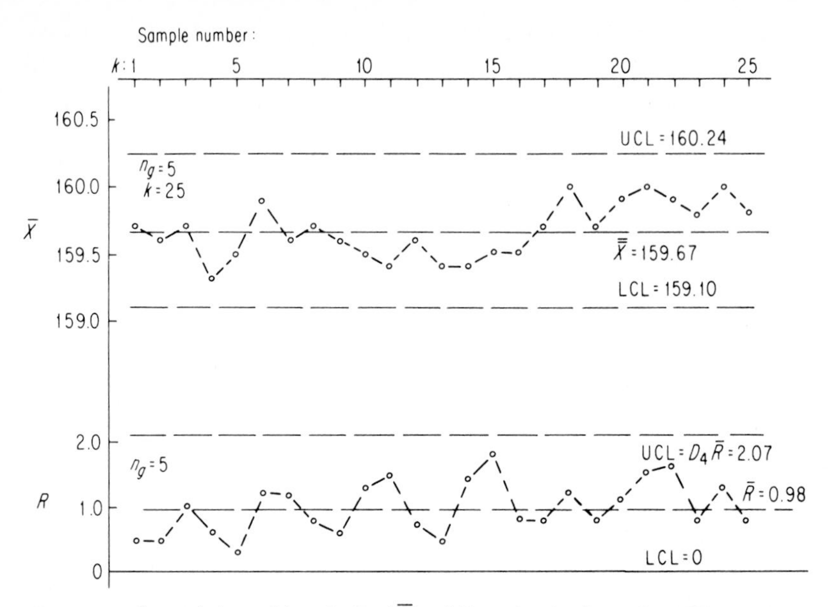

Figure 2.7 Control chart (historical) of $\bar{X}$ and R on depth of cut (Case History 2.1; data from Table 2.4).

Since all range points fall below 2.07, we proceed to (4).

4. $\text{UCL}(\bar{X}) = 159.67 + (0.58)(0.98) = 160.24.$

   $\text{LCL}(\bar{X}) = 159.67 - 0.57 = 159.10.$
5. See Fig. 2.7 for UCL and LCL.
6. *Possible evidence of assignable causes;*
   a. Runs: Either the run of eight points below $\bar{\bar{X}}$ or the following run of nine points above $\bar{\bar{X}}$ is evidence of a process operating at a level other than $\bar{\bar{X}}$. See discussion in (7) below.
   b. Are there any points outside $\bar{\bar{X}} \pm 3\hat{\sigma}_{\bar{x}} = \bar{\bar{X}} \pm A_2\bar{R}$? No. Any pair of consecutive points above $\bar{\bar{X}} \pm 2\hat{\sigma}_{\bar{x}}$? No, but three points in the last eight points are "close."
   c. Observations have been plotted on cumulative normal probability paper, see Fig. 1.7 and Sec. 1.7.
7. The run evidence is conclusive that some fairly abrupt drop in $\bar{X}$ occurred at the ninth or tenth point and $\bar{X}$ increased at the seventeenth or eighteenth point (from production of about 2 P.M.). After looking at the data, the supervisor was asked what had happened at about 2 P.M. "Nothing, no change." "Did you change the cutting tool?" "No." "Change inspectors?" "No."

The supervisor finally looked at the lathe and thought "perhaps" the chuck governing the depth of cut might have slipped. Such a slip might well explain the increase in depth of cut (at the seventeenth or eighteenth sample) but *not* the smaller values at the ninth and tenth samples. The supervisor got quite interested in the control-chart procedure and decided to continue the charting while waiting for the new lathe to arrive and learned to recognize patterns resulting from a chip broken out of the cutting tool—an abrupt drop; the gradual downward effect of tool wear; effects from changing stock rod. It was an interesting experience.

Eventually the new lathe arrived, but it had been removed a few weeks later and the old lathe back in use. "What happened?" The supervisor explained that with the control chart as a guide, the old lathe was shown to be producing more uniform depth of cut than they could get from the new one.

2.6 Probabilities Associated with an $\overline{X}$ Control Chart: Operating-Characteristic Curves

Identifying presence of assignable causes

Troubleshooting is successful when it gives us ideas of *when* trouble began and *what* may be causing it. It is important to have different sources which suggest sensible ideas about when and what to investigate. Specialists in data analysis can learn to cooperate with the engineer or scientist in suggesting the general type of trouble to consider. Data presented in the form of control-chart criteria, patterns of runs, the presence of outliers in the data; these will often suggest areas of investigation (hypotheses). The suggested hypotheses will ordinarily evolve from joint discussions between the scientist and the specialist in data analysis. The objective is to identify the physical sources producing the unusual data effects and to decide whether the cost of the cure is economically justified. The role of identifying causes rests principally with the process specialists.

The role of the Shewhart control chart in signaling production to make standard *adjustments* on a process is an important one. It also has the role of signaling opportune times to *investigate* the system. The risks of signals occurring just by chance (without the presence of an assignable cause) are quite small. When a process is stable, the probabilities (α) that the following criteria *will signal erroneously* a shift are small. They are:

1. That a single point will fall outside 3-sigma limits just by chance: about three chances in a thousand, that is $\alpha \cong 0.003$.
2. That a single point will fall outside 2-sigma limits just by chance: about one time in 20, that is, $\alpha \cong 0.05$.
3. That the last two points will both fall outside 2-sigma limits on the same side just by chance: about one time in 800,[10] that is, $\alpha \cong 0.001$.

Thus, the two-consecutive-points criterion is evidence of a change in the process at essentially the same probability level as a single point outside 3-sigma limits.

Operating-characteristic curves of $\overline{X}$ charts

When working to improve a process, our concern is not so much that we shall investigate a process without justification; rather it is that we shall miss a

[10] The probability that the first point will be outside is approximately 1/20; the probability that the next point will then be outside on the *same* side is essentially 1/40; thus the probability that two consecutive points will be outside, on the basis of chance alone, is about (1/20)(1/40) = 1/800.

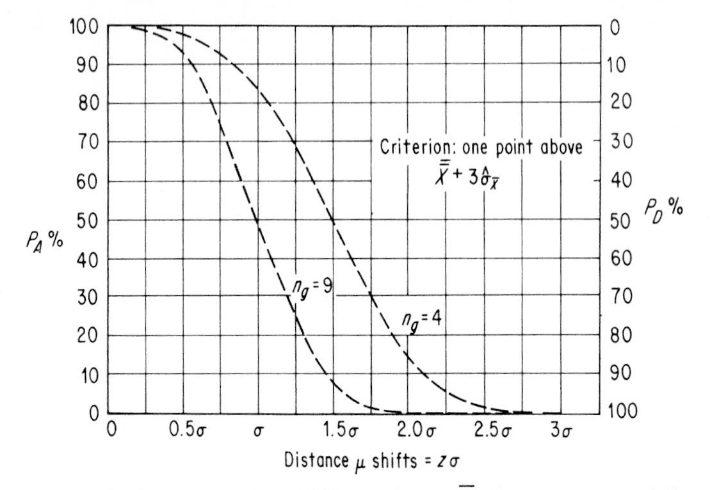

Figure 2.8 Comparing sensitivities of two $\bar{X}$ charts, $n_g = 4$ and $n_g = 9$ with operating-characteristic curves.

worthwhile opportunity to discover something important. As a rule of thumb, we may assume that a shift in process average of *one standard* deviation (one sigma) is of practical interest in troubleshooting. Just how sensitive is an $\bar{X}$ chart in detecting a shift of 1σ? Or in detecting a shift of 1.5σ? Or in detecting a shift of $z\sigma$? The operating-characteristic curves (OC curves) of Figs. 2.8 and 2.10 provide some answers.[11] It shows the probability of accepting the process as in control. P_A plotted against the size of shift to be detected. Clearly, the probability of detecting the shift is $P_D = (100 - P_A)$.

The two OC curves in Fig. 2.8 have been computed for averages of $n_g = 4$ and $n_g = 9$; the criterion for detecting a shift of $z\sigma$ in average is one point above $\bar{X} + A_2\bar{R} = \bar{\bar{X}} + 3\sigma_{\bar{X}}$. The abscissa represents the amount of shift in $\bar{X}$; the probabilities P_A of missing such shifts are shown on the left vertical scale, while the probability of detecting such shifts P_D is shown on the right.

Consider first the OC curve for $n_g = 4$

- A shift of σ has a small probability of being detected: $P_D \cong 16$ or 17 percent.
- A shift of 1.5σ has a 50 percent chance of being detected.
- A shift of 2σ has $P_D \cong 85$ percent.
- Shifts of more than 3σ are almost certain to be detected.

Consider now the OC curve for $n_g = 9$

- Except for very small and very large shifts in $\bar{X}$, samples of $n_g = 9$ are much more sensitive than with $n_g = 4$.

[11]The method of deriving OC curves is outlined below.

- A shift of 1σ has a 50 percent chance of being detected.
- A shift of 1.5σ has a 92 or 93 percent chance of detection.

Discussion: Samples of $n_g = 9$ are appreciably more sensitive than samples of $n_g = 4$ in detecting shifts in average. Every scientist knows this, almost by instinct.

This may suggest the idea that we should use samples of nine rather than the recommended practice of $n_g = 4$ or 5. And sometimes in nonroutine process-improvement projects, one may elect to do this. Even then, however, we tend to hold to the smaller samples and take them more frequently. During ordinary production, experience has shown that assignable causes which produce a point out of 3-sigma limits with samples of $n_g = 4$ or 5 can usually be identified by an engineer or production supervisor provided investigation is begun promptly. If they are not detected on the first sample, then usually on the second or third, or by one of the earlier run criteria of this chapter.

Samples as large as $n_g = 9$ or 10 frequently indicate causes which do not warrant the time and effort required to investigate them during regular production.

Some computations associated with OC curves[12]

The following discussion will consider samples of $n_g = 4$. Figure 2.9 represents four locations of a production process.

In position 1, the "outer" curve (the wider one) represents the process centered at $\overline{\overline{X}}$; the "inside" curve portrays the distribution of samples of $n_g = 4$, which is just one-half the spread of the process itself ($\sigma_{\bar{x}} = \sigma/\sqrt{n}$).

In position 2, the process has shifted $1.5\sigma = 3\sigma_{\bar{x}}$ and 50 percent of the shaded distribution of averages is now above $\overline{\overline{X}} + A_2\overline{R}$. ($P_D = 50$ percent.)

In position 4, the process has shifted $3\sigma = 6\sigma_{\bar{x}}$; "all" the distribution of averages is above the original control limit. That is, $P_D \cong 100$ percent.

In the general position 3, the process has shifted $z\sigma = 2z\sigma_{\bar{x}}$. The distance of the new process average below the original control limit is $(3 - 2z)\sigma_{\bar{x}}$. The area of the shaded tail above $\overline{\overline{X}} + A_2\overline{R}$ may be obtained from Table A.1. The distribution of samples even as small as $n_g = 4$ are essentially normally distributed, even when the process distribution is nonnormal as discussed in Sec. 1.8, Theorems 1 and 3.

The three figures on the right of Fig. 2.9 represent a process with averages increased by 1.5σ, $z\sigma$, and 3σ, respectively. The probabilities P_D that a single $\overline{X}$ point will fall above the upper control limit, after the shift in process, are indicated by the shaded areas. The unshaded areas represent the probability of acceptance of control, P_A.

Using the principles of Fig. 2.9, computation of the OC curve for control charts with samples of size $n_g = 4$ is summarized in Table 2.4A.

---

[12]This section may be omitted without seriously affecting the understanding of subsequent sections.

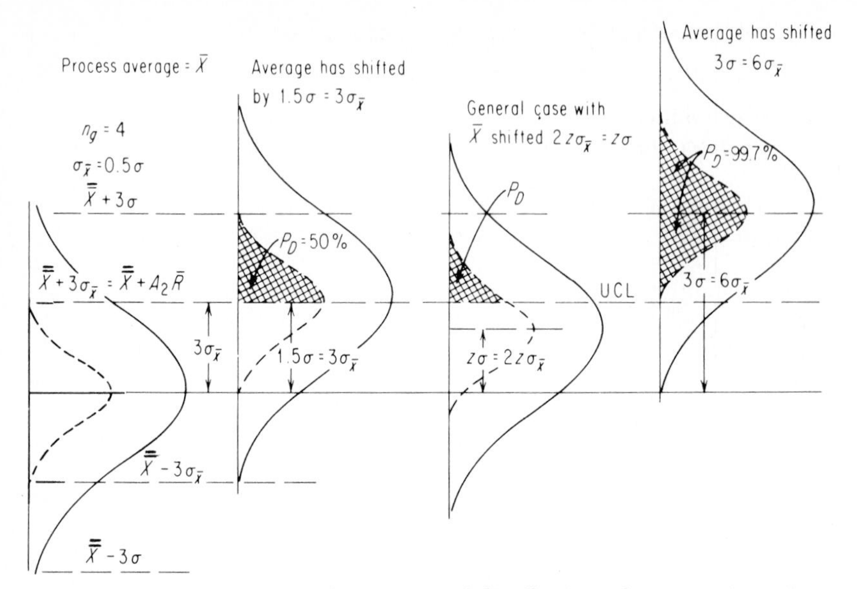

Figure 2.9 Distributions with their associated distributions of averages (n_g = 4).

TABLE 2.4A Computation of OC Curve for Shewhart $\bar{X}$ Control Chart with Sample Size n_g = 4

Shift in mean in units of σ z	Shift in mean in units of $\sigma_{\bar{x}}$ $z\sqrt{n_g}$	Distance from mean to control limit $z_{\bar{x}} = 3 - z\sqrt{n_g}$	Probability of detection $P_D = P_r$ $(z \geq z_{\bar{x}})$	Probability of acceptance $P_A = 1 - P_D$
0	0	3	0.135	99.865
0.5	1.0	2	2.28	97.72
1.0	2.0	1	15.87	84.13
1.5	3.0	0	50.00	50.00
2.0	4.0	− 1	84.13	15.87
2.5	5.0	− 2	97.72	2.28
3.0	6.0	− 3	99.865	0.135

Some OC curves associated with other criteria

Figure 2.10 shows the increased sensitivity when using 2-sigma decision limits over 3-sigma limits in troubleshooting projects. Both the one-point and two-point criteria of plans (2) and (3) are more sensitive to change than plan (1) with 3-sigma limits. These plans are often to be used when looking for ways to improve a process.

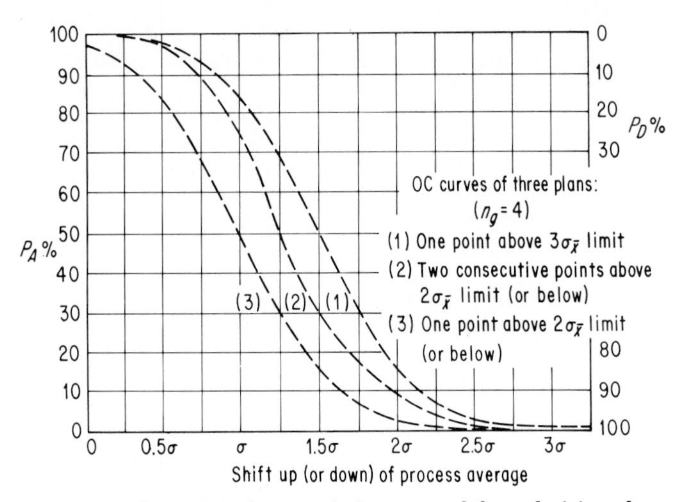

Figure 2.10 Operating-characteristic curves of three decision plans associated with an $\overline{X}$ control chart, $n_g = 4$.

CASE HISTORY 2.2

Excessive Variation in Chemical Concentration

Figure 2.11 shows some measurements (coded) of the chemical *concentration* obtained by sampling from a continuous production line in a large chemical company; samples were obtained at hourly intervals over a period of two weeks. During this period, every effort was made to hold the manufacturing conditions at the same levels; whatever variation there was in the process was unintentional and was considered inherent to the process. Although the process average was excellent, the variation being experienced was greater than could be tolerated in subsequent vital steps of the chemical process. This distribution of concentration was presented (by the scientists) as conclusive evidence that the process could not be held to closer variation than 263 to 273.

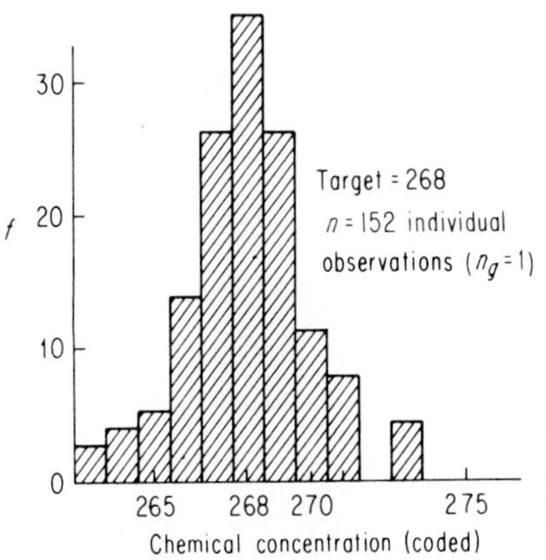

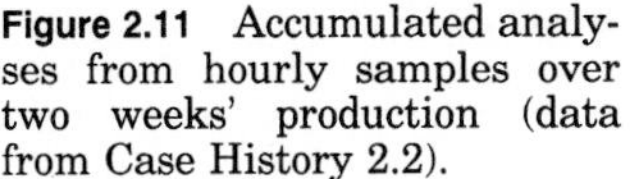

Figure 2.11 Accumulated analyses from hourly samples over two weeks' production (data from Case History 2.2).

But such a picture (histogram) of the process *does not necessarily* represent the *potential capability* of the process. Perhaps unsuspected changes occurred during the two weeks, caused by factors which could be controlled once their presence was recognized.

In Fig. 2.12 we show the data (used to prepare Fig. 2.11) as averages of four consecutive readings, i.e., covering a 4-hr production period. Each point on the $\overline{X}$ chart represents the average of four consecutive readings, and each point on the R chart represents the range of these four readings. This is a record over time. We see conclusive evidences of certain important changes having taken place during these two weeks—changes whose existence were not recognized by the very competent chemists who were guiding the process.

The R chart increases abruptly about May 14. Its earlier average is $\overline{R} \cong 2.5$, and the later average about $\overline{R} = 6$. Something happened quite abruptly to affect the 4-hr average of the process.

On the $\overline{X}$ chart the median overall is just about 267. The total number of runs above and below the median is 11, which is convincing evidence of a nonstable process average (risk less than 0.01; the critical value is 12). Although no more evidence is needed, one can draw tentative control limits on the $\overline{X}$ chart and see some further evidence of a nonstable average. Control limits drawn over the first half of the data (which averages about 267) are

$$\overline{\overline{X}} \pm A_2\overline{R} \cong 267 \pm (0.73)(2.5)$$

$$\text{UCL} \cong 268.8 \qquad \text{and} \qquad \text{LCL} \cong 265.2$$

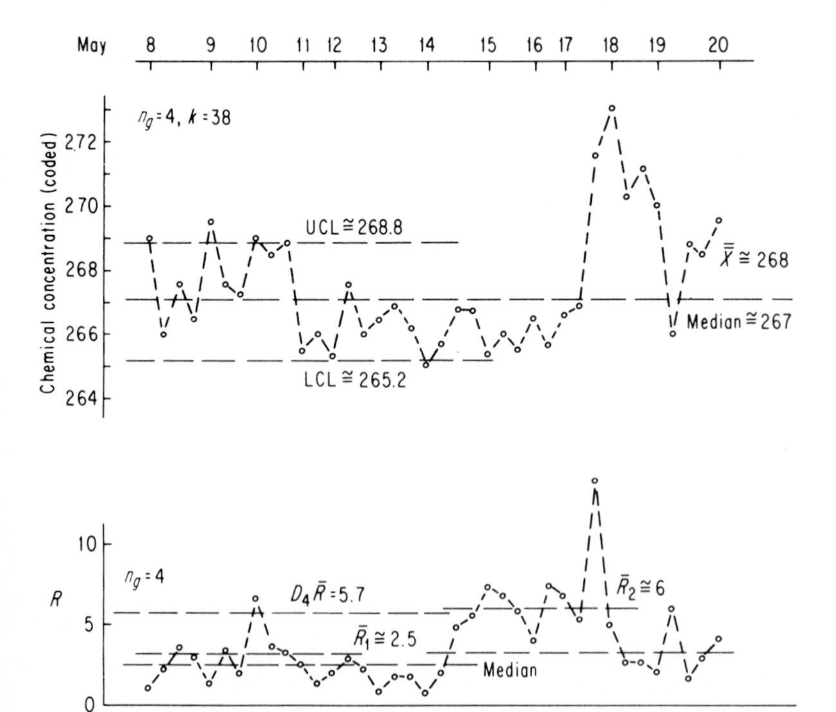

Figure 2.12 A control chart (historical) of chemical concentration of data taken about once an hour over a 2-week period (sample averages and ranges of four consecutive analyses).

These lines have been drawn in Fig. 2.12. Even during this first week, there are outages on the upper control chart limit on May 8, 9, and 10 and below the LCL during the next three or four days. There was an abrupt drop in average on May 11.

We could compute control-chart limits on both the R chart and $\overline{X}$ chart over the entire set of data; this would not mean much because of the obvious shifting in R and $\overline{X}$. The process has been affected by several assignable causes. It was agreed by production that points should be plotted on the control chart when available (every 4 hr or so) and evidences of assignable causes investigated to improve the process.

This particular process history is typical of those in every industry. Samples of size $n = 1$ are relatively common in industries where measurements are on finished batches, samples from continuous processes, complicated items (missiles, electronic test equipment), or monthly sales records, as examples. Grouping the data into subgroups of $n_g = 3$, 4, or 5 for analysis will usually be beneficial. It is usually best if there is a rationale for the grouping. But even an arbitrary grouping will often be of supplemental value to a histogram.

Recognition of the existence of important changes in any process is a necessary prerequisite to a serious study of causes affecting that process.

CASE HISTORY 2.3

Filling Vanilla Ice-Cream Containers

A plant was manufacturing French-style vanilla ice cream. The ice cream was marketed in 2.5-gal containers. Specified gross weight tolerances were 200 ± 4 oz. Four containers were taken from production at 10-min intervals in a special production study. The gross weights for $k = 24$ subgroups of $n_g = 4$ containers are shown in Table 2.5. What are appropriate ways[13] of presenting these observations for analysis?

Computation of control limits

The 24 sample averages and ranges are shown in Table 2.5; they have been plotted in Fig. 2.13.

$$\overline{\overline{X}} = 203.95, \overline{R} = 5.917; \text{ also the median, } \tilde{X} = 204.12$$

$$\text{For} \quad n = 4, \text{UCL}(R) = D_4\overline{R} = (2.28)(5.917) = 13.49$$

The fifth range point 14 exceeds the UCL. We recommend the exclusion of this range point; it represents more variation than expected from a stable, controlled process. Then we compute the average range and get $\overline{R} = 5.57$.

$$\text{Then} \quad \text{UCL} = D_4\overline{R} = (2.28)(5.57) = 12.70$$

When we compare the short-term inherent process variability of $6\hat{\sigma}_2 = 6(5.57)/2.06 = 16.2$ oz with the specification tolerance of ±4 oz = 8 oz, we find the short-term process is twice as variable as specifications allow. In other words, the inherent process

[13]Runs about the median for averages $\overline{X}$ from this set of data were considered in Sec. 2.4.

TABLE 2.5 Data: Gross Weight of Ice-Cream Fill in 2.5-Gal Container

Samples of $n_g = 4$ in order of production in 10-min intervals

Subgroup number					R	$\bar{X}$
1	202	201	198	199	4	200.00
2	200	202	212	202	12	204.00
3	202	201	208	201	7	203.00
4	201	200	200	202	2	200.75
5	210	196	200	198	14	201.00
6	202	206	205	203	4	204.00
7	198	196	202	199	6	198.75
8	206	204	204	206	2	205.00
9	206	204	203	204	3	204.25
10	208	214	213	207	7	210.50
11	198	201	199	198	3	199.00
12	204	204	202	206	4	204.00
13	203	204	204	203	1	203.50
14	214	212	206	208	8	210.00
15	192	198	204	198	12	198.00
16	207	208	206	204	4	206.25
17	205	214	215	212	10	211.50
18	204	208	196	196	12	201.00
19	205	204	205	204	1	204.50
20	202	202	208	208	6	205.00
21	204	206	209	202	7	205.25
22	206	206	206	210	4	207.00
23	204	202	204	207	5	204.25
24	206	205	204	202	4	204.25
						$\bar{\bar{X}} = 203.95$

SOURCE: Data by courtesy of David Lipman, then a graduate student at the Rutgers University Statistics Center.

variability is not economically adequate even if the process average were stable at $\bar{\bar{X}} = 200$ oz.

Since the 23 remaining range points fall below this revised UCL(R), we also proceed to calculate control limits on $\bar{X}$. For $n_g = 4$

$$\text{UCL}(\bar{X}): \bar{\bar{X}} + A_2\bar{R} = 203.95 + (0.73)(5.57) = 203.95 + 4.07 = 208.02$$

$$\text{LCL}(\bar{X}): \bar{\bar{X}} - A_2\bar{R} = 203.95 - 4.07 = 199.88$$

See Fig. 2.13.

Evidence from the $\bar{X}$ control chart

Before these data were obtained, it was known that the variability of the filled 2.5-gal ice cream containers was more than desired. In every such process where there is excessive variability, there are two possibilities:

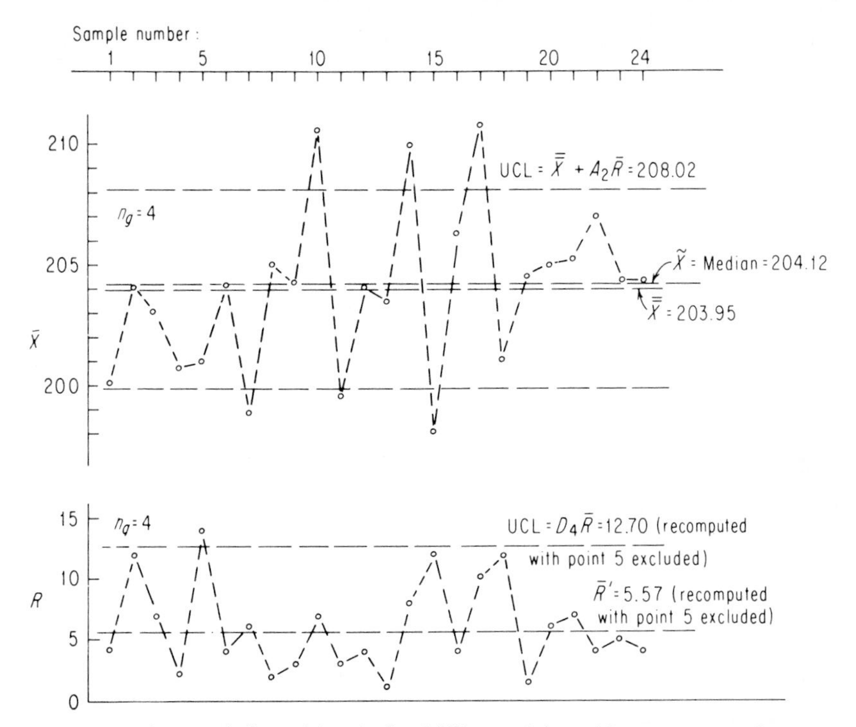

Figure 2.13 A control chart (historical) of filling weights of ice-cream containers (data from Table 2.5).

1. Excessive average short-term variation represented by $6\hat{\sigma}_2 = 6\bar{R}/d_2$ as discussed above. Even with an R chart in control and an $\bar{X}$ chart in control, the process just is not constructed to meet the desired specifications.
2. A *process average not controlled.* The two general methods of this chapter (runs and control charts) are applicable to consider this question of average process stability.
   a. The control chart for $\bar{X}$ in Fig. 2.13 provides evidence that the process average was affected by *assignable causes* on several occasions:
   b. Three separate points above UCL: points 10, 14, 17
   c. Three separate points below LCL: points 7, 11, 15
   d. Consider: (1) the group of four points 7, 8, 9, 10 (over a 30-min period); then (2) the group of four points 11, 12, 13, 14 (over 30 min); then (3) the group of three points 15, 16, 17 (over 20 min) and; then (4) the group of five points 18, 19, 20, 21, 22 (over 40 min).

The pattern of these sequences *suggests* that the process average would creep upward over a 25- to 30-min period, then was probably adjusted downward; then the adjustment cycle was repeated for a total of four such cycles. This is a supposition (hypothesis) worth investigation.

The total number of runs about the median is 8; since the small critical value is 8 for one-sided $\alpha = 0.05$, this is statistically significant at the 0.10 level of risk.

The long run of 7 below the median at the beginning and the run of 6 at the end are additional evidence of a nonstable process.

Summary of evidence The process average was quite variable; the 24 samples averaged 2 percent higher than specified. Even so, there is some danger of underfilled containers.

The inherent process variability is not economically acceptable even if process average is stabilized at $\bar{\bar{X}} = 200$ oz.

Evidence from frequency distribution analysis

The individual fill weights are shown in Table 2.6. Several containers were overfilled; there is some underfill. The histogram does not provide information about process stability or about process capability.

We can go through the routine of computing $\bar{X}$ and $\hat{\sigma}_1$ as in Table 1.3, although this cannot be expected to add to the analysis

$$\bar{X} = A + \frac{m\Sigma x_i^2}{n} = 203.5 + 2(0.094) = 203.69$$

$$\hat{\sigma}_1 = s = (2)\sqrt{5.01} = 4.48$$

This is much larger than $\hat{\sigma}_2 = \bar{R}/d_2 = 2.70$. This will almost always be the case when the control chart shows lack of control as discussed in Sec. 1.13.

Summary—Case history 2.3

- The control chart shows the process was not operating at any fixed level; a series of process changes occurred during the 4-hr study.

TABLE 2.6 Gross Weight of Ice Cream Fill in 2.5-gal Containers

Individual fill weights grouped in a histogram

Cell interval	f	d	fd	fd^2	
215–216	1	6	6	36	
213–214	4	5	20	100	
211–212	3	4	12	48	
209–210	3	3	9	27	
207–208	10	2	20	40	
205–206	17	1	17	17	USL = 204
203–204	21	0	0	0	
201–202	18	−1	−18	18	
199–200	7	−2	−14	28	
197–198	7	−3	−21	63	
195–196	4	−4	−16	64	
193–194	0	−5	0	0	
191–192	1	−6	−6	36	
$n =$	96		+9	477	

$$\bar{X} = 203.5 + (2)(0.094) = 203.69$$

$$\hat{\sigma}_1 = (2)\sqrt{\frac{96(477) - (9)^2}{96(95)}} = 4.48$$

- The average overfill of the 96 filled containers was about 2 percent; also, 38/96 = 39.6 percent of the containers were filled in excess of the USL of 204 oz (Table 2.6).
- The *inherent process capability* is estimated as follows:

$$\hat{\sigma}_2 = \overline{R}/d_2 = 5.57/2.06 = 2.70$$

and

$$6\hat{\sigma}_2 = 16.20 \text{ oz}$$

This means that the inherent variability of the process is about twice what is specified.

- Two major types of investigation will be needed to reduce the process variability to the stated specifications:

1. *What are the important assignable causes producing the shifting process average shown by the $\overline{X}$ control chart?* Once identified, how can improvements be effected? The control chart can be continued and watched by production personnel to learn how to control this average; investigations by such personnel made right at the time an assignable cause is signaled can usually identify the cause. Identifications are necessary to develop remedies.
2. *What are possible ways of reducing the inherent variability ($\hat{\sigma} = \overline{R}/d_2$) of the process?* Sometimes relationships between recorded adjustments made in the process and changes in the *Range chart* can be helpful. For example, there is a *suggestion* that the process variation increased for about 40 min (points 14 to 18) on the R chart. This suggestion would usually be disregarded in routine production; however, when trying to improve the process, it would warrant investigation. A serious process-improvement project will almost surely require a more elaborately designed study. Such studies are discussed in later chapters.

CASE HISTORY 2.4

An Adjustment Procedure for Test Equipment

Summary

The procedure of using items of production as standards to compare test-set performance indirectly with a primary standard (such as a bridge resistance in electrical testing) is discussed. The procedure is not limited to the electronic example described below; the *control chart of differences is recommended* also in analytical chemistry laboratories and in other analytical laboratories.

Introduction

In measuring certain electrical characteristics of high-intensity street lamps, it was necessary to have a simple, efficient method of adjusting (calibrating) the test sets which are in continuous use by production inspectors on the factory floor. No fundamental standard could be carried from one test set to another. The usual procedure in the industry had been to attempt comparisons between individual test sets and a standard bridge by the intermediate use of "standard" lamps. The bridge itself was calibrated by a laborious method; the same technique was not considered practical for the different test sets located in the factory.

A serious lack of confidence had developed in the reliability of the factory test sets which were being used. Large quantities were involved, and the situation was serious. Inaccurate or unstable test sets could approve some nonconforming lamps from the daily production; also, they could reject other lamps at one inspection which conformed to specifications on a retest. The manufacturing engineers chided the test-set engineers for poor engineering practice. Conversely, of course, those responsible for the test sets said: "It's not our fault." They argued that the lamps were unstable and were to blame for the excessive variations in measurements. It became a matter of honor and neither group made serious efforts to substantiate their position or exerted effort to improve the performance either of the lamps or the test equipment. Large quantities of lamps were being tested daily; it became urgent to devise a more effective criterion to make adjustments.

The comparison procedure used previously to adjust the test sets was to select five "standard" lamps which had been aged to ensure reasonable stability and whose readings were within or near specification limits. The wattage of the five standard lamps was read on the bridge and the readings recorded on form sheets, one for each test set. The lamps were then taken to each of the floor test sets where the lamps were read again and the readings recorded on its form sheet. From a series of these form sheets, the responsible persons attempted to see some signals or patterns to guide the adjustment and maintenance of the test equipment. Whether it was the adjustment procedure on test equipment which was ineffective or the instability of lamps could not be established.

A modified approach

It was generally agreed that there might be appreciable variation in a test set over a period of a few days and that operating conditions could gradually produce increasing errors in a set. Also, changes in temperature and humidity were expected to produce fluctuations. It was also known that internal variations within a lamp would produce appreciable variations in wattage at unpredictable times. Consequently, it was decided to use a control chart with averages and ranges *in some ways.*

After some discussion, the original comparison technique between the bridge and the test sets with five standard lamps was modified to permit a control-chart procedure.

The bridge was accepted as a working plant standard since there were data to indicate that it varied by only a fraction of a percent from day to day.

In the modified procedure, each of five standard lamps was read on the bridge and then on each floor test set, as before; the readings were recorded on a modification of the original form sheet (see Table 2.7). The difference

$$\Delta_i = S_i - B_i$$

between the set reading and the bridge reading for each lamp was then determined, and the average $\bar{\Delta}$ and the range R of these five differences were computed. Plus and minus signs were used to indicate whether the set read higher or lower than the bridge.

In the initial program a control chart of these differences was recorded at one test set; readings of the standard lamps were taken on it every two hours. Within a few days, control charts of the same type were posted at other test sets. An example of one of the control charts is shown in Fig. 2.14. After a short experience, several things became apparent. Certain test sets were quite variable; it was not possible to keep them in satisfactory adjustment. But a much improved performance of some was obtained easily by making minor systematic adjustments with the control charts as guides.

One of the early interests was in comparing the performance of the test sets under the control charts' guidance with the previous performance. Data from previous months

TABLE 2.7 Computations Basic to a Control-Chart–Test-Set Calibration

Standard no.	Reading on bridge (B)	Reading on test set (S)	Difference $\Delta = S - B$
1	1820	1960	+140
2	2590	2660	+ 70
3	2370	2360	− 10
4	2030	1930	−100
5	1760	1840	+ 80
			$\bar{\Delta} = +36$
			$R = 240$

SOURCE: Ellis R. Ott, An Indirect Calibration of an Electronic Test Set, *Ind. Qual. Control*, January, 1947. (Reproduced by consent of the editor.)

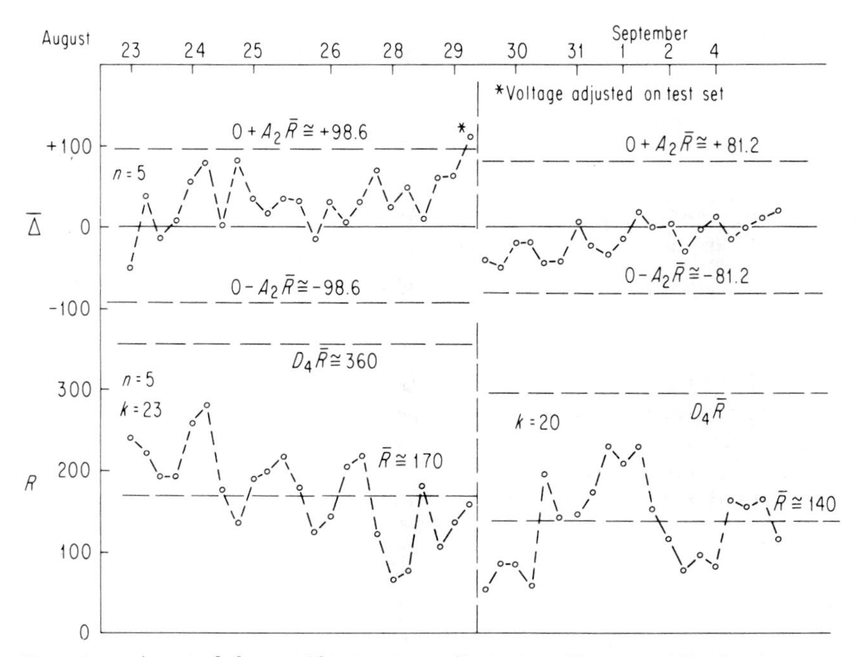

Figure 2.14 A control chart guide to test set adjustments. The central line has been set at a desired value of $\bar{\Delta} = 0$. Besides the one outage on August 29, there are other evidences of nonrandomness about $\bar{\Delta} = 0$ on the chart of averages and on the R charts; too few runs on each part of the R chart and on the August record on the $\bar{\Delta}$ chart.

were available to make control charts and a two-week period in June was selected to compare *before* and *after*. A comparison of variabilities of six test sets in June with a period in August (a few days after the start of the control charts) and then with a later two-week period in November is shown in Table 2.8.

TABLE 2.8 A Performance Comparison of Six Test Sets over Three Time Periods

Set no.	June 1–15 $\bar{\Delta}$	June 1–15 $\bar{R}$	Aug. 15–30 $\bar{\Delta}$	Aug. 15–30 $\bar{R}$	Nov. 1–15 $\bar{\Delta}$	Nov. 1–15 $\bar{R}$
1	−63	181	−6	72	17	74
3	−86	164	−1	68	−1	68
5	−62	216	−12	61	−13	61
6	−47	202	9	74	7	75
7	−136	138	16	86	17	140
8	−92	186	2	81	3	92

SOURCE: Ellis R. Ott, An Indirect Calibration of an Electronic Test Set, *Ind. Qual. Control*, January, 1947. (Reproduced with the consent of the editor.)

Note: The average $\bar{\Delta}$ indicates the amount of bias (inaccuracy); the desired average is $\bar{\Delta} = 0$. Small values of $\bar{R}$ indicate less variability.

The immediate improvements effected in August are apparent: the average differences (bias) are reduced on every test set; the variability $\bar{R}$ is reduced on several sets. It was soon discovered that set 7 was in need of a major overhauling.

No allowable limits of variation of the test sets with respect to the bridge had been established, but it was agreed that it would be a decided improvement to maintain the average of five lamps within 5 percent of the bridge readings. Table 2.8 showed that the variation had been in excess of 10 percent during the first two weeks in June. It was found during the first weeks of the experiment that the average of five lamps could be held within 4 percent of the bridge for most sets, and even closer agreements were obtained subsequently. Three-sigma control-chart limits were projected in advance on both $\bar{\Delta}$ and R charts and used as criteria for adjustment. Figure 2.14 shows a test set averaging about 40 units high during the last week in August. The voltage adjustment at the end of August centered it nicely.

It was surprising to find such large values of the range in Table 2.8. On the basis of logic, it was possible to explain this variability by any one of the following explanations which revert to the original "lamp versus test set" controversy:

1. *The five standard lamps were stable:* The variations in readings arose from the inability of a test set to duplicate its own readings. This assumption now had some support; $\bar{R}$ for set 7 in November was significantly larger than for other sets.

   Causes which produced a shift in the $\bar{\Delta}$ chart or in the R chart of one test set but not in others were assignable to the test set.

2. *The test sets were reliable:* The variations in readings resulted from internal variations within the standard lamps. The most obvious variations in all test sets would require the replacement of one of the five standard lamps by a new one. A reserve pool of standard lamps was kept for the purpose. A need for replacement was evidenced by an upward trend of several different R charts. Such a trend could not necessarily be attributed to the standard lamps, but analysis of recent bridge readings from the form sheet on individual lamps would show whether a particular lamp was the assignable cause.

3. *A combination of assumptions 1 and 2:* It was most convenient to have data from at least three test sets in order to compare their behavior. The test-set engineers started

control charts from data used in previous calibrations. They posted control charts at each test set, learned to predict and prevent serious difficulties as trends developed on the charts, and were able to make substantial improvements in test-set performance.

The advantages of control charts using differences in any similar indirect calibration program are essentially the same as the advantages of any control chart over the scanning of a series of figures.

The control chart of differences is applicable to a wide variety of calibration techniques in any industry where a sample of the manufacturer's product can be used as an intermediary. Reliable data are an important commodity.

2.7 Practice Exercises

1. Paul Olmstead is quoted as classifying the nature of assignable causes into five categories. In addition to these five, one might recognize a category known as "erratic jumps from one level to another for very short periods of time." For each of the six categories, identify or suggest a scenario which could lead to such results in actual practice. For instance, for category 1, gross error or blunder, we might consider that an inspector misplaces a decimal point when recording the result of an inspection.
2. The authors mention that there are two types of risks in scientific study just as in all other aspects of life. They mention specifically the acceptance of a new position, beginning a new business, hiring a new employee, or buying stock on the stock exchange. In addition, we can recognize situations such as the courtroom trial of an individual who pleads guilty to a crime; the release of a person from a facility for the criminally insane; the evaluation of information provided by an espionage agent. For several such situations, or others of your own choosing, identify the "null hypothesis" which is involved, and what constitutes the alpha risk and beta risk. Explain what actions might be taken to reduce either or both of these risks in the given situation.
3. Apply the runs criteria of Sec. 2.4 to the x-bar and R chart of Fig. 2.5. State your conclusions.
4. Use data of Table 2.4 to plot an x-bar and R chart for samples of size $n_g = 4$. Do this by excluding the fifth reading in each sample (159.7, 160, etc.). Compare your results with the authors' results using samples of size 5. Note that Table 2.4 is coded data. Interpret your results in terms of the original data (see Table 1.8). In this problem, apply all nine criteria listed in Sec. 2.5.
5. Extend Fig. 2.8 for $n_g = 2$ and $n_g = 5$. Draw sketches comparable to those of Fig. 2.9 to illustrate the probabilities involved. (This must be based on the applicable formulas.)
6. Redraw Fig. 2.9 for $n_g = 1$, $n_g = 4$, $n_g = 16$. Draw the OC curves for these sample sizes on the same chart.
7. Draw the OC curve for a control chart with $n_g = 3$.

Chapter

3

Ideas from Outliers[1]—Variables Data

3.1 Introduction

Since data with questionable pedigrees are commonplace in every science, it is important to have objective signals or clues to identify them. Strong feelings exist among scientists on the proper uses to be made of those which are suspected: one school of thought is to "leave them in"; other schools have different ideas. If the outlier represents very good or very bad quality, perhaps it represents evidence that some important, but unrecognized, effect in process or measurement was operative. Is this a signal which warrants a planned investigation? Is it a typical blunder warranting corrective action? Some excellent articles have been written on the implications of outliers and methods of testing for them.[2]

There are important reasons why the troubleshooter may want to detect the reasons for an outlier:

1. It may be an important signal of unsuspected important factor(s) affecting the stability of the process or testing procedure.
2. A maverick occurring in a small sample of four or five may have enough effect to cause $\overline{X}$ to fall outside[3] one of the control limits on the mean. The range will also be affected.

   We would not want to make an adjustment on the process average which has been signaled by a maverick.

[1]Other terms besides *outlier* in common usage include *maverick* and *wild-shot.*

[2]Frank E. Grubbs, "Procedures for Detecting Outlying Observations in Samples," *Technometrics,* vol. 11, no. 1, February 1969, pp. 1–21. Also, Frank Proschan, "Testing Suspected Observations," *Ind. Qual. Control,* January 1957, pp. 14–19.

[3]Any point falling outside control limits will be called an "outage."

3. A maverick (outlier) left in a relatively small collection of data may have a major effect when making comparisons with other samples.

CASE HISTORY 3.1

A Chemical Analysis—An *R* Chart As Evidence of Outliers

The percent by chemical analysis of a specific component A in successive batches of a plastic monomer became important evidence in a patent-infringement dispute. The numbers in column 1 of Table 3.1 represent the chemical analysis of individual, consecutively produced, several-ton batches of monomer. A crucial question was whether the chemical analyses on certain batches were reliable. Many more analyses than those shown in Table 3.1 were in evidence.

The recommended sequence of steps taken in any statistical analysis of data begins with plotting; in this set by plotting $\overline{X}$ and R charts where subsets were formed from five consecutive batches.[4] Table 3.1 lists only subsets 11 through 18. This analysis will assume there is no other evidence. In Fig. 3.1, we see that the control limits on the $\overline{X}$ chart include all the $\overline{X}$ points; the R chart has an outage in subset 14 and another in 17. What are some possible reasons for these two outages? Is it a general increase in variability, or is it the presence of mavericks?

We decide to retain the two subgroups with outages when computing $\overline{R}$, although we then obtain an estimate of σ which is probably large; this procedure is conservative and will be followed at least for now.

$$\hat{\sigma} = \overline{R}/d_2 = 1.085/2.33 = 0.47$$

Discussion

- The 3-sigma control limits on *individual batches* have been drawn, using this conservatively large estimate (see Fig. 3.2). We see that one batch analysis in subset 14 is below the lower 3-sigma limit; also, one low batch analysis is seen in subset 17. We must conclude that the analyses on these two batches are significantly different from those of their neighbors. There are two possible explanations for the two outages (based on logic alone):

  1. The chemical content of these two batches is indeed significantly lower than others (and dangerously close to a critical specification).
  2. There was an error (blunder) either in the chemical analysis or in the recording of it.

It is not possible to determine, now, which of the two possibilities is completely responsible. The preceding analysis does present evidence of important errors.

- At the time that these batches were produced and analyzed, chemical procedures were available which might have established conclusively which was the actual source of error. Such methods would include reruns of the chemical analysis, visual and physical tests on other batch properties, checks on the manufacturing log sheets in the material balance calculation, discussions with plant personnel.

---

[4]It is important to use a control-chart analysis, but it is not important here whether to choose subsets of five or of four. The maintenance of $\overline{X}$ and R charts as a part of the production system would probably have prevented the difficulties which arose, provided the evidence from the chart had been utilized.

TABLE 3.1 Record of Chemical Analyses (Column 1) Made on Consecutive Batches of a Chemical Compound

Column 2 shows a "material balance" content calculated for the same batches as column 1

Subset no.	(1) Chemical analysis, %	$\bar{X}$ $n_g = 5$	Range $n_g = 5$	(2) Material balance, %
11	2.76			4.12
	3.66			3.69
	3.47			3.92
	3.02			4.14
	3.55	3.29	0.90	3.70
12	3.55			3.74
	3.10			3.74
	3.28			3.65
	3.13			3.63
	3.21	3.25	0.45	3.92
13	3.66			3.95
	3.40			3.96
	3.25			3.95
	3.36			3.95
	3.59	3.45	0.41	3.76
14	1.32			4.10
	3.32			3.71
	2.91			4.12
	3.81			3.77
	3.47	2.97	2.49	4.15
15	3.70			3.79
	3.59			3.75
	3.85			3.72
	3.51			3.83
	4.12	3.75	0.61	3.77
16	4.08			3.65
	3.66			3.76
	3.66			3.65
	3.47			5.39
	4.49	3.87	1.02	3.96
17	3.85			3.78
	3.47			3.58
	3.32			5.35
	2.94			4.52
	1.43	3.00	2.42	5.46
18	3.51			4.15
	3.74			3.89
	3.51			3.96
	3.63			3.63
	3.36	3.55	0.38	5.27

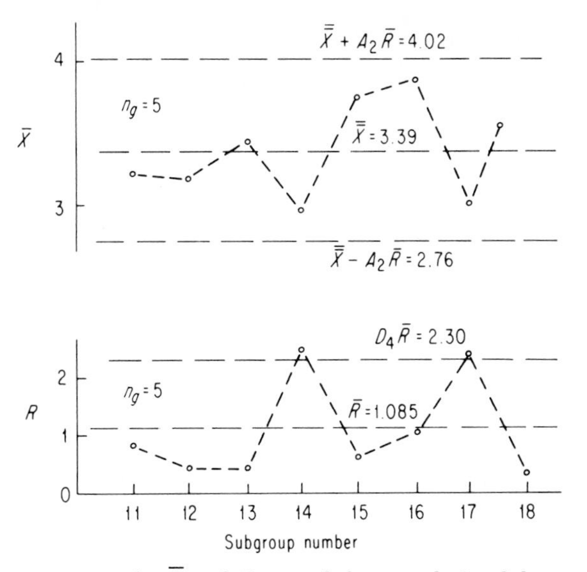

Figure 3.1 An $\bar{X}$ and R control-chart analysis of data (Table 3.1., column 1). Subsets of $n_g = 5$, $\hat{\sigma} = \bar{R}/d_2 = 0.47$.

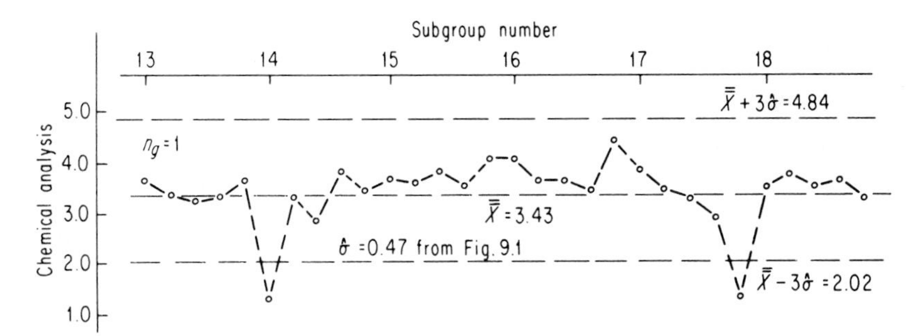

Figure 3.2 Individual batch analyses showing two outages.

- The column 2 figures were obtained on the basis of batch entries in the process log book. The materials balance, column 2, is computed for each batch on the assumption that the amounts of ingredients shown in the log book are actually correct. This assumption is very critical. Whenever there is a substantial discrepancy between columns 1 and 2 on a batch, it is advisable to make *immediate* and careful checks on the reasons for the discrepancy.

3.2 Other Objective Tests for Outliers

In Case History 3.1, eight different subgroups $n_g = 5$, were analyzed. The R chart of Fig. 3.1 showed two outages; individual batches responsible for the outages were easily identified. This R chart identification of outliers is applicable when the amount of data is "large."

When we have only a few observations, other criteria to test for possible outliers can be helpful. Two such criteria are the following:

1. An *R* chart of moving ranges as a test for outliers

In studying a process, observations should usually be recorded in the order of production. Data from Table 3.2 have been plotted in Fig. 3.3*a*—these are individual observations. There seems to be no obvious grouping at different levels, no obvious shift in average, nothing unusual when the number of runs is counted, no cycle; but the second observation is "somewhat" apart from the others. Possibly an outlier?

A moving range, MR, $n_g = 2$, is the *positive difference between two consecutive observations.* Moving ranges, $|X_{i+1} - X_i|$, $n_g = 2$, behave like ordinary ranges, $n = 2$. In Table 3.2, the moving ranges have been written in the column adjacent to the original X_1 observations; except that the tenth entry is $|X_1 - X_{10}|$. Then

$$\overline{MR} = \frac{\Sigma(MR_i)}{10} = 0.0144$$

The upper control limit on the moving-range chart, Fig. 3.3*b*, is

$$\text{UCL} = D_4(\overline{MR}) = (3.27)(0.0144) = 0.047$$

TABLE 3.2 Estimating σ from a Moving Range

	X_i	MR
1.	0.110	
2.	0.070	0.040
3.	0.110	0.040
4.	0.105	0.005
5.	0.100	0.005
6.	0.115	0.015
7.	0.100	0.015
8.	0.105	0.005
9.	0.105	0.000
10.	0.098	0.007
		$0.012 = X_1 - X_{10}$

$$\overline{MR} = 0.0144$$
$$D_4(\overline{MR}) = (3.27)(0.0144) = 0.047$$
$$\hat{\sigma} = \overline{MR}/d_2 = 0.0144/1.13 = 0.0127$$

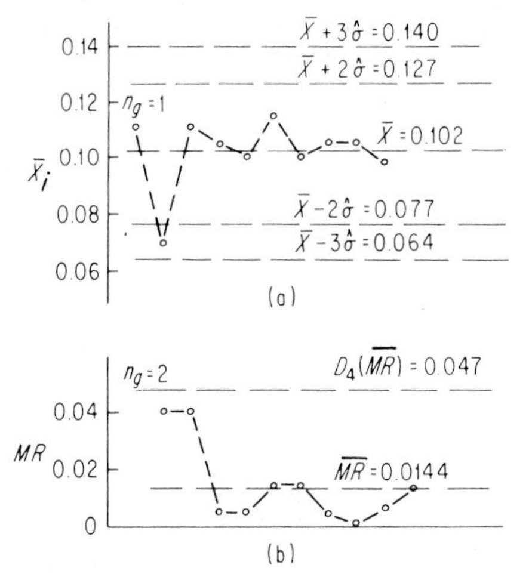

Figure 3.3 A chart check for an outlier (data of Table 3.2).

where D_4 is the ordinary factor for ranges, Table 2.3 or Table A.4.

The second observation in Table 3.2 is responsible for the first two points being near the UCL in Fig. 3.3*b*; thus $X_2 = 0.070$ is suspected of being an outlier.

Also, control limits have been drawn in Fig. 3.3*a* for $n = 1$ at

$$\bar{X} \pm 3\hat{\sigma} \qquad \text{and at} \qquad \bar{X} \pm 2\hat{\sigma}$$

where $\hat{\sigma} = 0.013$. The point corresponding to $X_2 = 0.070$ is between the $2\hat{\sigma}$ and $3\hat{\sigma}$ lines. This is additional evidence, consistent with that of Fig. 3.3*b*, indicating this one observation to be an outlier with risk between $\alpha = 0.05$ and 0.01. Consequently, an investigation for the reason is recommended.

2. Dixon's test for a single outlier in a sample of *n*

Consider the *ordered* set of *n* random observations from a population presumed to be normal:

$$X_1, X_2, X_3, \ldots, X_{n-1}, X_n$$

Either end point, X_1 or X_n, may be an outlier. Dixon studied[5] various ratios and recommends

$$\mathbf{r}_{10} = \frac{X_2 - X_1}{X_n - X_1}$$

[5] W. J. Dixon, "Processing Data for Outliers," *Biom.*, vol. 9, 1953, pp. 74–89.

to test the smallest X_1 of being an outlier in a sample of $n = 3, 4, 5, 6$, or 7. If the largest X_n is suspect, simply reverse the order of the data. When $n > 7$, similar but different ratios are recommended in Table A.9.

Example 3.1 Consider the previous data of Table 3.2, $n = 10$. From Table A.9 we are to compute:

$$\mathbf{r}_{11} = \frac{X_2 - X_1}{X_{n-1} - X_1} = \frac{0.098 - 0.070}{0.110 - 0.070} = 0.70$$

This computed value of $\mathbf{r}_{11} = 0.70$ exceeds the tabular entry 0.597 in Table A.9, corresponding to a very small risk of 0.01. We decide that the 0.070 observation does not belong to the same universe as the other nine observations. This conclusion is consistent with the evidence from Figs. 3.3*a* and *b*.

Whether the process warrants a study to identify the reason for this outlier and the expected frequency of future similar low mavericks is a matter for discussion with the engineer.

3.3 Two Suspected Outliers on the Same End of a Sample of *n* (Optional)

Besides the control chart, the following two procedures are suggested as tests for a pair of outliers:

1. *Dixon's test after excluding the more extreme of two observations:* Proceed in the usual way by testing the one suspect in the $(n - 1)$ remaining observations. If there are three extreme observations, exclude the two most extreme and proceed by testing the one suspect in the $(n - 2)$ remaining observations using Appendix Table A.9.

Consider the 10 measurements of Grubbs in the example below, where the two smallest suggest the possibility of having a different source than the other eight; see Fig. 3.4.

Analysis with Dixon's criterion. Exclude the lowest observation. Then in the remaining nine: $X_1 = 2.22$, $X_2 = 3.04$, $X_8 = 4.11$, and $X_9 = 4.13$.

Form the ratio

$$\mathbf{r}_{11} = \frac{X_2 - X_1}{X_8 - X_1} = \frac{0.82}{1.89} = 0.43$$

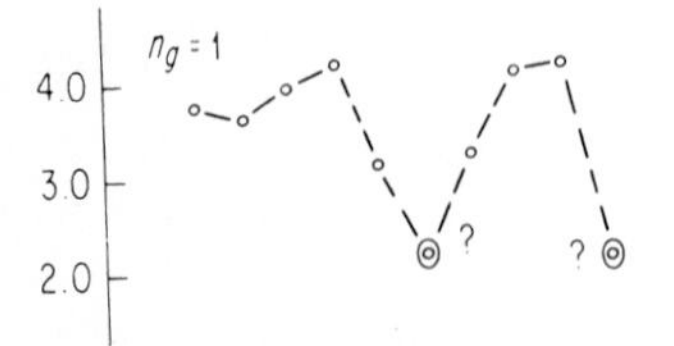

Figure 3.4 Data with two suggested outliers on the same end (see Example 3.2).

Analysis: This ratio is between the critical value of 0.352 for $\alpha = 0.20$ and 0.441 for $\alpha = 0.10$.

Decision: Then we consider both of the suspected observations to be from a different source if we are willing to accept a risk between 0.10 and 0.20.

2. A test for *two outliers on the same end provided by Grubbs*[6] is based on the ratio of the sample sum of squares when the doubtful values are excluded to the sum when included. This is illustrated by example.

Example 3.2 Following are ten measurements of percent elongation at break test on a certain material: 3.73, 3.59, 3.94, 4.13, 3.04, 2.22, 3.23, 4.05, 4.11, and 2.02. Arranged in ascending order of magnitude[7] these measurements are: 2.02, 2.22, 3.04, 3.23, 3.59, 3.73, 3.94, 4.05, 4.11, 4.13. We can test the two lowest readings simultaneously by using the criterion $S_{1,2}^2/S^2$ of Table A.10. For the above measurements

$$S^2 = \sum_{i=1}^{n}(X_i - \bar{X})^2 = \frac{n\Sigma X_i^2 - (\Sigma X_i)^2}{n} = \frac{10(121.3594) - (34.06)^2}{10} = (n - 1)s^2$$

$$S^2 = 5.351$$

and $$S_{1,2}^2 = \sum_{i=3}^{n}(X_i - \bar{X}_{1,2})^2 = \frac{(n-2)\sum_{i=3}^{n} X_i^2 - (\sum_{i=3}^{n} X_i)^2}{(n-2)} = (n-1)s_{1,2}^2$$

where $$\bar{X}_{1,2} = \sum_{i=3}^{n} X_i/(n - 2)$$

$$= \frac{8(112.3506) - (29.82)^2}{8} = 1.197$$

Then, $$S_{1,2}^2/S^2 = 1.197/5.351 = 0.224$$

From Table A.10, the critical value of $S_{1,2}^2/S^2$ at the 5 percent level is 0.2305. Since the calculated value is *less* than this, we conclude that *both* 2.02 and 2.22 are outliers, risk 5 percent. This compares with a risk between 10 percent and 20 percent by the previous analysis. Note that this text rejects the hypothesis of randomness when the test statistic is *less* that the critical value.

3.4 Practice Exercises

1. In Table 3.1, there is an apparent outlier in subset 14 ($X = 1.32$). Use Dixon's test to objectively determine if this is an outlier. Do this with subset 14 only ($n_g = 5$); subsets 13 and 14 ($n_g = 10$), subsets 12, 13, 14 ($n_g = 15$), and subsets 11, 12, 13, and 14 ($n_g = 20$). By interpolation in Table A.9, determine the approximate p value (the value of the statistic for

[6]Frank E. Grubbs, "Procedures for Detecting Outlying Observations in Samples," *Technometrics,* vol. 2, no. 1, February 1969, pp. 1–20.
[7]Also see Fig. 3.4.

which the result would just barely be significant for each case). Explain the different p values.

2. Use the data of subsets 11–14 of Table 3.1 column 1 to plot an x chart and a moving range chart. Compute the upper control limit on the MR chart based on subsets 11 and 12. Then continue to plot, noting whether or not the apparent outlier is detected by the chart.
3. In using an MR chart to check for an outlier in a set of data (where all the data is in hand before the check is made) do you think the suspect should be included or excluded in the computation of MR-bar. Explain your reasoning.
4. For the case of two suspected outliers, the author provides two methods of analysis: Dixon's and Grubbs'. Notice that the p value for Dixon's method is roughly 0.14, while for Grubbs' method it is about 0.04. Examine the two methods and suggest a theoretical reason why the latter p value is so much lower.
5. A third method has been proposed for the case of two suspected outliers—a t test. Let the two suspected units form one sample and the remaining units form the other sample. Compute t. What is your opinion of the validity of this method? Justify your answer. (*Note:* the t test is shown in Chap. 13).
6. Test the data in the sample below for outliers using both the Dixon test and the Grubbs test using $\alpha = 0.05$ as the criterion.

   56, 46, 43, 57, 70, 50, 43, 40, 41, 40, 51, 55, 72, 53, 42, 44.

7. Suppose the data of Table 3.1 is split into two parts with subset 11–14 in one part and 15–18 in the other. Test whether the means of these subsets are the same using $\alpha = 0.05$. Exclude any outliers and retest. Are the results the same? (*Note:* the t test is shown in Chap. 13).
8. Combine the $\overline{X}$ and R chart in Fig. 3.3 into one chart.

Chapter

4

Variability—Estimating and Comparing

4.1 Introduction

The variability of a stable process may be different on one machine than another, or under one set of conditions than under another. Such variabilities in a process can introduce difficulties in the comparison of averages or other measures of central tendency.

Comparisons of variability and of averages will depend upon estimates of variation. Statistical tables will be used in making some of these comparisons. The computation of some of these tables has depended upon the number of *degrees of freedom,* df, associated with estimates of the *variance* σ^2 and standard deviation σ. Degrees of freedom may be thought of, crudely, as "effective sample size," and depends on how much of the data is "used up" in making estimates necessary for the calculation of the statistic involved. The number of degrees of freedom may be different when computing estimates in different ways. The number of df associated with different methods of computation will be indicated.

Section 4.2 has been included for those who enjoy looking at extensions and ramifications of statistical procedures. It may be omitted without seriously affecting the understanding of subsequent sections.

4.2 Statistical Efficiency and Bias in Variability Estimates

Two terms used by statisticians have technical meanings suggested by the terms themselves. They are *unbiased* and *statistically efficient.* As the sample size is increased, a consistent statistic approaches a fixed value with errors that tend to be normally distributed. If the expected value of the statistic is equal to the population parameter, the estimate is called *unbiased.* A statistic which is the least variable for a given sample size is said to be the most *statistically efficient.*

TABLE 4.1 Factors c_4 to Give an Unbiased Estimate:

$$\hat{\sigma} = \frac{s}{c_4} = \frac{1}{c_4}\sqrt{\frac{\Sigma(X_i - \overline{X})^2}{n-1}}$$

n	c_4
2	0.80
3	0.88
4	0.92
6	0.95
10	0.97
25	0.99
∞	1.00

NOTE: $$c_4 = \frac{4n-4}{4n-3}$$

In the definition of variance, Eq. (4.1), the denominator $(n - 1)$ is used; this provides an unbiased estimate of the unknown population statistic σ^2. We might also expect the square root of s^2 to be an unbiased estimate of σ; actually this is "not quite" the case. However, it is unusual for anyone to make an adjustment for the slight bias.

An unbiased estimate is

$$\hat{\sigma} = \frac{s}{c_4} = \frac{1}{c_4}\sqrt{\frac{\Sigma(X_i - \overline{X})^2}{n-1}} \tag{4.1}$$

where some values of c_4 are given in Table 4.1 but are seldom used.

Note that $\bar{s}$ can be used in place of s in the above relationship.

The concept of statistical efficiency is discussed in texts on theoretical statistics. It permits some comparisons of statistical procedures especially under the assumptions of normality and stability. For example, the definition of the variance s^2 in Eq. (4.1*a*) would be the most efficient of all possible estimates *if* the assumptions were satisfied. However, the statistical efficiency of $\hat{\sigma}$ based on the *range,* $\hat{\sigma} = \overline{R}/d_2$, is only slightly less than that obtained in relation of Eq. (4.1) even when the assumptions are satisfied. (See Table 4.2.) When they are not satisfied, the advantages often favor $\hat{\sigma} = \overline{R}/d_2$.

Statistical methods can be important in analyzing production and scientific data, and it is advisable that such methods be as statistically efficient as possible. But cooperation between engineering and statistical personnel is essential; both groups should be ready to compromise on methods so that the highest overall simplicity, feasibility, and efficiency are obtained in each phase of the study.

4.3 Estimating σ and σ^2 from Data: One Sample of Size *n*

In Chap. 1, Table 1.3, the mechanics of computing $\overline{X}$ and $\hat{\sigma}$ from *grouped data* with large n were presented as follows:

TABLE 4.2 Statistical Efficiency of $\hat{\sigma} = \overline{R}/d_2$ in Estimating the Population Parameter from k Small Samples

n_g	Statistical efficiency
2	1.00
3	0.992
4	0.975
5	0.955
6	0.93
10	0.85

$$\hat{\sigma} = m \sqrt{\frac{n\Sigma fd^2 - (\Sigma fd)^2}{n(n - 1)}}$$

$$\overline{X} = \frac{\Sigma f_i d_i}{n}$$

The frequency distribution procedure as given serves a dual purpose: (1) it presents the data in a graphical form which is almost invariably useful, and (2) it provides a simple form of numerical computation which can be carried through without a desk calculator or other computer.

In some process-improvement studies, we shall have only *one set of data* which is too small to warrant grouping. Let the n independent, random observations from a stable process be $X_1, X_2, X_3, \ldots, X_n$. Then the variance σ^2 of the process can be estimated by

$$s^2 = \hat{\sigma}^2 = \frac{\Sigma(X_i - \overline{X})^2}{n - 1} \qquad \text{df} = n - 1 \tag{4.1a}$$

so

$$s = \sqrt{\hat{\sigma}^2} \Bigg/ \sqrt{\frac{n\Sigma X_i^2 - (\Sigma X_i)^2}{n(n - 1)}} \tag{4.1b}$$

The following sections show that the comparison of two process variabilities is based on variances rather than standard deviations.

4.4 Data from *n* Observations Consisting of *k* Subsets of $n_g = r$: Two Procedures

Introduction

Important procedures are presented in this section which will be applied continually throughout the remaining chapters of this book. They are summarized in Sec. 4.6 and Table 4.6. It is recommended that you refer to them as you read through this chapter.

In our industrial experiences, we often obtain repeated (replicated) observations from the same or similar sets of conditions. The small letter r will be used to represent the number of observations in each combination of conditions. There will usually be k sets of conditions each with $n_g = r$ replicates. The letter n will usually be reserved to use when two or more samples of size r are pooled. Thus,

$r = n_g$ = Number of observations in each combination of conditions

k = Number of subgroups

n = Sample size when two or more subgroups are pooled

so

$$n = rk$$

This distinction is important in this chapter which compares estimates based on k subgroups of size r with those using an overall sample size of $n = rk$.

- The *mechanics* of computing $\hat{\sigma}$ from a series of small rational subgroups of size n was discussed in Sec. 1.12. We usually considered k to be as large as 25 or 30. Then

$$\hat{\sigma} = \overline{R}/d_2 \tag{4.2}$$

where d_2 is a constant (see Table A.4) with values depending only on n_g (or r). An important advantage of this control-chart method is its use in checking the stability of process variation from the R chart.

This estimate $\hat{\sigma}$ in Eq. (4.2) is *unbiased.* However, squaring to obtain $\hat{\sigma}^2 = (\overline{R}/d_2)^2$ has the seemingly peculiar effect of producing a bias in $\hat{\sigma}^2$. The bias can be removed by the device of replacing d_2 by a slightly modified factor d_2^* (read "d-two star") depending on both the number of samples k and the number of replicates, r. See Table A.11. That is, the variance

$$\hat{\sigma}^2 = (\overline{R}/d_2^*)^2 \tag{4.3a}$$

is unbiased. Also

$$\hat{\sigma} = \overline{R}/d_2^* \tag{4.3b}$$

is slightly biased much as s in Eq. (4.1*b*) is biased. We shall sometimes use this biased estimate $\overline{R}/d_2^*$ especially when k is less than say 4 or 5 later in connection with certain statistical tables based on the bias of s in estimating σ. (This is somewhat confusing; the differentiation *is not critical,* as can be seen by comparing values of d_2 and d_2^*.) The degrees of freedom *df* associated with the estimate Eq. (4.3*a* or *b*) are also given in Table A.11 for each value of k and r. However, a simple comparison indicates that there is a loss of essentially 10 percent when using this range estimate, i.e.,

$$\text{df} \cong (0.9)k(r - 1) \qquad k > 2 \tag{4.4}$$

There is an alternate method of computing $\hat{\sigma}^2$ from a series of rational subgroups of varying sizes $r_1, r_2, r_3, \ldots, r_k$. Begin by computing a variance s_i^2 for each sample from Eq. (4.1*a*) to obtain: $s_1^2, s_2^2, s_3^2, \ldots, s_k^2$. Then

$$\hat{\sigma}^2 = s_p^2 = \frac{(r_1-1)s_1^2 + (r_2-1)s_2^2 + \ldots - (r_k-1)s_k^2}{r_1 + r_2 + \ldots + r_k - k} \tag{4.5a}$$

Each sample contributes $(r_i - 1)$ degrees of freedom for the total shown in the denominator of Eq. (4.5*a*). This estimate $\hat{\sigma}^2$ in Eq. (4.5*a*) is unbiased. When $r_1 = r_2 = r_3 = \ldots = r_k = r$, the denominator becomes simply

$$\text{df} = k(r - 1)$$

Equation (4.5*a*) is applicable for either large or small sample sizes r_i.

Note: When $k = 2$ and $r_1 = r_2 = r$, Eq. (4.5*a*) becomes simply the average

$$s_p^2 = \hat{\sigma}^2 = \frac{s_1^2 + s_2^2}{2} \qquad \text{df} = 2(r - 1) \tag{4.5b}$$

4.5 Comparing Variabilities of Two Populations

Consider two machines, for example, producing items to the same specifications. The product may differ with respect to some measured quality characteristic either because of differences in variability or because of unstable average performance.

Two random samples from the same machine (or population) will also vary. We would not expect the variability of the two samples to be exactly equal. Now if many random samples are drawn from the same machine (population, process), how much variation is expected in the variability of these samples? When is there enough of a difference between computed variances of two samples to indicate that they are *not* from the same machine or not from machines performing with the same basic variability?

This question can be answered by two statistical methods: the *variance ratio test* (*F test*) and the *range-square-ratio test* (F_R *test*).

Variance ratio test (*F* test)

This method originated with Professor George W. Snedecor who designated it the "*F* test" in honor of the pioneer agricultural researcher and statistician, Sir Ronald A. Fisher. The method is simple in mechanical application.

Method: Given two samples of sizes n_1 and n_2, respectively, *considered to be from the same population;* compute s_1^2 and s_2^2 and *designate the larger* value by s_1^2. What is the expected "largest ratio," with risk α, of the *F ratio*

$$F = s_1^2 / s_2^2 \tag{4.6}$$

To answer, we will need *degrees of freedom (df)* for

$$\textit{Numerator } (s_1^2)\text{:df}_1 = n_1 - 1, \textit{ and}$$

$$\textit{Denominator } (s_2^2)\text{:df}_2 = n_2 - 1$$

The two degrees of freedom will be written as

$$F(\text{df}_1, \text{df}_2) = F(n_1 - 1, n_2 - 1)$$

Critical values, F_α, are given in Tables A.12, corresponding to selected values of α and $F(n_1 - 1, n_2 - 1)$. The tables are constructed so that the df across the top of the tables applies to df_1 of the numerator (s_1^2); the df along the left side applies to df_2 of the denominator (s_2^2). Note that the tables are one-tailed. When performing a two-sided test, such as the above, the risk will be twice that shown in the table.

Example 4.1 The 25 tests on single-fiber yarn strength from two different machines are shown in Table 4.3; they are plotted in Fig. 4.1. The graph suggests the possibility that machine 56 is basically more variable than machine 49. A formal comparison can be made by the *F* test, Eq. (4.6), using

$$\hat{\sigma}^2 = s^2 = \frac{n(\Sigma X^2) - (\Sigma X)^2}{n(n-1)}$$

Computations *Machine 56*

$$n_1(\Sigma X^2) = 25(496.1481) = 12{,}403.7025$$

$$(\Sigma X)^2 = (110.71)^2 \qquad = \underline{12{,}256.7041}$$

$$146.9984$$

$$n_1(n_1 - 1) = 600$$

$$s_1^2 = 146.9984/600 = 0.245 \qquad \text{df} = 24$$

Machine 49

$$n_2(\Sigma X^2) = 25(456.1810) = 11{,}404.525$$

$$(\Sigma X)^2 = (106.56)^2 \qquad = \underline{11{,}355.034}$$

$$49.491$$

$$n_2(n_2 - 1) = 600$$

$$s_2^2 = 49.491/600 = 0.082 \qquad \text{df} = 24$$

TABLE 4.3 Data: Breaking Strength of Single-Fiber Yarn Spun on Two Machines

Machine 49	Machine 56
3.99	5.34
4.44	4.27
3.91	4.10
3.98	4.29
4.20	5.27
4.42	4.24
5.08	5.12
4.20	3.79
4.55	3.84
3.85	5.34
4.34	4.94
4.49	4.56
4.44	4.28
4.06	4.96
4.05	4.85
4.34	4.17
4.00	4.60
4.72	4.30
4.00	4.21
4.25	4.16
4.10	3.70
4.35	3.81
4.56	4.22
4.23	4.25
4.01	4.10

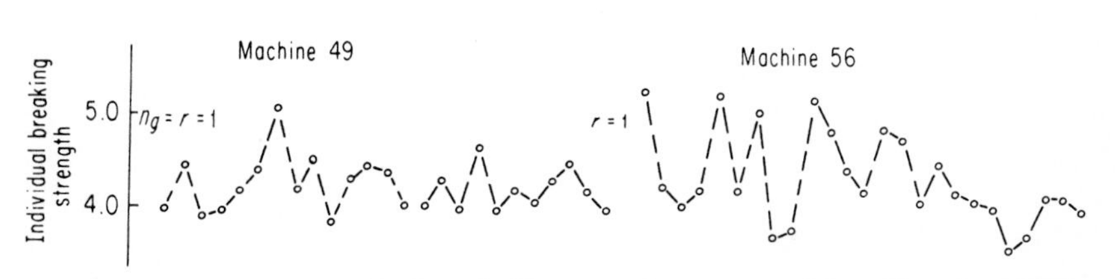

Figure 4.1 Breaking strength of single-fiber yarn from two machines (data from Table 4.3).

We now compare the two variabilities by the F ratio, Eq. (4.6).

$$F = s_1^2/s_2^2 = 0.245/0.082 = 2.99 \qquad \text{df} = F(24,24)$$

Critical Value of *F*

$$\text{for one-sided } \alpha = 0.01, F_{0.01} = 2.66$$

Since our test ratio 2.99 is larger than 2.66, we declare that the variability of machine 56 is greater than machine 49, with very small risk, $\alpha < 2(0.01) = 0.02$.

Example 4.2 The variability of an electronic device was of concern to the engineer who was assigned that product. Using cathode sleeves made from one batch of nickel (melt A), a group of 10 electronic devices was processed and an electrical characteristic (*transconductance*, G_m) was measured as in Table 4.4. Using nickel cathode sleeves from a new batch (melt B), a second test group of 10 devices was processed and G_m was read. Was there evidence that "the population variability" from melt B is significantly different from melt A?

The first step was to plot the data (Fig. 4.2).

The data were typical; neither set has an entirely convincing appearance of randomness. One obvious possibility is that the 1,420 reading in melt B is an outlier. The Dixon test for an outlier, $n = 10$, is

$$\mathbf{r}_{11} = \frac{X_2 - X_1}{X_{k-1} - X_1} = \frac{3770 - 1420}{6050 - 1420} = 0.508$$

TABLE 4.4 Data: Measurements of Transconductance of Two Groups of Electronic Devices Made from Two Batches (Melts) of Nickel

Melt A	Melt B
4,760	6,050
5,330	4,950
2,640	3,770
5,380	5,290
5,380	6,050
2,760	5,120
4,140	1,420
3,120	5,630
3,210	5,370
5,120	4,960
$\bar{A} = 4,184.0$	$\bar{B} = 4,861.0$
$(n_1 = 10)$	$(n_2 = 10)$
	$\bar{B}' = 5,243.3$
	$(n_2' = 9)$

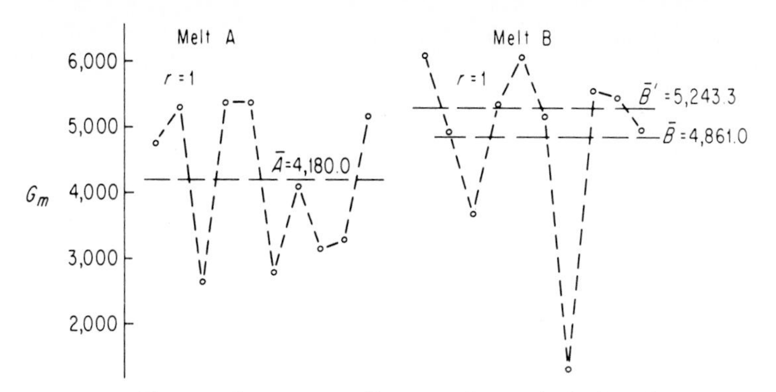

Figure 4.2 Transconductance readings on electronic devices from two batches of nickel (data from Table 4.4).

This computed value, 0.508, exceeds the critical value of 0.477, for $\alpha = 0.05$ and $n = 10$, Table A.9. This was "reasonably" convincing evidence that something peculiar occurred with the seventh unit in melt B either during production or testing to obtain the 1,420 reading.

Much of the variability of melt B is contributed by the suspected outlier observation. What would happen if the 1,420 observation were removed and F computed? (Note: B' is used to indicate the nine observations with 1,420 deleted.)

$$s_1^2 = s_A^2 = 1{,}319{,}604 \qquad n_1 = 10$$

$$s_2^2 = s_{B'}^2 = 477{,}675 \qquad n_2 = 9$$

Then

$$F = s_1^2/s_2^2 = (1{,}319{,}604)/(477{,}675) = 2.77 \qquad \text{df} = (9, 8)$$

Now this F exceeds $F_{0.10} = 2.56$ but not $F_{0.05} = 3.39$. This "suggested" that product from melt A may be more variable than product from melt B′ (one maverick removed), but is not entirely convincing, since the respective two-sided risks are 0.20 and 0.10.

There are important engineering considerations which must be the basis for deciding whether to consider melt B to be basically less variable than melt A, and whether to investigate reoccurrences of possible mavericks in either group.

On the other hand, what are the consequences, statistically, if the 1,420 observation is *not* considered to be a maverick? When the variance of melt B is recomputed with all 10 observations,

$$s_B^2 = 1{,}886{,}388$$

The F ratio becomes

$$F = s_1^2/s_2^2 = (1{,}886{,}388)/(1{,}319{,}604) = 1.43 \qquad \text{df} = (9, 9)$$

Even if the engineer were willing to assume a risk of $\alpha = 0.20$ there would still not be justification in considering the variability of the two samples to be different when the 1,420 (suspected maverick) is left in the analysis, since $F = 1.43$ is less than either $F_{0.10} = 2.44$ or even $F_{0.25} = 1.59$, with two-sided risks 0.20 and 0.50.

Conclusion The reversal which results from removing the 1,420 observation is not entirely unexpected when we observe the patterns of melt A and melt B; melt A gives an appearance of being more variable than melt B. These above statistical computations tend to support two suppositions:

1. That the observation 1420 is an outlier
2. That melt A is not a single population but a bimodal pattern having two sources, one averaging about 5000 and the other about 3000. This suspicion may justify an investigation either into the processing of the electronic devices or the uniformity of melt A in an effort to identify two sources "someplace in the system" and make appropriate adjustments.

(See Case History 13.4 for more discussion of this set of data.)

Range-square-ratio test, F_R

In this chapter, we have considered two methods of computing unbiased estimates of the variance, $\hat{\sigma}^2$: the mean-square, $\hat{\sigma}^2 = s^2$, and the range-square estimate, $\hat{\sigma}^2 = (\bar{R}/d_2^*)^2$. In the preceding section, two process variabilities were compared by forming an F ratio and comparing the computed F with tabulated critical values F_α. When data sets are available from two processes, or from one process at two different times, in the form of k_1 sets of r_1 from one and as k_2 sets of r_2 from a second, then we may use the *range-square-ratio*[1] to compare variabilities.

$$F_R = \frac{(\bar{R}_1/d_2^*)^2}{(\bar{R}_2/d_2^*)^2} \tag{4.7}$$

If the sample sizes $n = r_1k = r_2k$ are the same for both data sets, the ratio simply becomes

$$F_R = \frac{\bar{R}_1^2}{\bar{R}_2^2}$$

with

$$\mathrm{df}_1 \cong (0.9)k_1(r_1 - 1) \qquad \mathrm{df}_2 \cong (0.9)k_2(r_2 - 1) \tag{4.8}$$

The statistical significance of F_R is then compared to critical values in the F table (A.12) for degrees of freedom in relation (4.8). Details of the procedure are given below in Examples 4.3 and 4.4.

Example 4.3 This example concerns Case History 2.2 on chemical concentration. A visual inspection of Fig. 4.3 shows 11 range points representing a period of May 14

[1]Acheson J. Duncan, "The Use of Ranges in Comparing Variabilities," *Ind. Qual. Control*, vol. 11, no. 5, February 1955, pp. 18, 19, and 22. Values of r_1 and r_2 usually should be no larger than 6 or 7.

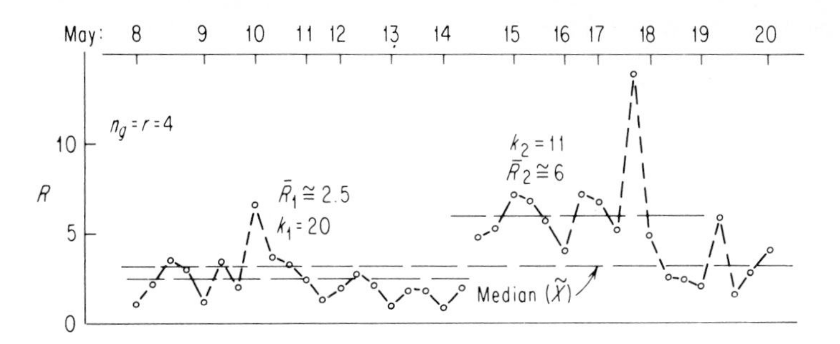

Figure 4.3 Evidence of increased process variability (Example 4.3).

into May 18 to be above the median. This long run is suggestive of a shift in the variability of the process during this period. Does the range-square-ratio test offer any evidence on variability?

We see an estimated average of $\overline{R}_1 = 2.5$ for the $k_1 = 20$ points during the period from May 8 into May 14, then a jump to an estimated $\overline{R}_2 = 6$ for the next 11 points, and possibly a drop back to the initial average during the period of the last 7 points. This is a conjecture, i.e., a tentative hypothesis.

Range-Square-Ratio Test *(Data of Fig. 4.3)*

$$\overline{R}_1 = 2.5, k_1 = 20, r_1 = 4, \text{ with } \text{df}_1 \cong (0.9)(20)(3) = 54$$

$$\overline{R}_2 = 6.0, k_2 = 11, r_2 = 4, \text{ with } \text{df}_2 \cong (0.9)(11)(3) = 30$$

Then

$$F_R = \frac{(\overline{R}_2/d_2^*)^2}{(\overline{R}_1/d_2^*)^2} \cong 36/6.25 = 5.76 \qquad \text{df} \cong (30, 54)$$

In Table A.12, no critical value of F is shown for 30 and 54 df, but one is shown for 30 and 60 df; it is $F_{0.01} = 2.03$. Since our test ratio, 5.76, is much larger than any value in the vicinity of $F_{0.01} = 2.03$, there is no need or advantage in interpolating. We declare there is a difference with risk $\alpha < 0.02$.

The range-square-ratio test supports the evidence from the long run that a shift in variability did occur about May 14. The process variability was not stable; an investigation at the time would have been expected to lead to physical explanations.

Control charts, $\overline{X}$, R, are helpful in process-improvement studies: these often indicate shifts in $\overline{R}$ (as well as in $\overline{X}$). Then the range-square-ratio test can be applied easily as a check, and investigations made into possible causes.

Example 4.4 Another use of the range-square-ratio test is in comparing variabilities of two analytical chemical procedures. The data of Table 4.5 pertain to the ranges of three determinations of a chemical characteristic by each of four analysts on (blind) samples from each of four barrels. In the first set, determinations are made by method A and in the second by method B. The question is whether the experimental variability of the second method is an improvement (reduction) over the first. It is assumed that the experimental variability is independent of analyst and sample so that the 16 ranges can be averaged for each set. Here we have

$$\overline{R}_1 = 1.37 \qquad \overline{R}_2 = 0.75$$

TABLE 4.5 Variability (As Measured by Ranges, $r = 3$) of Two Methods of Chemical Analysis Using Four Analysts

Method	Barrels	Analysts I	II	III	IV	
A	1	2.0	3.1	0	1.5	
	2	0.5	2.0	2.5	0.3	
	3	1.5	1.9	1.5	0.8	
	4	2.5	0.5	1.0	0.3	$\bar{R}_A = 1.37$
B	1	0.5	1.3	1.0	1.0	
	2	0.5	1.4	0	0.9	
	3	1.0	0.8	0.5	0.3	
	4	1.0	0.7	1.0	0.3	$\bar{R}_B = 0.75$

$$r_1 = 3 \qquad r_2 = 3$$
$$k_1 = 16 \qquad k_2 = 16$$

and from Table A.11, $df_1 = 29.3$ and $df_2 = 29.3$, or $df \cong (0.9)k(r - 1) = 29$ from Eq. (4.4). Then

$$F_R = \frac{(1.37)^2}{(0.75)^2} = 3.3 \qquad df \cong (29, 29)$$

From F table (A.12), we find $F_{0.05} = 1.87$. Since our computed $F_R = 3.3$ is in excess of the one-sided critical (0.05) value, we conclude that the variability of the second method is an improvement over the variability of method A with risk $\alpha = 0.05$ since this is a one-sided test.

4.6 Summary

This chapter has presented different methods of computing estimates of σ and σ^2 from data. These estimates are used in comparing process variabilities under operating conditions. Frequent references will be made to the following estimates:

- $\hat{\sigma} = \bar{R}/d_2$: used when the number of subgroups k is as large as 25 or 30, and is conservative for smaller numbers of subgroups—(see Note below). See Eq. (1.13). Gives unbiased estimate for all k.
- $\hat{\sigma} = \bar{R}/d_2^*$: used when the number of subgroups k is quite small (less than 30) and with computations in association with other tables designed for use with the slightly biased estimate s.

We frequently use this estimate in association with factors from Table A.15 which was designed for use with the similarly biased estimate s.

Note: In Table A.11, it may be seen that $d_2 < d_2^*$ in each column; thus, the two estimates above will differ slightly, and

$$\overline{R}/d_2 > \overline{R}/d_2^*$$

Consequently, the use of d_2 with a *small number* of subgroups will simply produce a slightly more conservative estimate of σ in making comparisons of k means. We shall usually use d_2^* in the case histories of the following chapters.

Two methods of comparing process variability have been discussed in Sec. 4.5: the F test and the range-square-ratio test (F_R).

An outline of some computational forms is given in Table 4.6. The different procedures for computing $\hat{\sigma}$ are quite confusing unless used frequently. Those forms which will be most useful in the following chapters are marked.

CASE HISTORY 4.1

Vial Variability

In an effort to evaluate two different vendors, data is taken on the weights (in grams) of 15 vials from firm A and 13 vials from firm B. (See Table 13.4.) Standard deviations were calculated to determine whether there was a difference in variation between the vendors. The results were $s_A = 2.52$ and $s_B = 1.24$. An F test was used to determine whether the observed difference could be attributed to chance, or was real, with an $\alpha = 0.10$ level of risk.

TABLE 4.6 Summary: Estimating Variability

Different procedures for computing $\hat{\sigma}$ are quite confusing unless used frequently. Forms which will be used most often in the following chapters are marked with the superscript#.

Computing measures of variation from k sets of r each			
#1. $\hat{\sigma} = \overline{R}/d_2$	df $\cong (0.9)k(r-1)$	unbiased	Eq. (1.13)
#2. $\hat{\sigma} = \overline{R}/d_2^*$	Also Table A.11	slightly biased	Eq. (4.3)
3. $\hat{\sigma}^2 = (\overline{R}/d_2^*)^2$		unbiased	Eq. (4.3)
4. $\hat{\sigma}^2 = \dfrac{(r_1-1)s_1^2 + (r_2-1)s_2^2 + \ldots + (r_k-1)s_k^2}{r_1 + r_2 + \ldots + r_k - k}$	df $= r_1 + r_2 + \ldots + r_k - k$	unbiased	Eq. (4.5)
#5. $\hat{\sigma}_{\bar{x}} = \hat{\sigma}/\sqrt{r}$			Eq. (1.7)
Computing measures of variation from one set of $n = rk$			
#6. $\hat{\sigma} = s = m\sqrt{\dfrac{n\Sigma fd^2 - (\Sigma fd)^2}{n(n-1)}}$		where m = cell width and df$=n-1$	Eq. (1.4b)
7. $\hat{\sigma}^2 = s^2 = \dfrac{\Sigma(X_i - \overline{X})^2}{n-1}$	df$=n-1$	unbiased	Eq.(4.1a)
8. $\hat{\sigma}^2 = s^2 = \dfrac{n\Sigma X_i^2 - (\Sigma X_i)^2}{n(n-1)}$		unbiased	Eq. (4.1b)
9. $\hat{\sigma} = s = \sqrt{\dfrac{\Sigma(X_i - \overline{X})^2}{n-1}}$		slightly biased	Eq. (4.1)

Here

$$F = \frac{2.52^2}{1.24^2} = 4.13$$

And, since this is a two-sided test with 14 and 12 degrees of freedom, the ratio of the larger over the smaller variance is compared to a critical value

$$F_{0.05} = 2.64$$

to achieve the $\alpha = 0.10$ risk. This would indicate that firm A has the greater variation, assuming the assumptions of the F test are met. This is further discussed in Case History 13.3.

4.7 Practice Exercises

1. Apply the range-square-ratio test to the first and second halves of the mica data shown in Fig. 2.5. Is there a difference in variation undetected by the control chart?
2. Given the data of Example 4.1, compute $\overline{R}$ for machine 49, using subgroups of $r = 5$. Also compute $\overline{R}$ for machine 56 by the same method. Compute for each machine using Eq. 4.3*a*, and compare the results with those obtained by the authors. Explain the reason for the differences. Which is more efficient? Why?
3. Assume that the following additional data are collected for the comparison of Example 4.1:

49:	4.01	5.08	4.4	4.25	4.20
56:	5.30	4.10	3.78	4.25	4.33

   Recompute s^2 for each machine.
   Perform and analyze an F test. State your conclusion, including the approximate p value. Define and state the value of n, k, and r. Use $\alpha = 0.05$.

4. If you have a calculator or computer with a built-in function, use it to compute s for each machine using Eq. 4.1*a* and then pool, using Eq. 4.5*a*. (Check to be sure that your method uses $n - 1$ rather than n in the denominator of Eq. 4.1*a*.) Compare your s_p with s obtained by the author and explain the likely cause of the difference.
5. Consider Example 4.2. Analyze the authors' statement that melt A is actually from two sources. How can this assertion be proven statistically? Perform the appropriate test(s) and draw conclusions. (Note: You may need to refer to Chapter 3 or 13 or an appropriate statistics text.)
6. Demand for part no. XQZ280 is such that you have had to run them on two process centers. Because of an engineering interest in the 2.8000 cm dimension you have taken the following five samples of five dimensions each from the two process centers (in order of their production occurrence).

a. Eliminate any outliers ($\alpha = 0.05$) based on all 25 unit values from each process center.
b. Formulate hypothesis H_0 of no difference and use the F test to attempt to reject H_0.
c. Plot the data.

	Sample No.				
Process	1	2	3	4	5
Center A:	2.8000	2.8001	2.7995	2.8014	2.8006
	2.8001	2.8012	2.8006	2.8000	2.8009
	2.8006	2.8015	2.8002	2.8005	2.7996
	2.8005	2.8002	2.8003	2.8003	2.8000
	2.8005	2.8010	2.8009	2.7992	2.7997
Center B:	2.7988	2.7985	2.7995	2.8004	2.8001
	2.7980	2.7991	2.7993	2.8001	2.8004
	2.7989	2.7986	2.7995	2.8002	2.8007
	2.7987	2.7990	2.7995	2.8004	2.8003
	2.7985	2.7994	2.7995	2.7997	2.8012

Chapter

5

Attributes or Go No-Go Data

5.1 Introduction

In every industry there are important quality characteristics which cannot be measured, or which are difficult or costly to measure. In these many cases, evidence from mechanical gauges, electrical meters used as gauges, or visual inspection may show that some units of production conform to specifications or desired standards and that some do not conform. Units which have cracks, missing components, appearance defects or other visual imperfections, or which are gauged for dimensional characteristics and fail to conform to specifications may be recorded as *rejects, defectives,* or *nonconforming* items.[1] They may be of a mechanical, electronic, or chemical nature. The *number* or *percentage* of such units is referred to as *attributes data;* each unit is recorded simply as having or not having the attribute.

Process improvement and troubleshooting with attributes data have received relatively little attention in the literature. In this book, methods of analyzing such data receive major consideration; they are of major economic importance in the great majority of manufacturing operations. (See especially Chap. 11.)

5.2 Three Important Problems

The ordinary manufacturing process will produce some defective[2] units. When random samples of the same size are drawn from a stable process, we expect variation in the number of defectives in the samples. Three important problems (questions) need consideration:

[1]Some nonconforming items may be sold as "seconds"; others reworked and retested; others scrapped and destroyed.

[2]If a unit of production has at least one nonconformity, defect, or flaw, then the unit is called nonconforming. If it will not perform its intended use, it is called defective. This is in conformity with the National Standard ANSI/ASQC A2 (1978). In this book, the two terms are often used interchangeably.

1. What *variation* is expected when samples of size n are drawn from a stable process?
2. Is the process *stable* in producing defectives? This question of stability is important in process-improvement projects.
3. How *large* a sample is needed to estimate the *percent defective* in a warehouse or in some other type of population?

Discussion of the three questions

What can be predicted about the sampling variation of the *number* of defectives found in random samples of size n from a stable process? There are two possibilities to consider: *first,* the percent[3] defective is assumed known. This condition rarely occurs in real-life situations. *Second,* the process percent defective is not known but is estimated from k samples, $k \geq 1$, where each sample usually consists of more than one unit. The samples may or may not all be of the same size. This is the usual problem we face in practice.

Binomial theorem

Assume that the process is stable and that the probability of each manufactured unit being defective is known to be p. Then the *probability of exactly x defectives* in a random sample of n units is known from the Binomial Theorem. It is

$$\Pr(x) = \frac{n!}{x!(n - x)!} p^x q^{n - x} \tag{5.1}$$

where p is the probability of a unit being defective and $q = 1 - p$ is the probability of it being nondefective. It can be proved that the *expected number* of defectives in the sample is np: there will be variation in the number which actually occur.

Example 5.1 When $n = 10$ and $p = q = 0.5$, for example, this corresponds to tossing 10 ordinary coins[4] and counting the number of heads or tails on any single toss of the 10.

Probabilities have been computed and are shown in Table 5.1 and plotted in Fig. 5.1. The expected or most probable number of defectives in a sample of 10 with $p = q = 0.5$ is $np = 5$. Also, it is almost as likely to have 4 or 6 heads as to have 5 and about *half* as likely to have 3 or 7 heads as 5.

The sum of all probabilities from Pr(0) to Pr(10) is *one,* i.e., certainty. The combined probability of 0, 1, or 2 heads can be represented by the symbol: $\Pr(x \leq 2)$. Also, $\Pr(x \geq 8)$ represents the probability of eight or more. Neither of these probabilities is large:

$$\Pr(x \leq 2) = \Pr(x \geq 8) = 0.055$$

These represent two tails or extremes in Fig. 5.1. We can state for example:

When we make a single toss of 10 coins, we predict that we shall observe between three and seven heads inclusive, and our risk of being wrong is about

---

[3]The percent defective P equals $100p$ where p is the *fraction* nonconforming in the process.
[4]Or of tossing a single coin 10 times.

TABLE 5.1 Probabilities Pr(*x*) of Exactly *x* Heads in 10 Tosses of an Ordinary Coin

$$\begin{aligned}
\Pr(0) &= \Pr(10) = (.5)^{10} = 0.001 \\
\Pr(1) &= \Pr(9) = 10(.5)^{10} = 0.010 \\
\Pr(2) &= \Pr(8) = 45(.5)^{10} = 0.044 \\
\Pr(3) &= \Pr(7) = 120(.5)^{10} = 0.117 \\
\Pr(4) &= \Pr(6) = 210(.5)^{10} = 0.205 \\
\Pr(5) &= 252(.5)^{10} = 0.246
\end{aligned}$$

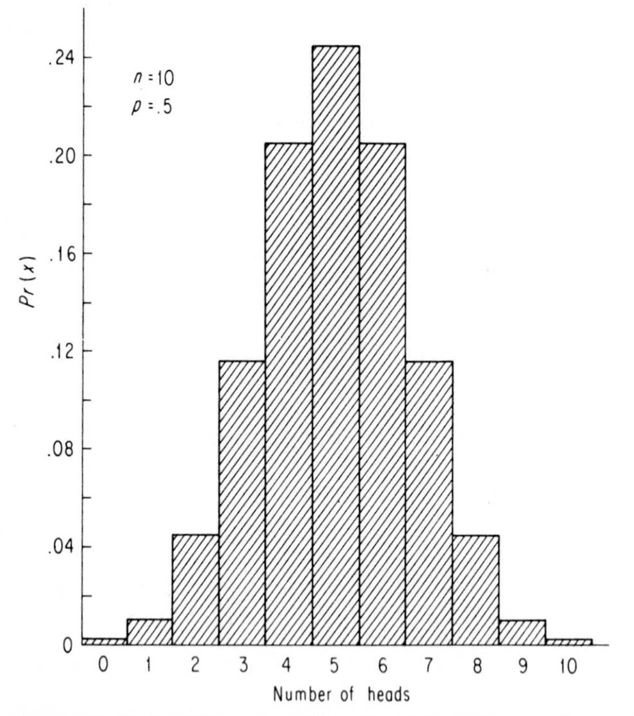

Figure 5.1 Probabilities of exactly x heads in 10 tosses of an ordinary coin ($n = 10, p = 0.50$).

$$0.055 + 0.055 = 0.11 \text{ or } 11 \text{ percent}$$

We expect to be right 90 percent of the time.

Binomial probability tables for selected values of *n*

Decimal values of probabilities are tedious to compute from Eq. (5.1) even for small values of n. Consequently, a table of binomial probabilities is included

(Table A.5) for selected values of n and p. Values in the table are probabilities of c or fewer defective units in a sample of n when the probability of occurrence of a defective is p on each item. The sum of the values I and J shown as the row and column headings for a given probability give c. Values in the table corresponding to c are *accumulated* values; they represent Pr $(x \le c)$. For example, the probability of 3 or fewer heads (tails) when $n = 10$, $p = 0.5$ is

$$\Pr(\le 3) = \Pr(0) + \Pr(1) + \Pr(2) + \Pr(3) = 0.172$$

The value of $c = I + J = 3 + 0$ for the entry 0.172.

Example 5.2 Assume a stable process has been producing 3 percent defectives; when we inspect a sample of $n = 75$ units, we find six defectives.

Question Is finding as many as six defectives consistent with an assumption that the process is still at the 3 percent level?

Answer Values of $\Pr(x \le c)$ calculated from the binomial are shown in Table 5.2 and again in Fig. 5.2. The probability of finding as many as six is seen to be small; it is represented by the symbol

$$\Pr(x \ge 6) = 1 - \Pr(x \le 5) = 1 - 0.975 = 0.025$$

Using the value $\Pr(c) = \Pr(x \le c) - \Pr(x \le c - 1)$ given in Table A.5.

Each individual probability above may have a rounding discrepancy of ± 0.0005. Also cumulative probabilities may have a rounding discrepancy; such discrepancies will be common but of little importance. Thus it is very unlikely that the process is still at its former level of 3 percent. There is only a 2.5 percent risk that an investigation would be unwarranted. If a process average greater than 3 percent is economically important, then an investigation of the process should be made.

When we obtain k samples of size n_g from a process, we count the number of defectives in each sample found by inspection or test. Let the numbers be

$$d_1, d_2, d_3, \ldots, d_k$$

Then the percent defective in the process, which is assumed to be stable, or in a population assumed to be homogeneous, is *estimated* by dividing the total number of defectives found by the total number inspected

TABLE 5.2 Probabilities of x Occurrences in $n = 75$ Trials and $p = 0.03$

Pr(0) = .101
Pr(1) = .236
Pr(2) = .270
Pr(3) = .203
Pr(4) = .113
Pr(5) = .049
Pr(6) = .018
Pr(7) = .005
Pr(8) = .001

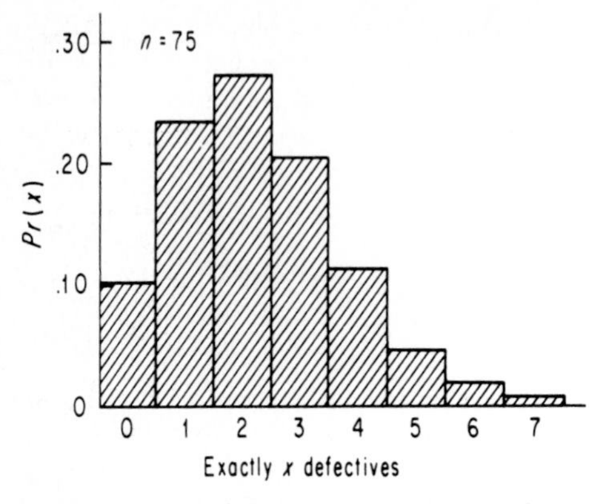

Figure 5.2 Probabilities of exactly x defectives in a sample of 75 from an infinite population and $p = 0.03$.

$$\hat{p} = \bar{p} = \frac{\Sigma d_i}{\Sigma n_g} \qquad \text{and} \qquad \hat{P} = 100\hat{p} \tag{5.2}$$

We do not expect these *estimates*[5] of the process average to be exactly equal to it, nor shall we ever know exactly the "true" value of P.

A measure of variability for the binomial distribution

When n is large and p and q are each larger than say 5 percent and

$$np \geq 5 \text{ or } 6 \qquad \text{and} \qquad nq \geq 5 \text{ or } 6 \tag{5.3}$$

then the binomial distribution closely approximates the continuous normal curve (Fig. 1.2). The values in Eq. (5.3) are guideposts and not exact requirements.

The computation of $\hat{\sigma}$ for the binomial (attribute data) can be a much simpler operation than when computing one for variables (Table 1.3). After the value of $\hat{p}$ is obtained as in Eq. (5.2), the computation is made from the following formulas.[6]

$$\hat{\sigma}_p = \sqrt{\frac{p(1-p)}{n}} \qquad \text{for proportion (fraction) defective} \tag{5.4}$$

$$\hat{\sigma}_p = \sqrt{\frac{P(100-P)}{n}} \qquad \text{for percent defective} \tag{5.5}$$

[5]We usually will not show the "hat" over either p or P.

[6]This is proved in texts on mathematical statistics.

$$\hat{\sigma}_{np} = \sqrt{np(1 - p)} \qquad \text{for number defective} \tag{5.6}$$

Thus knowing only the process level p and the sample size n, we compute standard deviations directly from the appropriate formula above.

Example 5.3 A stable process is producing 10 percent nonconforming items. Samplings of size $n = 50$ are taken at random from it. What is the standard deviation of the sampling?

Answer Any one of the following, depending upon the interest.

$$\hat{\sigma}_p = \sqrt{\frac{(0.10)(0.90)}{50}} = 0.0424 \qquad \hat{\sigma}_p = 4.24\% \qquad \hat{\sigma}_{np} = 2.12$$

Question 1. *Expected variation*

Assuming a stable process with $\bar{p} = 0.10$, the expected variation in the number of defectives in samples of $n = 50$ can be represented in terms of np and $\hat{\sigma}_{np}$. Just as in Eq. (1.10) for the normal curve, when the conditions of Eq. (5.3) are applicable, the amount of variation is predicted to be between

$$np - 3\sigma_{np} \quad \text{and} \quad np + 3\sigma_{np} \quad (99.7\%)$$

$$np - 2\sigma_{np} \quad \text{and} \quad np + 2\sigma_{np} \quad (95.4\%) \tag{5.7}$$

$$np - \sigma_{np} \quad \text{and} \quad np + \sigma_{np} \quad (68.3\%)$$

Thus in Example 5.3 with $n = 50$ and $p = 0.10$, the average number expected is $np = 5$ and the standard deviation is $\hat{\sigma}_{np} = 2.12$. From Eq. (5.7), we can predict the variation in samples to be from

$$5 - 6.36 \text{ to } 5 + 6.36 \qquad \text{i.e., 0 to 11 inclusive } (99.7\%)$$

or

$$5 - 4.24 \text{ to } 5 + 4.24 \qquad \text{i.e., 1 to 9 inclusive } (95.4\%)$$

Just as easily, the expected sampling variation in p or P can be obtained from Eqs. (5.4) and (5.5).

Question 2. *But is the process stable?*

A simple and effective answer is available when we have k samples of n_g each by making a control chart for fraction or percent defective. This chart is entirely analogous to a control chart for variables.

In routine production, control limits are usually drawn using 3-sigma limits. Points outside these limits (outages) are considered evidence of assignable causes. Also evidence from *runs* (Chap. 2) is used jointly with that from outages, especially in process improvement studies, when 2-sigma limits are usually the basis for investigations.

We expect "almost all" points which represent samplings from a *stable process* to fall inside 3-sigma lines. If they *do not*, we say "The process is *not* in control" and there is only a small risk of the conclusion being incorrect. If they *do* all fall inside, we say "The process appears to be in statistical control" or "The process appears stable"; this is not the same as saying "It *is* stable." There is an analogy. A person is accused of a crime: The evidence may (1) convict the defendant of guilt, and we realize there is some small chance that justice miscarried, or (2) fail to convict, but this is not the same as believing in the person's innocence. The "null hypothesis" (1) is that of no difference.

Example 5.4 (Data of Table 11.18, machine 1 only.) Final inspection of small glass bottles was showing an unsatisfactory reject rate. A sampling study of bottles was made over several days to obtain evidence regarding stability and to identify possible major sources of rejects. A partial record of the number of rejects, found in samples of $n_g = 120$ bottles taken three times per day from one machine, has been plotted in Fig. 5.3. The total number of rejects in the 21 samples was 147; the total number inspected was $n = 7(3)(120) = 2520$. Then $P = 0.0583$ or 5.83 percent. When considering daily production by shifts, the sample inspected was $n_g = 8(15) = 120$. Then

$$\hat{\sigma}_p = \sqrt{\frac{(5.83)(94.17)}{120}} = \sqrt{4.575\%} = 2.14\%$$

The upper 3-sigma limit is: 5.83 percent + 3(2.14) = 12.25 percent. There is no lower limit since the computation gives a negative value.

Discussion Was the process stable during the investigation?

Outages: There is one on August 17.

Runs: The entire set of data is suggestive of a process with a gradually increasing P. This apparent increase is not entirely supported by a long run. The run of five above at the end (and its two preceding points exactly on the median and average) is

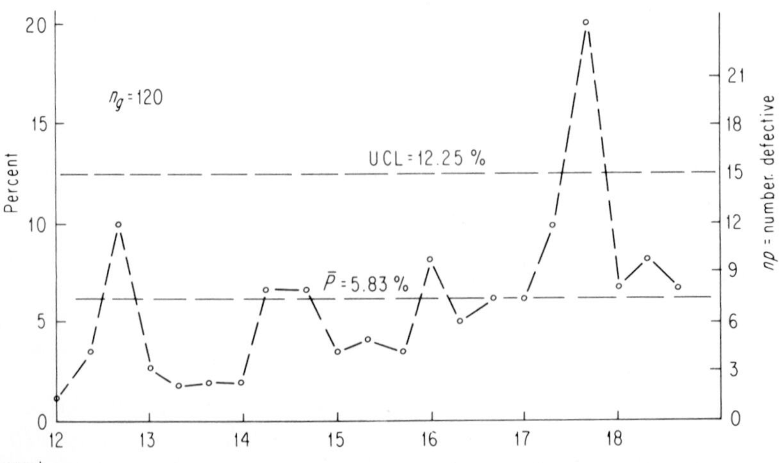

Figure 5.3 A control-chart record of defective glass bottles found in samples of 120 per shift over a 7-day period (see Example 5.4).

"suggestive support" (Table A.3); and the six below the median out of the first seven strengthens the notion of an increasingly defective process.

Both the outage and the general pattern indicate an uptrend.

Question 3. *How large a sample?*

How many are needed to estimate the percent defective in a warehouse or in some other population or universe? A newspaper article presented forecasts of a scientific survey of a national presidential election on the basis of 1400 individual voter interviews. The survey was conducted to determine the expected voting pattern of 60 to 80 million voters. The chosen sample size is not determined on a percentage basis.

The question of sample size was discussed in Sec. 1.11 for variables data. The procedure with attributes is much the same but differs in detail.

From a statistical viewpoint, we must first adopt estimates of three quantities:

1. The *magnitude* of allowable error Δ in the estimate we will obtain for P. Do we want the answer to be accurate within 1 percent? Or within 3 percent? *Some* tentative estimate must be made.
2. What is a rough guess as to the value of P? We shall designate it here by $\hat{P}$.
3. With what assurance do we want to determine the region within which P is to be established? Usually about 95 or 90 percent assurance is reasonable.

These two assurances correspond to $\pm 2\hat{\sigma}$ and $\pm 1.65\hat{\sigma}$.

Answer: The basic equations to determine sample size in estimating P in a population, allowing for possible variation of $\pm\Delta$, are

$$\pm\Delta = \pm 3\hat{\sigma}_p \qquad \text{if we insist on 99.7\% confidence}$$

$$\pm\Delta = \pm 2\hat{\sigma}_p \qquad \text{if about 95\% confidence is acceptable}$$

$$\pm\Delta = \pm 1.65\hat{\sigma}_p \qquad \text{if we accept about a 90\% confidence level}$$

The second of these equations may be rewritten as

$$\Delta = 2\sqrt{\frac{\hat{P}(100 - \hat{P})}{n}}$$

which simplifies to

$$n = \frac{4\hat{P}(100 - \hat{P})}{\Delta^2} \qquad \text{(95\% confidence)} \qquad (5.8a)$$

Similarly from the third and first equations above, we have

$$n = \frac{(1.65)^2 \hat{P}(100 - \hat{P})}{\Delta^2} \quad \text{(90\% confidence)} \tag{5.8b}$$

$$n = \frac{9\hat{P}(100 - \hat{P})}{\Delta^2} \quad (99.7\%) \tag{5.8c}$$

In general

$$n = \frac{Z_{\alpha/2}{}^2 \hat{P}(1 - \hat{P})}{\Delta^2}$$

When $\hat{P}$ is unknown use $P = 50$ for a conservatively large estimate of sample size. Note α is the complement of the confidence.

Other factors are very important, too, in planning a sampling to estimate a percent defective within a population. We must plan the sampling procedure so that whatever sample is chosen it is as nearly *representative* as possible. How expensive is it to obtain items from the process or other population for the sample? What is the cost of providing test equipment and operators? We would be reluctant to choose as large a sample when the testing is destructive as when it is nondestructive. These factors may be as important as the statistical ones involved in Eq. (5.8). However, values obtained from Eq. (5.8) will provide a basis for comparing the reasonableness of whatever sample size is eventually chosen.

Example 5.5 How large a sample is needed to estimate the percent P of defective glass bottles in a second warehouse similar to the one discussed in Example 5.4? Preliminary data suggest that $\hat{P} \cong 5$ percent or 6 percent. Now it would hardly be reasonable to ask for an estimate correct to 0.5 percent, but possibly to 1 percent or 2 percent. If we choose $\Delta = 1$ percent, a confidence of 95 percent, and $\hat{P} = 5$ percent, then

$$n = \frac{4(5)(95)}{1} = 1900$$

If we were to increase Δ to 2 percent, then

$$n = \frac{4(5)(95)}{4} = 475$$

Decision A representation of 1900 is probably unnecessary, too expensive, and too impractical. A reasonable compromise might be to inspect about 475 initially. Then whatever percent defective is found, compute $P \pm 2\hat{\sigma}_p$. Be sure to keep a record of all different *types* of defects found; a sample of 475 will probably provide more than enough information to decide what further samplings or other steps need be taken.

5.3 On How to Sample

There are different reasons for sampling

Estimating the percent P in a warehouse or from a process are examples just discussed. Now, we shall discuss the frequency and size of samples necessary to *monitor* a production process. This is a major question of process control.

- When daily production from any one shift is less than say 500, *and* 100 percent inspection is already in progress, it may be sensible to initiate a control chart with the entire production as the sample.[7] The control limits UCL and LCL can be computed and drawn on the chart for an *average* n_g as a basis for monitoring daily production. The control limits can be adjusted (recomputed) for any point corresponding to a substantially larger or smaller sample; this would be warranted only for points near the computed limits for average n_g. A factor of 2 in sample size is a reasonable basis for recomputing.
- When a daily shift production is more than say 500 or when evidence suggests process trouble, smaller random samples—checked by a special inspector—can provide valuable information. Samples of $n_g = 50$ or 100 can provide much useful information; sometimes even smaller ones will be adequate as a starter. The special inspector will not only record whether each item is defective but will inspect for *all* defect categories and will record the major types of defects and the number of each. See Case Histories 6.1, 6.2, and Sec. 6.13.
- How large a sample is necessary when starting a control chart with attributes? In Case History 5.1, a chart was plotted daily from records of 100 percent inspection. In Fig. 5.6 we see that variation of daily production defectives during January was as much as 4 or 5 percent and more above and below average. If a variation of $\Delta = 2$ percent about $p = 5.6$ percent were now accepted as a reasonable variation to signal a daily shift, and we choose 3-sigma limits to use in production, then from Eq. (5.8c) it follows that

$$n = \frac{9(5.6)(94.4)}{4} = 1189 \cong 1200$$

But a sample of about 1200 would be adequate for a *total* day's production. When samples are to be inspected hourly, then $n_g = 1200/8 = 150$ becomes a reasonable choice. If samples were to be inspected bihourly, then $n_g = 1200/4 = 300$ would be indicated from a statistical viewpoint. However a decision to choose a smaller sample would be better than no sampling.

The question of whether a sampling system should be instituted, and the

---

[7]The entire production will be considered the population at times; but it is usually more profitable to consider it as a sample of what the process *may produce in the future,* if it is a stable process.

details of both size and frequency of samples, should consider the potential savings compared to the cost of sampling.

Conclusion: If the potential economic advantages favor the start of a sampling plan, then initial samples of no larger than 150 are indicated. This would provide a feedback of information to aid production. The sample size can later be changed—either smaller or larger—as suggested by experience with this specific process.

There is another reason for sampling in production. Consider a shipment of items—either outgoing or incoming. How large a random sample shall we inspect and how many defectives shall we allow in the sample and still approve the lot? This is the problem of *acceptance sampling* (see Chap. 6) for lot inspection as opposed to its use in monitoring production.

5.4 Attributes Data Which Approximate a Poisson Distribution

In Table A.5, binomial tables are provided for selected values up to $n = 100$. In this section, we consider count data in the form of number of nonconformities counted in a physical unit (or units) of a given size.

Example 5.6 A spinning frame spins monofilament rayon yarn;[8] it has over a hundred spinarets. There are occasional spinning stoppages at an individual spinaret because of yarn breakage. A worker then corrects the breakage and restarts the spinaret. An observer records the number of stoppages on an entire spinning frame during a series of 15-min periods. The record over 20 time periods shows the following number of stoppages:

6, 2, 1, 6, 5, 2, 3, 5, 6, 1, 5, 6, 4, 3, 1, 3, 5, 7, 4, 5...

with an average of four per period.

This type of data is attribute or countable data. However, it differs from our preceding attribute data in the following ways:

1. There is no way of knowing the "sample size n"; the number of *possible* breaks is "very large" since each spinaret may have several stoppages during the same time period. Although we do not know n, we know it is potentially "large," at least conceptually.
2. There is no way of knowing the "probability p" of a single breakage; we *do* know that it is "small."
3. We do not have estimates on either n or p, but we have a good estimate of their *product:* in fact, the average number of stoppages per period is $\hat{\mu} = np = 4$.

There are many processes from which we obtain attribute data which satisfies these three criteria: n is "large" and p is "small" (both usually unknown) but their average product $\hat{\mu}_2 = np$ is known. Data of this sort are called *Poisson* type. A classical example of such data is the number of deaths caused by the kick of a horse among different

[8]See Case History 11.3.

Prussian army units. In that case the number of opportunities for a kick and the probability of a kick were unknown, but the average number of fatalities turned out to be $\mu = 0.61$ per unit per year.

How much variation is predicted in samples from a Poisson distribution?

The question can be answered in two ways, both useful to know:

1. *By computing* $\hat{\sigma}_c$. From Eq. (5.6), with $\hat{\sigma}_c = \sqrt{np(1 - p)}$ we have, for the Poisson with p small, and therefore $(1 - p) \cong 1$

$$\hat{\sigma}_c \cong \sqrt{np} = \sqrt{\hat{\mu}} \qquad (5.9)$$

where $\hat{\mu}$ is the average number of defects per unit.

Discussion: In the artificial data of Example 5.6, we are given $\hat{\mu} = np = 4$. Then from Eq. (5.9), $\hat{\sigma} = \sqrt{4} = 2$. Consequently, variation expected from this Poisson type of process, assumed stable, can be expected to extend from

$$4 - 2\hat{\sigma} \text{ to } 4 + 2\hat{\sigma} \qquad \text{(about 95\% confidence)}$$

i.e., from 0 to 8.

2. *From Poisson curves (Table A.6).* These very useful curves give probabilities of c or *fewer* defects for different values of $\mu = np$. The value of $\Pr(\le 8)$, for the previous example, is estimated by first locating the point $\mu = np = 4$ on the base line; follow the vertical line up to the curve for $c = 8$. Then using a plastic ruler locate $\Pr(\le 8)$ on the left-hand vertical scale; it is slightly more than 0.98.

The Poisson formula for the probability of x outcomes when the mean is μ becomes

$$\Pr(x) = \frac{\mu^x e^{-\mu}}{x!}$$

where $e = 2.71828$

Example 5.7 The question of whether the process which generated the defects data is stable is answered by the c control chart. To construct such a chart it is, of course, necessary to estimate the average number of defects c found as

$$\hat{\mu} = \bar{c}$$

Limits then become

$$\text{UCL}_c = \bar{c} + 3\sqrt{\bar{c}}$$
$$\text{CL} = \bar{c}$$
$$\text{LCL}_c = \bar{c} - 3\sqrt{\bar{c}}$$

so for the data on stoppages of the spinning frame we have

$$\hat{\mu} = \bar{c} = 4$$

and

$$\text{UCL}_c = 4 + 3\sqrt{4} = 10$$
$$\text{CL} = 4$$
$$\text{LCL}_c = 4 - 3\sqrt{4} = -2 \sim 0$$

Note that no sensible practitioner would construct a chart showing −2 as a lower limit in the count of defects. In such a case as this, we simply do not show a lower limit, since a value of 0 is obviously within the $3\hat{\sigma}$ spread. The chart is shown in Fig. 5.4.

Now the stoppage data showed the number of stoppages of the spinning frame itself which consisted of 100 spinarets. Sometimes, it is desirable to plot a chart for defects per unit, particularly when the number of units n_g involved changes from sample to sample. In such a case, we plot a $\bar{u}$ chart where

$$\bar{u} = \frac{\bar{c}}{n} = \text{defects per unit}$$

and the limits are

$$\text{UCL}_u = \bar{u} + 3\sqrt{\frac{\bar{u}}{n_g}}$$
$$\text{CL} = \bar{u}$$
$$\text{LCL}_u = \bar{u} - 3\sqrt{\frac{\bar{u}}{n_g}}$$

Obviously, when $n_g = 1$, the limits reduce to those of the c chart. If we adjust the data for the $n = 100$ spinarets involved in each count we obtain the following data in terms of stoppages per spinaret, u.

0.06, 0.02, 0.01, 0.06, 0.05, 0.02, 0.03, 0.05, 0.06, 0.01, 0.05, 0.06, 0.04, 0.03, 0.01, 0.03, 0.05, 0.07, 0.04, 0.05

and the limits are

$$\text{UCL}_u = 0.04 + 3\sqrt{\frac{0.04}{100}} = 0.04 + 0.06 = 0.10$$

$$\text{CL} = 0.04$$

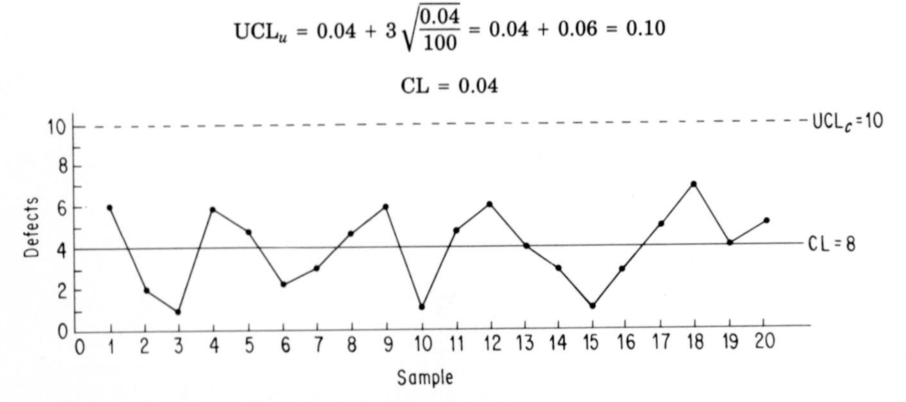

Figure 5.4 c chart on stoppages of spinning frame.

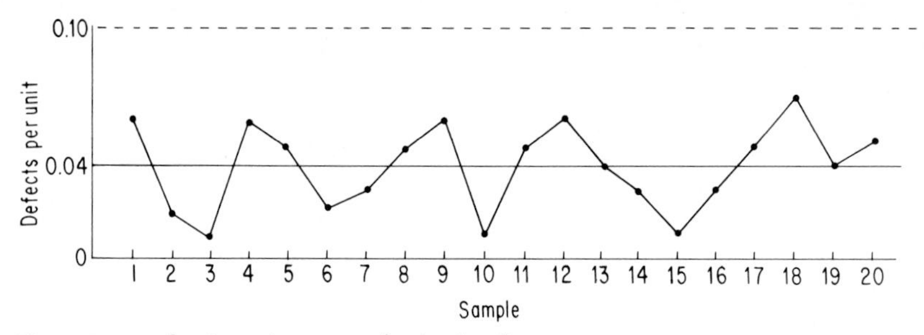

Figure 5.5 u chart on stoppages of spinning frame.

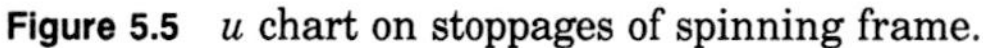

$$LCL_u = 0.04 - 3\sqrt{\frac{0.04}{100}} = 0.04 - 0.06 = -0.02$$

which produces a chart similar to the c chart since n is constant. (See Fig. 5.5.)

Discussion: Processes and events representable by Poisson distributions are quite common.

CASE HISTORY 5.1

Defective Glass Stems in a Picture Tube for a Color TV Set[9]

A 100 percent inspection was made following the molding process on machine A. A record of lot number, and then the number defective, and the percent defective for each lot are shown in Fig. 5.6 at the bottom. A chart of daily P values is shown on the same vertical line above the computed P values below. The inspector (only one) was instructed to note operating conditions and any problems observed. It was learned that it was possible to explain why things were bad more often than why things were good.

Two types of visual defects predominated:

1. *Cracked throat,* the major problem, caused by stems sticking in a mold, probably from "cold" fires.
2. *Bubbles,* caused by "hot" fires.

All reject types have been combined for this present analysis and discussion. The control limits in Fig. 5.6 have been computed for $\bar{n}_h = 66{,}080/26 \cong 2{,}540$ and $\bar{P} = 100(3{,}703/66{,}080) = 5.60$ percent.

Discussion The day-to-day variation is large; the percent of rejects was substantially better than the month's average on at least 11 different days. When possible reasons for these better performances were explored, not much specific evidence could be produced:

- It is not surprising to have excessive rejects on the day after New Year's nor at the startup of the process.
- It was believed that the fires were out of adjustment on January 9, producing almost 20 percent rejects. Fires were adjusted on the tenth, and improved performance was

[9]Courtesy of Carl Mentch, General Electric Company.

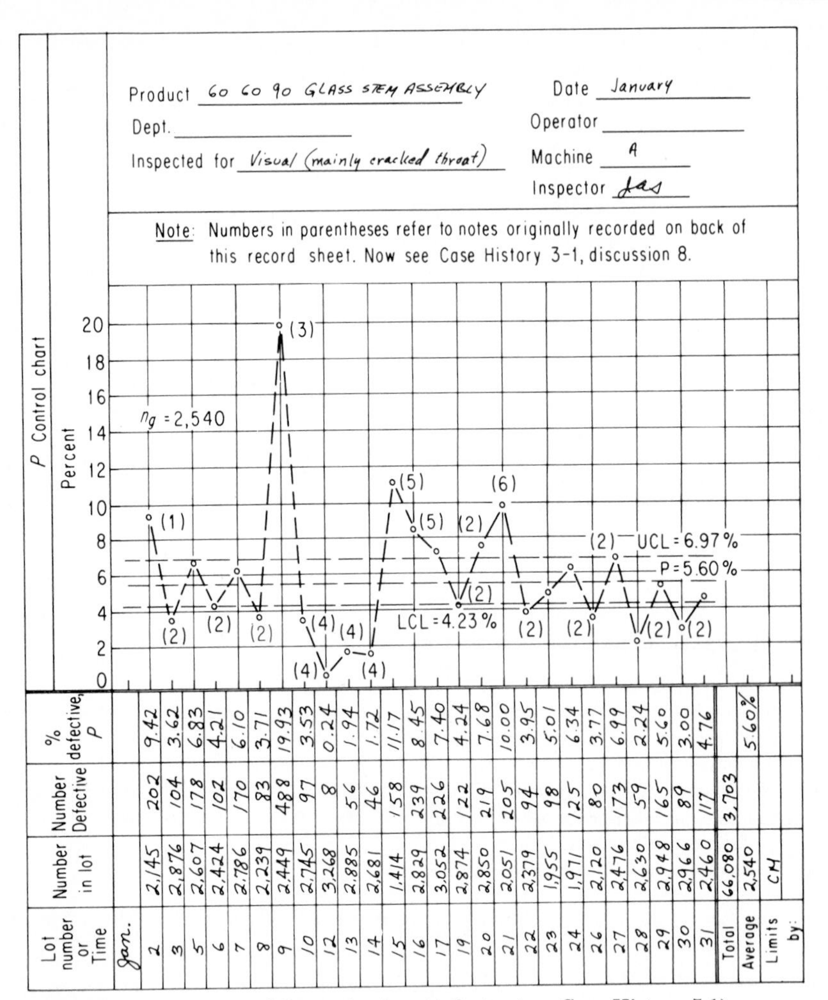

Product 60 60 90 GLASS STEM ASSEMBLY
Date January
Dept.
Operator
Inspected for Visual (mainly cracked throat)
Machine A
Inspector Jas

Note: Numbers in parentheses refer to notes originally recorded on back of this record sheet. Now see Case History 3-1, discussion 8.

Lot number or Time	Number in lot	Number Defective	% defective, P
Jan.			
2	2,145	202	9.42
3	2,876	104	3.62
5	2,607	178	6.83
6	2,424	102	4.21
7	2,786	170	6.10
8	2,239	83	3.71
9	2,449	488	19.93
10	2,745	97	3.53
12	3,268	8	0.24
13	2,885	56	1.94
14	2,681	46	1.72
15	1,414	158	11.17
16	2,829	239	8.45
17	3,052	226	7.40
19	2,874	122	4.24
20	2,850	219	7.68
21	2,051	205	10.00
22	2,379	94	3.95
23	1,955	98	5.01
24	1,971	125	6.34
26	2,120	80	3.77
27	2,476	173	6.99
28	2,630	59	2.24
29	2,948	165	5.60
30	2,966	89	3.00
31	2,460	117	4.76
Total	66,080	3,703	
Average	2,540		5.60%
Limits by:	CM		

Figure 5.6 Form to record inspection by attributes (see Case History 5.1).

evident on the next four days (January 10 through 14).

- Notes and their number as recorded originally by inspector on back of Fig. 5.6.
    1. Day after New Year's shutdown
    2. Reason that these points were out of control is unknown
    3. Four fires known to be out of adjustment
    4. Fires readjusted to bogie settings on 10th
    5. Stems sticking in 4 mold
    6. Fire position 6 too hot
- On January 15, rejects jumped again (to 11 percent). An investigation showed stems sticking in mold 4. This condition was improved somewhat over the next three days.
- On January 21, fire position 6 on the machine was found to be too hot. An adjustment

was made early on January 22, and the process showed an improvement in the overall average and with less erratic performance for the remainder of January.

- When this study was begun, the stem machine was considered to be in need of a major overhaul. This chart shows that either the machine easily lost its adjustment or possibly it was overadjusted. At any rate, the machine was shut down and completely rebuilt during February, and then started up again.
- Since the methods used to adjust the machine during January were not very effective, perhaps it would now be helpful to inspect a sample of stems hourly or bihourly and post the findings. This systematic feedback of information to production would be expected to do several things: prevent a machine operating unnoticed all day at a high reject rate; signal both better and worse performance allowing production and engineering to establish reasons for the difference in performance.
- Sometimes, large variations as in Fig. 5.6 are found to be a consequence of differences in inspectors. Since only one inspector was involved here, it is doubtful that this type of variation from day to day was a major contribution to the variation.
- Calculation of control limits in Fig. 5.6 for average $n_g \cong 2540$, attribute data. From Eq. (5.5)

$$\hat{\sigma}_p = \sqrt{\frac{(5.60)(94.40)}{2540}} = 0.456\%$$

and

$$3\hat{\sigma}_p = 1.37\%$$

$$\text{UCL} = \bar{P} + 3\hat{\sigma}_P = 6.97\%$$

$$\text{LCL} = \bar{P} - 3\hat{\sigma}_P = 4.23\%$$

CASE HISTORY 5.2

Incoming Inspection of a TV Component[10]

A relatively inexpensive glass component (a mount) was molded at one plant of a company. After a 100 percent inspection, it was transported to a second plant of the same company. After a 100 percent inspection at the receiving plant, it was sealed into a TV picture tube.

The two principal reasons for the second 100 percent inspection were:

1. The transportation was a possible source of defective mounts, but a defective mount was easily repaired.
2. Defective mounts were the principal cause of defective picture tubes. A defective picture tube was an expensive item and was nonrepairable.

The supplier plant was notified immediately by phone of any problems found at the receiver plant. Also, a monthly report of performance was given the supplier plant by type of defect and this was passed along to production. (All defects have been combined in Fig. 5.7.) In the course of a year, this operation dropped from over 5 percent to less than 1 percent defective. The system was considered to be successful.

[10]Courtesy of Carl Mentch, General Electric Company.

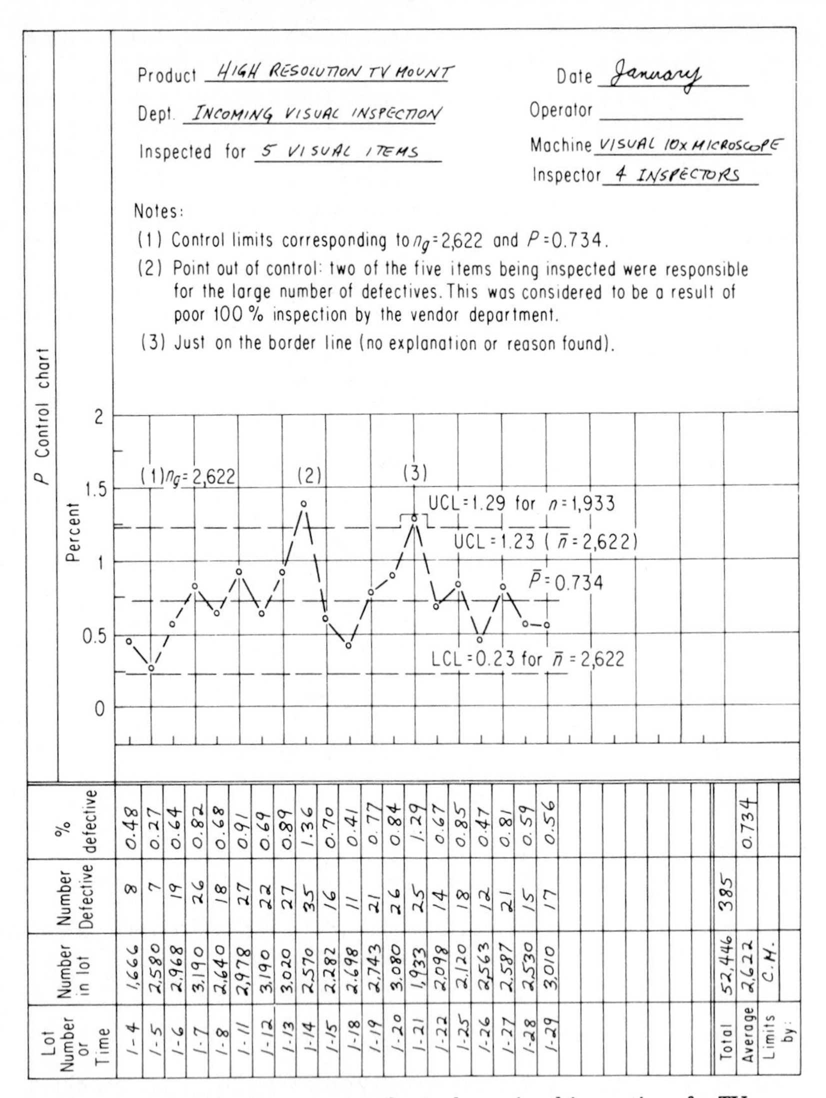

Lot Number or Time	Number in lot	Number Defective	% defective
1-4	1,666	8	0.48
1-5	2,580	7	0.27
1-6	2,968	19	0.64
1-7	3,190	26	0.82
1-8	2,640	18	0.68
1-11	2,978	27	0.91
1-12	3,190	22	0.69
1-13	3,020	27	0.89
1-14	2,570	35	1.36
1-15	2,282	16	0.70
1-18	2,698	11	0.41
1-19	2,743	21	0.77
1-20	3,080	26	0.84
1-21	1,933	25	1.29
1-22	2,098	14	0.67
1-25	2,120	18	0.85
1-26	2,563	12	0.47
1-27	2,587	21	0.81
1-28	2,530	15	0.59
1-29	3,010	17	0.56
Total	52,446	385	
Average	2,622		0.734
Limits by:	C.H.		

Figure 5.7 A control chart using attributes data; visual inspection of a TV component.

Discussion of numbered notes on the chart:

1. The sample size of 2580 on January 5 is smaller than n_g = 2622. No recomputation was made since it would only lower the LCL. No investigation was made.
2. The number of mounts inspected on January 14 was 2570. This is less than n = 2622 by so little that a recomputation is not warranted; it would not affect the decision that the day's lot was significantly worse than average. The record of defects showed two particular items to be the reason for the increase. An investigation established

that inadequate 100 percent inspection by the supplier was the cause, and not the transportation.

3. The number inspected on January 21 was 1933; when the UCL is recomputed, it is found to be 1.29 percent. On this date, $P = 1.293$ percent; although this is "less" than UCL = 1.294 percent, it indicates a high probability that there is a findable cause. (Actually, no substantial investigation was made.)

Computation of adjusted control limits on Fig. 5.7: $n = 2622$

$$\hat{\sigma}_p = \sqrt{\frac{(0.734)(99.266)}{2622}} = 0.1665\% \qquad \text{and} \qquad 3\hat{\sigma}_p = 0.500\%$$

$$\text{UCL} = 1.234\%$$

$$\text{LCL} = 0.234\%$$

5.5 Practice Exercises

1. Table 5.2 shows the probabilities of x occurrences in $n = 75$ trials, given $p = 0.03$.
   a. Find the probability that x will exceed $np + 3$ sigma and compare this result with the statement of Eq. 5.7. Explain the discrepancy.
   b. Repeat this comparison for $n = 50$ and $n = 25$.
   c. For $n = 50$, what fraction defective level would have at least a 90 percent chance of being detected on a single sample, using the control limit found in Exercise 1.*b*?
2. Adapt Eq. 5.8 to find a sample size to give with 50 percent confidence that $p = 20$ percent defective material is estimated with a possible error of ± 5 percent. Rework for $p = 10$ and 50 percent.
3. Consider Example 5.6. Use the Poisson probability curves to find the following probabilities.

$$P(0 \le x \le 8 \text{ given } \mu = 4) \qquad (\mu + 2\,\sigma)$$

$$P(2 \le x \le 6 \text{ given } \mu = 4) \qquad (\mu + \sigma)$$

Compare these probabilities with the normal approximation with the same mean and variance. Explain the discrepancy if any.

4. Show numerically that the Poisson is the limit of the binomial (Eq. 5.1) as np remains constant, n approaches infinity and p approaches 0. For $X = 1$, start with $n = 10, p = 0.1$, then evaluate $n = 100, p = 0.01$, and finally use the Poisson limit formula. Hint: Table A.5 gives sufficient accuracy in evaluating Eq. 5.1.
5. In Case History 5.1, the average sample size, n bar, is 2540. Recompute the limits assuming the percent defective shown in column 4 of Fig. 5.6 remains the same, but the average size is reduced from 2540 to 100. Which points are now out of control? Why is it changed? What does this say about sample size requirements for p charts?
6. In Case History 5.2, there are two points for 1-14 and 1-21, outside the UCL

of 1.23. However, the point of 1-21 has a special little control limit of 1.29 just for it. Explain why this is and show how it was calculated.

7. In analyzing experiments, an alternative to simply accepting or rejecting the null hypothesis is to compute the "p value" (also called "prob. value" or "significance level") of an experimental outcome. This is the probability, given that the null hypothesis is true, that the observed outcome or one more extreme would occur. The computation of a p value can often aid in deciding what action to take next. For example, see Fig. 5.7 on 1-21, percent defective was 1.293 percent based on $n = 1933$.

$$\hat{\sigma} = \sqrt{\frac{(0.734)(99.266)}{1933}} = 0.1941 \qquad \text{and } p = 0.734\%$$

$$Z = \frac{1.2934 - 0.734}{0.1941} = 2.88$$

$$\Pr(2.88) = 0.9980 \qquad p \text{ value} = 0.002$$

Using this same logic, compute p values for the following situations:

Binomial process, $n = 30, p = 0.05, x = 5$ (use Table A.5)

Poisson process, $np = 2, x = 1, x = 4$ (use Table A.6)

Normal process, $\mu = 50, \sigma = 10, x = 72$ (use Table A.1)

Student's t, $\mu = 50, \sigma = 40, n = 16, x = 72$ (interpolate in Table A.15)

8. Compare actual probability of 3σ limits with that obtained from the normal approximation to the binomial for small n and p.
9. Compare limits obtained by the Poisson and binomial distributions when the approximation is poor.

Part 2

Statistical Process Control

Chapter

6

On Sampling to Provide a Feedback of Information

6.1 Introduction[1]

Incoming inspection departments traditionally decide whether to accept an entire lot of a product submitted by a vendor. Factors such as the reputation of the vendor, the urgency of the need for the purchased material, and the availability of test equipment influence this vital decision—sometimes to the extent of eliminating all inspection of a purchase.

In contrast, some companies even perform 100 percent inspection for nondestructive characteristics and accept only those individual units which conform to specifications. However, 100 percent screening inspection of large lots does not ensure 100 percent accuracy. The inspection may fail to reject some nonconforming units and/or reject some conforming units. Fatigue, boredom, distraction, inadequate lighting, test-equipment variation and many other factors introduce substantial errors into a screening inspection.

Sampling offers a compromise in time and expense between the extremes of 100 percent inspection and no inspection. It can be carried out in several ways. First, a few items, often called a "grab sample," may be taken from the lot indiscriminately and examined visually or measured for a quality characteristic or group of characteristics. The entire lot may then be accepted or rejected on the findings from the sample. Another procedure is to take some fixed percentage of the lot as the sample. This was once a fairly standard procedure. However, this practice results in large differences in protection for dif-

[1]This chapter remains essentially unchanged from the first edition and reflects the philosophy on acceptance sampling of Ellis R. Ott. The co-author is in complete agreement with Ott's approach. See E. G. Schilling, "An Overview of Acceptance Control," *Quality Progress,* April 1984, pp. 22–24 and E. G. Schilling, "The Role of Acceptance Sampling in Modern Quality Control," *Communications in Statistics—Theory and Methods,* vol. 14, no. 11, 1985, pp. 2769–2783.

ferent sized lots. Also, the vendor can "play games" by the choice of lot size, submitting small lots when the percent defective is large, and thus increasing the probability of their acceptance.

This chapter presents and discusses the advantages and applications of scientific acceptance sampling plans.[2] These plans designate sample sizes for different lot sizes. If the number of defective units found in a sample exceeds the number specified in the plan for that lot and sample size, the entire lot is rejected. A rejected lot may be returned to the vendor for reworking and improvement before being returned for resampling, it may be inspected 100 percent by the vendor or vendee as agreed, or it may be scrapped. Otherwise the entire lot is accepted except for any defectives found in the sample.

Historically, the primary function of acceptance sampling plans was, naturally enough, acceptance-rejection of lots. Application of acceptance sampling plans was a police function; they were designed as protection against accepting lots of unsatisfactory quality. However, the vendor and customer nomenclature can also be applied to a shipping and receiving department within a single manufacturing plant or to any point within the plant where material or product is received for further processing. Variations in the use of scientific sampling plans are, for instance:

1. At incoming inspection in a production organization
2. As a check on a product moving from one department or process of a plant to another
3. As a basis for approving the startup of a machine
4. As a basis for adjusting an operating process or machine before approving its continued operation
5. As a check on the outgoing quality of product ready for shipment to a customer

6.2 Other Scientific Sampling Plans

The control chart for attributes discussed in Chap. 5 is one possible system of surveillance in any of these situations. Another method, similar to the Military Standard Acceptance Plans by Attributes, is a single or double sampling plan or set of plans for nondestructive testing of attributes, such as the Dodge–Romig[3] sampling inspection tables.

Consider the following single sampling plan applied to a process.

$$\text{Plan: } n = 45, c = 2$$

This notation indicates that:

---

[2]Various sampling plans in usage are referenced. The emphasis of this discussion is on some basic ideas related to acceptance sampling plans.

[3]Harold F. Dodge and Harry G. Romig, *Sampling Inspection Tables, Single and Double Sampling*, 2d ed., John Wiley and Sons, Inc., New York, 1959.

- A random sample of $n = 45$ units is obtained—perhaps taken during the last half-hour or some other chosen period, possibly from the last 1000 items produced, or even the last 45 units if interest on the present status of the process is primary. The quantity from which the sample is taken is called a lot.
- The 45 units are inspected for quality-characteristic A which may be a single quality characteristic or a group of them. If it is a group of them, the characteristics should usually be of about equal importance and be determined at the same inspection station.
- If not more than two ($c = 2$) defective units are found in the sample, the entire lot (or process) except for defectives is accepted for characteristic A; i.e., the entire lot is accepted if 0, 1, or 2 defectives are found. When more than 2 defectives are found in the sample, the *lot* is *not* acceptable.

In addition to this decision to accept or reject, a second important use of a plan is a *feedback system,* providing information to help production itself (or the vendor) improve the quality of subsequent lots as produced. In any case such plans often exercise a healthy influence on the control of a process. This will be discussed in Secs. 6.11, 6.12, and 6.13.

6.3 A Simple Probability

What may we expect to happen (on the average) if many successive samples from the process are examined under the above plan, $n = 45$, $c = 2$? What fraction of samplings will approve the process for continuance? To provide an answer, additional information is required. Let us assume first, for example, that the actual process is stable and producing 5 percent defective, i.e., the probability that any single item is defective is $p = 0.05$ or $P = 5$ percent.

Discussion

From Table A.5 of binomial probabilities, the probability[4] P_A of no more than $c = 2$ defectives in a sample of 45, with $p = 0.05$ is

$$P(x \leq 2) = 0.608 \text{ or } 60.8\%$$

6.4 Operating-Characteristic Curves of a Single Sampling Plan

There are two areas of special interest in practice.

1. What happens when lots with a very small percentage of defective units are submitted for acceptance? A *good* plan would usually accept such lots on the basis of the sample.
2. What happens when lots with a "large" percentage of defective units are

[4]The symbol P_A (read "P sub A") represents the "probability of acceptance." When the acceptance plan relates to an entire lot of a product, the P_A is the probability that any particular lot will be accepted by the plan when it is indeed the stated percent defective. Thus we shall talk about *accepting or rejecting a process* as well as accepting or rejecting a lot.

submitted? A *good* plan ought to, and usually will, reject such lots on the basis of the sample.

As in Sec. 6.3, values of P_A have been obtained for selected values of P and have been tabulated in Table 6.1 and graphed in Fig. 6.1. The resulting curve is called the *operating-characteristic curve* (OC curve) of the sampling plan, $n = 45$, $c = 2$.

TABLE 6.1 Probabilities P_A of Finding $x \leq 2$ in a Sample of $n = 45$ for Different Values of p

Values from Table A.5, Binomial Probability Tables

p, in percent	$P_A = P_r(x \leqq 2)$
0	1.00
1	.99
2	.94
3	.85
4	.73
5	.61
6	.49
8	.29
10	.16
15	.03

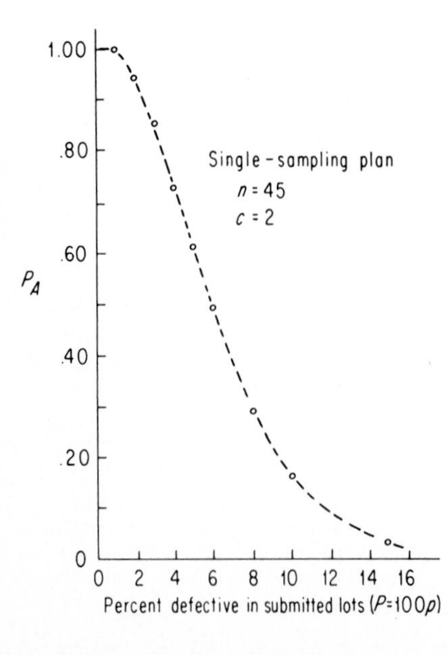

Figure 6.1 Operating-characteristic curve of a single sampling plan for attributes ($n = 45$, $c = 2$). The probability of a submitted lot being accepted, P_A, is shown on the vertical axis while different percent defective in submitted lots is shown on the horizontal axis (data from Table 6.1).

Discussion: regarding Fig. 6.1

- When lots with less than 1 percent defective are submitted to the plan, only occasionally will they be rejected; $P_A = 0.99$.
- When lots with more than 10 percent defective are submitted, the probability of acceptance is small; $P_A \cong 0.10$.
- Lots with P between 2 percent and 10 percent defective have a probability of acceptance which drops sharply.
- Whether the plan $n = 45$, $c = 2$, is a reasonable plan to use in a particular application has to be given consideration.

6.5 But Is It a Good Plan?

Whether the plan $n = 45$, $c = 2$ is a sensible, economical plan for a specific application involves the following points:

1. Sampling plans for process control and improvement should provide signals of economically important changes in production quality, a deterioration or improvement in the process, or other evidence of fluctuations. Are signals important in this application during ordinary production or in a process improvement project? Should a set of plans be devised to detect important differences between operators, shifts, machines, vendors? Improvements of industrial processes are often relatively inexpensive.
2. What are the costs of using this plan versus not using any?
   a. What is the cost to the company if a defective item is allowed to proceed to the next department or assembly? If it is simple and inexpensive to eliminate defectives in subsequent assembly, perhaps no sampling plan is necessary. If it is virtually impossible to prevent a defective from being included in the next assembly, and if this assembly is expensive and is ruined thereby, failure to detect and eliminate an inexpensive component cannot be tolerated.
   b. What is the cost of removing a defective unit by inspection? What is the cost of improving it by reworking it? What is the possibility of improving the process by reducing or eliminating defective components? When the total cost of permitting P percent defective items to proceed to the next assembly is equal to the cost of removing the defectives or improving the process, there is an economic standoff. When the costs of sampling and the possible consequent 100 percent screening are less than the alternative costs of forwarding P percent defectives, *some* sampling plan should be instituted.
3. Does this sampling plan minimize total amount of inspection? The Dodge–Romig tables provide plans for different lot sizes and AOQLs and LTPDs (see Secs. 6.6, 6.7, and 6.8). These plans were designed to minimize the total inspection resulting from the inspection of the sample(s) and whatever 100 percent screening inspection is required on lots which fail sampling.

Note: In process control we often use smaller sample sizes than required by a plan when our dependence on quality is only the acceptance-rejection aspect of the plan. Convenience sometimes stipulates the use of smaller sample sizes, referred to as *convenience samples. Any sampling plan* which detects a deterioration of quality and sends rejected lots or records of them back to the producing department can have a most salutary influence on production practices.

Many companies use one type of acceptance procedure on purchases and another on work in process, or outgoing product. In contracting the purchase of materials, it is common practice to specify the sampling procedure for determining acceptability of lots. Agreement to the acceptance procedures may be as critical to the contract as price or date of delivery.

6.6 Average Outgoing Quality (AOQ) and Its Maximum Limit (AOQL)

These two concepts will be discussed here with reference to a single-sampling rectification inspection plan for attributes (nondestructive test) which includes the following steps:

1. A lot rejected by the plan is given a 100 percent screening inspection. All defectives are removed and replaced by nondefectives; i.e., the lot is rectified. It is then resubmitted for sampling before acceptance.
2. Defectives are always removed when found and replaced by nondefectives even in the samples.

As a consequence of (1) and (2), the *average outgoing quality* (AOQ) of lots passing through the sampling station will be improved. Since very good lots will usually be accepted by the sampling plan, their AOQ will be improved only slightly (see Fig. 6.2). Lots with a larger percent of defectives will be rejected more often and their defectives removed in screening. Their AOQ will be improved substantially (Fig. 6.2).

The *worst possible* average situation is represented by the height of the *peak*

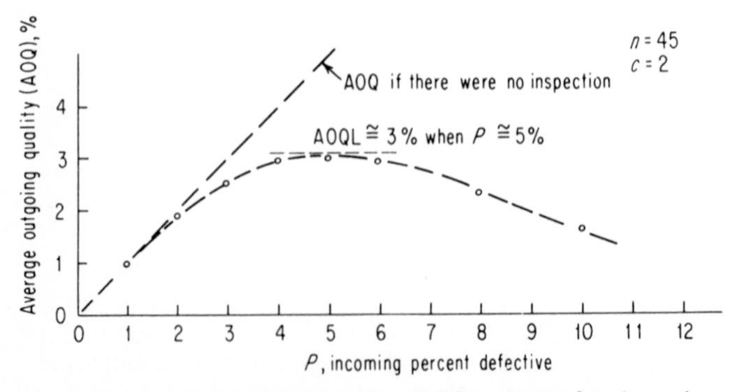

Figure 6.2 Average outgoing quality (AOQ) compared to incoming percent defective P for the plan $n = 45$, $c = 2$. (AOQ is after any lots that fail the inspection plan have been 100 percent inspected; see Table 6.2, columns 1 and 3.)

of the AOQ curve. This maximum value is called the *average outgoing quality limit* (AOQL). In Fig. 6.2, AOQL ≅ 3 percent. This very useful AOQL concept forms the basis for a system of sampling plans. Several plans are available in Dodge–Romig for each AOQL from 0.1 to 10 percent.

Any sampling plan will provide information on the quality of a lot as it is submitted—certainly more information[5] than if no inspection is done. But some plans require much too large a sample for the need; some specify ridiculously small samples. The AOQL concept is helpful in assessing the adequacy of the plan under consideration and/or indicating how it could be improved. When a 3 percent AOQL system is used, the worst possible long-term average accepted will be 3 percent. But this could occur only if the producer always fed 5 percent defective to the sampling station. Except in this unlikely event, the average outgoing quality[6] will be less than 3 percent.

6.7 Computing the Average Outgoing Quality (AOQ) of Lots from a Process Producing *P* Percent Defective

When a sequence of lots is submitted to an acceptance plan $n = 45$, $c = 2$, what is the AOQ for different values of p?

Consider first the case for $p = 0.05$; the number of defectives in an average lot of size $N = 2000$, for example, is $N_p = 100$. The average number of defectives *removed* from the lot as a consequence of the sampling plan come from two sources:

1. From the *sample.* Average number removed is

$$np = (0.05)(45) = 2.25$$

2. From *lots* which are rejected by the plan. We designate the probability of a lot being rejected by

$$P_R = 1 - P_A = 1 - 0.61 = 0.39 \tag{6.2}$$

Then the additional average number of defectives removed from these rejected lots is

$$p(N - n)P_R = (0.05)(1955)(0.39) = 38.12 \tag{6.3}$$

[5]Other valuable information on its quality could be provided by a production control chart(s). These charts are usually available to us only when the vendor is a department of our own organization; but this is not necessarily so. A basis for control-chart information on incoming material can often be established with outside vendors.

[6]There is a hint of a spurious suggestion from Fig. 6.2; namely, that one way to get excellent product quality is to find a supplier who provides a large percent of defectives! This is not really a paradox. The only way to ensure good quality product is to provide good manufacturing practices; dependence upon 100 percent inspection is sometimes a short-term "necessary evil."

It is told that three friends at a company's country club for dinner ordered clams on the half shell. After an exceptionally long wait, three plates of clams were brought with apologies for the delay: "Am very sorry, but we had to open and throw out an awful lot of clams to find these good ones!" Would you eat the clams served?

So the total average number of defectives *removed* from a lot is the sum of these two:

$$np + p(N - n)P_R = 2.25 + 38.12 = 40.37 \tag{6.4}$$

The average number of defectives remaining in a lot divided by N is the AOQ: 100 − 40.37 = 59.63 and

$$\text{AOQ} = 59.63/2000 = 0.0298 = 2.98\%$$

Consider now the general case. The average number of defectives in a lot is Np. From Eq. (6.4), the number removed is

$$np + p(N - n)P_R$$

Then

$$\text{AOQ} = \frac{Np - [np + p(N - n)P_R]}{N} = pP_A\left(1 - \frac{n}{N}\right)$$

Since the value of n/N is usually very small,

$$\text{AOQ} \cong pP_A \tag{6.5}$$

Equation (6.5) was used[7] to compute values of AOQ in Table 6.2.

Also, the operating-characteristic curves of two commonly used systems of sampling plans have been computed and appear along with the sampling tables. For those who have an interest in mathematical consideration of probabilities, the following is presented.

The probability P_A of a lot of size N being accepted on the basis of a random sample of 45 drawn from 2000 units of a lot is the sum of the probabilities of there being 0, 1, or 2 defectives in the sample:

---

[7]Anyone intending to become expert in devising sampling plans should become familiar with the mechanics of computing probabilities such as the following, based on sampling from lots of size N and with changing probabilities as each item is drawn from the lot.

When an item is drawn from a lot with Np defectives, the probability that the *next* item will be defective is *not quite* equal to p:

- If the first item drawn is defective, then $(Np - 1)$ defectives remain, and the probability that the second item will be defective is

$$\frac{Np - 1}{N - 1} = \frac{(N - 1)p - (1 - p)}{N - 1} = p - \frac{1 - p}{N - 1}$$

This is evidently only slightly *less* than p.

- However, if the first item withdrawn is *not* defective, then the probability that the second item will be defective is

$$\frac{Np}{N - 1} = \frac{(N - 1)p + p}{N - 1} = p + \frac{p}{N - 1}$$

This is only slightly *more* than p.

For ordinary practical purposes, it is adequate to regard the characteristics of sampling plans as sampling from a production process with fixed fraction defective p. An expert in the field will want to learn how computers can be used to calculate probabilities.

TABLE 6.2 Average Outgoing Quality (AOQ) of Lots Proceeding Past an Acceptance Sampling Station Using the Plan $n = 45$, $c = 2$

Lots which fail to pass sampling are submitted to 100 percent screening with replacement of defectives. See Eq. (6.5) for method of approximating AOQ

P	P_A	$\text{AOQ} = P \cdot P_A$
0	1.00	0
1	.99	.99
2	.94	1.88
3	.85	2.55
4	.73	2.92
5	.61	3.05
6	.49	2.94
8	.29	2.32
10	.16	1.60
15	.03	0.45

$$P_A = \Pr(x = 0) + \Pr(x = 1) + \Pr(x = 2) \tag{6.6}$$

Each of these probabilities can be expressed in terms of the number of possible combinations. In a lot of $N = 2000$ units having 5 percent defective, the number of defective units is 100, and the number of nondefective units is 1900. The Eq. (6.6) can be written explicitly

$$\begin{aligned} P_A &= [\mathrm{C}(100, 0) \cdot \mathrm{C}(1900, 45) + \mathrm{C}(100, 1) \cdot \mathrm{C}(1900, 44) \\ &\quad + \mathrm{C}(100, 2) \cdot \mathrm{C}(1900, 43)] \div \mathrm{C}(2000, 45) \\ &= \left(\frac{1900!}{45!1855!} + (100)\frac{1900!}{44!1856!} + \frac{(50)(99)1900!}{43!1857!}\right) \div \frac{2000!}{45!1955!} \end{aligned} \tag{6.7}$$

The representation of this as decimal fraction requires a programmed computer, or a table of logarithms of factorials.

6.8 Other Important Concepts Associated with Sampling Plans

Minimum average total inspection. Includes inspection of the samples and the 100 percent screening of those lots which are rejected by the plan. Both Dodge–Romig systems, AOQL and LTPD, include this principle.

Acceptable-quality level (AQL). Represents the largest average percent defective which is still considered acceptable. Sometimes an attempt is made to formalize the concept as "that quality in percent defective which the consumer is willing to accept about 95 percent of the time such lots are submitted." This definition has been the basis for some heated arguments.

Point of indifference[8]. Represents a percent defective in lots which will be accepted *half the time* when submitted (P_A = 50 percent).

Lot tolerance percent defective plans. To many consumers, it seems that the quality of *each* lot is so critical that the average outgoing quality concept does not offer adequate protection. The customer often feels a need for a lot-by-lot system of protection. Such systems have been devised and are called *lot tolerance percent defective plans* (LTPD). These are provided in Dodge–Romig for values of P from 0.5 to 10 percent.

Good quality is a consequence of good manufacturing. Good quality is not the result of inspection; inspection is often considered a "necessary complement." Most important is the role that acceptance sampling plans can play in providing useful information to help manufacturing improve its processes (see Sec. 6.11).

6.9 Risks

The vendor/producer wants reasonable assurance (a small producer's risk) of only a small risk of rejection when lots are submitted having a small percent defective. In Fig. 6.1, the *producer's risk* is about 5 percent for lots with 2 percent defective and less than 5 percent for better lots. This may be reasonable and acceptable to the producer on some product items and not on others; negotiation is normally required.

The vendee/consumer wants reasonable assurance (a small consumer's risk) that lots with a large percent defective will usually be rejected. In Fig. 6.1, the consumer's risk is seen to be about 16 percent for lots submitted with 10 percent defective; less than 16 percent for larger percents defective.

Compromises between the consumer and producer are necessary. Two systems in wide use provide a range of possibilities between these two risks; the Dodge–Romig plans and the MIL-STD-105 series.

6.10 Military Standard 105E[9]

Tightened and reduced inspection MIL-STD-105E is used extensively in acceptance sampling for government contracts and for many nongovernmental applications as well. The plans are based on an *acceptable quality level* (AQL) concept. The *producer's risk* is emphasized when AQL plans are chosen. Plans

[8]Hugo C. Hamaker, "Some Basic Principles of Sampling Inspection by Attributes," *Appl. Stat.*, vol. 7, 1958, pp. 149–159.

[9]United States Department of Defense, *Military Standard Sampling Procedures and Tables for Inspection by Attributes,* (MIL-STD-105E), U.S. Government Printing Office, Washington, D.C., 1989.

are intended to protect the producer when producing at or better than the AQL level, unless there is a previous history or other basis for questioning the quality of the product. When product is submitted from a process at the AQL level, it will be accepted "almost always." Since OC curves drop only gradually for percent defectives slightly larger than the AQL value, such product will have a fairly high probability of acceptance. The producer's interest is protected under criteria designated as *normal inspection.*

How then is the consumer protected? Whenever there is reason to doubt the quality level of the producer, MIL-STD-105E plans provide stricter criteria for acceptance. These plans are called *tightened inspection.* Criteria are provided in the plans to *govern switching from normal to tightened* inspection. Proper use of MIL-STD-105E demands that the rules for shifting from normal to tightened inspection be observed.

When the producer has an excellent record of quality on a particular item, the MIL-STD-105E plans permit a reduction in sample sizes by switching to *reduced inspection.* This shift to *reduced inspection* is not designed to maintain the AQL protection, but to allow a saving in inspection effort by the consumer.

For use with individual lots, specific plans can be selected by referring to OC curves printed in the standard. For more details on MIL-STD-105E, see *Quality and Reliability Handbook* (H53A), available from the Superintendent of Documents, Washington, D.C.

6.11 Feedback of Information

Problems, problems, everywhere

Problems always abound when manufacturing any product; they may be found both during processing and in the finished product. Problems may result from product design, vendor quality, testing inadequacies, and on and on. It is tempting to blame the problem on factors outside our own immediate sphere of responsibility. As a matter of fact, there are occasions when a vendor is known to be supplying low-quality items; there are also occasions when we have examined our process very carefully without finding how to improve it.There are two standard procedures that, though often good in themselves, can serve to postpone careful analysis of the production process:

1. On-line inspection stations (100 percent screening). These can become a way of life.
2. On-line acceptance sampling plans which prevent excessively defective lots from proceeding on down the production line, but have no feedback procedure included.

These procedures become bad when they allow or encourage carelessness in production. It gets easy for production to shrug off responsibility for quality and criticize inspection for letting bad quality proceed.

More than a police function

No screening inspection should simply separate the good from the bad, the conforming from the nonconforming, the sheep from the goats. No on-line acceptance sampling system should serve merely a police function by just keeping unsatisfactory lots from continuing on down the production line. Incorporated in any sampling system should be procedures for the recording of important detailed information on the number and types of production defects. It is a great loss when these data are not sent back to help production improve itself. A form for use in reporting such information is vital although preparing an effective one is not always a simple task.

Any systematic reporting of defects which can trigger corrective action is a step forward. Contentions that the start of a system should be postponed—"we aren't ready yet"—should be disregarded. Get started. Any new information will be useful in itself and will suggest adjustments and improvements.

Defect classification

Any inspection station has some concepts of "good" and "bad." This may be enough to get started. But corrective action on the process cannot begin until it is known what needs correction. At a station for the visual inspection of enamel bowls, items for a sampling sheet (Table 6.6) were discussed with the regular inspector, a foreman, and the chief inspector. Table 6.10 was similarly devised for weaving defects in cloth. Some general principles can be inferred from them. Figure 6.3 and the discussion below offer some ideas for record sheets associated with single-sampling acceptance plans.

- Give some consideration to the *seriousness* of defects. Table 6.7 uses two categories, *serious* and *very serious*. Categories can be defined more carefully after some experience with the plan. (More sophisticated plans may use three or four categories.)
- Characterize defects into *groups* with some regard for their *manufacturing source*. This requires advice from those familiar with the production process. In Table 6.6, for example, *black spots* are listed both as A.4 and A.5, and B.8 and B.9. They were the result of different production causes. Corrective action is better indicated when they are reported separately. Also note Metal Exposed, A.6, A.7, and A.8, and B.5, B.6, and B.7.
- Do not list too many different defect types; limit the list to those which occur most often; then list "others." When some other defect appears important, it can be added to the list.
- Eventually, information relating to the natural sources of defects may be appropriate; individual machines, operators, shifts. Even heads on a machine or cavities in a mold may perform differently.

Sampling versus 100 percent inspection

Information from samples is usually more helpful than from 100 percent inspection because:

Department: *Mounting*
Tube Type: *6AK5*
Item: *Grid*
Test: *Visual*

Sampling Plan: $n = 45$, $c = 2$

Date	Time	n	Inspector	Spacy	Taper	Damage	Slant	Other		Total	Action	Comments
2/5/73	10^{00}	45	mB	2	1	0	1			4	R	
	12^{00}	45	mB	3	0	1	2			6	R	
	2^{00}	45	aR	1	0	0	1			2	A	
	4^{00}	45	aR	2	0	0	1			3	R	
Daily Total		180		8	1	1	5			15		

Circulation:
Jma WCF
FFR RMA

Figure 6.3 Lot by lot record acceptance sampling (single sampling).

1. In 100 percent screening, inspection is often terminated as soon as *any* defect is found in the unit. This can result in undercounting important defects. In a sample, however, inspection of a unit can usually be continued until *all* quality characteristics have been checked and counted.

2. Much 100 percent inspection is routine and uninspiring by its very nature. Records from such inspection are often full of inaccuracies and offer little useful information for improvement. Small samples give some release from the boredom and allow more careful attention to listed defect items. They also permit recognition and attention to peculiarities which occur.

3. With the use of small, convenient, fixed-size samples information can be fed back as illustrated in the two case histories below. The resulting improvements in those situations had been thought impossible. Also see Case History 5.2.

Teeth and incentives

Firmness and tact are important when persuading people that they are at fault and that they can correct it. There are various possibilities.

1. Some major companies physically return defectives back to the erring department. Others have them repaired by a repair department, but charge the repair back to the erring department. A department can often improve itself if suitable information is fed back.

2. The physical holdup of product proceeding down the line has a most salutary effect. When no complaints or records on bad quality are made, but instead bad product continues on down the line, it is almost certain to induce carelessness. It says, loud and clear, "Who cares?"

3. Even a control chart on percent defective (a p chart) posted in the manufacturing department *can* provide encouragement. Used carefully, this can have as much interest and value as a golf score to an individual or a department.

6.12 Where Should a Feedback Begin?

There is no *one* answer, but there are some guidelines.

1. An acceptance plan may already be operating but serving only as a police function. Attach a feedback aspect, organized so as to suggest important manufacturing problems.

2. *Sore thumb.* Sometimes a large amount of scrap or a failure to assemble will indicate an obvious problem. Often no objective information is available to indicate its severity, the apparent sources, or whether it is regular or intermittent. Start a small-scale sampling with a feedback. This may be a formal acceptance sampling plan or a *convenience* sample large enough to provide some useful information. (A sample of $n = 5$ will not usually be large enough[10] when using attributes data, but frequently, a sample of 25 or 50 taken at reasonable time intervals will be very useful.)

3. *Begin at the beginning?* It is often proposed that any project to improve should start at the beginning of the process, making any necessary adjustments at each successive step. Then at the end of the process, it is argued, there will be no problems. This approach appeals especially to those in charge of the manufacturing processes. Sadly, it is often not good practice.

First, there is rarely an opportunity to complete such a well-intentioned project. A "bigger fire" develops elsewhere, and this one is postponed, often indefinitely.

Second, most of the steps in a process are usually right. In the process of following operations step by step, and in checking each successive operation, much time is lost unnecessarily. Usually it proves better to start at the *back* end; find the major problems occurring in the final product. Some will have arisen in one department, some in another. The method of the following section, 6.13, was designed to pinpoint areas in manufacturing which warrant attention, whether from raw materials or components, process adjustment, engineering design, inspection, or other reason. Pareto analysis allows prioritization of these aspects of any process study.[11]

---

[10]For an exception, see Case History 11.1 (spot welding).

[11]J. M. Juran, "Pareto, Lorenz, Cournot, Bernoulli, Juran and others," *Ind. Qual. Control,* vol. 17, no. 4, October 1960, p. 25.

6.13 Outgoing Product Quality Rating (OPQR)

Introduction

This is a program which gets to the source of difficulties in a hurry. Further, it enlists the cooperation of various departments. The method starts by *rating* small samples of the outgoing product. This outgoing product quality rating program[12] was suggested by a plan[13] developed in connection with complicated electronic and electrical equipment. A well-known pharmaceutical house utilized this system for a major drive on package quality. It is equally applicable in many other industries.

In the OPQR program, each product is sampled daily just before being placed in shipping cases. The sample items are then inspected visually and a demerit rating assigned according to the type and magnitude of defects observed. The rating is done by a department which is independent of manufacturing. The procedure is primarily for *information* and feedback purposes and is *not* a police function:

1. *It is an information service* to top management, providing an objective evaluation of outgoing product. In many companies, careful information is provided to assure management that the organization is geared to produce high quality at a reasonable cost. But how many such managements can claim accurate information on the actual product quality as it goes to the customer? Many *assume* good quality because the product has "had 100 percent inspection." But how many really know, for example, how many inoperative radios leave their plant; how many refrigerators leave with scratched doors; how many phonograph records have audio imperfections; how many of their solder joints or welds are defective? Reliable answers would usually provide important and surprising information.

2. Another important function of an OPQR rating program is to *pinpoint areas in manufacturing* which warrant attention. These may include raw materials (or components), process adjustment, engineering design, inspection, and so forth. It provides a much faster and surer method of getting to trouble areas than does the alternative method of following production step by step from the beginning on through successive stages.

Outline of the program

1. Members of this study committee were named from top management, sales, manufacturing, and quality control.

[12]William C. Frey, "A Plan for Outgoing Quality," *Mod. Packag.*, October 1962. Besides special details in our Table 6.3, Fig. 6.4, and the classification of defects, other ideas and phrases from the article are included here. Permission for these inclusions from the author and publisher are gratefully acknowledged.

[13]Harold F. Dodge and Mary N. Torrey, "A Check Inspection and Demerit Rating Plan," *Ind. Qual. Control*, vol. 13, no. 1, July 1956.

2. Two standard products were selected for study.
3. For this initial study 50 sample bottles of each product were collected at stated intervals each day for a period of one week from each manufacturing line and product. The usual sample should be smaller, perhaps 20, and never more than 40 per line per day.
4. These samples were examined carefully by an independent checker. Each bottle with any deviation from perfection was tagged, the defects recorded as in the attached OPQR weekly summary from Table 6.3 and saved for the committee.
5. The study committee examined the visual defects found and added others which it was felt might occur.
6. Defects were classified into four categories.[14] A demerit rating was attached to each of the four.
7. Regular meetings were held to discuss findings and plan future progress.

Defect classification

What is a defect? What is a nonconforming unit? Invariably, these are difficult questions to answer especially with visual defects. An upside down label is clearly a defect; dirty bottles and wrinkled labels are not desirable, but there are all degrees of dirty bottles and wrinkles. A discussion of the types of defects which might possibly be found in one's product is rather academic until actual facts are obtained. Thus, one of the first steps agreed upon by the planning committee, appointed to activate the program, was to obtain such information.

The four defect categories accepted were the following:

1. *Minor.* These are items which are not perfect; they are objectionable, but so slightly objectionable that management would agree that these items should not be removed from the production line, if observed. However, it is desired that corrective efforts be made to reduce the number of subsequent defects of this category. (Slightly crooked labels are typical.)

2. *Moderately serious.* These are nonfunctional package defects of a somewhat more serious nature. They would probably be noticed by the customer, although they would probably not incur a loss of good will. They are sufficiently objectionable, however, to require line inspectors to cull them out of production since they should not reach the customer (e.g., an upside-down, dirty, or wrinkled label).

3. *Serious.* These are appearance defects or functional defects which would certainly result in loss of good will and possibly result in loss of a customer. (A broken plastic cap or a short fill are examples.)

4. *Very serious.* This category is reserved for those rare defects which could conceivably result in bodily injury.

[14]Harold F. Dodge and Mary N. Torrey, "A Check Inspection and Demerit Weighting Plan," *Ind. Qual. Control*, vol. 13, no. 1, July 1956, pp. 5–12.

In this program the demerit rating values assigned to each category in accordance with its estimated importance were; minor = 2 demerits; moderately serious = 10; serious = 25; and very serious = 50. Control charts for the overall demerit rating were established (per hundred units of each product).

In an initial discussion, it often proves difficult to get agreement on levels of defects which are tolerable. Every company official, whether representing manufacturing, sales, or management, is reluctant to accept quality standards which are short of perfection or to agree that even a single serious defective can be tolerated. To refute this concept, consider this question: "What percent of airplane accidents are tolerable in commercial flights in the U.S.?" The immediate answer, of course is "none"; however, we recognize that to guarantee this standard, all flights would have to be grounded. This approach helped make it possible to establish initial working standards of quality for the pharmaceutical program.

Within each defect category were grouped defects which were related to each machine within the plant. As shown in an attached OPQR weekly summary, Table 6.3, many defects came from the labeler. This relationship to a machine is helpful in providing an effective information feedback procedure. In this plant, outgoing product quality rating reports are published weekly. The report consists of a rating control chart (Fig. 6.4) with a new point plotted each week plus a detailed listing of the defectives which led to the overall rating value. Limits for the control chart are obtained as shown in Table 6.4. Occasionally, when the weekly reports stimulate interest in certain production areas, daily reports are furnished on critical items in that area so that the interested department has an independent detailed analysis of difficulties being found.

TABLE 6.3 OPQR: Outgoing Product Quality Rating—Weekly Summary

		50 Demerits D Very Serious			25 Demerits C Serious			(10 Demerits) B Moderately serious: Labeller										A Minor (2 Demerits): Labeller							Materials									
								b	b	b	b	a	c	d				b	b	b	b	b	b	b	a	a	a	a	c	d	d			
Week	n	Chipped Thread		Total Demerits	Brcken cap		Total Demerits	Loose Label	Very Crooked Label	Serious Wrinkle	Very Dirty Label	Free Fluid	Cracked Cap	Scorched Cello.		Miscellaneous	Total Demerits	Illegal Code	Crooked Label	Wrinkled Label	Sl. Loose Label	Scuffed Label	Ripped Label	Dirty Label	Print Imperfection	Cap Imperfection	Holiday in Cap	Slight Fr. Fl.	Dirty Cap	Torn Cello.	Damaged carton	Miscellaneous	Total Demerits	Total Demerits per 100 Units
9/15	120											1					10	10		3	3		1		2			2					42	52×.83 = 43
9/22	200									1		1	1				30	2		17	3		3		1		4	3		3	5		82	112×.5 = 56
9/29	200							2					1				30		2	17	2		3		3			3			1	1	64	94×.5 = 47
10/6	200							1				2	1				40	3	1	4	4	2			4			4			2	2	54	94×.5 = 47
10/13	180								1	2		1	1	1		2	80	4		8	1	1	1		8			4		2		2	70	150×.56 = 84
10/20	160							1		9						1	110			17	2	1	1		3	1	1	3		2	3	3	74	184×.625 = 115
10/27	195									2							20	1		9	4	9	2		2	2	3	2	2		5	4	96	116×.51 = 59

a: Materials b: Labeller c: Filler d: Cartoner

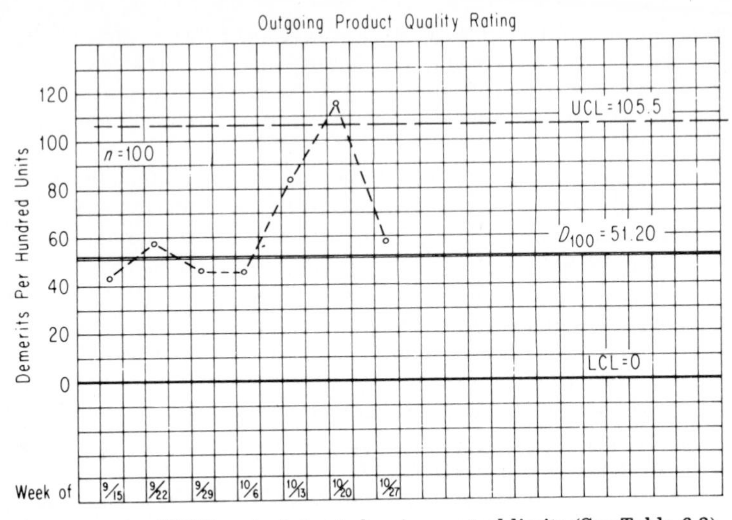

Figure 6.4 An OPQR control chart showing control limits (See Table 6.3).

Additional applications

One important function of this program is in comparing the production quality with previous production. Is the quality from line 1 similar to line 2? If there is a marked difference, where is the major departure? The OPQR program provides answers to these and other similar questions.

Occasionally a defect is found on the shelf of a drugstore. It is important to know whether this was something which might represent a sporadic occurrence or an epidemic at the time the product left the plant, or whether it is a characteristic which developed subsequently. The production superintendent who was sometimes skeptical of the program initially, found it very useful to quote the OPQR records which provide objective testimony about defects which developed after the product left the plant. It is also important to know systematically and regularly about how many off-standard items are being shipped; when, for example, a field complaint is received, you can tell the vice president that not all the toothpaste tubes went out empty. (He actually bought some empty ones in a drug store.)

Summary: OPQR

We have found the outgoing product quality rating procedure adaptable to cosmetic and pharmaceutical products and an appropriate means to initiate and extend a quality control program. It very quickly pinpoints areas where quality improvement is warranted and is both an insurance policy and an information service to the production management. This approach to an extension of a total quality program merits your consideration.

TABLE 6.4 Computation of Standard Quality Demerit Level and Control Limits ($\overline{QD}_n$ and σ_n) per n Units

Assigned Class Demerit Weights in this project:

Class D defect: 50 = w_D
Class C defect: 25 = w_C
Class B defect: 10 = w_B
Class A defect: 2 = w_A

Now

$$\overline{QD} = \frac{(w_A)d_A + (w_B)d_B + (w_C)d_C + (w_D)d_D}{n}$$

where d_A, d_B, d_C, and d_D represent the number of each class of defect found in the n items inspected. In this example, $n = 2400$, $d_A = 302$, $d_B = 60$, $d_C = 1$, and $d_D = 0$. Then

$$\overline{QD} = 0.512 \quad \text{and} \quad \overline{QD}_{100} = 51.2$$

This value is the central line in Fig. 6.4. Now when we assume a Poisson distribution of demerits we obtain:

$$\hat{\sigma}_{QD} = \sqrt{\frac{\Sigma w_i^2 d_i}{n}}$$

$$\hat{\sigma}_{100} = \sqrt{\frac{100[(50)^2(0) + (25)^2(1) + (10)^2(60) + (2)^2(302)]}{2400}}$$

$$= \sqrt{\frac{783{,}300}{2400}} = 18.1 \textit{ demerits per } 100 \textit{ units}$$

In Fig. 6.4:

$$\text{UCL} = \overline{QD}_{100} + 3\hat{\sigma}_{100}$$
$$= 51.2 + 54.3 = 105.5$$
$$\text{LCL} = 51.2 - 54.3 \cong \text{zero}$$

CASE HISTORY 6.1

Metal Stamping and Enameling

Many different enameled items were made in a plant in India—such as basins, trays, cups. The manufacture of each product began with punching blanks from large sheets of steel and cold forming them to shape. The enamel was then applied by dipping the item into a vat of enamel slurry and by firing in an oven. (This enameling process consisted of two or three coating applications.) See Table 6.5 and Fig. 6.6 for steps in producing an enameled basin.

As we made our initial tour of the plant, we saw two main visual inspection (sorting) stations: (1) after metal forming (before enameling) and (2) after final enameling (before shipment to the customer). Either station would be a logical place to collect data.

It is never enough just to collect data. The sensitivities of various production and inspection groups must be recognized and their participation and support enlisted. Many projects produce important, meaningful data, which are useless until their interpretations are implemented. The sequence of steps shown below was important to success in enlisting support of the production study.

TABLE 6.5 Steps in Producing an Enameled Basin

STEP 1. *Metal Fabrication:*

- *a.* Metal punching (one machine with one punching head); Produced circular blanks from a large sheet of steel.
- *b.* Stampings (three-stage forming): one machine with its dies at each stage to produce rough-edged form.
- *c.* Trimming; a hand operation using large metal shears.
- *d.* Cold spinning (a hand operation on a lathe to roll the edge into a band).
- *e.* Sorting inspection (100%) (no records).

STEP 2. *Acid Bath*

STEP 3. *Enameling* (Blue and white coats):

- *a.* Mixing enamel.
- *b.* Apply blue enamel coating (by dipping).
- *c.* Fire coating (in ovens).
- *d.* Apply white enamel coating (by dipping).
- *c.* Paint border (hand operation).
- *f.* Final firing.

STEP 4. *Final Inspection:*

Product classified but no record kept of defects found.

To begin

We arrived at the factory about 8 A.M. and our meeting with the plant manager ended about 12:30 P.M. When asked "When can you begin?" the answer was "Now."

Since we had agreed to begin with visual defects on bowls (Fig. 6.5), the chief inspector and our young quality control people sketched out an inspection sheet which allowed a start of *sampling* at final inspection. Regular final inspection continued to classify items as:

- First quality—approved for export.
- Second quality—with minor defects; these sold at a slightly reduced price.
- Third quality—with some serious defects; these sold at a substantial reduction in price.

The daily inspection sheet (Table 6.6) was used.

Random samples of 30 units from each day's production were inspected for the 16- and 40-cm basins and the numbers of defects of each type recorded. A single basin might have more than one type of defect; an inspection record was kept of all defects found on it. The same inspector was used throughout the workshop to reduce differences in standards of inspection.

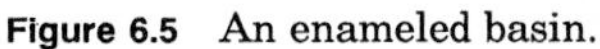

Figure 6.5 An enameled basin.

Progress

Daily meetings of the workshop team were held to look at the data on the daily sampling inspection sheets. They led to discussions on ways to reduce high-defect items. Beginning with the first day, information was given to production via the production supervisor and was a major factor in several early corrections. Then after four days, a summary of the causes of defects was prepared and discussed at a meeting of the workshop team (Table 6.7).

Sequence of steps in the workshop investigation

1. A team was formed; it included the chief inspector, the production supervisor, two young experienced quality control people, and one of the authors (Ellis Ott).
2. A 2-hr tour of the plant was made with the team to identify potential stations to gather information.
3. A meeting with members of the team and the works manager was held following the tour. The discussion included:
   a. Types of problems being experienced in the plant.
   b. Important cooperative aspects of the project.
   c. Various projects which might be undertaken. Management suggested that we emphasize the reduction of visual defects, especially in their large volume 16-cm enameled bowl (see Fig. 6.5). The suggestion was accepted.
4. Two locations were approved to begin the workshop:
   a. At final visual inspection: step 4 in Table 6.5.
   b. At the point of 100 percent sorting after metal fabrication and forming (step 1*e*) and just before an acid bath which preceded enameling. No records were being kept of the number of actual defects found.
5. A final oral summary with the works manager, which included an outline presentation of findings and ideas for improvements suggested by the data.

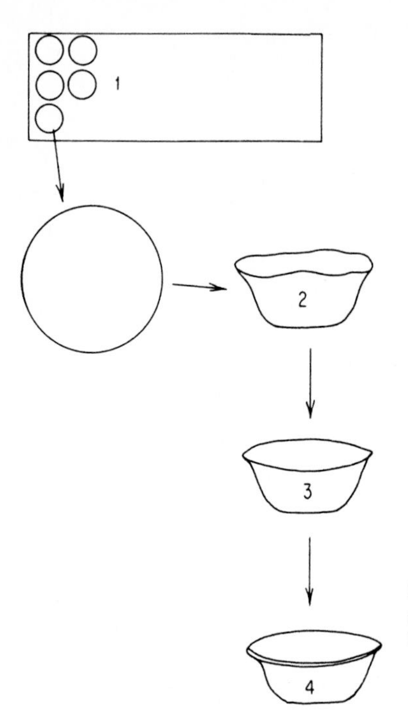

Figure 6.6 Representation of steps in metal fabrication to form an enameled basin.

Some findings

A quick check of Table 6.7 shows that four types of defects accounted for about 80 percent of all defects found during the first four days (a typical experience):

Defect	40 cm	16 cm
1. Blue and black spots	62%	46%
2. Nonuniform border	16%	38%
3. Metal exposed	6%	35%
4. Sheet blister	23%	6%

Many good ideas came from a discussion of this four-day summary sheet. It was noticed, for example, that the smaller 16-cm basin had a record of 35% defects for "metal exposed" while the 40-cm basin—over 6 times as much area—had only 6 percent! "What would explain this peculiar result?" The supervisor said, "Wait a minute," left us abruptly and returned with a metal tripod used to support both the 16- and 40-cm basins during the firing of the enamel coating, suddenly realizing what might be happening. On the small basin, the exposed metal was on the basin rim. The small basin nestled down inside the tripod, letting the edges touch the supporting tripod during firing, and the glaze (Fig. 6.7) often adhered to the tripod as well as to the basin; when the basin was removed from the tripod, the enamel pulled off the edge and left metal ex-

TABLE 6.6 Daily Inspection Sheet (Sampling)

Date ________ Sample Size ________

Product: Basin size: 16 cm or 40 cm

Stage: Final inspection (after firing)

Defects	Number inspected	Classification First	Classification Second	Classification Third	Number inspected	Summary First	Summary Second	Summary Third	Notes
A. Serious									
1. Jig mark									
2. Lump									
3. Nonuniform border									
4. Black spot inside									
5. Black spot outside									
6. Metal exp. (rim)									
7. Metal exp. (border)									
8. Metal exp. (body)									
9. Bad coating									
10. Others									
B. Very serious									
1. Very nonuniform border									
2. Chip									
3. Sheet blister									
4. Dented									
5. Metal exp. border									
6. Metal exp. rim									
7. Metal exp. body									
8. Black spot inside									
9. Black spot outside									
10. Lumps									
11. Very bad coating									
12. Dust particles									
13. Others									

Inspector Signature

posed (a serious defect). The large basin sat on top of the tripod, and any exposed metal was on the bottom in an area where it was classified as minor.

In this case, the solution was simple to recognize and effect, once the comparison between the two basins was noted, because the supervisor was an active member of the team.

Some subsequent summaries

Four summaries at 3 to 5 day intervals were prepared during the two weeks. The one in Table 6.8 compares the progress on the four major defects which had been found.

The defects in periods 3 and 4 have decreased considerably except for blue and black spots (40 cm) and sheet blister. The summary of inspection results was discussed each period with the production people who took various actions, including the following, to reduce defects:

1. Pickling process—degreasing time was increased.

TABLE 6.7 Enamel Basins—Defect Analysis After Four Days

Classification of defects	Defects observed, Number inspected = 4 × 30 = 120											
	40-cm basins						16-cm basins					
	Serious		Very serious		Total		Serious		Very serious		Total	
	no.	%	no.	%	no.	%	no.	%	no.	%	no.	%
Nonuniform border	19	16	—	—	19	16	34	28	12	10	46	38
Blue, black spot	51	42	24	20	75	62	45	37	11	9	56	46
Metal exposed	5	4	3	2	8	6	10	8	32	27	42	35
Sheet blister	—	—	28	23	28	23	—	—	7	6	7	6
Jig mark	1	1	—	—	1	1	9	8	—	—	9	8
Lump	14	12	5	4	19	16	2	2	—	—	2	2
Bad coating	15	13	5	4	20	17	—	—	—	—	—	—
Chips	—	—	3	2	3	2	—	—	—	—	—	—
Dented	—	—	2	2	2	2	—	—	1	1	1	1
Dust particles	—	—	12	10	12	10	—	—	—	—	—	—
Others	—	—	—	—	—	—	—	—	—	—	—	—
Total defects	105 + 82 = 187						100 + 63 = 163					

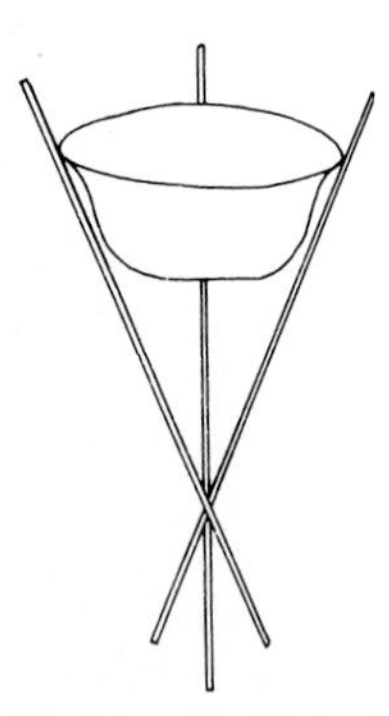

Figure 6.7 Tripod supporting 16-cm enameled basin during firing.

2. Change in enamel solution.
3. Better supervision on firing temperature.

These and other changes were effective in reducing defects.

The record (for 16-cm basins, from Table 6.9) is shown graphically in Fig. 6.8. A sharp decrease is evident in the critical third-grade quality and an increase in the percent of first quality items.

TABLE 6.8 Summary Showing Percentage of Major Defects over Four Time Periods

Quality classification	40-cm basins			16-cm basins			
	Period*			Period			
	1	2	3	1	2	3	4
First	18.3%	14.2%	23.3%	20.0%	27.5%	48.0%	44.0%
Second	37.5%	58.3%	58.9%	45.0%	49.2%	46.0%	53.3%
Third	44.2%	27.5%	17.8%	35.0%	23.3%	6.0%	2.7%
Total	100.0%	100.0%	100.0%	100.0%	100.0%	100.0%	100.0%

* Note: No production of 40-cm basin in period 4.

TABLE 6.9 Changes in Quality Classification over Four Time Periods
Probably most important was the decrease in third quality

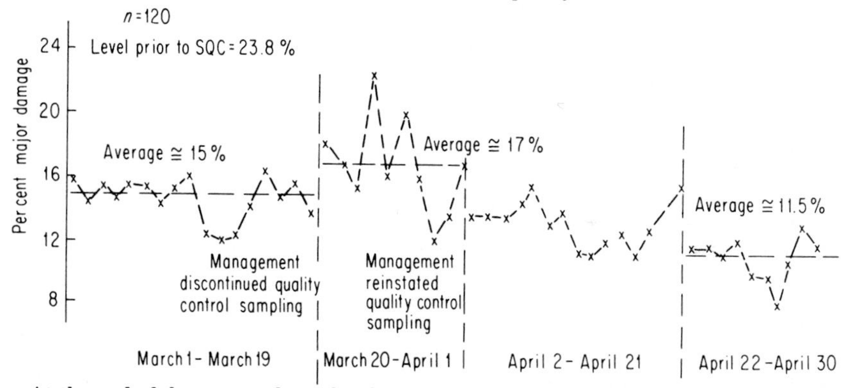

At the end of the two-week study, the team met again with top management. Types of accomplishments and problems were discussed (in nonstatistical terms). Everyone was pleased with the progress attained and in favor of extending the methods to other product items.

A summary of some proposed plans for extensions was prepared by the team and presented orally and in writing to management. It is outlined below.

A plan for the development and extension of a quality control system at an enamel works

1. *Official quality control committee.* Although quality is the concern of everybody in the organization, it is usually found to be the responsibility of none. It is always important to have a small committee to plan and review progress; and this committee should have as secretary the person who will be charged with the responsibility of implementing the program at the factory level. The committee should meet at least once a week to study the results achieved and plan future action.

   The committee should be composed of a management representative, production personnel in charge of the manufacturing and of the enameling sections, and the chief inspector (chairperson), assisted by the quality control person.

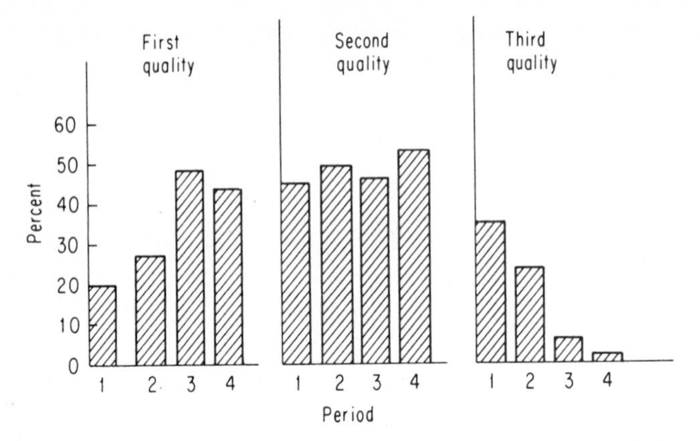

Figure 6.8 Summary of percent classification of 16-cm enameled basins over four sampling periods (data from Table 6.9).

2. *Control of visual defects in production.* Initially, systematic sampling on a routine basis should be done on all items produced every day and at least once a week after conditions have been stabilized. Inspect for visual defects after machining and after enameling. These data should be kept in a suitable file in an easily distinguishable way, product by product.

   All data collected should be maintained on a control chart to be kept in the departments concerned, and quality control should bring to the notice of the appropriate personnel any abnormalities that need to be investigated or corrected.

   Weekly summaries would be discussed by the quality control committee.
3. *Control of outgoing quality (quality assurance).* Starting with exported products, a regular check of about 20 items per day should be made on firsts, seconds, and thirds (quality). Based on appropriate demerit scores for the type and intensity of defects, a demerit chart can be kept in each case for groups of similar products.
4. *Control of inspector differences.* Establish "just acceptable" and "just not acceptable" standards for each defect. Make it available to all inspectors. Keep the data of quality assurance for each inspector (sorter). Summarize the information on a monthly basis to study the extent of misclassification per inspector and take corrective measures to improve poor inspectors (sorters).
5. *Control of nonvisual defects.* In addition to the visual defects some important quality characteristics that require control are:
   a. Weight of the product
   b. Weight of the enamel on the product
   c. Uniformity of enameling
   d. Chipping strength of enameling

   The processes have to be studied with regard to the performance on these characteristics and, where required, simple experimentation should be planned for effecting improvements and thereafter routine control with the help of control charts.
6. *Specification.* In due course the committee should concern itself with laying down realistic specifications. The data of paragraphs 2 to 5 would be of immense help.
7. *Incoming material: acceptance sampling.* The quality of the materials accepted has a vital bearing on the quality of manufacture. Acceptance sampling plans may be started on a few vital items and based on experience can be gradually extended. In

each case, it is necessary to *devise suitable forms* so that it will be possible to analyze each product and vendor without too much trouble.

8. *Training.* There should be a 1-hr talk every week with the help of data and charts collected above to a group of workers and supervisors on different product types; the group will be different for different weeks. The talk should pertain to the data in which the group itself would be interested so as to ensure responsive cooperation.
9. *Regular reports.* On all quality-control data, daily reports should be available to the concerned person in charge of production as well as the works manager and technical adviser.

In each case reports will be short and should pinpoint the achievements as well as points needing attention.

CASE HISTORY 6.2

An Investigation of Cloth Defects in a Cotton Mill (Loom Shed)

Some background information

The advanced age and condition of the looms and loom shed had led the management of a factory to schedule a complete replacement of looms.

The factory had had some previous helpful experience using quality-control methods to increase machine utilization in the spinning department. Consequently, there was a climate of cooperation and hopefulness in this new venture.

There were two shifts, A and B; the same looms were studied on the two shifts.

The 500 looms were arranged in 10 lines of 50 looms each. There were 10 supervisors per shift, one supervising each line of 50 looms. There were 25 loom operators per line (per shift), each operator servicing two looms (in the same line).

The regular inspection practice was to 100 percent inspect each piece of cloth (which was 18 ft long and 45 in wide), rating it as good, second, or poor. No record had been kept of the number, type, or origin of defects. Thus inspection provided no feedback of information to guide improvement of the production process. This is quite typical of 100 percent inspection procedures, unfortunately.

The principles of exploratory investigation used in this in-plant workshop study are general ones, applicable in studying defects in many types of technical operations.

Summary of workshop

1. *Planning.* The essential sequence of steps was:
   a. A team was formed representing supervision, technology, and quality control.
   b. A tour of the plant was made for obtaining background information to formulate methods of sampling and to prepare sampling inspection records.
   c. An initial meeting was held by members of the team with plant management to select the projects for study and outline a general plan of procedure.
   d. Frequent meetings of the team with appropriate supervisory and technical personnel were held as the study progressed.
2. *Data form.* A standard form was prepared in advance; it listed the types of major and minor defects which might be expected (Table 6.10). This form was used by the in-

TABLE 6.10 Record of Weaving Defects—Major and Minor—Found in Cloth Pieces over Two Days (from Five Looms on Two Shifts)

All five looms in line 2

Loom Number	Number of Pieces Inspected	Major Defects: No Head	Imperfect Head	Crack	Float Wft.	Wrong Wft.	Others Wft.	Float Wp.	Long End 6"–18"	Long End > 18"	Smash	Oily	Rusty	Thick	Others	Border Selvedge	Border Flange Cut	Border Others	Total Major	Minor Defects: Float Wft.	Wrong Wft.	Crack Wft.	Others	Float Wp.	Long End 3 ends > 18 in.	Long End 6 in.	Long End 6–18 in.	Fluff	Others	Oily	Thin	Thick	Other Spots	Total Minor
SHIFT A																																		
210	13			1				2		1							1	1	6			4	1	1	8		2					2		18
211	15		1	1		1							2						5		1	3					2			3	1	1		11
223	13							2		1			1	1					5			5			7		4							16
251	15			2									1			1			4			4		1	3		2		1	1		6		18
260	7		2	1										1					4			4		1	5				1			5		16
Σ	63		(3)	(5)		1		(4)		2			(4)	2		1	1	1	24		1	(20)	1	3	(23)		(10)		2	4	1	(14)		79
SHIFT B																																		
210	14							4	1										5			3			5	2	3					1		14
211	14			1				2	1		2	1							7						4				1			2		7
223	12																					2			3	1		1				1		8
251	14													1					1			1		1	1							1		4
260	8																					1			7							1		9
Σ	62			1				(6)	2		2	1		1					13			(7)		1	(20)	3	3	1	1			(6)		42
A & B SHIFTS COMBINED																																		
Σ			3	6		1		(10)	2	2	2	1	4	3		1	1	1	37		1	(27)	1	4	(43)	3	13	1	3	4	1	(20)		121

spector to record the defects found in the initial investigation and in subsequent special studies.

The factory representatives on the team suggested that certain types of defects were operator-induced (and could be reduced by good, well-trained supervision) and that other types were machine-induced (some of which should be reduced by machine adjustments, when recognized).

3. *Sampling versus 100 percent in study.* It was not feasible to examine cloth from each of the 500 loom-operator combinations; but even had it been technically feasible, we would prefer to use a *sampling of looms* for an initial study. Information from a sample of looms can indicate the general types of existing differences, acquaint supervision with actual records of defect occurrences, and permit attacks on major problems more effectively and quickly than by waiting for the collection and analysis of data from all 500 looms.

   What specific method of sampling should be used? Any plan to sample from those groupings which might be operating differently is preferred to a random sampling from the entire loom shed. Differences among the 10 line supervisors is an obvious possibility for differences. Within each line, there may be reasons why the sampling should be stratified:

   a. Proximity to humidifiers, or sunny and shady walls

   b. Different types of looms, if any

c. Different types of cloth being woven, if any

4. *The proposed scheme of sampling looms.* Management was able to provide one inspector for this study. This made it possible to inspect and record defects on the production of about 15 looms on each shift. (One inspector could inspect about 200 pieces of cloth daily.) Each loom produced 6 or 7 pieces of cloth per shift or 12 to 14 pieces in the two shifts. Then an estimate of the number n of looms to include in the study is: $n \cong 200/12 \cong 16$. After discussion with the team, it was decided to select five looms from each of lines 1, 2, 3 on the first day of the study and repeat it on the second day; then five looms from lines 4, 5, 6 on the third day and repeat on the fourth day; then lines 7, 8, 9 on two successive days.

   This sampling scheme would permit the following comparisons to be made:

   a. Between the 10 line supervisors on each shift, by comparing differences in numbers of defects between lines.
   b. Between the two shifts, by comparing differences in numbers of defects of the same looms on the two shifts. (Shift differences would probably be attributed either to supervision or operator differences; temperature and humidity were other possibilities.)
   c. Between the looms within lines included in the study.

   Each piece of cloth was inspected completely, and every defect observed was recorded on the inspection form (see Table 6.10). The technical purpose of this sampling-study workshop was to determine major sources and types of defects in order to indicate corrective action and reduce defects in subsequent production. After an initial determination of major types of differences, it was expected that a sampling system would be extended by management to other looms and operators on a routine basis.

5. *A final oral summary-outline presentation* of findings and indicated differences was held between the team and management. Findings were presented graphically to indicate major effects; specific methods of improving the manufacturing and supervisory processes were discussed. Also, suggestions were made on how to extend the methods to other looms, operators, supervision, and possible sources of differences.

6. *Some findings.* Many different types of useful information were obtained from the two-week study, but improvement began almost immediately. Table 6.10, for example, includes the first two-days' record of defects of five rather bad looms (line 2) on shifts A and B.

   a. *The vital few.* Five types of defects accounted for 70 percent of all major defects found during the first two days; it is typical that a few types of defects account for the great majority of all defects.
   b. *Difference between shifts.* Management was surprised when shown that almost twice as many defects, major and minor, came from shift A as from shift B. (See Table 6.10.) This observed difference was the reason for a serious study to find the reasons; important ones were found by plant personnel.
   c. *Differences between loom-operator combinations in line 2.* All five of these bad looms are in line 2; they have the same line supervisor but different operators. No statistically significant difference was determined between loom-operator combinations. It was found that more improvements could be expected by improving line supervision than from operator or machine performance.

Discussion

Management maintained a mild skepticism, initially, for this "ivory tower" sampling study but soon became involved. It interested them, for example, that the number of defects on the *second* day of sampling was substantially lower than on the *first* day! Supervisors had obtained evidence from the loom records of between loom-operator differences and could give directions on corrective methods. Improvements came from better operator attention and loom adjustments; operators readily cooperated in making improvements.

The average number of major and minor defects (causing downgrading of cloth) had been 24 percent prior to the workshop study (i.e., about 0.24 major defects per piece since major defects were the principal cause of downgrading). During this workshop in December, the system of recording and charting the percent of pieces with major defects was begun and continued.

The data from five loom-operators per line immediately showed important differences between lines (supervisors) and loom-operators; the word was passed from supervisor to supervisor of differences being found and suggestions for making substantial improvements.

No record is available showing the improvements made by the end of the second week when a presentation to management outlined the major findings; but management arranged for the sampling procedure to be continued and posted charts of the sampling defects.

Figure 6.9 shows a daily record of the major defects per piece during March and April; this study began the previous December. It shows several apparent levels of performance; an explanation of two of them is given in a letter written the following July:

> Damages which came down from 24 percent to 16 percent a little while after December 19 have now further reduced to 11 percent as a result of additional sampling checks for quality at the looms. You will see some periods when management lifted the controls in the hope that they would not be necessary. But as soon as things started worsening, they reinstated the procedures laid down earlier. Progress in this plant is quite satisfactory, and it provides a method immediately applicable to other cotton mills.

6.14 Practice Exercises

1. Copy Fig. 6.1 onto a sheet of graph paper, using Table 6.1 for assistance. Extend the chart by plotting similar curves for $c = 1$ and $c = 0$, holding $n = 45$ constant.

	40-cm basins			16-cm basins			
	Period*			Period			
Major defects	1	2	3	1	2	3	4
Blue and black spots	62.5%	52.3%	58.8%	56.7%	37.2%	27.3%	30.0%
Nonuniform border	15.8%	13.3%	3.3%	38.3%	38.3%	14.6%	23.0%
Metal exposed	6.7%	8.4%	3.3%	35.0%	11.6%	7.3%	2.0%
Sheet blister	23.3%	12.4%	18.9%	5.8%	8.4%	6.7%	2.0%

* Note: No production of 40-cm basin in period 4.

Figure 6.9 Record of percent major damaged cloth in March and April following start of quality control program. Average prior to the program was about 24 percent.

2. Copy Fig. 6.1 onto a second sheet of paper and extend the chart by plotting similar curves for $n = 25$ and $n = 100$, holding $c = 2$ constant.

3. Write a short essay on the effect of varying n and c, generalizing from the results of the above exercises. Prepare for oral presentation.

4. Using $P_A = 0.95$ (producer risk of 5 percent) for AQL, $P_A = 0.50$ for IQL, and $P_A = 0.10$ (consumer risk of 10 percent) for LTPD, set up a table of the AQL, IQL, and LTPD for the five plans in 1 and 2 above. (*Note:* Current ANSI standards use LQ, meaning "limiting quality" instead of LTPD; IQL means indifference quality level.)

5. Find a sample size which will assure that if the process remains in control at $p = 0.03$, there is a 99.7 percent confidence that any given sample will be within 3-sigma control limits, but if the process defectiveness suddenly changes to $p = 0.06$, there is a 90 percent chance that the first sample after the change will be outside the upper control limit. (*Hint:* Use normal approximation and trial-and-error.)

Chapter 7

Narrow-Limit Gauging in Process Control

7.1 Introduction

Go no-go gauging has *advantages* over measurements which are made on a variables scale; less skill and time are usually required. But also, the tradition of gauging has been established in many shops even when it may be entirely feasible to make measurements.

However, there are real *disadvantages* associated with gauging. Large sample sizes are required to detect important changes in the process. This is expensive when the test is nondestructive; it becomes exorbitant when the test is destructive. The function of inspection is a dual one: (1) it separates the sheep from the goats, of course; but (2) it should provide a warning feedback of developing trends. Go no-go gauges made to the specifications provide little or no warning of approaching trouble. When the process produces only a small percent of units out of specifications, a sample selected for gauging will seldom contain any out-of-spec units. By the time out-of-spec units are found, a large percentage of the product may already be out of specifications, and we are in real trouble.

Hopefully there is a procedure which retains the important advantages of gauging but improves its efficiency. Narrow-limit gauging is such a method. It is a gauging procedure; it is versatile; it is applicable to chemical as well as to mechanical and electrical applications. Required sample sizes are only nominally larger than equivalent ones when using measurements.

The discussion in this chapter centers about *narrowed* or *compressed gauges* used to guide a process. They function to prevent trouble rather than to wait until sometime after manufacture to learn that the process has not been operating satisfactorily. A process can be guided only if information from gauging is made available to production in advance of a substantial increase in defectives.

The variation of some processes over short periods is often less than permitted by the specifications (tolerances). It is then economical to allow some shift in the process, provided a system is operating which detects the *approach* of rejects.

7.2 Outline of an NL-Gauging Plan

At the start, narrowed or compressed gauges (NL gauges) must be specified and prepared. They may be mechanical ones made in the machine shop; they may simply be limits computed and marked on a dial gauge or computed and used with any variables measurements procedure.

NL gauges are narrowed by an amount indicated by $t\sigma$; see Fig. 7.1, where the mean is (arbitrarily) taken a distance 3σ from the lower specification limit.

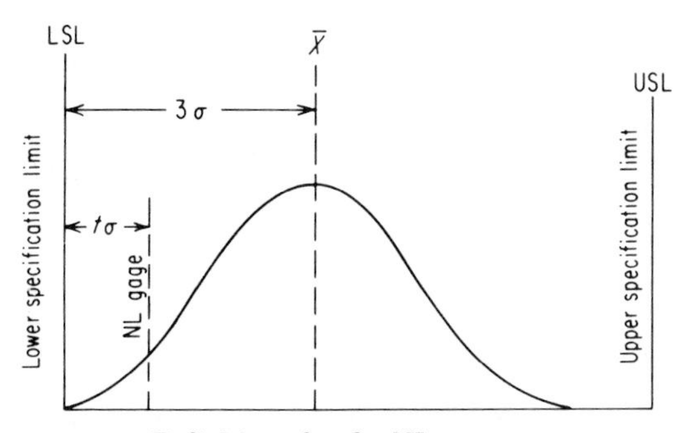

Figure 7.1 Definition of $t\sigma$ for NL gauge.

Small samples of size n are usually gauged at regular time intervals; n may be as small as 4 or 5 and is not usually greater than 10. (The sample is of the most recent production when being used for process control.)

A specified number c of items in a sample of n will be allowed to "fail[1]" the NL gauges. If there are c or fewer items nonconforming to the NL gauges in a sample of n, the process is continued without adjustment. If more than c nonconforming items are found, then the process is to be adjusted.

Separate records are kept of the number of units which fail the smaller NL gauge and of the number failing the larger NL gauge.

In the applications discussed here, there are upper and lower specifications, and the distance between them is larger than the process capability. That is, $(\text{USL} - \text{LSL}) > 6\sigma$.

It is assumed that the process produces a product whose quality characteristic is a reasonably normal distribution (over a short time interval). See "Hazards," Sec. 7.6.

[1]Any unit which fails an NL gauge can be regauged at the actual specifications to determine salability.

Basic assumptions for NL gauging

1. An estimate of the basic process variability is available or can be made. Perhaps an estimate can be made from a control chart on ranges or perhaps from experience with a similar application.
2. The difference between the upper and lower specification limits (USL, LSL) is greater than 6σ; i.e., $USL - LSL > 6\sigma$.
   This assumption means that some shifting in the process average is acceptable. It also means that only one tail of the distribution at a time need be considered when computing operating characteristic curves (OC curves).
3. The distribution of the quality characteristic should be reasonably normal over a short time.

7.3 Selection of a Simple NL-Gauging Sampling Plan

Those who have used $\bar{X}$, R charts are familiar with the usefulness of small samples of $n_g = 4$ or 5. Samples as large as 9 or 10 are not often used; they are too sensitive and indicate many shifts and peculiarities in the process which are not important problems.

Then those of you who are familiar with $\bar{X}$, R charts will approve of an NL-gauge plan which gives process guidance comparable to that of an $\bar{X}$, R chart for $n_g = 4$ or 5. Four such plans are the following:

A. $n_g = 5, c = 1, t = 1.0$

C. $n_g = 10, c = 2, t = 1.0$

D. $n_g = 4, c = 1, t = 1.2$

F. $n_g = 10, c = 2, t = 1.2$

Two OC curves are shown in Fig. 7.2, along with that of an $\bar{X}$ control-chart plan, $n_g = 4$. OC curves are shown for plans A and C (data from Table 7.2). The OC curves of plans A and D are very close to that of an $\bar{X}$ chart, $n_g = 4$. For that reason, the curve for plan D has not been drawn. A discussion of the construction of operating characteristic curves (OC curves) of some plans is given in Sec. 7.5.

When guiding a process, we may believe that there is need for more information than provided by an $\bar{X}$, R chart using $n_g = 5$. If so, we usually take samples of five *more frequently* in preference to increasing the size of n_g. The same procedure is recommended for an NL-gauge system.

The effectiveness of NL-gauging in helping prevent trouble is improved when we *chart* the information from successive samples. These charts indicate the approach of trouble in time to take preventive action. Plans which have zero acceptance numbers are all or nothing in signaling trouble.

Since an indication of approaching trouble is not possible in a plan with $c = 0$, the smallest recommended value is $c = 1$. A frequent preference is a plan with $n_g = 4$ or 5, $c = 1$, and $t = 1$.

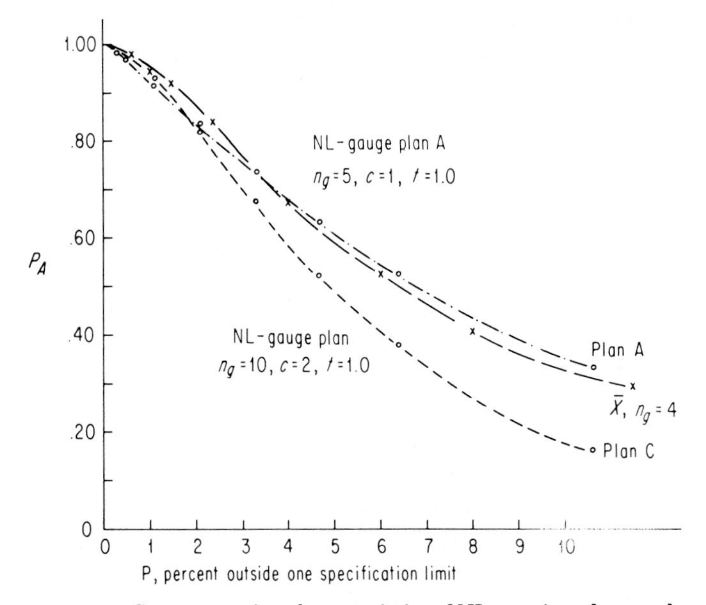

Figure 7.2 Some operating characteristics of NL-gauging plans and a variables control chart on $\bar{X}$, n_g = 4. OC curves are shown for plans A and C (data from Table 7.2). The OC curves of plans A and D are very close to that of an $\bar{X}$ chart, n_g = 4. For that reason, the curve for plan D has not been drawn.

CASE HISTORY 7.1

Extruding Plastic Caps and Bottles

The mating fit of a semiflexible cap on a bottle depends on the *inside diameter* (ID) of the cap and the *outside diameter* (OD) of the bottle as shown in Fig. 7.3.

Some pertinent information

1. The molded plastic caps and bottles shrink after leaving the mold and during cooling. We found it reasonable to immerse them in cold water before NL-gauges were used.

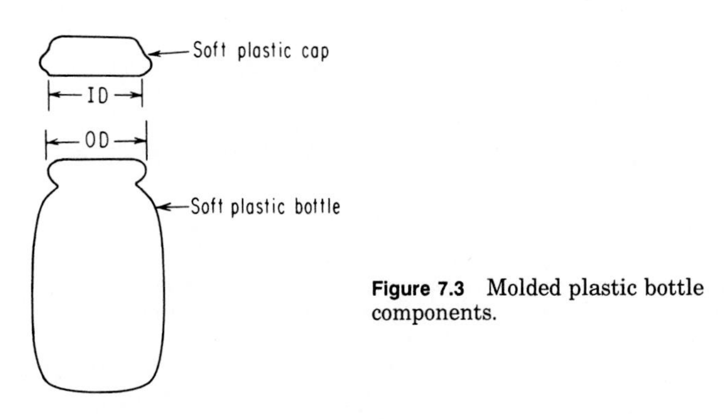

Figure 7.3 Molded plastic bottle components.

Sometimes they are held in a plastic bag during immersion. Gauging could then be done shortly after production.

2. The ID and OD dimensions can be adjusted by certain temperature ranges in the machine molds and by other machine adjustments. It requires specific knowledge of the process to effect these machine changes on diameter without introducing visual defects into the product.
3. The usual production control check on these two diameters is by gauging; a plug gauge for ID and a ring gauge for OD. It is traditional to use plug gauges made to the maximum and minimum specifications. Then by the time they find a reject, large numbers of out-of-spec caps or bottles are already in the bin. Since it is rarely economical to make a 100 percent inspection of caps or bottles, the partial bin of the product must be scrapped.
4. It is possible to purchase dimensional equipment to measure the OD and the ID. However they are slow to use, not as accurate as one would expect, require more skill than gauging, and the plotting of data is more elaborate.
5. Each cap or bottle has the cavity of origin molded in it because differences between cavities are common. However, it requires excessive time to establish the cavity numbers represented in a sample and then to measure and record diameters.

Determining the variability of cap ID

On one type of plastic cap, the specifications were

$$\text{USL} = 1.015 \text{ in} \qquad \text{and} \qquad \text{LSL} = 1.00 \text{ in}$$

Samples of three caps from each of the 20 cavities were gauged with a series of six gauges having diameters of 1.000 in, 1.003 in, 1.006 in, 1.009 in, 1.012 in, and 1.015 in. The department participated in measuring diameters with these plug gauges. It was first determined that the ID of a cap could be determined to within 0.001 in or 0.002 in with this set of gauges. Then the ID of samples of three caps were "measured" with them and ranges of $n_g = 3$ computed. From these measurements we computed

$$\hat{\sigma} = \overline{R}/d_2 \cong 0.003 \text{ in}$$

We already had plug gauges to measure 1.003 in and 1.012 in; these corresponded to NL gauges compressed by $t\hat{\sigma} \cong 0.003$ ($t = 1.0$). We decided to use the plan: $n_g = 10$, $t = 1.0$, and $c = 1$; random samples of 10 would allow more representative sampling of the 20 cavities.[2] Other plans using different values of n_g and c might have been equally effective, of course.

The NL-gauging plan was used to guide the process, as planned originally. But also, several *special studies* were made under different process adjustments; the NL gauges were used to measure and compare the effects of these changes. Data from a typical study were tabulated as in Table 7.1. The usual study used a 2^2 or a 2^3 design as in Chap. 11.

Consequences

Much of the art of adjusting very expensive and complicated equipment was replaced by cause-and-effect relationships which were developed as in Table 7.1. A production pro-

[2]See item 5 above.

TABLE 7.1 A Sampling Method of Estimating Process Relation to Specifications (1.000–1.015 in.) Using a Sequence of 5 Go, No-Go Gauges*

	Classification steps						
	LSL						USL
(n_g = 10) Sample no.	< 1.000	1.000–1.003	1.003–1.006	1.006–1.009	1.009–1.012	1.012–1.015	> 1.015
1					1	8	1
2					7	3	
3			1	8	1		
4			3	7			
5			5	5			
6			4	6			
7			etc.				
8							

*The numbers in the columns suggest a possible sequence of classifications from samples of 10 using gauges milled to low and high dimensions as follows:

	Low, in	High, in
G_1	1.000	1.003
G_2	1.003	1.006
G_3	1.006	1.009
G_4	1.009	1.012
G_5	1.012	1.015

These five gauges permit the seven classifications shown at the top of columns above.

cess which had been a source of daily rejects, aggravation, and trouble began gradually to improve.

Then the department was expanded by purchasing equipment for the manufacture of plastic bottles; the critical dimension was now the OD of the neck. The same general approach which had been used for caps was still applicable; ring gauges when narrowed by 0.003 in from the LSL and USL proved satisfactory NL gauges.

Soft plastic caps and bottles are produced in very large quantities and the cost of a single unit is not great. There are serious reasons why traditional manufacturing controls discussed in (3) and (4) above are not satisfactory.

Industrial experiences with NL gauges have shown them to be practical and effective.

CASE HISTORY 7.2

Chemical Titration

Introduction

Several multihead presses were producing very large quantities of a pharmaceutical product in the form of tablets. It was intended to double the number of presses which would require an increase in the amount of titration and the number of chemists; this was at a time when analytical chemists were in short supply. It was decided to explore the possibility of applying gauging methods to the titration procedure. The following

NL-gauging plan was a joint effort of persons knowledgeable in titration procedures and applied statistics.

Some pertinent information

1. Data to establish the variability of tablet-making presses were already in the files; they were used to provide the estimate $\hat{\sigma} \cong 0.125$ grains.
2. Specifications on individual tablets were LSL = 4.5 grains, USL = 5.5 grains.
3. Then $6\hat{\sigma} = 0.75$ is less than the difference USL–LSL of 1.0 grains. A choice of $t = 1.2$ with $\hat{\sigma} = 0.125$ gives $t\hat{\sigma} \cong 0.15$. The chemists agreed that this was feasible. The lower NL-gauge value was then 4.65 grains, and the upper NL-gauge value was 5.35 grains.

Sampling

It was proposed that we continue the custom of taking several samples per day from each machine; then an OC curve of our NL-gauge plan was desired which would be comparable to an $\bar{X}$, R chart, $n_g = 4$ or 5. Some calculations led to the plan; $n_g = 4$, $t = 1.2$, and $c = 1$. (See Table 7.3 for the computation of the OC curve.)

A semiautomatic charging machine was adjusted to deliver the titrant required to detect by color change:

1. Less than 4.65 grains on the first charge
2. More than 5.35 grains on the second charge

Consequences

With minor adjustments, the procedure was successful from the beginning. The accuracy of the gauging method was checked regularly by titration to endpoint.

Shifts on individual presses were detected by the presence of tablets outside the NL gauges, but usually within specifications. The presses could be adjusted to make tablets to specifications. The required number of chemists was not doubled when the increased number of tablet presses began operation—in fact the number was reduced by almost half.

7.4 Sequential NL-Gauging Plans

Introduction

Almost everyone accepts the premise that visual records should be kept of process performance. Almost everyone is a bit careless in the matter, too. Consider a process with samples taken frequently and a visual record being kept. Then it may be satisfactory to permit a shift in process to go undetected on a single sample *provided* it is detected on a second or third sample. A double-phased plan has proved to be practical to use on a production floor; the second phase is based on the *accumulated evidence from the last three samples.* Since a visual chart is being kept, the accumulated total can be done mentally.

For example, the double plan ($n_1 = 5, c = 1; n_2 = 15, c = 2$) with $t = 1.0$ provides joint criteria for machine adjustment on a nonconforming characteristic. This is referred to as plan E.

Adjustment of process

Check with the NL gauge a sample of five pieces of the most recent production at regular time intervals (quarter-hourly, half-hourly, hourly, or as experience indicates a need).

- Record the number of pieces which are high and the number low on the chart (see Fig. 7.4).
- The machine shall be adjusted on the nonconforming characteristic (to NL gauge) when and only when either

  1. Two or more pieces fail the NL gauges in the last *single* sample ($n_1 = 5$, $c = 1$ criterion), or
  2. Three or more pieces fail the NL gauges in the *accumulation* of the last three samples ($n_2 = 15$, $c = 2$ criterion).

- Continue otherwise to operate the machine without adjustment.
- When a rejection occurs on a single sample (and the process is then adjusted), only the single-sample criterion is applied until two successive conforming samples have been obtained. Then the criterion, $n_2 = 15$, $c = 2$, also is operative.

Note: The probability of detecting a defective process can be increased by including the pooled information from the last three samples with the $n_2 = 15$, $c = 2$ criterion. Some probabilities are given in Table 7.3, and OC curves are shown in Fig. 7.7.

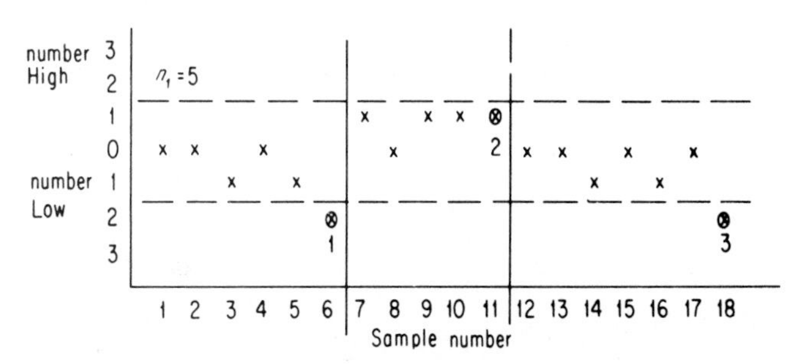

Figure 7.4 An NL-gauge chart for spraying cathodes. Plan: $n_1 = 5$, $c = 1$ and $n_2 = 15$, $c = 2$ with gauge compressed one standard deviation, i.e., $t = 1.0$. *Notes:* (1) Sprayer cleaned after failing $n_1 = 5$, $c = 1$ criterion; (2) number of spray passes reduced after a conference with supervisor (process failed the three combined sample criterion $n_2 = 15$, $c = 2$); (3) number of spray passes increased after conference with supervisor (process failed $n_1 = 5$, $c = 1$ criterion).

CASE HISTORY 7.3

Thickness of Oxide-Coated Metal

Trays holding 25 nickel cathodes were sprayed with an oxide coating. A critical characteristic was the thickness of the coating being applied by several passes across each tray.

Some pertinent information

1. The process was too variable. It was not possible to use 100 percent inspection to select the good ones; the softness of the coating made any gauging or measurement a destructive process.
2. Not only was it destructive to measure; it was difficult because the oxide-coated surface was very rough and uneven. Gauging was easier.
3. The cost of rejecting a tray of sprayed cathodes and reworking them was not much, but it reduced the output and there was no assurance that reworked trays would be any better. Good cathodes required that the process itself be good.
4. Sprayed cathodes were assembled into components, then tested for electronic characteristics which were dependent upon the thickness of the cathode coating. A rejected unit was not reworkable and meant a major loss.

Procedure

It was decided to replace the existing process control procedure by NL gauging. The gauges were compressed by 1.0σ ($t = 1.0$). The double NL-gauging plan was: $n_1 = 5$, $c = 1$ and $n_2 = 15$, $c = 2$.

The OC curves of this plan corresponding to the single sample and the three accumulated samples ($n_2 = 15$) are shown in Fig. 7.7.

From each tray of cathodes which had been sprayed, five were selected and checked with the NL gauges. The number of oversize (high) and undersize (low) cathodes were marked on the chart as in Fig. 7.4.

By studying the behavior of the spraying, it was soon learned when and how the process should be adjusted: by cleaning the spray guns, by increasing or decreasing the number of passes, by instructing the operator as to the speed of passes.

The improvements in this process were of different kinds:

1. It was possible to use NL gauging as an acceptance sampling plan on trays of sprayed cathodes and thus eliminate a holdup on the production line.
2. Because of rapid determination of the state of control available by go no-go gauging, the spray operator had current information to use in adjusting the process. This plan quickly reduced the number of trays sent back for additional spray passes (cathodes undersized), and simultaneously reduced the number of cathodes which were scrapped for being oversize from an average of 20 percent to less than 1 percent. With less rework flowing through the production line, it was possible to reduce the inspection personnel while increasing the yield of good cathodes.
3. Even more important, however, was the uniformity of cathode thickness. It was possible to reduce the overall spread of the sprayed cathodes by a factor of 50 percent using NL gauging. This improvement was so radical and so important in subsequent operations that the completed product of which the cathodes were a subassembly was improved from being mediocre to being a successful product on the market.

CASE HISTORY 7.4

Machine-Shop Dimensions

Some aircraft navigational instruments use parts machined with great precision. It was customary in this shop to measure critical dimensions to the nearest ten-thousandth. The specifications on one part were 0.2377 to 0.2383 in (a spread of 0.0006 in). Data on hand indicated a machine capability of ±0.0002, i.e., an estimated $\hat{\sigma} = (0.0004)/6$. The realities of measurement would not permit using limits compressed less than 0.0001 in; this corresponded to $t \cong 1.5$. The double plan was: $n_1 = 5$, $c = 1$ and $n_2 = 15$, $c = 2$.

The toolmaker would measure a machined piece with a toolmaker's micrometer and indicate by a check the reading which he observed as in Fig. 7.5. This chart is a combination between a variables chart and NL-gauge chart. It is a form which was accepted and maintained by "old-line" machine operators.

The machine operators were willing to make checkmarks above and below the NL-gauge lines. Previously, their practice had been to flinch and make a sequence of three or four consecutive checkmarks at UCL = 0.2383. Then rechecks later of production at these times would show oversized parts. The psychology of making checks outside NL gauges but within specifications resulted in resetting the tool before rejects were machined.

Although no physical NL gauges were made for this particular machine shop application, the entire concept of adjusting the process was exactly that of NL gauging.

7.5 OC Curves of NL-Gauge Plans[3]

It is not easy at first to accept the apparent effectiveness of NL gauging with small samples. True, the OC curves in Fig. 7.2 do indicate that some NL-gauging plans are very comparable to ordinary $\bar{X}$, R control charts. But actual experience is also helpful in developing a confidence in them.

We used samples of $n_g = 4$, 5, 10, and 15 in preceding examples. It will be seen that OC curves comparable to $\bar{X}$, R charts, $n_g = 4$, can be obtained from NL-gauging plans using

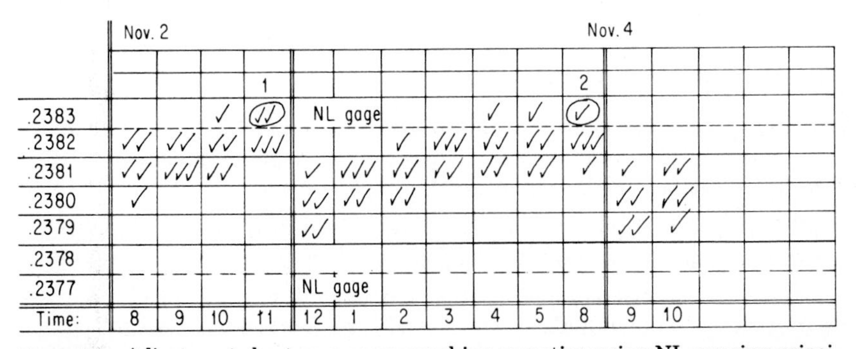

Figure 7.5 Adjustment chart on a screw machine operation using NL-gauging principles. Plan: $n_1 = 5$, $c = 1$ and $n_2 = 15$, $c = 2$ with $t \cong 1.5$. *Notes:* (1) Sample failed $n_1 = 5$, $c = 1$ criterion; tool was reset. (2) Sample failed $n_2 = 15$, $c = 2$ criterion; tool was reset.

[3]May be omitted by reader.

$$n_g = 4 \text{ or } 5 \qquad t = 1.0 \text{ to } 1.2 \qquad \text{and } c = 1$$

We show the method of deriving OC curves for $n_g = 5$ and 10 in Table 7.2. Curves for many other plans can be derived similarly.

Two different types of percents will be used in this discussion:

1. P will represent the *percent outside the actual specification.* It appears as the abscissa in Fig. 7.7. It is important in assessing the suitability of a particular sampling plan. It appears as column 1 in Table 7.2.
2. P' will represent the *percent outside the NL gauge* corresponding to each value of P. It appears as column 4 in Table 7.2. It is an auxiliary percent used in Table A.5 to determine probabilities P_A.

Derivation of OC curves

On the vertical scale in Fig. 7.7 we show the probability of acceptance, P_A; it represents the probabilities that the process will be *approved* or *accepted without adjustment* under the NL-gauging plan for different values of P. To obtain P_A, we use the binomial distribution (Table A.5) in conjunction with the normal distribution as in Fig. 7.6.

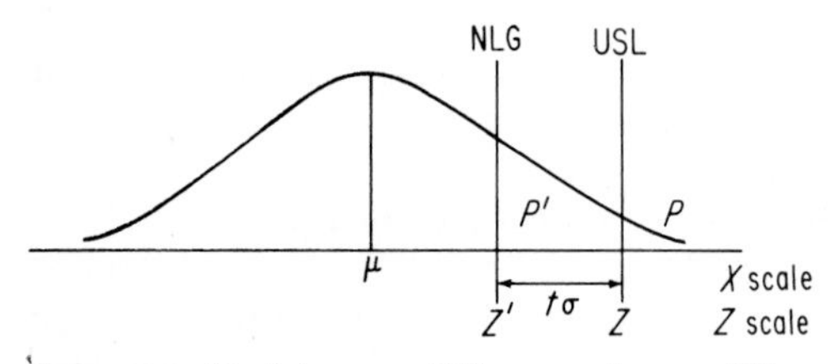

Figure 7.6 Deriving an OC curve for an NL-gauging plan (general procedure). (Detail given in Tables 7.2 and 7.3.)

The steps in filling out Table 7.2 are as follows:

1. Select appropriate percents, P, out of specification and list them in column 1. These are directly related to the position of the process mean μ relative to the specification. For P to change, μ must shift.
2. From the normal table (Table A.1) obtain the standard normal deviate Z corresponding to the percent out of specification, P. List these in column 2.
3. Determine the corresponding standard normal deviate for the narrow limit gauge. This will always be $Z' = Z - t$ since the narrow limit gauge will be a distance $t\sigma$ from the specification limit. List the values of Z' in column 3.
4. Find the percent outside the narrow limit gauge P' from the value of Z' shown using the standard normal distribution (Table A.1). List these in column 4.

TABLE 7.2 Derivation of Operating Characteristic Curves (OC Curves) for Some NL-Gauge Plans with Gauge Compressed by 1.0σ(t = 1.0)

(1) P = percent outside spec.	(2) Z value of P Z_p	(3) Z value of P' $Z' = Z_p - t$	(4) P' = percent outside NLG	(5) P_A Plan A $n = 5$ $c = 1$	(5a) P_A Plan B $n = 5$ $c = 2$	(5b) P_A Plan C $n = 10$ $c = 2$
0.1	3.09	2.09	1.83	0.9968	0.9999	0.9993
0.135	3.00	2.00	2.28	0.9950	0.9999	0.9987
1.0	2.33	1.33	9.18	0.9302	0.9933	0.9432
2.0	2.05	1.05	14.69	0.8409	0.9749	0.8282
5.0	1.64	0.64	26.11	0.6094	0.8844	0.4925
10.0	1.28	0.28	38.97	0.3550	0.7002	0.1845
15.0	1.04	0.04	48.40	0.2081	0.5300	0.0669

5. Find the probability of acceptance for the NLG plan at the specified percent, P, out of specification by using the binomial distribution (Table A.5) with percent nonconforming P', and the sample size n and acceptance number c given. List these values in column 5. Note that, given column 4, P_A can be calculated for any other values of n and c. Additional plans are shown in columns 5*a* and 5*b*.

6. Plot points corresponding to column 1 on the horizontal axis and column 5 on the vertical axis to show the OC curve (Fig. 7.7).

Table 7.3 utilizes the results of Table 7.2 to compute the probability of acceptance for sequential plan E ($n_1 = 5$, $c = 1$; $n_2 = 15$, $c = 2$; $t = 1.0$) referred to in Sec. 7.4.

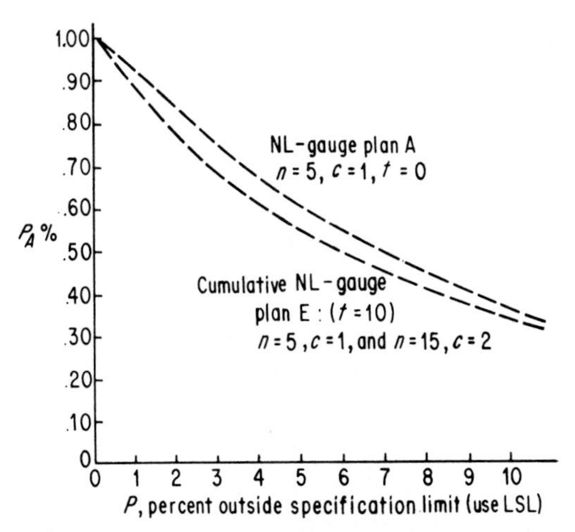

Figure 7.7 OC curves of NL-gauge plans (data from Table 7.3).

TABLE 7.3 Some Probabilities for Sequential NLG Plan* ($n_1 = 5$, $c = 1$; $n_2 = 15$, $c = 2$; $t = 1.0$)

		Plan A	Plan A	Plan A	Plan A	Plan E
		Probability of acceptance on last sample	Probability of rejection on last sample	Probability of exactly one outside NLG	Probability of last three with exactly one outside NLG	Probability of acceptance
P	P'	$P_A = P_r(x \leq 1)$	$P_R = 1 - P_A$	$P_1 = P_r(x = 1)$	P_1^3	$P_A = 1 - P_R - P_1^3$
0.1	1.83	0.9968	0.0032	0.0850	0.0006	0.9962
0.135	3.36	0.9895	0.0105	0.1465	0.0031	0.9864
1.0	9.18	0.9302	0.0698	0.3123	0.0305	0.8997
2.0	14.69	0.8409	0.1591	0.3890	0.0589	0.7820
5.0	26.11	0.6094	0.3906	0.3892	0.0590	0.5504
10.0	38.97	0.3550	0.6450	0.2703	0.0197	0.3353
15.0	48.40	0.2081	0.7919	0.1716	0.0051	0.2030

*NOTE: The probability of rejecting of plan E is the sum of the probability of rejecting on the last sample P_R plus the probability of exactly one item outside the narrow limit gauge on each of the last three samples of five, P_1^3.

TABLE 7.4 Percent of Normally Distributed Product Outside 3σ Specification from Nominal Mean of Control Chart for Comparison of NLG to other Control Chart Procedures

Shift in mean (Z)	Distance to spec. in σ units ($3 - Z$)	Percent nonconforming
Nominal 0.0	3.0	0.135
0.5	2.5	0.62
1.0	2.0	2.28
1.5	1.5	6.68
2.0	1.0	15.87
2.5	0.5	30.85
3.0	0.0	50.00

Narrow-limit plans have been used in acceptance sampling (e.g., Schilling and Sommers[4]) and as a process-control device. To compare the OC curve of the narrow-limit procedure to that of a standard control chart, it is usually assumed that the process is in control with a mean value positioned 3σ from the specification limits with $P = 0.135$ percent. Shifts in the process mean from that position will result in changes in P. The relationship between the shift in the mean and the percent outside 3σ is shown in Table 7.4. Thus, a shift in the mean of $(3 - Z)\sigma$ will result in the percents defective shown in column 1 of Table 7.5 for either type of chart. The probability of acceptance can then be calculated and the charts compared (see Fig. 2.4*a*).

TABLE 7.5 Deriving an OC Curve for the NL-Gauge Plan $n = 4$, $t = 1.2$, $c = 1$. The OC curve of the plan is very similar to that of an $\overline{X}$ chart, $n = 4$ (See Fig. 7.2 and Table 7.4.)

(1) P = percent outside spec.	(2) Z value of P Z_p	(3) Z value of P' $Z' = Z_p - t$	(4) P' = percent outside NLG	(5) P_A Plan D $n = 4$ $t = 1.2$ $c = 1$	P_A Shewhart $\overline{X}$ chart (See Table 7.4)
0.135	3.0	1.8	3.59	99.26	99.87
0.62	2.5	1.3	9.68	95.08	97.72
2.28	2.0	0.8	21.19	80.07	84.13
6.68	1.5	0.3	38.21	50.63	50.00
15.87	1.0	−0.2	57.93	20.39	15.87
30.85	0.5	−0.7	75.80	4.64	2.28
50.00	0.0	−1.2	88.49	0.56	0.135

[4]E. G. Schilling and D. J. Sommers, "Two Point Optimal Narrow Limit Plans with Applications to MIL-STD-105D," *J. of Qual. Tech.*, vol. 13, no. 2, April 1981, pp. 83–92.

Note that the binomial distribution (rather than the Poisson approximation) is used here because, for NLG applications, very often nP' is greater than 5.

7.6 Hazards

There is a potential error in values of P_A if our estimate of σ is in substantial error. In practice, our estimate of σ might be in error by 25 percent, for example. Then instead of operating on a curve corresponding to $t = 1.0$, we either operate on the curve corresponding to $t = 1.25$ when our estimate is too large, or we operate on the curve corresponding to $t = 0.75$ when our estimate is too small. In either case, a difference of this small magnitude does not appear to be an important factor.

Suppose that the portion of the distribution nearest the specification is not normally distributed. In most instances this is more of a statistical question than a practical one, although there are notable exceptions in certain electronic characteristics, for example. In machining operations, we have never found enough departure from a normal distribution to be important except when units produced from different sources (heads, spindles) are being combined. Even then, that portion of the basic curve nearest the specification limit (and from which we draw our sample) is typically normal.

If the distribution is *violently* nonnormal, or if an error is made in estimating σ, the NL-gauging system still provides control of the process but not necessarily at the predicted level.

In discussing a similar situation, Tippett[5] remarks that there need not be too much concern about whether there was an "accurate and precise statistical result, because in the complete problem there were so many other elements which could not be accurately measured."

It should be noted that narrow-limit plans need not be limited to normal distribution of measurements. Since the OC curve is determined by the relationship of the proportion of product beyond the specification limit, P, to the proportion of product outside the narrow limit gauge, P', any distribution can be used to establish the relationship. This can be done from probability paper or from existing tables of nonnormal distributions. For example, when a Pearson Type III distribution is involved, the tables of Salvosa[6] can easily be used to establish the relationship, given the coefficient of skewness α_3 involved. The procedure would be similar to that described here (Table 7.3), using the Salvosa table in place of the normal table for various values of Z.

Discussion

We have used NL gauges in a variety of process-control applications over the last several years. Both from experience and the underlying theory, we find

[5]L. H. C. Tippett, *Technological Applications of Statistics*, John Wiley & Sons, Inc., New York, 1950.

[6]Luis R. Salvosa, "Tables of Pearson's Type III Function," *Annals of Mathematical Statistics*, vol. 1, May 1930, pp. 191ff. See also Albert E. Waugh, *Elements of Statistical Method*, McGraw-Hill Book Company, New York, 1952, pp. 212–215.

that NL gauges offer a major contribution to a study of industrial processes even when it is possible to use $\overline{X}, R$ charts. There are different reasons which recommend NL gauges:

1. Only gauging is required—no measurements.
2. Even for samples as small as *five,* the *sensitivity* is comparable to control charts with samples of *four* and *five.*
3. *Record keeping* is simple and effective. Charting the results at the machine is simpler and faster than with $\overline{X}, R$ charts. The number of pieces which fail to pass the NL gauges can be charted by the operator quickly. Trends in the process and shifts in level are often detected by the operator before trouble is serious. Operator comprehension is often better than with $\overline{X}, R$ charts.
4. NL-gauge plans may be applied in many operations where $\overline{X}, R$ charts are not feasible, thereby bringing the sensitivity of $\overline{X}, R$ charts to many difficult problems.

7.7 Selection of an NL-Gauge Plan

The selection of an NL-gauge plan is similar to the selection of an $\overline{X}, R$ plan. The same principles are applicable. A sample size of five in either is usually adequate. A sample of ten will sometimes be preferred with NL gauging; it provides more assurance that the sample is representative of the process. We recommend the following plans, or slight modifications of them, since their sensitivity corresponds closely to an $\overline{X}, R$ chart with $n_g = 5$:

Plan A	Plan F	Plan D
$n_g = 5$ $t = 1.0$ $c = 1$	$n_g = 10$ $t = 1.2$ $c = 2$	$n_g = 4$ $t = 1.2$ $c = 1$

Various charts and ideas presented in this chapter are reproduced from an article by Ellis R. Ott and August B. Mundel, "Narrow-limit Gaging," *Ind. Qual. Control*, vol. 10, no. 5, March 1954. They are used with the permission of the editor.

The use of these plans in process control has been discussed by Ott and Marash.[7] Optimal narrow limit plans are presented and tabulated by Edward G. Schilling and Dan J. Sommers[8] in their paper.

[7]Ellis R. Ott and S. A. Marash, "Process Control with Narrow Limit Gauging." *Transactions of the 33rd ASQC Annual Technical Conference,* Houston, Tex., 1979, p. 371.

[8]E. G. Schilling and D. J. Sommers, "Two Point Optimal Narrow Limit Plans with Applications to MIL-STD-105D," *J. Qual. Tech.*, vol. 13, no. 2, April 1981, pp. 83–92.

7.8 Practice Exercises

1. State the two important requirements for narrow-limit gauging to work.
2. What should be done with individual units which fail the NL gauge?
3. Derive OC curves for $n = 10$, $c = 1$, $t = 1$, and $n = 10$, $c = 1$, $t = 1.5$. Plot them on a graph together with the three plans presented in Table 7.2. Use different colors to clearly distinguish between the five curves. Use a French curve if necessary to obtain smooth curves. From study of these five curves, write a general discussion of the effect of n, c, and t on an NL-gauge plan, and how they interact.
4. Given, NLG plan $n = 5$, $c = 1$, $t = 1$, find probability of acceptance (P_A) for a fraction defective (p) of 0.03 (3 percent).
5. A process has been monitored with a control chart and shows good evidence of control, with X double bar 180 and R bar = 5.4. The upper specification limit on this process is 188. Given a sample size of 7, find (a) an upper limit on X bar that will provide a 90 percent chance of rejecting a lot with 10 percent defective, and (b) a comparable NLG limit using $n = 7$, $c = 1$. Illustrate the relationships involved with sketches.
6. Sketch the OC curve of a conventional Shewhart X bar chart with $n = 5$. For X bar charts assume that the control limit is set for mu = USL − 3 sigma: Then derive the OC curve of an NLG plan with $n = 4$, $t = 1.2$, $c = 1$. Compare these two OC curves and discuss tradeoffs involved in using the Shewhart X bar chart and the NLG technique for ongoing process control. Consider the need for maintaining an R chart concurrent with an NLG process control technique, and consider as an alternative a double NLG plan which monitors both tails of the process. (Note that such a procedure gives ongoing information about process variability as well as fraction defective, and can be used in conjunction with a two-sided specification limit.)

Chapter

8

On Implementing Statistical Process Control

8.1 Introduction

Process quality control is not new. It had its genesis in the conviction of Walter Shewhart that "constant systems of chance causes do exist in nature" and that "assignable causes of variation may be found and eliminated."[1] That is to say that controlled processes exist and the causes for shifts in such processes can be found. Of course, the key technique in doing so is the control chart, whose contribution is often as much in terms of a physical representation of the philosophy it represents as it is a vital technique for implementation of that philosophy.

In its most prosaic form we think of process quality control in terms of a continuing effort to keep processes centered at their target value while maintaining the spread at prescribed values. This is what Taguchi[2] has called online quality control. But process quality control is, and must be, more than that, for to attain the qualities desired all elements of an organization must participate and all phases of development from conception to completion must be addressed. Problems must be unearthed before they blossom and spread their seeds of difficulty. Causes must be found for unforeseen results. Action must be taken to rectify the problems. And finally controls must be implemented so the problems do not reoccur. This is the most exciting aspect of process control: not the control, but the conquest. For nothing is more stimulating than the new. New ideas, new concepts, new solutions, new methods—all of which can come out of a well-organized and directed program of process quality control.

[1] W. A. Shewhart, *Economic Control of Quality of Manufactured Product,* D. Van Nostrand Company, Inc., New York, 1931.

[2] G. Taguchi, "On-Line Quality Control during Production," *Jap. Stand. Assoc.,* Tokyo, 1981.

8.2 Key Aspects of Process Quality Control

Process quality control may be addressed in terms of three key aspects:

1. *Process control.* Maintaining the process on target with respect to centering and spread.
2. *Process capability.* Determining the inherent spread of a controlled process for establishing realistic specifications, use for comparative purposes, and so forth.
3. *Process change.* Implementing process modifications as a part of process improvement and troubleshooting.

These aspects of process control are exhibited in Fig. 8.1.

Naturally, these aspects work together in a coordinated program of process quality control in that achievement of statistical control is necessary for a meaningful assessment of capability and the analysis of capability of a process against requirements is often an instrument of change. But changes may necessitate new efforts for control and the cycle starts over again.

There is much more to process quality control than statistics. Yet statistics plays a part all along the way. Interpretation of data is necessary for studies of capability and for achieving control. The statistical methodology involved in troubleshooting and design of experiments is essential in affecting change. After all, processes are deaf, dumb, blind, and usually not ambulatory. In other words, they are not very communicative. Yet they speak to us over time through their performance. Add bad materials and the process will exhibit indigestion. Tweak the controls and the process will say "ouch!" Yet in the presence of the variation common to industrial enterprise, it is difficult to interpret these replies without amplification and filtering by statistical methods designed to eliminate the "noise" and focus on the real shifts in level or spread of performance.

One approach to the analysis of a process is continued observation so that, when a variable changes affecting the process, the resulting change in the performance of the process will identify the cause. This is the essence of interpretation of data as a means for determining and achieving process capability and control. Alternatively, deliberate changes can be made in variables thought to affect the process. The resulting changes in performance identify

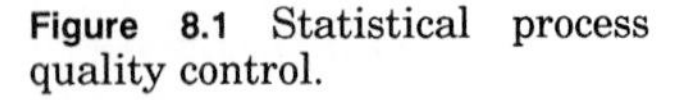

Figure 8.1 Statistical process quality control.

and quantify the real effect of these changes. This is a basic approach in troubleshooting and use of design of experiments in process improvement. In any event, statistical analysis is necessary because it provides a communication link between the process and the investigator which is unbiased and which transcends the variation that bedevils interpretation of what the process is saying.

It is possible, then, to distinguish two types of investigation in application of statistics to process control:

- *Interpretation:* Listen to the process; detect signals from process as to when variables change.
- *Experimentation:* Talk to the process; perturb variables by experimental design.

The key is communication with the process through statistics.

8.3 Process Control

There are many ways to control a process. One way is through experience, but that takes too long. Another is through intuition, but that is too risky. A third approach (all too common) is to assume the process is well-behaved and not to bother with it, but that may lead to a rude awakening. All these have their place but should be used judiciously in support of a scientific approach to achieving and maintaining statistical control through control charts.

It is the philosophy of use of control charts that is so important. The search for an assignable or "special" cause and the measurement of inherent variation brought about by "common" causes are at the heart of that philosophy. The purpose of control is to identify and correct for the assignable causes as they occur and so to keep variation in the process within its "natural" limits. In so doing, the control chart is used to test whether the data represents random variation from stable sources and, if not, to help infer the nature of the source(s) responsible for any nonrandomness.

There are many types and uses of control charts. The chart may be used for "standards given," that is, to maintain future control when previous standards for the mean and standard deviation have been established. Alternatively, they may be used to investigate and establish control from past and current data with "no standards given." Control limits and appropriate factors for variables and attributes charts for use in these situations are shown in Table 8.1. Values of the factors are given in Table A.4.

8.4 Uses of Control Charts

Control charts may be used in making judgments about the process, such as establishing whether the process was in a state of control at a given time. This is useful in determining the capability of the process. Again, they may be used in an ongoing effort to maintain the centering and spread of the process, that is, in maintaining control. The charts may also be used to detect clues for process change. This is at the heart of process improvement and troubleshooting.

TABLE 8.1 Factors for Shewhart Charts, $n = n_g$

	Standards given			
Plot	Mean $\bar{X}$ with μ, σ given	Standard deviation s or range R with σ given	Proportion $\hat{p}$ or number defects $n\hat{p}$ with p given	Defects $\hat{c}$ or defects per unit $\hat{u}$ against c or μ given
Upper control limit	$\mu + 3\sigma/\sqrt{n}$	$s: B_6\sigma$	$\hat{p}: p + 3\sqrt{\frac{p(1-p)}{n}}$	$\hat{c}: c + 3\sqrt{c}$
	$= \mu + A\,\sigma$	$R: D_2\sigma$	$n\hat{p}: np + 3\sqrt{np(1-p)}$	$\hat{u}: \mu + 3\sqrt{\frac{\mu}{n}}$
Center-line	μ	$s: c_4\sigma$ $R: d_2\sigma$	$\hat{p}: p$ $n\hat{p}: np$	$\hat{c}: c$ $\hat{u}: \mu$
Lower control limit	$\mu - 3\sigma/\sqrt{n}$	$s: B_5\sigma$	$\hat{p}: p - 3\sqrt{\frac{p(1-p)}{n}}$	$\hat{c}: c - 3\sqrt{c}$
	$= \mu - A\,\sigma$	$R: D_1\sigma$	$n\hat{p}: np - 3\sqrt{np(1-p)}$	$\hat{u}: \mu - 3\sqrt{\frac{\mu}{n}}$

	No standards given			
Plot	Mean $\bar{X}$ of past data using s or R against past data	Standard deviation s or range R against past data	Proportion $\hat{p}$ or number defective $n\hat{p}$ against past data	Defects $\hat{c}$ or defects per unit $\hat{u}$ against past data
Upper control limit	$s: \bar{\bar{X}} + A_3\bar{s}$	$s: B_4\bar{s}$	$\hat{p}: \bar{p} + 3\sqrt{\frac{\bar{p}(1-\bar{p})}{n}}$	$\hat{c}: \bar{c} + 3\sqrt{\bar{c}}$
	$R: \bar{\bar{X}} + A_2\bar{R}$	$R: D_4\bar{R}$	$n\hat{p}: n\bar{p} + 3\sqrt{n\bar{p}(1-\bar{p})}$	$\hat{u}: \bar{u} + 3\sqrt{\frac{\bar{u}}{n}}$
Center-line	$s: \bar{\bar{X}}$ $R: \bar{\bar{X}}$	$s: \bar{s}$ $R: \bar{R}$	$\hat{p}: \bar{p}$ $n\hat{p}: n\bar{p}$	$\hat{c}: \bar{c}$ $\hat{u}: \bar{u}$
Lower control limit	$s: \bar{\bar{X}} - A_3\bar{s}$	$s: B_3\bar{s}$	$\hat{p}: \bar{p} - 3\sqrt{\frac{\bar{p}(1-\bar{p})}{n}}$	$\hat{c}: \bar{c} - 3\sqrt{\bar{c}}$
	$R: \bar{\bar{X}} - A_2\bar{R}$	$R: D_3\bar{R}$	$n\hat{p}: n\bar{p} - 3\sqrt{n\bar{p}(1-\bar{p})}$	$\hat{u}: \bar{u} - 3\sqrt{\frac{\bar{u}}{n}}$

In all aspects of process control it is desirable to distinguish, with W. E. Deming[3] between two types of study:

- *Enumerative study:* Aim is to gain better knowledge about material in a population.

[3]W. E. Deming, *Some Theory of Sampling,* Dover Publications, Inc., New York, 1966. See especially Chap. 7, "Distinction between Enumerative and Analytic Studies."

- *Analytic study:* Aim is to obtain information by which to take action on a cause system that has provided material in the past and will produce material in the future.

Process control studies are by nature analytic. The objective is to characterize the process at any point in time and not necessarily the product which is being produced. Therefore the sampling procedures are not necessarily those of random sampling from the population or lot of product produced over a given period. Rather, the samples are structured to give sure and definitive signals about the process. This is the essence of rational subgrouping as opposed to randomization. It explains why it is reasonable to take regular samples at specific intervals regardless of the quantity of product produced. It also indicates why successive units produced are sometimes taken as a sample for process control, rather than a random selection over time.

8.5 Rational Subgroups

Use of the control chart to detect shifts in process centering or spread requires that the data be taken in so-called rational subgroups. These data sets should be set up and taken in such a way that variation within a subgroup reflects nonassignable random variation only, while any significant variation between subgroups reflects assignable causes. Experience has shown that the reason for an assignable cause can be found and will give insight into the shifts in process performance that are observed.

Rational subgrouping must be done *beforehand.* Control charts are no better than the effort expended in setting them up. This requires technical knowledge about the process itself. It requires answers to such questions as:

- What do we want the chart to show?
- What are the possible sources of variation in the process?
- How shall nonassignable or random error be measured?
- What should be the time period between which samples are taken?
- What sources can be combined in one chart and which sources should be split among several charts?

The answers to these and similar questions will determine the nature of sampling and the charting procedure.

8.6 Special Control Charts

Table 8.1 shows how to compute control limits for the standard charts with which most people engaged in industrial use of statistics are familiar. There are some charts, however, that are well-suited to specific situations in which process control is to be applied.

8.7 Median Chart

Of particular importance for in-plant control is the median chart, in which the median is plotted in lieu of $\overline{X}$ on a chart for process location. While special methods have been developed for the construction of such charts, using the median range, for example,[4,5] it is very simple to convert standard $\bar{X}$ chart limits to limits for the median. In so doing, familiar methods are used for constructing the $\bar{X}$ limits, they are then converted to median limits, and the person responsible for upkeep of the chart simply plots the median $\tilde{X}$ along with the range R sample by sample. Calculation of the limits is straightforward since it is by the standard methods, which may be available on a calculator or computer. The calculation of the limits is transparent to the operator who plots statistics which are commonplace and rich in intuitive meaning. Even when displayed by a computer terminal at a work station, the median and range charts are meaningful in that they display quantities which are well known to the operator, allowing concentration on the philosophy rather than the mechanics of process control.

The conversion of an $\bar{X}$ chart to a median chart is simplified by Table 8.2 which presents three forms of conversion.

TABLE 8.2 Factors for Conversion of $\bar{X}$ into Median Chart

$n = n_g$	Widen by W	Factor Z_M	Alternate sample size n_M	Efficiency E
2	1.00	3.00	2	1.000
3	1.16	3.48	5	0.743
4	1.09	3.28	5	0.828
5	1.20	3.59	8	0.697
6	1.14	3.41	8	0.776
7	1.21	3.64	11	0.679
8	1.16	3.48	11	0.743
9	1.22	3.67	14	0.669
10	1.18	3.53	14	0.723
11	1.23	3.68	17	0.663
12	1.19	3.56	17	0.709
13	1.23	3.70	20	0.659
14	1.20	3.59	21	0.699
15	1.23	3.70	23	0.656
16	1.20	3.61	24	0.692
17	1.24	3.71	27	0.653
18	1.21	3.62	27	0.686
19	1.24	3.72	30	0.651
20	1.21	3.64	30	0.681
∞	1.25	3.76	$1.57n$	0.637

SOURCE: Computed from efficiencies, E, given by W. J. Dixon and F. J. Massey, Jr., *Introduction to Statistical Analysis,* 2d ed., McGraw-Hill Book Company, New York, 1957. Table A.8*b*4 using the relationships

$$W = \frac{1}{\sqrt{E}} \qquad Z_M = \frac{3}{\sqrt{E}} \qquad n_M = \frac{n}{E}$$

[4] E. B. Farrell, "Control Charts Using Midranges and Medians," *Ind. Qual. Control,* vol. 9, no. 5, March 1953, pp. 30–34.

[5] P. C. Clifford, "Control Charts Without Calculations," *Ind. Qual. Control,* vol. 15, no. 11, May 1959, pp. 40–44.

1. Widen the $\bar{X}$ limits by a multiple W. That is, if the limits are $\bar{\bar{X}} \pm 3\hat{\sigma}/\sqrt{n_g}$, widen them to $\bar{\bar{X}} \pm W(3\hat{\sigma}/\sqrt{n_g})$. Keep the sample size the same.
2. Use the factor Z_M in the place of 3 in the limits for $\bar{X}$. That is, if the limits are $\bar{\bar{X}} \pm 3\hat{\sigma}/\sqrt{n_g}$, use $\bar{\bar{X}} \pm Z_M\hat{\sigma}/\sqrt{n_g}$ and plot medians on the new chart. Keep the sample size the same.
3. Keep the limits the same as for the $\bar{X}$ chart but increase the sample size from n_g to n_M. This is useful when the location of the limits have special significance, such as in converting modified control limits or acceptance control limits.

It should be emphasized that median charts assume the individual measurements upon which the chart is based to have a normal distribution.

Sometimes it is desirable to plot the midrange, that is, the average of the extreme observations in a subgroup rather than the median. This procedure actually has greater efficiency than the median for sample size 5 or less. Standard $\bar{X}$ charts may be converted to use of the midrange by using the factors shown below in Table 8.2A in a manner similar to the conversion of the median.

Conversion of an $\bar{X}$ chart to a median chart may be illustrated using the statistics of mica thickness data compiled in Table 8.3 for $k = 40$ samples of size $n_g = 5$. For each sample the mean $\bar{X}$, median $\tilde{X}$, range R and standard deviation s are shown. Using the means and ranges, the control limits for an $\bar{X}$ and R chart are found to be

$$UCL_{\bar{X}} = \bar{\bar{X}} + A_2\bar{R} = 11.15 + 0.58(4.875) = 13.98$$

$$CL_{\bar{X}} = \bar{\bar{X}} = 11.15$$

$$LCL_{\bar{X}} = \bar{\bar{X}} - A_2\bar{R} = 11.15 - 0.58(4.875) = 8.32$$

and

$$UCL_R = D_4\bar{R} = 2.11(4.875) = 10.29$$

$$CL_R = \bar{R} = 4.875$$

TABLE 8.2A Factors for Conversion of $\bar{X}$ into Midrange Chart

$n = n_g$	Widen by W	Factor Z_{MR}	Alternate sample size n_{MR}	Efficiency E
2	1.00	3.00	2	1.000
3	1.04	3.13	4	0.920
4	1.09	3.28	5	0.838
5	1.14	3.42	7	0.767

SOURCE: Computed from efficiencies, E, given by W. J. Dixon and F. J. Massey, Jr., *Introduction to Statistical Analysis*, 2d ed., McGraw-Hill Book Company, New York, 1957 (Table A.8b4).

TABLE 8.3 Mean, Median, Range and Standard Deviation of Mica Thickness

Sample	1	2	3	4	5	6	7	8	9	10
$\overline{X}$	10.7	11.0	11.9	13.1	11.9	14.3	11.7	10.7	12.0	13.7
$\tilde{X}$	11.5	10.5	12.5	13.5	12.5	14.0	11.5	11.0	11.5	14.5
R	4.0	3.5	7.5	3.5	4.5	5.0	6.0	5.5	4.0	6.0
s	1.72	1.50	3.03	1.56	1.71	1.99	2.49	2.11	1.54	2.64
Sample	**11**	**12**	**13**	**14**	**15**	**16**	**17**	**18**	**19**	**20**
$\overline{X}$	9.8	13.0	11.7	9.6	12.0	11.9	11.7	11.1	10.0	11.0
$\tilde{X}$	9.5	13.0	10.0	9.5	12.5	12.0	11.5	10.0	10.5	10.5
R	3.5	3.0	6.0	8.0	4.5	7.5	4.0	7.5	3.5	5.0
s	1.30	1.12	2.82	3.31	1.73	2.75	1.60	3.05	1.41	1.84
Sample	**21**	**22**	**23**	**24**	**25**	**26**	**27**	**28**	**29**	**30**
$\overline{X}$	12.8	9.7	9.9	10.1	10.7	8.9	10.7	11.6	11.4	11.2
$\tilde{X}$	13.0	10.0	10.5	10.0	10.5	8.5	10.5	11.0	12.0	10.5
R	3.5	6.0	3.5	6.0	5.5	6.5	2.0	4.5	2.0	6.5
s	1.35	2.33	1.56	2.33	2.51	2.63	0.84	1.78	0.89	2.44
Sample	**31**	**32**	**33**	**34**	**35**	**36**	**37**	**38**	**39**	**40**
$\overline{X}$	11.1	8.6	9.6	10.9	11.9	12.2	10.3	11.7	10.1	10.0
$\tilde{X}$	11.0	8.0	10.0	10.5	13.0	12.0	10.0	12.0	9.5	10.0
R	3.5	4.5	5.0	8.5	5.0	7.0	1.5	3.5	4.0	4.5
s	1.29	1.92	1.98	3.27	2.27	3.05	0.67	1.44	1.82	1.70

$$\text{LCL}_R = 0$$

The median chart is plotted in Fig. 8.2.

Point 6 is out of control on the $\overline{X}$ chart while the R chart appears in control against its limits. The $\overline{X}$ chart may be converted to a median chart using

$$\text{UCL}_M = 11.15 + 1.20(0.58)(4.875) = 14.54$$

$$\text{CL}_M = \text{CL}_{\overline{X}} = 11.15$$

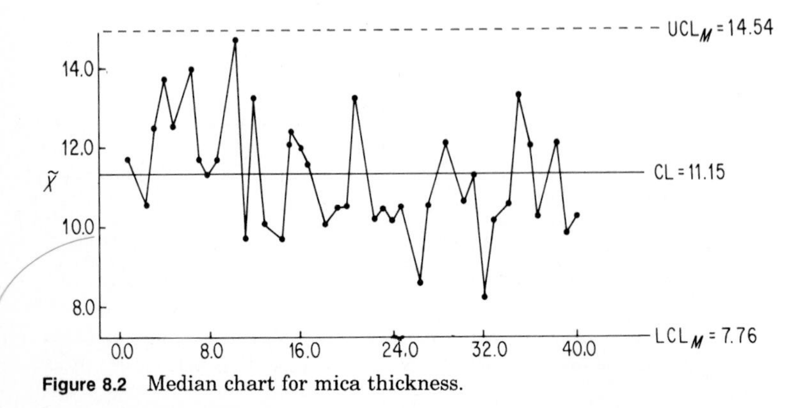

Figure 8.2 Median chart for mica thickness.

$$\text{LCL}_M = 11.15 - 1.20(0.58)(4.875) = 7.76$$

The tenth point is just in control on the median chart as it is on the $\overline{X}$ chart. The sixth point is also barely in control as is the thirty-second point. The fluctuations are roughly the same as in the $\overline{X}$ chart, but the median chart has an efficiency of 90 percent compared to the $\overline{X}$ chart, which accounts for the lack of indication of an out-of-control condition on the sixth point.

8.8 Standard Deviation Chart

As an alternative to the range chart, a standard deviation chart is sometimes calculated. Such charts are particularly well adapted to computation by the computer. Factors for the construction of an s chart are given in Table 8.1. Its construction may be illustrated with the mica data of Table 8.3 as follows:

$$\bar{s} = 1.98$$

$$\text{UCL}_s = B_4\bar{s} = 2.089(1.98) = 4.14$$

$$\text{CL}_s = \bar{s} = 1.98$$

$$\text{LCL}_s = B_3\bar{s} = 0$$

The s chart is shown in Fig. 8.3. The s chart appears in control as does the R chart.

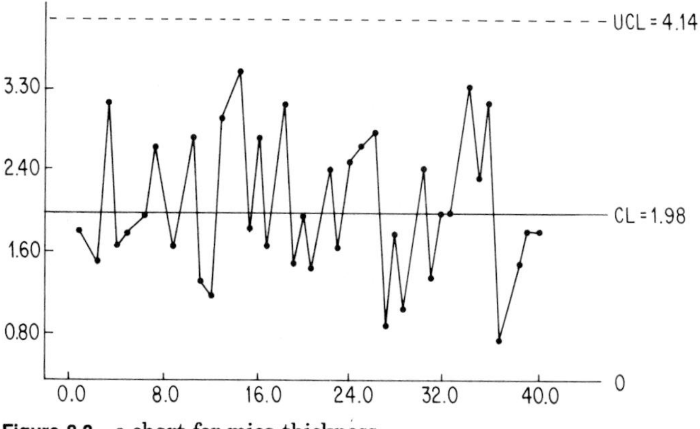

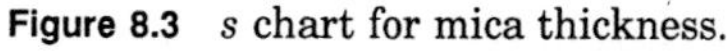

Figure 8.3 s chart for mica thickness.

8.9 Acceptance Control Chart

Acceptance control charts are particularly well adapted to troubleshooting in process control. Unlike the conventional Shewhart chart, the acceptance control chart fixes the risk β of missing a signal when the process mean goes beyond a specified rejectable process level (RPL). It also incorporates a specified risk α of a signal when the process mean is within a specified acceptable process level (APL). Since, in troubleshooting and improvement studies, it is essential that aberrant levels be detected, the acceptance control chart is a versatile tool in that β can be set at appropriate levels. Acceptance control charts can be used for continuing control of a process when it is desirable to fix the risks of a signal. They also serve as an acceptance control device when interest is centered on acceptance or rejection of the process which produced the product, that is, type B sampling for analytic purposes. These charts are ordinarily used when the standard deviation is known and stable. A control chart for R or s is ordinarily run with the acceptance control chart to assure the constancy of process variation.

An acceptance control chart is set up as follows:

1. Determine the sample size as

$$n_g = \left(\frac{(Z_\alpha + Z_\beta)\sigma}{\text{RPL} - \text{APL}}\right)^2$$

where Z_λ is the upper tail normal deviate for probability λ. A few values of λ are

Risk λ	Z_λ
0.10	1.282
0.05	1.645
0.025	1.960
0.01	2.326
0.005	2.576

Note that a risk of $\alpha/2$ should be used if the chart is to be two-sided. The β risk is not divided by 2 since it applies to only one side of a process at any time.

2. Determine the acceptance control limit, ACL, as follows:
   a. APL given

   Upper APL: $\text{ACL} = \text{APL} + Z_\alpha\sigma/\sqrt{n_g}$

   Lower APL: $\text{ACL} = \text{APL} - Z_\alpha\sigma/\sqrt{n_g}$

   b. RPL given

   Upper RPL: $\text{ACL} = \text{RPL} - Z_\beta\sigma/\sqrt{n_g}$

   Lower RPL: $\text{ACL} = \text{RPL} + Z_\beta\sigma/\sqrt{n_g}$

   c. A nominal centerline (CL) is often shown halfway between the upper and lower ACL for a two-sided chart. It is sometimes necessary to work with either the APL or the RPL, whichever is more important. The other value will then be a function of sample size and may be back calculated from the relationships shown. That is why two sets of formulas are given. In normal practice when both the APL and RPL with appropriate risks are set, the sample size is determined and either of the sets of formulas may be used to determine the acceptance control limit.

3. When $\alpha = 0.05$ and $\beta = 0.10$, these formulas become

   a. Single-sided limit

$$Z_\alpha = 1.645 \qquad Z_\beta = 1.282$$

$$n_g = \frac{8.567\sigma^2}{(\text{RPL} - \text{APL})^2}$$

   b. Double-sided limit

$$Z_{\alpha/2} = 1.960 \qquad Z_\beta = 1.282$$

$$n_g = \frac{10.511\sigma^2}{(\text{RPL} - \text{APL})^2}$$

These risks are quite reasonable for many applications of acceptance control charts. The development and application of acceptance control charts are detailed in the seminal paper by R. A. Freund.[6] Application of the procedure to attributes data is discussed by Mhatre, Scheaffer, and Leavenworth.[7]

Consider the mica data given in Table 8.3. For these data, purchase specifications were 8.5 to 15 mils with an industry allowance of 5 percent over and 5 percent under these dimensions. Thus

$$\text{Upper RPL} = 1.05(15) = 15.75$$

$$\text{Upper APL} = 15$$

$$\text{Lower APL} = 8.5$$

$$\text{Lower RPL} = 0.95(8.5) = 8.075$$

The previously calculated s chart indicates that the process is stable with $\bar{s} = 1.98$. But s is a biased estimate of σ as revealed from the factor for the centerline of a standards given s chart based on known σ. We see that

$$\bar{s} = c_4\sigma$$

so, an unbiased estimate of σ is

$$\hat{\sigma} = \frac{\bar{s}}{c_4} = \frac{1.98}{0.9400} = 2.11$$

For risks $\alpha = 0.10$, $\beta = 0.10$ the sample size required is

[6] R. A. Freund, "Acceptance Control Charts," *Ind. Qual. Control,* vol. 14, no. 4, October 1957, pp. 13–23.

[7] S. Mhatre, R. L. Scheaffer, and Richard S. Leavenworth, "Acceptance Control Charts Based on Normal Approximations to the Poisson Distribution," *J. Qual. Tech.,* vol. 13, no. 4, October 1981, pp. 221–227.

$$\text{Upper limit } n_g = \left(\frac{(1.282 + 1.282)}{15.75 - 15.00}2.11\right)^2 = 52$$

$$\text{Lower limit } n_g = \left(\frac{(1.282 + 1.282)}{8.075 - 8.5}2.11\right)^2 = 162$$

Suppose, for convenience we take $n_g = 50$. Then the acceptance control limits, using the RPL formulas are

$$\text{Upper ACL} = 15.75 - 1.282\frac{(2.11)}{\sqrt{50}} = 15.37$$

$$\text{Lower ACL} = 8.075 + 1.282\frac{(2.11)}{\sqrt{50}} = 8.46$$

The risks on the upper limit will be essentially held by this procedure, because we selected the sample size appropriate to that limit. Only the RPL will be held for the lower limit since the ACL was calculated from the RPL formula. The risk at the APL will have changed, but may be back-calculated as follows.

$$\text{ACL} = \text{APL} - Z_\alpha \frac{\sigma}{\sqrt{n_g}}$$

so with some algebra

$$Z_\alpha = \frac{(\text{APL} - \text{ACL})\sqrt{n}}{\sigma}$$

$$Z_\alpha = \frac{(8.50 - 8.46)\sqrt{50}}{2.11}$$

$$Z_\alpha = 0.13$$

and from the normal table

$$\alpha = 0.4483$$

The new lower APL having $\alpha = 0.10$ risk is

$$\text{APL} = \text{ACL} + Z_\alpha \sigma / \sqrt{n_g} = 8.46 + \frac{1.282(2.11)}{\sqrt{50}} = 8.84$$

rather than 8.50. Clearly some tradeoffs are in order. If we proceed with the chart, plotting the mean of successive samples of 50 represented by each row of Table 8.3, the chart would appear as in Fig. 8.4.

8.10 Modified Control Limits

It should be emphasized that acceptance control charts are *not* modified limit control charts in that they incorporate stated risks and allowable *process* lev-

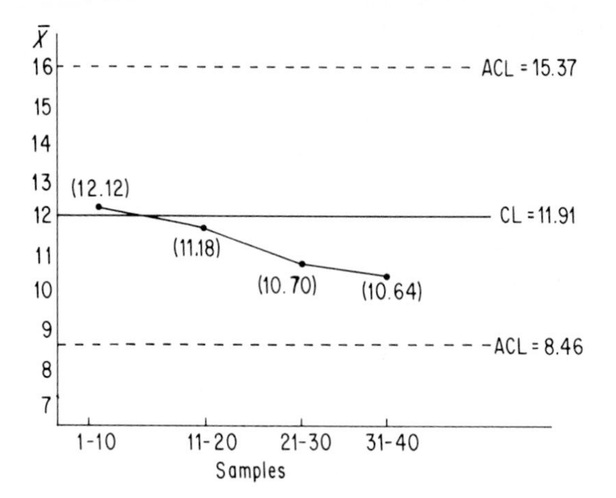

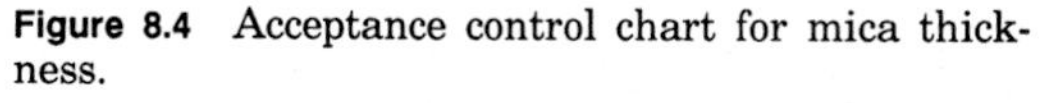

Figure 8.4 Acceptance control chart for mica thickness.

els. Acceptance control charts are oriented toward the process, whereas modified limit charts are designed to detect when nonconforming product is being produced. Modified limits are set directly from the specifications (USL and LSL) as in Fig. 8.5.

The process is assumed normally distributed. The nominal centerline of a modified limit chart is then set 3σ in (on the good side) from the specification limit. The control limit is set back toward the specification limit a distance $3\sigma_{\bar{x}}$. In this way signals will not be given unless the process is centered so close to the specification that nonconforming product may be produced. It will be seen that the modified control limits are simply

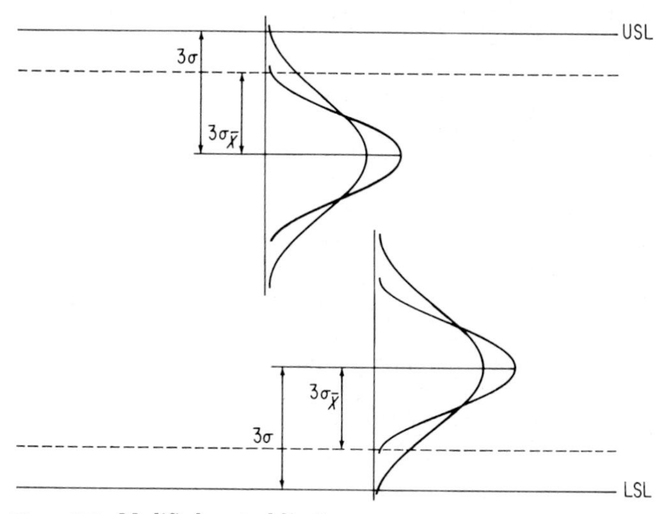

Figure 8.5 Modified control limits.

$$\text{Upper specification: USL} - 3\sigma + 3\frac{\sigma}{\sqrt{n_g}} = \text{USL} - 3\left(1 - \frac{1}{\sqrt{n_g}}\right)\sigma$$

$$\text{Lower specification: LSL} + 3\sigma - 3\frac{\sigma}{\sqrt{n_g}} = \text{LSL} + 3\left(1 - \frac{1}{\sqrt{n_g}}\right)\sigma$$

Clearly there is no consideration as to the selection of sample size or the process levels which might be regarded as acceptable or rejectable with certain risks as in acceptance control charts. If modified limits are used with the mica data of Table 8.3, for samples of five the limits are:

$$\text{Upper modified limit } 15.75 - 3\left(1 - \frac{1}{\sqrt{5}}\right)2.11 = 12.25$$

$$\text{Lower modified limit } 8.075 + 3\left(1 - \frac{1}{\sqrt{5}}\right)2.11 = 11.57$$

Inspection of the means shown for the 50 samples indicates that only 11 samples would produce averages inside the modified limits. This is because the process is incapable of meeting the specification limits, since

$$\frac{\text{USL} - \text{LSL}}{\sigma} = \frac{15.75 - 8.075}{2.11} = 3.64$$

instead of the 6 that would indicate marginal capability.

8.11 Arithmetic and Geometric Moving-Average Charts

Often data is not obtained in successive subsamples but comes naturally in a sequence of single observations. Sometimes the subsample data is lost or not recorded so that all that is available is a series of means or ranges. The daily temperatures listed in the newspaper hour by hour is of this form. Such data is often analyzed by arithmetic moving-average and range charts in which successive subgroups (often of size $k = 2$) are formed by deleting the earliest observation from a subgroup and appending the next available observation to obtain successive arithmetic moving averages. The resulting subgroups can be analyzed (approximately) by the standard methods.

An alternative is the geometric or exponentially weighted moving-average chart developed by Roberts[8] in which a cumulative score is developed which weights the earlier observations successively less than subsequent observations in such a way as to automatically phase out distant observations almost entirely. It is particularly useful for a continuing series of individual observations and is useful when the observations to be plotted on the chart are not independent as in the case of the temperatures.

To set up an exponentially weighted moving-average chart, proceed as follows:

[8]S. W. Roberts, "Control Chart Tests Based on Geometric Moving Averages," *Technometrics,* vol. 1, no. 3, August 1959, pp. 239–250.

1. For each point Z_t plotted at time t, calculate

$$Z_t = rx_t + (1 - r)Z_{t-1}$$

where Z_{t-1} is the weighted value at the immediately preceding time and r is the weight factor $0 < r < 1$ between the immediate observation x_t and the preceding weighted value. Typically, $r = 1/4$.

2. Set limits at

$$\mu_0 \pm 3\sigma\sqrt{\frac{r}{2-r}}\sqrt{1-(1-r)^{2t}}$$

where μ_0 is the target value for the chart. After the first few points the last factor in the formula effectively drops out and the limits become

$$\mu_0 \pm 3\sigma\sqrt{\frac{r}{2-r}}$$

Notice, if $r = 1$ we have a conventional control chart.

A study of Z_t will show that the influence of any observation decreases as time passes. For example

$$\begin{aligned} Z_3 &= rx_3 + (1-r)Z_2 \\ &= rx_3 + (1-r)(rx_2 + (1-r)Z_1) \\ &= rx_3 + (1-r)rx_2 + (1-r)(1-r)rx_1 \end{aligned}$$

This is a geometric progression which, in general, amounts to

$$Z_t = \sum_{i=0}^{t-1} r(1-r)^i x_{t-i}$$

Suppose the arithmetic moving-average and geometric moving-average methods are to be applied to the first 10 points of the $\overline{X}$ data of Table 8.3. We will take $\mu_0 = 11.15$ and $\sigma_X = 2.11/\sqrt{5} = 0.94$ so the limits for $\overline{X}$ are as follows:

Moving average

$$n_g = 2$$

$$\mu_0 \pm 3\sigma/\sqrt{n_g}$$

$$11.15 \pm 3\frac{(0.94)}{\sqrt{2}}$$

$$11.15 \pm 1.99$$

$$9.16 \text{ to } 13.14$$

Geometric moving average

$$r = 1/4$$

$$\mu_0 \pm 3\sigma\sqrt{\frac{r}{2-r}}\sqrt{1-(1-r)^{2t}}$$

$$11.15 \pm 3(0.94)\sqrt{\frac{0.25}{1.75}}\sqrt{1-(0.75)^{2t}}$$

$$11.15 \pm 1.07\sqrt{1-(.75)^{2t}}$$

Giving

t	Limits	UCL	LCL
1	11.15 ± 0.71	10.44	11.86
2	11.15 ± 0.88	10.27	12.03
3	11.15 ± 1.02	10.13	12.17
4	11.15 ± 1.06	10.09	12.21
≥ 5	11.15 ± 1.07	10.08	12.22

Then the moving averages are as follows:

Sample	$\bar{X}$	$\bar{X}$ $n = 2$	Z_t
1	10.7		
2	11.0	10.85	10.78
3	11.9	11.45	11.06
4	13.1	12.50	11.57
5	11.9	12.50	11.65
6	14.3	13.10	12.31
7	11.7	13.00	12.16
8	10.7	11.20	11.80
9	12.0	11.35	11.85
10	13.7	12.85	12.31

The resulting arithmetic moving-average chart shown in Fig. 8.6 does not detect an outage against its limits although sample 6 is just at the limit.

The geometric moving average does, however, detect outages at the sixth and tenth points because of its superior power relative to the simple moving average chart used. This can be seen from Fig. 8-7.

8.12 Cumulative Sum Charts

The cumulative sum (CUSUM) chart provides a very sensitive and flexible vehicle for the analysis of a sequence of individual observations or statistics. Such charts involve plotting the sum of the observations (in terms of devia-

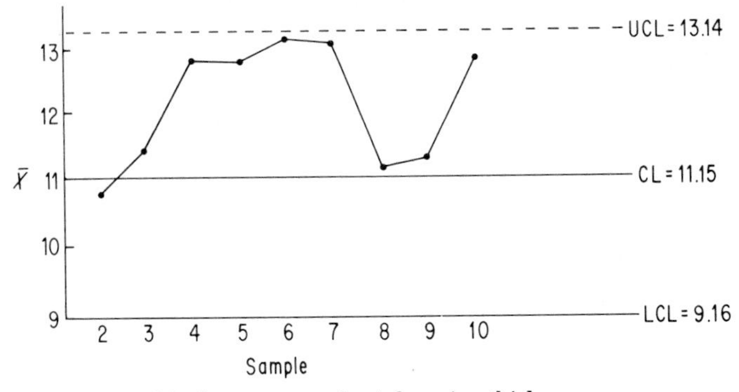

Figure 8.6 Moving average chart for mica thickness.

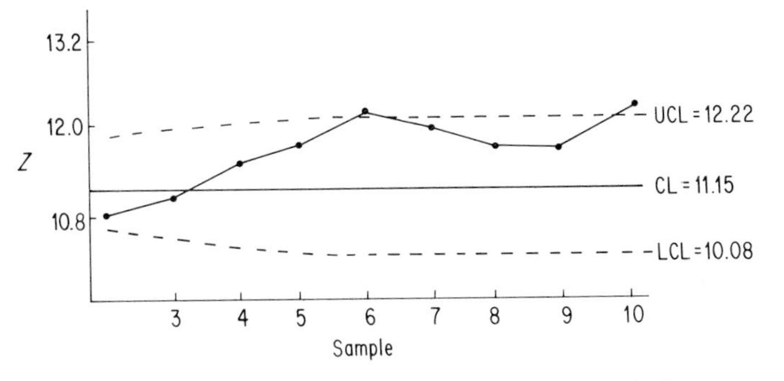

Figure 8.7 Geometric moving average chart for mica thickness.

tions from target) up to a given point against the number of samples taken. The resultant sums when plotted do not often have direct meaning, such as $\bar{X}$ in the Shewhart chart, but are used as an index of the behavior of some parameter of the process. Cumulative sum charts have been shown to be more sensitive than the Shewhart chart for detecting shifts of less than about $3\sigma_{\bar{X}}$ in the mean. The Shewhart chart is more sensitive than CUSUM in detecting departures greater than $3\sigma_{\bar{X}}$ as pointed out by Lucas.[9] Of course, the $3\sigma_{\bar{X}}$ level of sensitivity for sample sizes 4 or 5 was selected by Shewhart because he found that, at that level of a process shift, assignable causes could reasonably be expected to be found by the user. Nevertheless, for purposes of trouble-shooting and process improvement, greater sensitivity is often welcome and the cumulative sum chart has proven to be a parsimonious and reliable tool.

Let us first consider a two-sided approach, suggested by Barnard[10] which

[9]J. M. Lucas, "A Modified 'V' Mask Control Scheme," *Technometrics,* vol. 15, no. 4, November 1973, pp. 833–847.

[10]G. E. A. Barnard, "Control Charts and Stochastic Processes," *J. Roy. Stat. Soc.,* series B, vol. 21, 1959, pp. 239–271.

incorporates use of a V mask superimposed on the plot to assess the significance of any apparent change. The cumulative sum chart for testing the mean simply plots the sum of all the data collected up to a point against time. A V mask is constructed and positioned against the last point at the positioning point indicated on the mask with the bottom of the mask parallel to the x axis. As long as all the previous points remain visible within the cut out portion (or notch) of the mask, the process is regarded as in control. When a point is covered under the solid portion of the mask, or its extension, the process is regarded as out of control. Thus, Fig. 8.8 indicates an out-of-control condition.

To construct a V mask for the process it is necessary to determine the following:

1. μ_0 = APL = acceptable process level
2. α = risk of false signal that process has shifted from APL (use $\alpha = \alpha/2$ for two-sided test)
3. μ_1 = RPL = rejectable process level
4. $D = |\mu_1 - \mu_0|$ = change in process to be detected
5. β = risk of not detecting a change of magnitude D
6. $\sigma_{\bar{x}}$ = standard error of the points plotted ($\sigma_{\bar{x}} = \sigma$ when $n = 1$)
7. $\delta = D/\sigma_{\bar{x}}$ standardized change to be detected
8. w = scaling factor showing the ratio of the y to the x axis (distance of w units on ordinate corresponding to 1 unit length on abscissa) $w = 2\sigma$ is recommended.

Then the mask is determined by specifying its lead distance d and half angle θ as shown in Fig. 8.9.

These quantities are calculated using the following relationships

$$\tan\theta = \frac{D}{2w} = \frac{\delta\sigma_{\bar{x}}}{2w}$$

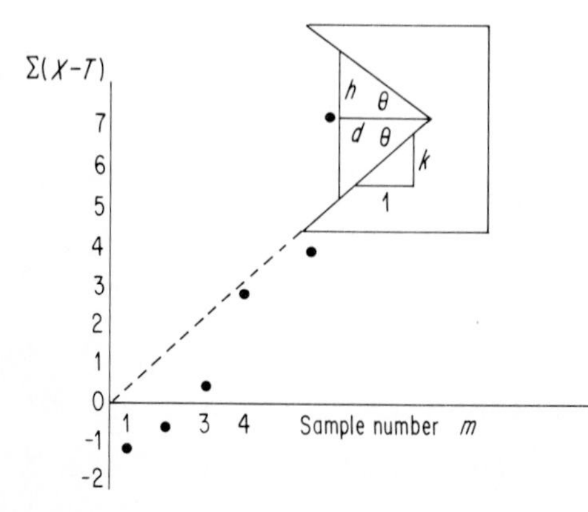

Figure 8.8 CUSUM chart for mica thickness, $d = 1.58$, $\theta = 45°$.

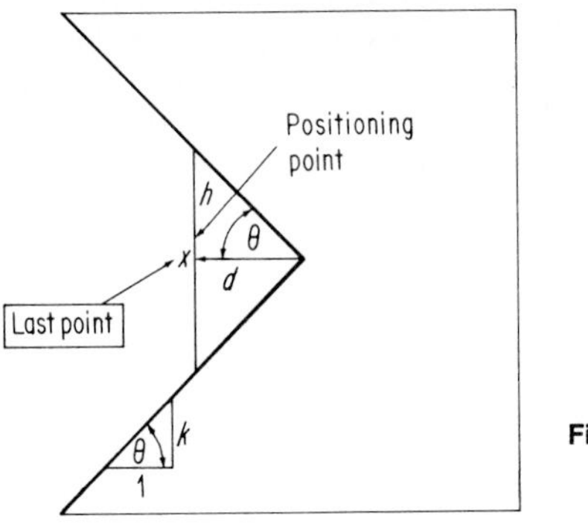

Figure 8.9 V mask.

$$d = \frac{2}{\delta^2} \ln \left(\frac{1 - \beta}{\alpha} \right)$$

Johnson and Leone[11] have noted that for β small (negligible):

$$d = \frac{-2}{\delta^2} \ln \alpha$$

It should be pointed out that the mask may be reparameterized in terms of two other quantities. Some authors use h and k as parameters of the CUSUM procedure

where h = decision interval (intercept of the V mask)
k = reference value (slope of the V mask)

The slope of the V mask, k, corresponds to the centerline of a Shewhart chart in the sense that the observations will climb at slope k if the process is at the target. A shift of k from the goal μ_0 will produce a process level $\mu_0 + k$ having approximate probability of acceptance $\beta \simeq 0.5$. The decision interval h corresponds to the control limits in the sense that for the first observation to give a signal, it must be at a distance $k + h$ above the positioning point of the chart.

These two systems of specifying cumulative sum charts are obviously related. The relationship is as follows

$$h = wd \tan \theta$$

$$k = w \tan \theta$$

so that, as indicated by Ewan[12]

[11]N. L. Johnson and F. C. Leone, "Cumulative Sum Control Charts—Mathematical Principles Applied to their Construction and Use," *Ind. Qual. Control,* part 1, vol. 18, no. 12, June 1962, pp. 15–20; part 2, vol. 19, no. 1, July 1962, pp. 29–36; part 3, vol. 19, no. 2, August 1962, pp. 22–28.

[12]W. D. Ewan, "When and How to Use CU-SUM Charts," *Technometrics,* vol. 5, no. 1, February 1963, pp. 4–22.

$$\tan \theta = k/w$$

$$d = h/k$$

Note that h and k correspond directly to the slope and intercept of sequential sampling plans with

$$k = \text{slope} = s$$

$$h = \text{intercept} = h_2$$

and as pointed out by Schilling,[13] it is possible to utilize this relationship with tables or computer programs for sequential plans by taking

$$\tan \theta = s/w$$

and

$$d = h_2/s$$

Commonly, deviations of the observations from a target value T for the chart are plotted, rather than the observations themselves. T is often taken as the acceptable process level μ_0 for CUSUM charts utilizing the Barnard procedure.

Of course, scaling of the chart is of great importance. If the equal physical distances on the y and x axes are in the ratio $y{:}x = w{:}1$, it is necessary to adjust the half angle so that its tangent is $1/w$ times the former value. This is shown in the formulas given above.

The plot of the cumulative sum can be used to estimate the process average from the slope of the points plotted. The estimate is simply

$$\hat{\mu} = T + (\text{slope})$$

where T is the target value for the chart. The slope can be determined by eye or, alternatively, from the average slope for the last r points of the cumulative sum. If, for a range of r plotted points, S_1 is the first cumulative sum and S_r the last, the process mean may be estimated as

$$\hat{\mu} = T + \frac{S_r - S_1}{r - 1}$$

The time of a process change may be estimated by passing a trend line through the points of trend and observing the sample number at which it intersects the previous stable process line.

To illustrate the use of the cumulative sum chart, consider the first ten means of Table 8.3. Take the target value as $T = \mu_0 = 11.15$ with a standard error of the means as $\sigma_{\bar{X}} = 2.11/\sqrt{5} = 0.94$ and use a scaling factor $w = 1$. The chart will be set up to detect a two-sided difference of $D = 2$ in the mean with $\alpha = 0.05$ and $\beta = 0.10$. Then

[13] E. G. Schilling, *Acceptance Sampling in Quality Control,* Marcel Dekker, Inc., New York, 1982.

$$D = 2, \quad \delta = \frac{2}{0.94} = 2.13$$

$$\frac{\alpha}{2} = 0.025, \quad \beta = 0.10$$

$$\theta = \tan^{-1}\left(\frac{\delta\sigma_{\bar{x}}}{2w}\right) = \tan^{-1}\left(\frac{0.94(2.13)}{2(1)}\right) = \tan^{1}(1) = 45°$$

$$d = \frac{2}{\delta^2}\ln\left(\frac{1-\beta}{\alpha}\right) = \frac{2}{(2.13)^2}\ln\left(\frac{1-0.10}{0.025}\right) = 1.58$$

The data is cumulated as follows

Sample	$\bar{X}$	$\bar{X} - \mu_0$	$\Sigma(\bar{X} - \mu_0)$
1	10.7	−0.45	−0.45
2	11.0	−0.15	−0.60
3	11.9	0.75	0.15
4	13.1	1.95	2.10
5	11.9	0.75	2.85
6	14.3	3.15	6.00
7	11.7	0.55	6.55
8	10.7	−0.45	6.10
9	12.0	0.85	6.95
10	13.7	2.55	9.50

The cumulative sum chart appears in Fig. 8.8. Clearly a shift is detected at the sixth point. A line through the fifth and sixth points when compared to a line through the remainder of the points indicates the shift occurred after the fifth point plotted. The new mean is estimated as

$$\begin{aligned}\hat{\mu} &= T + \frac{S_6 - S_5}{1} \\ &= 11.15 + \frac{6.00 - 2.85}{1} \\ &= 11.15 + 3.15 \\ &= 14.30\end{aligned}$$

It is possible to present the cumulative sum chart in computational form using the reparameterization suggested by Kemp.[14] This approach is directly suitable for computerization, see Lucas.[15] We see from Fig. 8.8 that if a point is to fall below the lower area of the V mask in a subsequence of r points, it must be a distance $h + rk$ below a projection of the lead distance line d at a distance $r + d$ from the positioning point of the mask. So, the difference be-

[14]K. W. Kemp, "The Use of Cumulative Sums for Sampling Inspection Schemes," *App. Stat.,* vol. 11, no. 1, March 1962, pp. 16–31.
[15]J. M. Lucas, "A modified 'V' Mask Control Scheme," *Technometrics,* vol. 15, no. 4, November 1973, pp. 833–847.

tween the cumulative sum at the rth point S_r and at the first point S_1 of the sequence must be such that

$$S_r - S_1 > h + rk$$

$$(S_r - S_1) - rk > h$$

$$\Sigma(\overline{X} - k) > h$$

for the cumulative sum chart to detect an increase in the mean.

Normally, the cumulative sum chart for the mean is plotted in terms of differences from a target value T for the chart. This target value may be thought of as composed of the goal G or aimed at value for the process μ_0, plus a factor k which, as Lucas[16] has pointed out, may be thought of as allowable slack in the process. In the Barnard procedure, $\Sigma(\overline{X} - G)$ is plotted directly and k is taken account of in the mask. In the reparameterization, k is taken account of in the points plotted, so that $\Sigma(\overline{X} - T) = \Sigma(\overline{X} - G - k)$ when plotting against an upper limit and $\Sigma(\overline{X} - T) = \Sigma(\overline{X} - G + k)$ when plotting against a lower limit. Goldsmith and Whitfield[17] have pointed out that for a two-sided procedure, Ewan and Kemp[18] suggest the use of cumulative sums from a goal taken halfway between either the two acceptable or the two rejectable quality levels. Similarly, to detect a decrease in the mean, the difference between S_r and S_1 must diminish by an amount less than $h + rk$. That is

$$S_r - S_1 < -(h + rk)$$

$$(S_r - S_1) + rk < -h$$

$$\Sigma(\overline{X} + k) < -h$$

These relations can be stated in tabular form, which is natural for analysis by the computer.

The computational method requires calculation of two quantities

$$S_H = \sum_{i=1}^{m} ((X_i - G) - k) \text{ for testing against an increase in process level}$$

$$S_L = \sum_{i=1}^{m} ((X_i - G) + k) \text{ for testing against a decrease in process level}$$

They are cumulated in such a way that, if

[16] J. M. Lucas, "The Design and Use of V-Mask Control Schemes," *J. Qual. Tech.*, vol. 8, no. 1, January 1976, pp. 1–12.

[17] P. L. Goldsmith and H. Whitfield, "Average Run Lengths in Cumulative Chart Quality Control Schemes," *Technometrics,* vol. 3, no. 1. February 1961, pp. 11–20.

[18] W. D. Ewan and K. W. Kemp, "Sampling Inspection of Continuous Processes with No Autocorrelation between Successive Results," *Biometrics,* vol. 47, 1960, pp. 363–380.

$S_H < 0$, set $S_H = 0$ until a positive value of $(X_i - G) + k$ is obtained. Then begin cumulating S_H again.

$S_L > 0$, set $S_L = 0$ until a negative value of $(X_i - G) + k$ is obtained. Then begin cumulating S_L again.

These quantities indicate lack of control if

$$S_H > h \text{ or } S_L < -h$$

Lucas[19] has suggested that the computational procedure also keeps track of the number of successive readings r showing the cumulative sum to be greater than zero. When an out-of-control condition is detected, an estimate of the new process average can then be obtained using the relation

Out-of-control high $(S_H > h)$

$$\hat{\mu} = G + \frac{(S_H + rk)}{r}$$

Out-of-control low $(S_L < -h)$

$$\hat{\mu} = G + \frac{(S_L - rk)}{r}$$

The procedure may be illustrated with a computational approach to the cumulative sum chart for the means of mica thickness given in Table 8.3. Recall

$$d = 1.58$$

$$\theta = 45°$$

So

$$\begin{aligned} h &= wd \tan \theta \\ &= 1(1.58) \tan 45° \\ &= 1(1.58)1 \\ &= 1.58 \end{aligned}$$

$$\begin{aligned} k &= w \tan \theta \\ &= 1 \tan 45° \\ &= 1(1) \\ &= 1 \end{aligned}$$

The cumulation, then, is as follows:

[19]J. M. Lucas, "The Design and Use of V-Mask Control Schemes," *J. Qual. Tech.*, vol. 8, no. 1, January 1976, pp. 1–12.

	Increase in mean				Decrease in mean			
Sample	$\bar{X}$	$\bar{X}$-μ_0	$\bar{X}$-μ_0-k	S_H	$\bar{X}$-μ_0+k	S_L	Number high	Number low
1	10.7	−0.45	−1.45	0	0.55	0	0	0
2	11.0	−0.15	−1.15	0	0.85	0	0	0
3	11.9	0.75	−0.25	0	1.75	0	0	0
4	13.1	1.95	0.95	0.95	2.95	0	1	0
5	11.9	0.75	−0.25	0.70	1.75	0	2	0
6	14.3	3.15	2.15	2.85*	4.15	0	3	0
7	11.7	0.55	−0.45	2.40*	1.55	0	4	0
8	10.7	−0.45	−1.45	0.95	0.55	0	5	0
9	12.0	0.85	−0.15	0.80	1.85	0	6	0
10	13.7	2.55	1.55	2.36*	2.55	0	7	0
11	9.8	−1.35	−2.35	0.01	−0.35	− 0.35	8	1
12	13.0	1.85	0.85	0.86	2.85	0	9	0
13	11.7	0.55	−0.45	0.41	1.55	0	10	0
14	9.6	−1.55	−2.55	0	−0.55	− 0.55	0	1
15	12.0	0.85	0.15	0.15	0.15	0	1	0

The procedure detects an out of control condition at points 6, 7 and 10, as shown by asterisks. An estimate of the process mean at point 6 would be

$$\mu = 11.15 + \frac{2.85 + 3(1)}{3} = 11.15 + 1.95 = 13.1$$

Kemp[20] has suggested plotting the above results in the form of a control chart. Such a chart could be one-sided (plotting S_h or S_L only) or two-sided as shown in Fig. 8.10.

A one-sided version of the Barnard chart can be developed using the top or bottom half of the V mask with α rather than $\alpha/2$ as the risk. Usually the arm of the mask is extended all the way down to the abscissa. A one-sided cumulative sum chart testing for an increase in mean of the mica data would be developed as follows with $\alpha = 0.05$, $\beta = 0.10$:

$$\theta = \tan^{-1}\left(\frac{\delta\sigma_{\bar{x}}}{2w}\right) = 45^\circ$$

$$\begin{aligned} d &= \frac{2}{\delta^2} \ln\left(\frac{1-\beta}{\alpha}\right) \\ &= \frac{2}{(2.13)^2} \ln\left(\frac{1-0.10}{0.05}\right) \\ &= 1.27 \end{aligned}$$

Figure 8.11 shows that if such a one-sided test were conducted the process would be found to be out of control at the sixth point.

[20] K. W. Kemp, "The Use of Cumulative Sums for Sampling Inspection Schemes," *Appl. Stat.*, vol. 11, no. 1, March 1962, pp. 16–31.

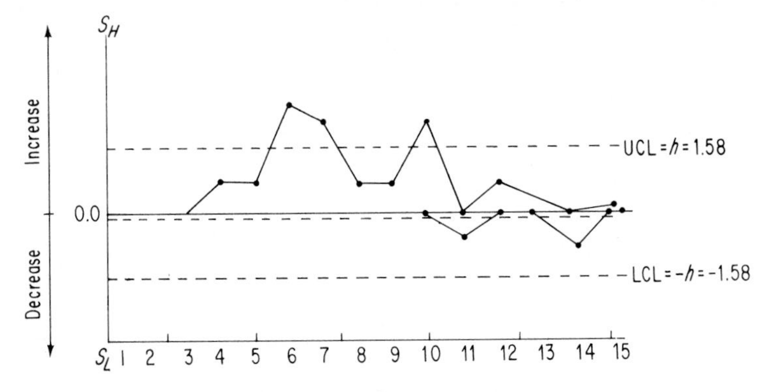

Figure 8.10 Kemp cumulative sum chart.

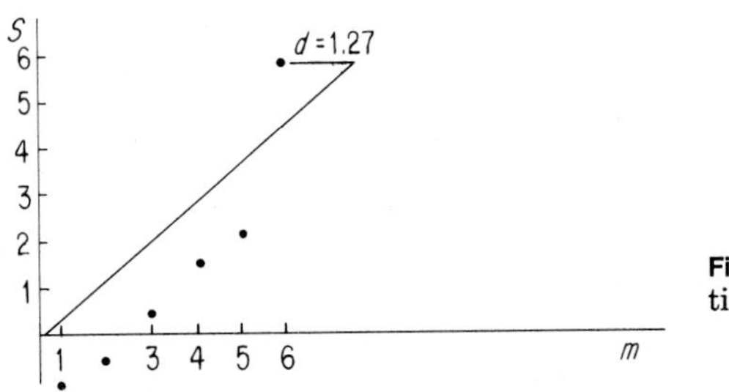

Figure 8.11 One-sided cumulative sum chart.

Lucas[21] has shown that a cumulative sum chart equivalent to the Shewhart chart has

$$h = 0$$

$$k = 3\sigma$$

which implies

$$\tan \theta = 3\sigma/w$$

$$d = 0$$

so that, when $w = 3\sigma$, $d = 0$ and $\theta = 45°$.

A CUSUM chart of this form would appear as Fig. 8.12 for any scaling.

Such a chart seems to have little advantage over a Shewhart chart. In fact Lucas[22] states.

> A V mask...designed to detect large deviations quickly is very similar to a Shewhart chart and if only very large deviations are to be detected, a Shewhart chart is best.

[21]J. M. Lucas, "A Modified 'V' Mask Control Scheme," *Technometrics*, vol. 15, no. 4, November 1973, pp. 833–847.

[22]Ibid.

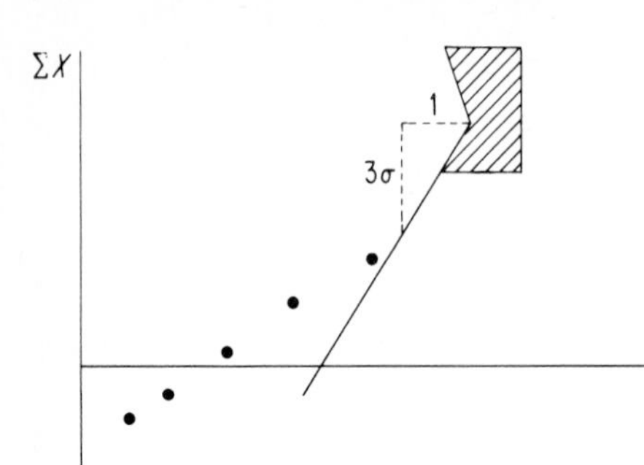

Figure 8.12 CUSUM chart equivalent to Shewhart chart.

Nevertheless, it has been proposed by Ewan[23] that the plan $h = 5$, $k = 0.5$ has properties similar to the Shewhart chart, but possesses more sensitivity in the region of δ less than 3σ and a higher average run length (ARL) at μ_0. This amounts to a Barnard chart with

$$d = 10$$

$$\theta = 26.57° = 26°34'$$

This plan may provide a useful substitute for the Shewhart chart when a cumulative sum chart is in order. It is a good match to the corresponding arithmetic and geometric moving-average chart.

8.13 Other Control Charts

A variety of other control-chart procedures are available for specific applications. Notable among them are adaptive control charts which provide a system of feedback from the data to achieve appropriate adjustment of the process. The interested reader should consult the seminal paper by Box and Jenkins[24] and subsequent literature.

Multivariate control charts are an essential part of the methodology of process control. They provide a vehicle for simultaneous control of several correlated variables. With, say, five process variables of concern, it is possible to run a single T^2 control chart which will indicate when any of them have gone out of control. This can be used to replace the five individual charts used if these variables are treated in a conventional manner. The T^2 chart also incorporates any correlation which may exist between the variables and thus overcomes difficulty in interpretation which may exist if the variables are treated separately. Multivariate methods are discussed in detail by Jackson.[25]

Finally, quality scores are often plotted on a demerits per unit or OPQR (outgoing product quality rating) chart. Such charts provide weights for var-

[23]W. D. Ewan and K. W. Kemp, "Sampling Inspection of Continuous Processes with no Autocorrelation between Successive Results," *Biometrics*, vol. 47, 1960, pp. 363–380.

[24]G. E. P. Box and G. M. Jenkins, "Some Statistical Aspects of Adaptive Optimization and Control," *J. Roy. Stat. Soc.*, vol. 24, 1962, pp. 297–343.

[25]J. E. Jackson, "Principle Components and Factor Analysis," *J. Qual. Tech.*, part 1, vol. 12, pp. 201–213; part 2, vol. 13, pp. 46–58; part 3, vol. 13, pp. 125–130.

ious nonconformities to give an overall picture of quality. See Dodge and Torrey[26] and Frey.[27]

An excellent comparison of control-chart procedures has been given by Freund.[28]

8.14 Precontrol

A process of any kind will perform only as well as it is set up before it is allowed to run. Many processes, once set up, will run well and so need be subject only to occasional check inspections. For these processes, control charts would be overkill. Precontrol[29] is a natural procedure to use on such processes. It is based on the worst-case scenario of a normally distributed process centered between the specifications and just capable of meeting specifications, that is, the difference between the specifications is assumed just equal to a 6σ spread. If precontrol lines are set in from the specifications a distance one-quarter of the difference between the specifications there would then be a 7 percent chance of an observation falling outside the precontrol (PC) lines on one side by normal theory. The chance of two successive points falling outside the precontrol lines on the same side would be

$$P(2 \text{ outside}) = 0.14 \times 0.07 \cong 0.01$$

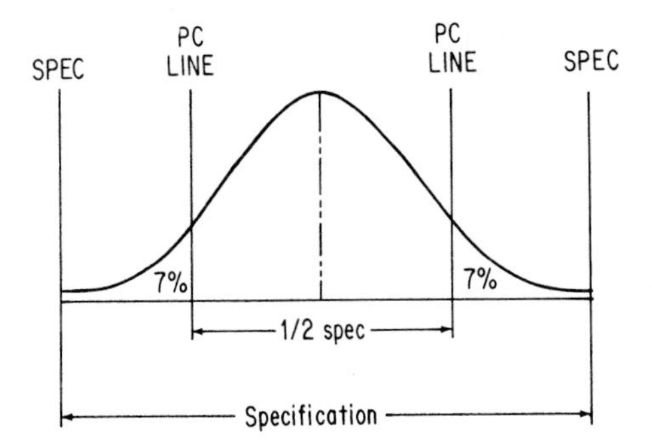

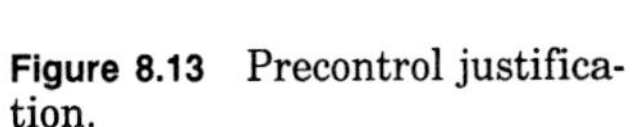

Figure 8.13 Precontrol justification.

This can be seen from Fig. 8.13. This principal is basic to the precontrol approach. A typical set of rules for application of precontrol are given in Table 8.4 and are intended for maintaining an AQL (acceptable quality level) of 1 to 3 percent when the specifications are about 8σ wide. Application of these rules will lead to the diagrammatic representation shown in Fig. 8.14.

As an example, consider the following sequence of initial mica thickness measurements in starting up the process

8.0, 10.0, 12.0, 12.0, 11.5, 12.5, 10.5, 11.5, 10.5, 7.0, 7.0

[26]H. F. Dodge and M. N. Torrey, "A Check Inspection and Demerit Rating Plan," *Ind. Qual. Control,* vol. 13, no. 1, July 1956, pp. 5–12.

[27]W. C. Frey, "A Plan for Outgoing Quality," *Mod. Pack.,* October 1962.

[28]R. A. Freund, "Graphical Process Control," *Ind. Qual. Control,* January 1962, pp. 15–22.

[29]D. Shainin, "Techniques for Maintaining a Zero Defects Program," *AMA Bull. 71,* 1965.

TABLE 8.4 Precontrol Rules

1. Set precontrol lines in 1/4 from the specifications.
2. Begin process.
3. If first piece outside specifications, reset.
4. If first piece outside PC line, check next piece.
5. If second piece outside same PC line, reset.
6. If second piece inside PC lines, continue process and reset only when two successive pieces are outside PC lines.
7. If two pieces are outside opposite PC lines, reduce variation immediately.
8. When 5 successive pieces are inside PC lines, go to frequency gauging and continue as long as average checks to reset is 25.

Frequency guidelines		
Process	Frequency	Process characterization
Erratic	1/50	Intermittently good and bad
Stable	1/100	May have drift
Controlled	1/200	In statistical control

9. During frequency gauging, do not reset until piece is outside PC lines. Then check next piece and go to step 5.

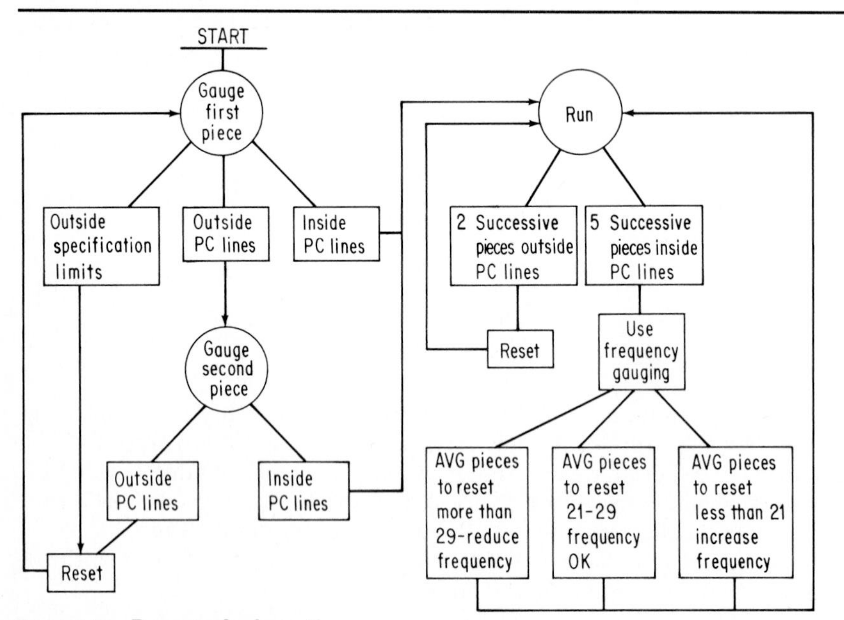

Figure 8.14 Precontrol schematic.

where the specifications are 8.5 and 15. The precontrol procedure would operate as follows:

1. Set precontrol lines at

$$8.5 + \frac{(15 - 8.5)}{4} = 10.1$$

$$15 - \frac{(15 - 8.5)}{4} = 13.4$$

2. Begin process
3. First piece is 8.0, outside specifications
   3*a*. Reset process and begin again
4. Next piece is 10.0, outside lower PC line
5. Second piece 12.0 is within PC lines, so let process run

. . . .

8. Next pieces are 12.0, 11.5, 12.5, 10.5, all within PC line so start frequency gauging, roughly 1 in 50 pieces
   8*a*. Next two sample pieces are 11.5, 10.5. Within PC line so continue sampling
9. Next sample piece is 7.0, outside lower PC line so check next piece
5. Next successive piece is 7.0, reset and start over

This procedure takes advantage of the principle of collapsing the specifications to obtain greater sensitivity in a manner similar to narrow-limit gauging. While it is sensitive to the assumption of a normally distributed process, it is not necessary to know σ as in narrow-limit gauging. Precontrol provides an excellent approach to check inspection which can be used after the control charts are removed.

8.15 Narrow-Limit Control Charts

Narrow-limit gauging (NLG) incorporates a compressed or narrow limit much like the PC line in precontrol. The narrow limit is set a distance $t\sigma$ inside the specification limit when the procedure is to be used for acceptance against the spec. The narrow limit may be set a distance $(3 - t)\sigma$ from the mean when it is to be used as a process-control device, ignoring any specification. The number of observations outside the narrow limit may be used to characterize the process. A control chart may be set up by plotting the resulting count (high or low) against an allowable number c for the sample size n used. The plan $n_g = 5, t = 1, c = 1$ corresponds to a Shewhart chart with sample size 4, while $n_g = 10, t = 1.2, c = 2$ corresponds to a Shewhart chart with sample size 5. When used with the specification limit, such a chart corresponds to a modified limits control chart, while when used with a crude estimate of process spread (3σ) such a chart may be used in tracking the process.

8.16 How to Apply Control Charts

There are a variety of control-chart forms and procedures. Some of these are summarized in Table 8.5 which shows the type of chart, the type of data to

TABLE 8.5 Use of Control Charts

Type	Data	Use	Level
X	Measurement	Rough plot of sequence	B*
$\bar{X}, R$	Measurement	In-plant by operator	B
$\bar{X}, s$	Measurement	In-plant by computer	A*
Median, R, s	Measurement	Excellent introductory tool	B
NLG	Measurement	In-plant ease of gauging with greater sensitivity	B
p	Proportion	Attributes comparison	B
np	Number defective	In-plant attributes	B
c	Defects	In-plant defects	B
u	Defects/unit	Defects comparison	B
CUSUM	Measurement proportion defects	One observation at a time; engineering analysis; natural for computer	A
Moving $\bar{X}, R$	Measurement	In-plant one observation at a time	B
Geometric moving avg.	Measurement	Continuing sequences; no definite period	A
Demerits/unit	Attributes characteristics	Audit	B
Adaptive	Real-time measurements	System feedback and control	A
Acceptance control	Measurements	Fixes risks; combines acceptance sampling and process control	A
T^2	Multiple correlated measurements	Combined figure of merit for many characteristics	A

*B—basic; A—advanced.

which it applies, its use, and the level of sophistication required for effective application.

Effective application of control charts requires that the chart selected be appropriate to the application intended. Some reasonable selections for various alternatives are shown in Table 8.6.

TABLE 8.6 Selection of Chart

Purpose	Data	
	Individuals	Subgroups
Overall indication of quality	CUSUM p,c	Shewhart p,c
Attain/maintain control of attributes	CUSUM p,c	Shewhart p,c
Attain/maintain control of measurement	Moving $\bar{X}$, R; Geometric moving average	Shewhart $\bar{X}$, R; NLG
Attain/maintain control of correlated characteristics	Multivariate control chart	Multivariate control chart
Feedback control	Adaptive	Adaptive
Investigate assignable causes	CUSUM	Analysis of means
Overall audit of quality	Demerits/unit	Demerits/unit
Acceptance with control	CUSUM	Acceptance control

Proper use of control charts requires that they be matched to the degree of control the process has exhibited, together with the extent of knowledge and understanding which has been achieved at a given time. In this way the sophistication and frequency of charting may be changed over time to stay in keeping with the physical circumstances of the process. This progression is shown in Fig. 8.15.

As a process or product is introduced, little is known about potential assignable causes or, in fact, the particular characteristics of the process which require control. At that time, it is appropriate to do 100 percent inspection or screening while data is collected to allow for implementation of more economic procedures. After the pilot plant and start-up phase of production process development, acceptance sampling plans may be instituted to provide a degree of protection against an out-of-control process while at the same time collecting data for eventual implementation of process control. Whenever dealing with a process, acceptance sampling should be viewed as an adjunct and precursor of process control, rather than as a substitute for it.

Sometimes acceptance sampling plans can be used to play the role of a process-control device. When this is done, emphasis is on feedback of information rather than simple acceptance or rejection of lots. Eventually enough information has been gathered to allow implementation of control charts and other process-control devices along with existing acceptance sampling plans. It is at this point that acceptance sampling of lots should be phased out in preference to expanded process control. In its turn, when a high degree of confidence in the process exists, control charts should be phased out in favor of check inspections, such as precontrol, and eventually process checking or no inspection at all. These ideas are illustrated in Fig. 8.16.

It will be seen that there is a life cycle in the application of control charts. Preparation requires investigation of the process to determine the critical variables and potential rational subgrouping. Motivational aspects should be considered in implementation. This is often accomplished by using a team approach while attempting to get the operators and supervisors as much in-

Control \ Process understanding	Little	Some	Extensive
Excellent	$\bar{X}, R$	NLG	Spot check
Average	p, c	$\bar{X}, R$	NLG
Poor	X chart	p, c	$\bar{X}, R$

Figure 8.15 Progression of control charts.

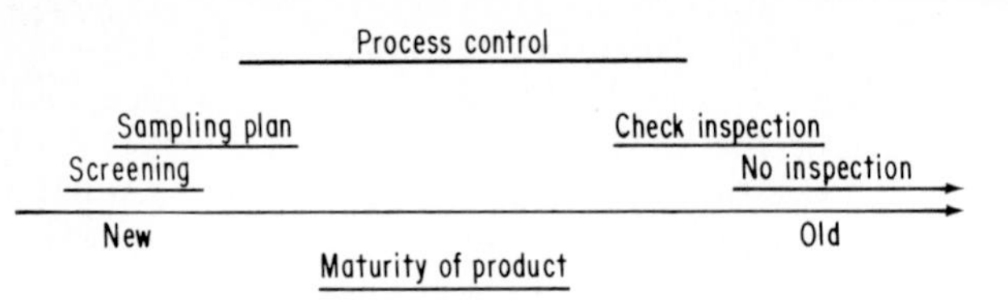

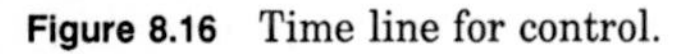

Figure 8.16 Time line for control.

volved as possible. Charts must be changed over the life of the application to sustain interest. A given application might utilize p charts, $\overline{X}$, R charts, median charts, narrow limit charts, and so forth, successively in an effort to draw attention to the application. Eventually, of course, with the assurance of continued control, the charts should be withdrawn in favor of spot checks as appropriate. This is seen in Fig. 8.17.

Certain considerations are paramount in initiation of a control chart including rational subgrouping, type of chart, frequency, and the type of study being conducted. A check sequence for implementation of control charts is shown in Fig. 8.18.

Stage	Step	Method
Preparatory	State purpose of investigation	Relate to quality system
	Determine state of control	Attributes chart
	Determine critical variables	Fishbone
	Determine candidates for control	Pareto
	Choose appropriate type of chart	Depends on data and purpose
	Decide how to sample	Rational subgroups
	Choose subgroup size and frequency	Sensitivity desired
Initiation	Insure cooperation	Team approach
	Train user	Log actions
	Analyze results	Look for patterns
Operational	Assess effectiveness	Periodically check usage and relevance
	Keep up interest	Change chart, involve users
	Modify chart	Keep frequency and nature of chart current with results
Phase out	Eliminate chart after purpose is accomplished	Go to spot checks, periodic sample inspection, overall p, c charts

Figure 8.17 Life cycle of control chart application.

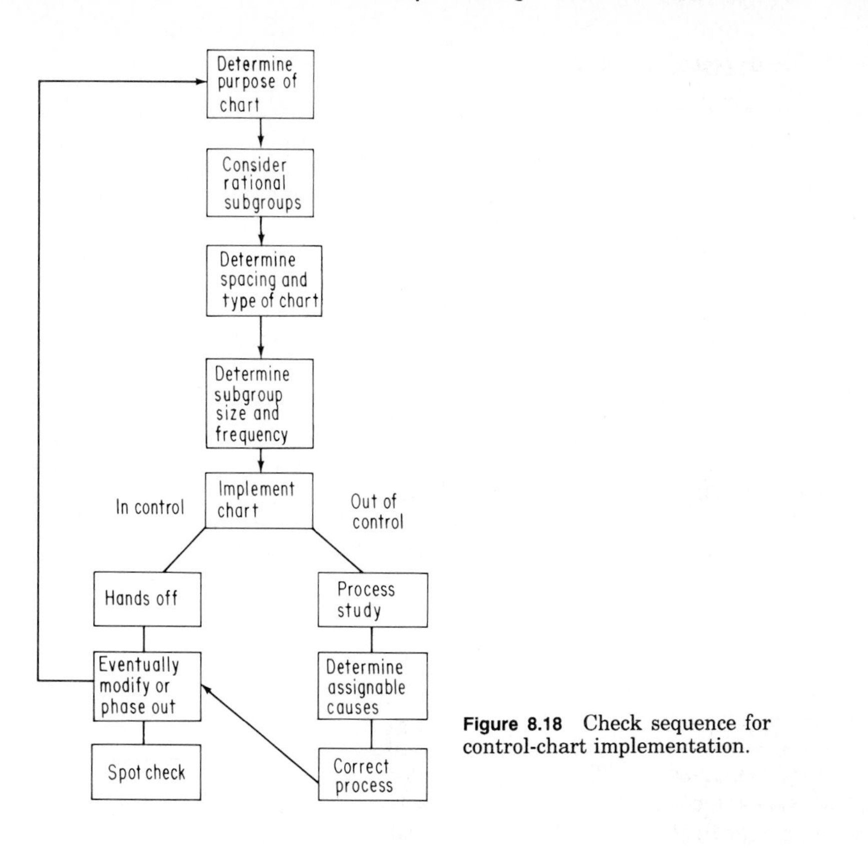

Figure 8.18 Check sequence for control-chart implementation.

Control charts are not a cure-all. It takes a great deal of time and effort to use them properly. They are not appropriate in every situation to which statistical quality control is to be applied. A retailer with a large number of small job-shop vendors is hard put to insist on process control at the source for acceptance of products, since few pieces are made and purchased at any given time. Here acceptance sampling is the method of choice. On the other hand a large firm dealing with a large amount of product from a few vendors is well advised to work with the vendors to institute process control at the source, thus relieving the necessity for extensive incoming inspection. These ideas were well-summarized by Shewhart[30] as follows in his description of the use of control charts in process control

> ...control of this kind cannot be reached in one day. It cannot be reached in the production of a product in which only a few pieces are manufactured. It can, however, be approached scientifically in a continuing mass production.

[30]W. A. Shewhart, *Economic Control of Quality of Manufactured Product,* D. Van Nostrand Company, Inc., New York, 1931.

8.17 Process Capability

What do we mean by the capability of a process? The ATT Statistical Quality Control Handbook[31] states, "The natural behavior of the process after unnatural disturbances are eliminated is called the 'process capability.'" The handbook emphasizes that a process capability study is a systematic investigation of the process by control charts to determine its state of control, checking any lack of control for its cause, and taking action to eliminate any nonrandom behavior when justified in terms of economics or quality. Process capability can, then, never be divorced from control charts or from the concepts of control which Shewhart envisaged. It is basic to the process and may be thought of as inherent process capability. This may be estimated from a range or standard deviation chart on past data, but it can be measured only when the process itself is in control (on the $\overline{X}$ chart also) because of possible synergistic effects on the spread which come about by bringing the average under control. It is not necessary for the process to be normally distributed to use control charts to affect control. Hence, measures of location and spread will not always be found to be independent of each other and so complete control is required to establish the capability of a process.

Again

THERE IS NO CAPABILITY WITHOUT CONTROL

It is important to observe that the inherent capability of a process has nothing to do with specifications. It is a property of the process, not of the print. It is a natural phenomenon which can be estimated with some effort and measured with even more. Process capability studies may be performed on measurements with $\overline{X}$, R charts but also can be accomplished with p or c charts to demonstrate a condition of control. A p chart in control at 3 percent nonconforming says that the inherent capability of that process is 3 percent unless something is done to change the process itself from what it is now.

8.18 Process-Optimization Studies

Process-capability studies are one type of four studies which may be performed in what Mentch[32] calls a process optimization program. These are

- *Process-performance check:* A quick check of the product produced by a process. Based on a small amount of data at a given time, it gives a snapshot of the process performance within a limited time frame. Output is short-run capability. Example: Calculate gas mileage from one tankful.
- *Process-performance evaluation:* A comprehensive evaluation of product produced by the process based on whatever historical data is available, it

[31] B. B. Small (ed.), *ATT Statistical Quality Control Handbook,* ATT, New York, 1956.

[32] C. C. Mentch, "Manufacturing Process Quality Optimization Studies," *J. Qual. Tech.,* vol. 12, no. 3, July 1980, pp. 119–129.

gives a moving picture of how the process has performed in the past. Output is the estimated process capability that could be achieved. Example: Study past records to estimate how good mileage could be.

- *Process-capability study:* An investigation undertaken to actually achieve a state of statistical control on a process based on current data taken in real time, including efforts to achieve control. It gives a live image of control. Output is inherent process capability of controlled process. Example: Study gas mileage and make adjustments until control is achieved.
- *Process-improvement study:* A comprehensive study to improve a process that is not capable of meeting specifications even though in statistical control. It gives a vision of what the process could be and sets out to attain it. Output (if successful) is target capability. Example: Modify car after study by changing exhaust system to increase gas mileage.

It is clear that these studies are sometimes performed individually although they can comprise the elements of a complete process-improvement program.

The process-performance check is conducted in a short time frame. From one day to one week is typical. Based on existing data, the primary tools are frequency distributions, sample statistics, or Pareto analysis of attributes data. It is simple to perform, usually by one person and may lead to quick corrective action.

The process-performance evaluation is a longer term study, typically a few weeks. It is based on existing historical data, usually enough for a control chart. The primary tools are $\overline{X}$, R charts for variables data; p, np, or c charts for attributes data; and sometimes Pareto analysis. Usually done by one person it can lead to relatively quick corrective action. Process-performance evaluations are, however, based on "what has been done" not on "what can be done." This can lead to underestimates of process capability when it is not practical to eliminate assignable causes or to underestimates when synergistic effects exist between level and spread.

The process-capability study is a much longer term study, usually of a month or more. It is conducted using data from current production in an effort to demonstrate inherent process capability, not to estimate or predict it. Tools include $\overline{X}$, R charts for variables and p, np, or c charts for attributes. Process capability is the best in-control performance that an existing process can achieve without major expenditures. Such studies are relatively inexpensive and while it is possible for a single person to perform them, it is normally conducted by a team.

A process-improvement study is usually recommended only after a process-capability study has shown the present process (equipment) to be inadequate. It requires participation by all interested parties from the very beginning. A cost analysis should be performed at the outset since such studies can be expensive. A working agenda should be drawn up and control charts kept throughout to verify improvements when they occur. Tools here include design of experiments, regression, correlation, evolutionary operations (EVOP), and other advanced statistical techniques. This is almost always a team project with management leadership.

Proper use of these studies will identify the true capability of the process.

They can be used progressively as necessary in an improvement program with each study leading to further process optimization. The studies are intended to pinpoint precise areas for corrective action and bring about cost savings through yield improvement. Such studies often result in cost avoidance by preventing unnecessary expenditures on new processes or equipment. Don't throw out the old process until you have a realistic estimate of what it can do!

8.19 Capability and Specifications

The capability of a process is independent of any specifications that may be applied to it. Capability represents the natural behavior of the process after unnatural disturbances are eliminated. It is an inherent phenomenon and is crudely measured by the 6σ spread obtained by using the estimate of standard deviation from an in-control chart for variation.

The thrust of modern quality control is toward reduction of variation. This follows the Japanese emphasis on quality as product uniformity about a target rather than simple conformance to specifications. Thus, process capability becomes a key measure of quality and must be appropriately and correctly estimated.

While it is true that a product with less variation around nominal is, in a sense, better quality, specifications will probably never be eliminated; for specifications tell us how much variation can be tolerated. They provide an upper limit on variation which is important in the *use* of the product, but which should be only incidental to its manufacture. The objective of manufacture should be to achieve nominal, for the same product can be subjected to different specifications from various customers. Also, specifications are not stable over time. They have a tendency to shrink. The only protection the manufacturer has against this phenomenon is to strive for product as close to nominal as possible in a constant effort toward improvement through reduction in variation. Otherwise, the best marketing plan can be defeated by a competitor who has discovered the secret of decreased variation.

When a duplicate key is made, it is expected to fit. If it is too thick, it will not fit. If it is too thin, it may snap off. If it is at nominal, the customer will be pleased by its smooth operation. If quality is measured by the tendency to repurchase, the user will avoid purchase when product has been found out of spec, but the customer will be encouraged to repurchase where product is made at nominal. Thus, specifications should be regarded as an upper bound or flag beyond which the manufacturer should not trespass. But nominal product is the hallmark of a quality producer.

The idea of relating specifications to capability is incorporated in the capability index, C_p, where

$$C_p = \frac{\text{Spread of specifications}}{\text{Process spread}} = \frac{\text{USL} - \text{LSL}}{6\sigma}$$

A process just meeting specifications has $C_p = 1$. Sullivan[33] has pointed out

[33]L. P. Sullivan, "Reducing Variability: A New Approach to Quality," *Qual. Prog.*, vol. 17, no. 7, July 1984, pp. 15–21.

that the Japanese regard $C_p = 1.33$ as a minimum, which implies an 8σ spread in the specifications, with $C_p = 1.66$ preferred.

Values of 3, 5, and 8 can be found. There is, in fact, a relation between the C_p index, acceptable quality level (AQL), and parts per million (PPM) as follows:

C_p	AQL(%)	AQL(PPM)
0.5	13.36	130000
0.75	2.44	24400
1.00	0.26	2500
1.25	0.02	200
1.33	0.003	30
1.50	0.001	10
1.63	0.0001	1.0
1.75	0.00001	0.1
2.00	0.0000002	0.002

This correctly implies that the way to achieve quality levels in the range of parts per million is to work on the process to achieve C_p in excess of 1.33. This can be done and is being done through the methods of statistical process control.

A rough guess at the C_p index will indicate the type of process optimization study that may be appropriate. We have

C_p	Study
< 1	Process improvement
1–1.3	Process capability
1.3–1.6	Performance evaluation
> 1.6	Performance check

Consider the mica thickness data. Since the s chart is in control, we estimate process capability as $\hat{\sigma} = s/c_4 = 2.11$. The spread in the specifications is $(15 - 8.5) = 6.5$ so

$$C_p = \frac{6.5}{6(2.11)} = 0.47$$

Clearly, the process is inferior. A process improvement study is definitely called for.

Note that use of C_p implies the ability to hold the mean at nominal, that is, a process in control. When the process is centered away from nominal, the standard deviation used in C_{pk} is sometimes calculated using nominal μ_0 in place of the average of the data so that

$$\hat{\sigma} = \sqrt{\frac{\Sigma(x - \mu_0)^2}{n - 1}}$$

This will give an inflated measure of variability and decrease C_p.

When a single specification limit is involved, or when the process is deliberately run off-center for physical or economic reasons, the C_{pk} index is used. Here

$$C_{pk} = \frac{\text{USL} - \bar{x}}{3\sigma}$$

and/or

$$C_{pk} = \frac{\bar{x} - \text{LSL}}{3\sigma}$$

When the process is offset and two-sided specification limits are involved, the capability index is taken to be the minimum of these two values.

CASE HISTORY 8.1

The Case of the Schizophrenic Chopper

Introduction

A plant was experiencing too much variation in the length of a part which was later fabricated into a dimensionally critical component of an assembly. The part was simply cut from wire which was purchased on spools. Two spools were fed through ports into a chopper, one on the left and one on the right, so that two parts could be cut at one blow. The parts then fell into a barrel, which was periodically sampled.

A histogram of a fifty-piece sample from the barrel showed $\overline{X}$ = 49.56 mils with a standard deviation of 0.93 mils. The specifications were 44.40 ± 0.20 mils. Clearly, this check showed process performance to be inadequate. A process-performance study of past samples showed several points out of control for the range and wide swings in the mean well outside the control limits.

A new supervisor was assigned to the area and took special interest in this process. It was decided to study its capability. A control chart for samples of size 5 was set up on the process and confirmed the previous results.

One day the control chart exhibited excellent control. The mean was well-behaved and the ranges fell well below the established centerline of $\overline{R}$ = 2.2 What had happened? The best place to find out was at the chopper. But this was a period of low productivity because wire was being fed in from one side only. The other side was jammed. Perhaps that had an effect. Perhaps it was something else.

It was then that the supervisor realized what had happened. Each side of the chopper was ordinarily set up separately. That would mean that any drift on either side would increase the spread of the product and, of course, shift the mean. What he had learned about rational subgrouping came back to him. It would be sensible to run control charts on each side separately. Then they could be adjusted as needed and be kept closer to nominal. Closer to nominal on the two sides meant less overall variation in the product and better control of the mean. This could well be the answer.

Control charts were set up on the two sides separately. They stayed in reasonable control and were followed closely so that adjustments would be made when they were needed (and not when they were not needed). Assignable causes were now easier to find because the charts showed which side of the chopper to look at. The control charts for the range eventually showed $\overline{R}$ = 0.067 for the right side and $\overline{R}$ = 0.076 for the left. The mixed product had $\hat{\sigma}$ = 0.045. This gave a capability index of C_p = 0.40/0.27 = 1.48. The sorting operation was discontinued and the product had attained a uniformity beyond the hopes of anyone in the operation. All this at a cost of an additional five samples plotted for the second chart. This is an example of what can be done with statistical process control when it is properly applied.

8.20 Process Improvement

Control charts are the method of choice in conducting process optimization programs. They sort out the assignable or special causes of variation in the process which can be corrected on the floor from those random or common causes which only redefinition of the process by management can correct.

Dr. Deming[34] has indicated that most people think the job of statistical process control has ended when the process is in control. But that is only the beginning, for he emphasizes that this is the time to concentrate on elimination of the common causes.

This is not as difficult as it may appear and, in some sense, comes naturally in the life cycle of a product or process. During the development phase of a process there are often violent swings in average and spread. Since process variation is usually more stable than the average, the spread can be expected to be controlled at some stable level sometime during the introduction of the product. Thereafter, as the process matures using process-control techniques, the average comes under control and the process is reasonably stable. This provides an excellent opportunity for innovation, for now the erratic state of lack of control has been eliminated, variation is stable, and meaningful process-improvement studies can be performed. It is at this point that changes in the process can be implemented that will lead to further reduced variation with still better process control.

8.21 Process Change

Statistical control of a process is not, in itself, an end of process control. The objective is, as pointed out by Shewhart, to obtain satisfactory, adequate, dependable, economic quality. It is sometimes the case that a process produces inadequate quality even though it is in a state of control. Natural variation may exceed the span of the specifications. The variation may be acceptable, but it may not be possible to center the mean because of tradeoffs with other variables. It is precisely in this situation that a process-improvement program is appropriate to bring about change in the process—that is, a new process—such that the desirable attributes of quality are achieved. This may include new equipment, improved process flow, better raw materials, personnel changes, and so forth, and may be quite expensive. No process-improvement program should be undertaken without a process-capability study to show the effort is justified. Such a program is usually a team effort consisting of representatives from manufacturing, quality, engineering, and

[34]W. E. Deming, "Quality Productivity and Competitive Position," MIT Center for Advanced Engineering Study, Cambridge, Mass., 1982.

headed by a member of management. Mentch[35] has outlined the steps in a process-improvement study as follows:

1. Develop a formal work agenda for the team selected to perform the study, including components to be worked on, priorities, responsibilities, and a completion schedule. Compile cost analysis at every stage.
2. Determine critical problem areas through cause-and-effect analysis and perform a Pareto analysis to show where effort should be directed.
3. Utilize statistically designed experiments, EVOP, and so forth, to show what actions will be required to correct problem areas.
4. Continue control charts from the previous process-capability study to show the effect of changes made.
5. Conclude the program when the process is in control, running at an acceptable rate and is producing product that meets specifications so that further expenditures are not justified.
6. Institute continuing controls to ensure that the problems do not reappear.

8.22 Problem Identification

Initial brainstorming sessions are aided by listing possible causes of a problem on a "cause-and-effect" or "fishbone" diagram developed by Professor Kaoru Ishikawa in 1950. Its Japanese name is Tokusei Yoinzu, or characteristics diagram. It displays the characteristics of a problem that lead casually to the effect of interest.

Often, in problem solving, the skeletal framework is laid out in terms of the generic categories of man (operator), machine, material, and method. Included in method is management, although this is sometimes split out.

As a somewhat prosaic example of use of this technique, suppose you are confronted with burned toast for breakfast. Consider the possible causes. They can be laid out using the cause-and-effect diagram as shown in Fig. 8.19. List-

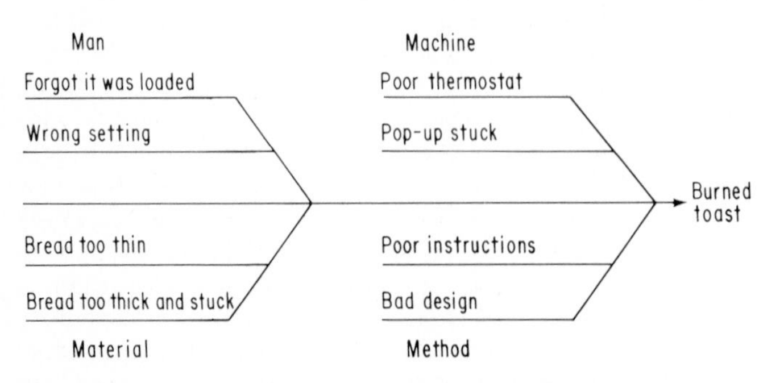

Figure 8.19 Cause-and-effect diagram for burned toast.

[35] C. C. Mentch, "Manufacturing Process Quality Optimization Studies," *J. Qual. Tech.*, vol. 12, no. 3, July 1980, pp. 119–129.

ing the causes in this way facilitates their identification and prepares for Pareto analysis to assess relative importance.

8.23 Prioritization

Pareto analysis addresses the frequency distribution associated with various causes. Since the causes are normally nominal variables, the frequency distribution is ordered from highest frequency to lowest. The resulting histogram and cumulative frequency distribution are plotted to give a visual representation of the distribution of causes. This will help separate out the most important causes to be worked on.

Vilfredo Pareto (1848–1923) studied the distribution of wealth in Italy and found that roughly 20 percent of the population had 80 percent of the wealth. But also, in marketing it was later found that 20 percent of the customers account for roughly 80 percent of the sales. In cost analysis, 20 percent of the parts contain roughly 80 percent of the cost, and so forth.

Juran[36] was the first to identify this as a universal principle which could be applied to quality and distinguish between what he called the "vital few" and the "trivial many."

A typical application is the number of defects found in pieces of pressed glass over the period of a month (see Table 8.7).

The resulting cumulative distribution of the causes is plotted in Fig. 8.20. Note that the bars shown correspond to the histogram of the causes.

8.24 A Statistical Tool for Process Change

The secret of process change is not only in analysis, but also in action. It is not enough to find the cause for a problem, the solution must be implemented. Controls must be set up to insure the problem does not occur again. There is no better way to get action than to have relevant data presented in a way that anyone can understand. And there is no better way to understand and present the data than with a graph. In other words, "plot the data."

A process-improvement program requires that sources of variability, normally concealed in random error, be identified. The search is for common causes. This would imply changes in level less than $3\sigma_{\bar{X}}$ for the sample size used in assessing capability. This search can be accomplished by observation and interpretation of data using cumulative sum charts or Shewhart charts made more sensitive through runs analysis, warning limits, increased sample size, and the like. But this is very difficult since it is not easy to associate a small change in level detected by such a chart with its cause. This is precisely why Shewhart recommended 3σ with small sample sizes of 4 or 5. This means that to achieve process improvement after control is attained one must resort to statistically designed experiments rather than interpretation of data.

[36]J. M. Juran, "Pareto, Lorenz, Cournot, Benoulli, Juran and Others," *Ind. Qual Control,* vol. 17, no. 4, October 1960, p. 25.

TABLE 8.7 Pressed-Glass Defects*

	Defects	Percent of all defects found
Man (operator)		
A-Jack	36	0.27
B-Lucille	69	0.51
C-Carl	66	0.50
D-Dan	3317	24.52
Machine		
A-left	1543	11.41
B-middle	95	0.70
C-right	120	0.89
Material		
A-supplier 1	1126	8.32
B-supplier 2	2822	20.86
C-supplier 3	225	1.66
Method		
A-design 1	3799	28.08
B-design 2	159	1.18
C-design 3	35	0.26
Miscellaneous (12 items)	116	0.86
Total	13528	

*Courtesy of C. C. Mentch.

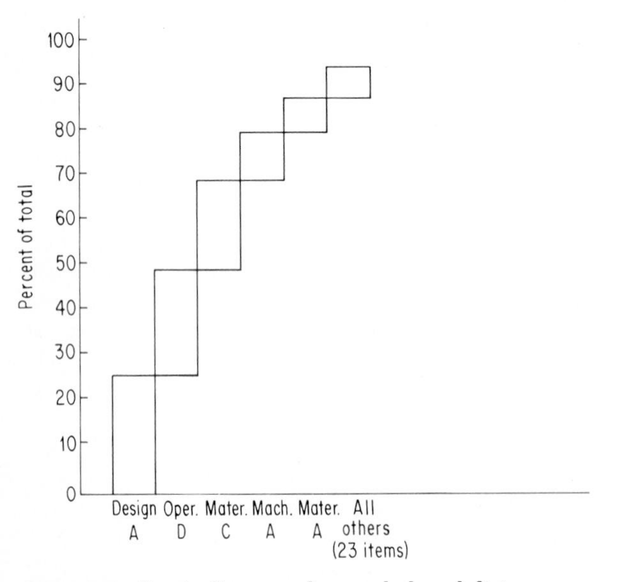

Figure 8.20 Pareto diagram of pressed-glass defects.

There are a variety of methods used in the analysis and presentation of the results of designed experiments, but none is more appropriate to the analysis and presentation of data to nonstatisticians than analysis of means. The reason for this is that it is graphical. It uses the Shewhart chart as a vehicle for the analysis and presentation of data. As such it has all the advantages of Shewhart procedure. The limit lines are slightly more difficult to calculate, but this difficulty is transparent to the observer of the chart. The simple statement that, if there is no real difference between the experimental treatments the odds are 1 to $(1 - \alpha)/\alpha$ that the points will plot within the limits, will suffice for statistical interpretation. From that point, the chart and its use are as familiar from plant to boardroom as a Shewhart control chart.

The control chart has often been used in the exposition and analysis of designed experiments. Excellent early examples may be seen in the work of Dr. Grant Wernimont.[37,38] The analysis of means as developed by Ott[39] is such a procedure, but differs from the traditional control-chart approach in that probabilistic limits are employed which adjust for the compound probabilities encountered in comparing many experimental points to one set of limits. Thus, this method gives an exact estimate of the α risk over all the points plotted.

8.25 Analysis of Means for Measurement Data

The basic procedure is useful in the analysis of groups of points on a control chart, in determining the significance and direction of response of the means resulting from main effects in an analysis of variance, as a test for outliers, and in many other applications. Steps in the application of analysis of means to a set of k subgroups of equal size, n_g, are as follows:

1. Compute $\overline{X}_i$, the mean of each subgroup, for $i = 1, 2, \ldots, k$
2. Compute $\overline{\overline{X}}$, the overall grand mean
3. Compute $\hat{\sigma}_e$, the estimate of experimental error as

$$\hat{\sigma}_e = \frac{\overline{R}}{d_2^*}$$

where $\overline{R}$ = the mean of the subgroup ranges
d^* = Duncan's adjusted d_2 factor given in appendix Table A.11

with degrees of freedom, df = v, as shown in Table A.11. For values of d_2^* not shown, the experimental error can be estimated as

$$\hat{\sigma}_e = \frac{\overline{R}}{d_2}\left[\frac{k(n_g - 1)}{k(n_g - 1) + 0.2778}\right]$$

[37] G. Wernimont, "Quality Control in the Chemical Industry II: Statistical Quality Control in the Chemical Laboratory," *Ind. Qual Control,* vol. 3, no. 6, May 1947, p. 5.

[38] G. Wernimont, "Design and Interpretation of Interlaboratory Studies of Test Methods," *Anal. Chem.,* vol. 23, November 1951, p. 1572.

[39] E. R. Ott, "Analysis of Means," *Rutgers U. Stat. Cent., Tech. Rep. no. 1,* August 10, 1958.

where d_2 = factor for calculating the centerline of an R control chart when σ is known

The degrees of freedom for the estimate may then be approximated by

$$v \approx 0.9\,k(n_g - 1)$$

Clearly other estimates of σ may be used if available.

4. Plot the means $\overline{X}_i$ in control chart format against decision limits computed as

$$\overline{\overline{X}} \pm H_\alpha \frac{\hat{\sigma}_e}{\sqrt{n_g}}$$

where H_α = Ott's factor for analysis of means given in appendix Table A.8

5. Conclude the means are significantly different if any point plots outside the decision limits.

The above procedure is for use when no standards are given, that is, when the mean and standard deviation giving rise to the data are not known or specified. When standards are given, i.e., in testing against known values of μ and σ, the H_α factor shown in Table A.8 in the row marked SG should be used. The limits then become:

$$\mu \pm H_\alpha \frac{\sigma}{\sqrt{n_g}}$$

The values for H_α shown in appendix Table A.8 are exact values for the studentized maximum-absolute deviate in normal samples as computed by L. S. Nelson[40] and were adapted by D. C. Smialek[41] for use here to correspond to the original values developed by Ott. Values for $k = 2$ and SG are as derived by Ott[42].

8.26 Example—Measurement Data

Wernimont[43] presented the results of a series of Parr calorimeter calibrations on eight different days. The results of samples of four in BTU/lb/°F are:

[40] L. S. Nelson, "Exact Critical Values for Use with the Analysis of Means," *J. Qual. Tech.*, vol. 15, no. 1, January 1983, pp. 40–44.

[41] E. G. Schilling and D. Smialek, "Simplified Analysis of Means for Crossed and Nested Experiments," *43d Ann. Qual. Control Conf.*, Rochester Section, ASQC, March 10, 1987.

[42] E. R. Ott, "Analysis of Means," Rutgers U. Stat. Cent. Tech. Rep., no. 1, August 10, 1958.

[43] G. Wernimont, "Quality Control in the Chemical Industry II: Statistical Quality Control in the Chemical Laboratory," *Ind. Qual. Control*, vol. 3, no. 6, May 1947, p. 5.

Day	$\overline{X}$	R
1	2435.6	13.5
2	2433.6	11.3
3	2428.8	10.0
4	2428.6	5.6
5	2435.9	18.7
6	2441.7	13.6
7	2433.7	9.9
8	2437.8	14.8
Average	2434.5	12.2

Analysis of means limits to detect differences between days, with a level of risk of $\alpha = 0.05$, may be determined as follows:

- An estimate of the experimental error is

$$\hat{\sigma}_e = \frac{\overline{R}}{d_2^*} = \frac{12.2}{2.08} = 5.87$$

$$\text{df} = 22.1 \sim 22$$

- The analysis of means limits are

$$\overline{\overline{X}} \pm H_\alpha \frac{\hat{\sigma}_e}{\sqrt{n_g}}$$

$$2434.5 \pm 2.80 \frac{5.87}{\sqrt{4}}$$

$$2434.5 \pm 8.22$$

The analysis of means plot is shown in Fig. 8.21 and does not indicate significant differences in calibration over the eight days at the 0.05 level of risk.

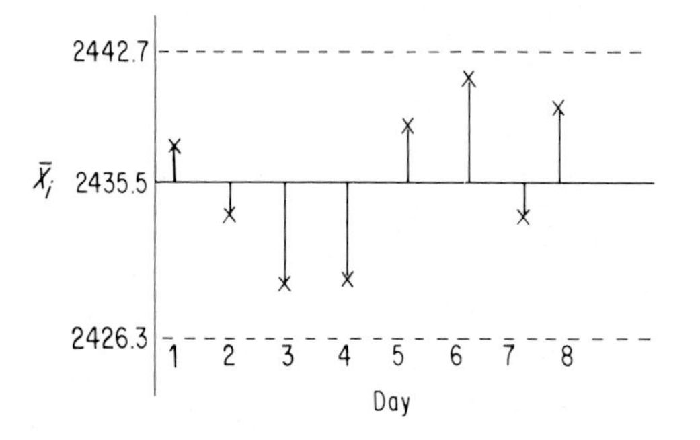

Figure 8.21 Analysis of means plot; Parr calorimeter determination.

8.27 Analysis of Means for Proportions

Lewis and Ott[44] applied analysis of means to binomially distributed data when the normal approximation to the binomial distribution applies. The procedure for k sample proportions from samples of equal size, n, when no standards are given, is as follows:

1. Compute $\hat{p}_i$, the sample proportion, for $i = 1, 2, \ldots, k$
2. Compute $\bar{p}$, the overall mean proportion
3. Estimate the standard error of the proportions by

$$\hat{\sigma}_e = \sqrt{\frac{\bar{p}(1 - \bar{p})}{n_g}}$$

Regard the estimate as having infinite degrees of freedom.

4. Plot the proportions, p_i, in control chart format against decision limits computed as:

$$\bar{p} \pm H_\alpha \sqrt{\frac{\bar{p}(1 - \bar{p})}{n_g}}$$

The H_α factor is obtained from appendix Table A.8 using df $= \infty$.

5. Conclude the proportions are significantly different if any point plots outside the decision limits.

When standards are given, i.e., when testing against a known or specified value of p, use the above procedure with $\bar{p}$ replaced by the known value of p and H_α taken from the row labeled SG in appendix Table A.8.

8.28 Example—Proportions

Hoel[45] poses the following problem:

> Five boxes of different brands of canned salmon containing 24 cans each were examined for high-quality specifications. The number of cans below specification were respectively 4, 10, 6, 2, 8. Can one conclude that the 5 brands are of comparable quality.

To answer the question by analysis of means, the procedure would be as follows, using the $\alpha = 0.05$ level of risk:

- The sample proportions are 0.17, 0.42, 0.25, 0.08, 0.33

[44] S. S. Lewis and E. R. Ott, "Analysis of Means Applied to Percent Defective Data," *Rutgers U. Stat. Cent. Tech. Rep. no. 2,* February 10, 1960.

[45] P. G. Hoel, *Introduction to Mathematical Statistics,* 3d ed., John Wiley & Sons, Inc., New York, 1962.

- The average proportion is $\bar{p} = 30/120 = 0.25$
- The decision lines are

$$\bar{p} \pm H_\alpha \sqrt{\bar{p}(1 - \bar{p})/n_g}$$

$$0.25 \pm 2.29 \sqrt{\frac{0.25(1 - 0.25)}{24}}$$

$$0.25 \pm 0.20$$

The analysis of means plot is as shown in Fig. 8.22 and leads to the conclusion that there is no evidence of a significant difference in quality at the $\alpha = 0.05$ level of risk.

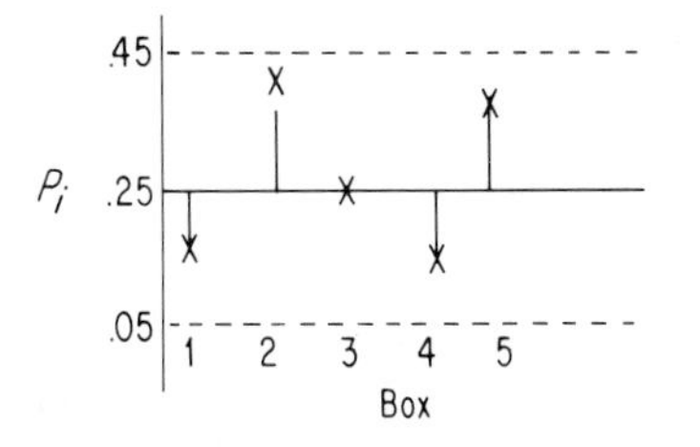

Figure 8.22 Analysis of means plot; proportion defective.

8.29 Analysis of Means for Count Data

Vaswani and Ott[46] used an analysis of means-type procedure on data which was distributed according to the Poisson distribution, when the normal approximation to the Poisson could be employed. For k units each with a count of c_i successes, when no standards are given, analysis of means is performed in the following manner:

1. Count the number of successes per unit, c_i, $i = 1, 2, \ldots, k$
2. Compute $\bar{c}$, the mean number of successes overall
3. Estimate the standard error of the counts as

$$\hat{\sigma}_e = \sqrt{\bar{c}}$$

   Regard the estimate as having infinite degrees of freedom
4. Plot the counts, $\hat{c}_i$, in control-chart format against decision limits computed as

[46]S. Vaswani and E. R. Ott, "Statistical Aids in Locating Machine Differences," *Ind. Qual. Control,* vol. 11, no. 1, July 1954.

$$\bar{c} \pm H_\alpha \sqrt{\bar{c}}$$

5. Conclude the counts are significantly different if any point falls outside the decision limits.

When standards are given, that is, when the mean of the Poisson distribution is known or specified to be μ, the value of μ is used to replace $\bar{c}$ in the above procedure and values of H_α are taken from appendix Table A.8 using the row labeled SG.

8.30 Example—Count Data

Brownlee[47] presents some data on the number of accidents over a period of time for three shifts. These were 2, 14, and 14 respectively. Analysis of means can be used to test if these data indicate an underlying difference in accident rate between the shifts. The analysis is performed as follows:

- The overall mean is $\bar{c} = 30/3 = 10$
- Limits for analysis of means at the $\alpha = 0.05$ level are

$$\bar{c} \pm H_\alpha \sqrt{\bar{c}}$$

$$10 \pm 1.91\sqrt{10}$$

$$10 \pm 6.0$$

The analysis of means plot is shown in Fig. 8.23 and shows the shifts to be significantly different at the 0.05 level of risk.

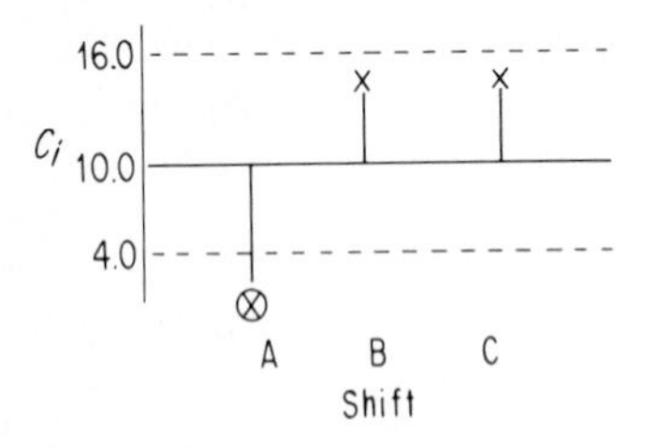

Figure 8.23 Analysis of means plot; accidents by shift.

8.31 Summary

We have looked at process quality control in the broad sense. By definition it encompasses all aspects of an operation. Its objective is what Shewhart called

[47] K. A. Brownlee, *Industrial Experimentation,* Chemical Publishing Company, Inc., New York, 1953.

SADE-Q, that is, satisfactory, adequate, dependable, economic, quality. Three important aspects of process quality control are process control, process capability, and process change. These are tied together with the control chart. It is used to track the process in process control, separating chance causes from assignable causes. A chart in control is necessary for any reasonable assessment of process capability. And it is an instrument of process change through the analysis of means. Thus, the control chart is the method of choice when dealing with the statistical side of process quality control.

8.32 Practice Exercises

Consider the following data for the exercises below, taken from Table 8.8.

1. Prepare a median chart for the data of Table 8.8.
2. What sample size would allow medians to be plotted on an $\overline{X}$ chart for samples of 6 with no change in the positioning of the limits.
3. Prepare a chart for the midrange using the data of Table 8.8 with samples of 5.
4. Prepare an s chart for the data of Table 8.8.
5. If the specifications for the data of Table 8.8 are LSL = 159.0 in, and USL = 160.0 in, and it is known that $\sigma = 0.4$, find limits for a modified-limit control chart for samples of $n_g = 5$. Why won't the chart work?
6. Using the specification limits of Exercise 5, set up an acceptance control chart with an upper and lower APL of 159.5 using $\sigma = 0.4$ and $\alpha = 0.05$ with samples of $n_g = 5$. Back-calculate to determine the RPL having $\beta = 0.10$. Note the APL and RPL are *process* levels and not specifications on individuals.
7. Plot a geometric moving average chart of the means from Table 8.8. Use $\sigma = 0.4$. Note $\sigma_{\overline{X}} = \sigma/\sqrt{n_g}$.
8. Plot a CUSUM chart for the means of the samples from Table 8.8. Use $\sigma = 0.4$. Remember to use $\sigma_{\overline{X}} = \sigma/\sqrt{n_g}$.
9. Convert the limits of Exercise 8 into values of h and k.
10. Plot a Kemp chart for the sample averages of Table 8.8. Use $\sigma = 0.4$. Note $\sigma_{\overline{X}} = \sigma/\sqrt{n_g}$.
11. Set up precontrol limits against the specifications of Exercise 5. Using the data of Table 8.8 in sequence, sample by sample, how many would be taken before a problem is detected?
12. Under what conditions should the process-performance check be as effective as a process-performance evaluation?
13. Using the data of Table 8.8 and the specifications of Exercise 5, compute the C_P index. What does it tell you?
14. Do a Pareto analysis of the demerits shown in Table 6.3.

TABLE 8.8 Data: Air-Receiver Magnetic Assembly (Depth of Cut)
Taken at 15-min intervals in order of production

						$\bar{X}$	Range R
1	160.0	159.5	159.6	159.7	159.7	159.7	0.5
2	159.7	159.5	159.5	159.5	160.0	159.6	0.5
3	159.2	159.7	159.7	159.5	160.2	159.7	1.0
4	159.5	159.7	159.2	159.2	159.1	159.3	0.6
5	159.6	159.3	159.6	159.5	159.4	159.5	0.3
6	159.8	160.5	160.2	159.3	159.5	159.9	1.2
7	159.7	160.2	159.5	159.0	159.7	159.6	1.2
8	159.2	159.6	159.6	160.0	159.9	159.7	0.8
9	159.4	159.7	159.3	159.9	159.5	159.6	0.6
10	159.5	160.2	159.5	158.9	159.5	159.5	1.3
11	159.4	158.3	159.6	159.8	159.8	159.4	1.5
12	159.5	159.7	160.0	159.3	159.4	159.6	0.7
13	159.7	159.5	159.3	159.4	159.2	159.4	0.5
14	159.3	159.7	159.9	158.5	159.5	159.4	1.4
15	159.7	159.1	158.8	160.6	159.1	159.5	1.8
16	159.1	159.4	158.9	159.6	159.7	159.5	0.8
17	159.2	160.0	159.8	159.8	159.7	159.7	0.8
18	160.0	160.5	159.9	160.3	159.3	160.0	1.2
19	159.9	160.1	159.7	159.6	159.3	159.7	0.8
20	159.5	159.5	160.6	160.6	159.8	159.9	1.1
21	159.9	159.7	159.9	159.5	161.0	160.0	1.5
22	159.6	161.1	159.5	159.7	159.5	159.9	1.6
23	159.8	160.2	159.4	160.0	159.7	159.8	0.8
24	159.3	160.6	160.3	159.9	160.0	160.0	1.3
25	159.3	159.8	159.7	160.1	160.1	159.8	0.8
						$\bar{\bar{X}} = 159.67$	$\bar{R} = 0.98$

15. Your doorbell doesn't work and you speculate on a cause. Draw up a cause-and-effect diagram.

16. Make an analysis of means chart for the results shown in Table 1.7 for 11 strips of ceramic. Use $\alpha = 0.05$ limits. Are the strips the same?

Part

3

Troubleshooting and Process Improvement

Chapter

9

Some Basic Ideas and Methods of Troubleshooting

9.1 Introduction

In Chaps. 2 and 5, a scientific process was studied by attempting to hold constant all variables which are thought to affect the process. Then data obtained in a time sequence from the process were examined for the presence of unknown causes (nonrandomness) by the number and length of runs and control charts. Experience in every industry has shown that its processes have opportunities for economic improvement to be discovered by this approach.

When evidence of nonrandomness has been observed, the assignable causes can sometimes be explained by standard engineering or production methods of investigation. Sometimes the method of investigation is to vary one factor or different factors suspected of affecting the quality of the process or the product. This should be done in a preplanned experimental pattern. This experimentation was formerly the responsibility of persons involved in research and development. More recently, process improvement and troubleshooting responsibilities have become the province of those engineers and supervisors who are intimately associated with the day-to-day operation of the plant processes. Effective methods of planning investigations have been developed and are being applied. Their adoption began in the electrical, mechanical, and chemical industries. However, the principles and methods are universal; applications into other industries may differ in detail.

The following sections will outline some procedures of designing and analyzing data from investigations (experiments). Examples from different sciences and industries will be presented to illustrate useful methods. We emphasize attribute data in Chap. 11 and variables in Chaps. 13, 14, and 15.

9.2 Some Types of Independent and Dependent Variables

Introductory courses in science introduce us to methods of experimentation. Time, temperature, rate of flow, pressure, and concentration are examples of variables often expected to have important effects in chemical reactions. Voltage, power output, resistance, and mechanical spacing are important in electronics and many laws involving them have been determined empirically. These laws have been obtained from many laboratory studies over long periods of time by many different experimenters. Such laws are often known by the names of the scientists who first proposed and studied them. We have special confidence in a law when some background of theory has been developed to support it, but we often find it very useful even when its only support is empirical.

In order to teach methods of experimentation in science courses, students are often assigned the study of possible effects of different factors. Different levels of temperature may be selected and the resultant responses determined. Hopefully, the response will behave like a *dependent* variable. After performing the experimental study, a previously determined relationship (law) may be shown to the student to compare with the experimental data.

As specialized studies in a science are continued, we may be assigned the project to determine which factors have major influence on a specific characteristic. Two general approaches are possible:

1. Recognized causative variables (factors)

We study the effects of many variables known to have been important in similar studies (temperature, light intensity, voltage, power output, as examples). This procedure is often successful, especially in well-equipped research laboratories and pilot plants. This is often considered basic to the "scientific method."

Frequently, however, those scientific factors which are expected to permit predictions regarding the new process are found to be grossly inadequate. This inadequacy is especially common when a process is transferred from the laboratory or pilot plant to production. The predicted results may be obtained at some times but not at others, although no known changes have been introduced. In these cases, the methods of Chaps. 2 and 5 are especially relevant to check on stability.

2. Omnibus-type factors[1]

Sometimes the results vary from machine to machine and from operator to operator. The following fundamental "law" has resulted from empirical stud-

[1]There is no term in common usage to designate what we mean by "omnibus-type" factors. Other terms which might be used are *bunch-type* or *chunky-type*. The idea is that of a classification-type factor which will usually require subsequent investigation to establish methods of adjustment or other corrective action. An omnibus-type factor deliberately confounds several factors; some may be known and others unknown.

ies in many types of industry; it is presented with only slight "tongue in cheek":

Consider k different machines assigned to the same basic operation:

- When there are three or four machines, one will be substantially better or worse than the others.
- When there are as many as five or six machines, at least one will be substantially better and one substantially worse than the others.

There are other important omnibus-type factors. We might study possible effects in production from *components* purchased from different vendors[2]; or differences between k *machines* intended to produce the same items or materials; or differences between *operators* or *shifts* of operators. This type of experimentation is often called "troubleshooting" or problem solving; its purpose is to improve *either the product or the process, or both.*

A troubleshooting project often begins by studying possible differences in the quality output of different machines, or machine heads, or operators or other types of variables discussed below. Then when important differences have been established, experience has shown that careful study of the sources of better and worse performance by the scientist and supervisor will usually provide important *reasons* for those differences.

A key to making adjustments and improvements is in *knowing that actual differences do exist* and in being able to *pinpoint the sources* of the differences.

It is sometimes argued that *any* important change or difference will be evident to an experienced engineer or supervisor; *this is not the case.* Certainly many important changes and improvements are recognized without resort to analytical studies, *but* the presence and identity of many economically important factors are not recognized without them. Several case histories are presented throughout the following chapters which illustrate this very important principle.

Summary on variables

Types of independent variables (factors) in a study

1. Continuous variables with a known or suspected association or effect on the process: temperature, humidity, time of reaction, voltage. Sometimes these variables can be set and held to different prescribed levels during a study—sometimes not.

2. Discrete omnibus-type factors. Several examples will be given relating to this type: different heads or cavities on a machine, different operators, different times of day, different vendors. Once it has been determined that important differences do exist, it can almost always lead to identification of specific adjustable causes and to subsequent process improvement.

---

[2]See Case History 11.9.

Types of quality characteristics (response variables, dependent variables, factors)

1. Measurable, variable factors: the brightness of a TV picture, the yield of a chemical process, the breaking strength of synthetic fibers, the thickness of a sheet of plastic, the life (in hours) of a battery.
2. Attribute or classification data (go no-go): the light bulb will or will not operate, the content of a bottle is or is not underfilled. There are occasions where the use of attributes data is recommended even though variables data are possible. In Chap. 7 the important, practical methods of narrow-limit gauging (NL gauging) were discussed.

Experimentation with variables response data is common in scientific investigations. Our discussion of experimentation in Chaps. 13, 14, and 15 will consider variables data. In practice, however, important investigations frequently begin with quality characteristics causing rejects of a go no-go nature. See Chap. 11 for discussions involving their use.

9.3 Some Strategies in Problem Finding, Problem Solving, and Troubleshooting

There are different strategies in approaching real-life experiences. The procedures presented here have been tested by many persons and in many types of engineering and production problems. Their effective use will sometimes be straightforward, but will always benefit from ingenuity in combining the art and science of troubleshooting.

It is traditional to study cause-and-effect relationships. However, there are frequently big advantages to studies which only identify areas, regions, or classification as the source of difference or difficulty. The pinpointing of specific cause and effect is thus postponed. The omnibus-type factor may be different areas of the manufacturing plant, or different subassemblies of the manufactured product. Several examples are discussed in the following chapters and in case histories.

Two important principles need to be emphasized:

Basic principle 1: plan to learn something initially—not everything

This is important especially in those many industrial situations where more data are rather easily attainable.

It is not possible to specify all the important rules to observe in carrying out a scientific investigation, but a second very important rule to observe, if at all possible, is:

Basic principle 2: be present at the time and place the data are being obtained, at least for the beginning of the investigation

- It often provides opportunities to observe possible error sources in the data acquisition. Compensations may be possible by improving the data-recording forms, or by changing the type of measuring instrumentation.

- Observance of the data being obtained may suggest causative relationships which can be suggested, questioned, or evaluated only at the time of the study.
- Observing the possibility of different effects due to operators, machines, shifts, vendors and other omnibus-type variables can be very rewarding.

CASE HISTORY 9.1

Black Patches on Aluminum Ingots[3]

Introductory investigations

While conducting some model investigations in different types of factories, an occasion to investigate a problem of excessive black-oxidized patches on aluminum ingots came into our jurisdiction. A general meeting was first held in the office of the plant manager. At this meeting the general purpose of two projects, including this specific one, was described. This problem had existed several months and competent metallurgists had considered such problems as contaminants in aluminum pigs and differences in furnace conditions.

Our study group had just two weeks to work on the problem; clearly, we could not expect to become better metallurgists than those competent ones already available. Rather than investigate the possible effects of such traditional independent variables as furnace conditions (temperature and time, etc.) we considered what omnibus-type variables might produce differences in the final ingots.

Planning the study

The ingots were cast in 10 different molds. A traveling crane carried a ladle of molten aluminum to a mold; aluminum was poured into the mold, where it was allowed to solidify before removal by a hoist. The plant layout in Fig. 9.1 shows the general location of the 10 molds (*M*), the electric furnace and track, doors (*D*), two walls, and two windows (*W*). It was considered that the location of these doors, windows, and walls might possibly affect the oxidation of black patches. The location of the patches was vaguely considered to be predominantly on the bottom of ingots.

It was decided to record the occurrence and location of the black patches on ingots from a selected sample of molds, one from each in the order M_1, M_{10}, M_3, M_6, M_5, M_8. Then the procedure was repeated once with these same six molds.

- This selection of ingots would indicate whether the location of black patches would occur and reoccur on some molds and not on others, whether it occurred in about the same location on all molds, and whether it would *reoccur* in about the same location on the same mold. If the reoccurrence of black patches was predictable, then the problem was *not* contamination in the molten aluminum or in furnace conditions, but would relate to some condition of the molds. If the black patches did not reoccur in the same areas but their locations appeared random, then the problem might be of a metallurgical nature. The problem might be contamination in the molten aluminum or in changing conditions of the molds.

[3]Ellis R. Ott, United Nations Technical Assistance Programme, report no. TAA/IND/18, March 25, 1958.

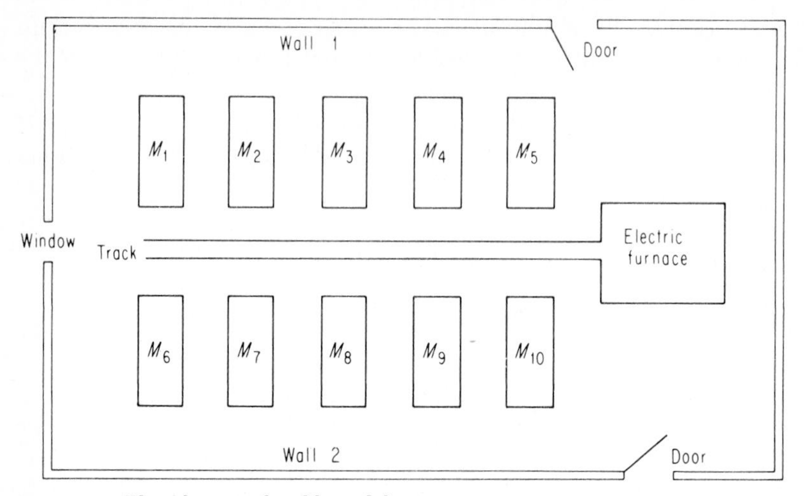

Figure 9.1 Plant layout of molds and furnace.

- A comparison of "inside locations" (M_3 and M_8) with "outside locations" (M_1, M_5, M_6, M_{10}) might also indicate possible effects related to distances from furnace, doors, and windows.

How were the location and intensity of black oxidation to be measured? There is no standard procedure: (1) Often a diagram can be prepared and the location of defects sketched or marked on it; Fig. 9.2 shows the blank diagrams which were prepared in

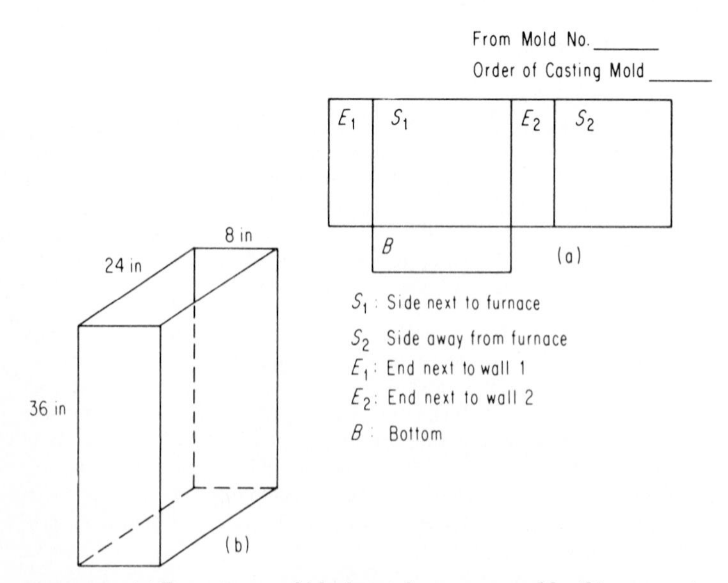

Figure 9.2 (*a*) Form to record black patch areas on molds; (*b*) representation of a mold.

advance on sheets of papers. (2) Included on each form was the order of casting to be followed.

Note: While getting organized, it was learned that the 10 molds had electrical heating in their walls and bottoms (the tops were open). The metallurgists agreed that differences in heating might have an effect, and it would be possible to measure the temperatures at different locations in a mold with a contact thermometer. Prior to pouring, the temperatures at designated locations of the mold were measured and recorded on the form (Fig. 9.2).

It was about 6 hours after the initial planning meeting that the data forms had been drawn, the plan formalized, and the first ingot poured. Then after solidifying, the ingot was withdrawn, its identity marked, and the procedure was continued.

Obtaining data

Now if possible, be present when the study begins—long enough, at least to observe *some data.* We examined the ingot from M_1; yes, there was a smallish 3-in irregular circle of oxide—*not* on the bottom, but on the side S_1. The location and size of the oxide were recorded as planned.

No clues were immediately available; the wall temperature in the area of the black patch was no different from the temperatures at locations lacking the oxide. Was there anything special about the condition of the mold wall at the origin of the black oxide? An immediate investigation "suggested" the *possibility* that the white-oxide dressing with which the molds were treated weekly "looked a bit different." It was of unlikely importance, but its existence was noted.

The casting of the first round of ingots was continued as planned; some of the ingots had black patches; some did not. Their location was indicated on the prepared forms. It was time to repeat molds beginning with M_1. When the second M_1 ingot was examined it showed a black patch in the *same* general location as the first M_1 ingot! And this was the general repeat pattern for the six molds. A careful examination in the area producing a black patch usually suggested a slight differing appearance: nothing obvious or very convincing.

It was the practice to dress the molds with white oxide every few days. When it was time to make a third casting on M_1, a redressing was applied (by brush) to the specific area of origin of the black patch.

Analysis

Then the next casting was made. Consequence? *No black patch.* It was found that this same procedure would repeatedly identify areas in other molds which needed redressing to prevent black oxidized patches.

9.4 Summary

The basic logic and method of this study are important. Repeat observations on selected single units of your process will demonstrate one of two things: either the performance of the unit will repeat; or, it will not.

It was established in this case history that the black patches came repeatedly from specific geometric areas within molds. The reason for the problem was thus unrelated to contaminants or other metallurgical properties of aluminum pigs, or to distances or relationships to the furnace or windows and walls. Temperature differences within a mold could have been a possible explanation. Being present and able to inspect the first mold illustrates the importance of the previously stated Principle 2.

It is not always that the correction of a process can be identified so readily; but the opportunity was provided for simple data to suggest ideas. In this case history, the retreatment of molds provided a complete solution to the problem.

- In some studies, the purpose of data collection is to provide organized information on relationships between variables. In many other instances, such as this one, the purpose is simply to find ways to eliminate a serious problem; the data themselves or a formal analysis of them are of little or no interest. It was the logic and informal analysis which was effective.
- In troubleshooting and process improvement studies, we can plan programs of data acquisition which offer opportunities for detecting types of important differences and repeat performances. The opportunity to notice possible differences or relations, such as the location of black patches and their origin within molds, comes much more surely to one who watches "data in the process of acquisition" than to one who sits comfortably in an office chair.

These ideas will be extended in subsequent case histories.

9.5 Practice Exercises

1. Identify an industrial process which you have had experience with or can study closely.
   a. Describe the physical nature of the process in detail.
   b. Identify all the continuous variable factors associated with this process. Be specific regarding the operating ranges of these variables and the degree of control which the operator has over them.
   c. Identify several omnibus-type factors.
2. Based on the author's discussion of Case History 9.1, write out a set of *short* specific guidelines for conducting a process troubleshooting project. Include from five to eight guidelines.

Chapter

10[1]

Some Concepts of Statistical Design of Experiments

10.1 Introduction

Success in troubleshooting and process improvement often rests on the appropriateness and efficiency of the experimental setup and its match to the environmental situation. Design suggests structure, and it is the structure of the statistically designed experiment that gives it its meaning. Consider a simple 2^2 factorial experiment laid out in Table 10.1 with measurements X_{11}, X_{12}, X_{21}, X_{22}. The subscripts i and j on X_{ij} simply show the machine (i) and operator (j), associated with a given measurement.

Here there are two factors or characteristics to be tested, operator and machine. There are two levels of each so that operator takes on the levels Dianne and Tom and the machine used is either old or new. The designation 2^p means two levels of each of p factors. If there were three factors in the experiment, say the addition of material (from two vendors), we would have a 2^3 experiment.

In a properly conducted experiment, the treatment combinations corre-

TABLE 10.1 Experiment Plan

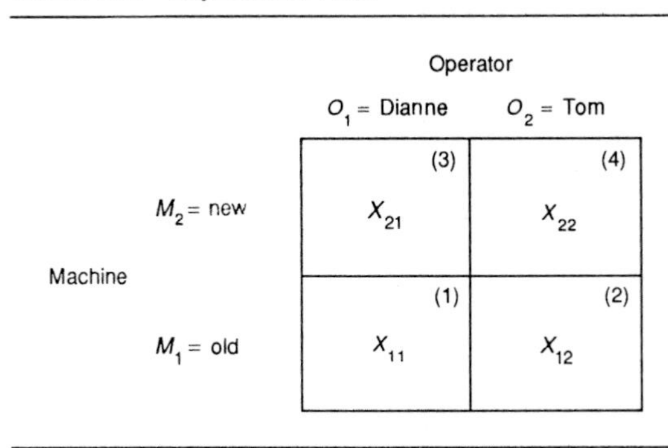

[1]Sections 10.3, 10.4, 10.5 and 10.6 are not vital to subsequent understanding and may be omitted by the reader. They are intended as a supplement for those already somewhat familiar with the topic.

sponding to the cells of the table must be run at random to avoid biasing the results. Tables of random numbers on slips of paper drawn from a hat can be used to set up the order of experimentation. Thus, if we numbered the cells as shown in the diagram and drew the numbers 3, 2, 1, 4 from a hat, we would run Dianne-new first followed by Tom-old, Dianne-old, and Tom-new in that order. This is done to insure than any external effect which might creep into the experiment while it is being run would affect the treatments in random fashion. Its effect would then appear as experimental error rather than biasing the experiment.

10.2 Effects

Of course, we must measure the results of the experiment. That measurement is called the response. Suppose the response is units produced in a given time, and that the results are as shown in Table 10.2.

The effect of a factor is the average change in response (units produced) brought about by moving from one level of a factor to the other. To obtain the machine effect we would simply subtract the average result for the old machine from that of the new. We obtain

$$\text{Machine effect} = \overline{X}_{2\cdot} - \overline{X}_{1\cdot} = \frac{(5 + 15)}{2} - \frac{(20 + 10)}{2} = -5$$

which says the old machine is better than the new. Notice that when we made this calculation, each machine was operated equally by both Dianne and Tom for each average. Now calculate the operator effect. We obtain

$$\text{Operator effect} = \overline{X}_{\cdot 2} - \overline{X}_{\cdot 1} = \frac{(15 + 10)}{2} - \frac{(5 + 20)}{2} = 0$$

The dots ($\cdot$) in the subscripts simply indicate which factor was averaged out in computing $\overline{X}$. It appears that operators have no effect on the operation. Notice that each average represents an equal time on each machine for each operator and so is a fair comparison.

But suppose there is a unique operator-machine combination that produces

TABLE 10.2 Experiment Results

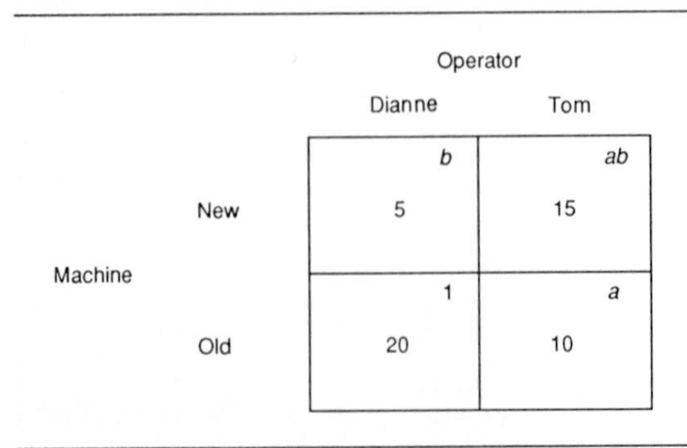

		Operator	
		Dianne	Tom
Machine	New	*b* 5	*ab* 15
	Old	1 20	*a* 10

a result beyond the effects we have already calculated. This is called an interaction. Remember, we averaged machines out of the operator effect and operators out of the machine effect. To see if there is an interaction between operators and machines we calculate the machine effect individually for each operator. If there is a peculiar relationship between operators and machine, it will show up as the average difference between these calculations. We obtain

$$\text{Machine effect for Dianne} = X_{21} - X_{11} = 5 - 20 = -15$$

$$\text{Machine effect for Tom} = X_{22} - X_{12} = 15 - 10 = 5$$

The average difference between these calculations is

$$\text{Interaction} = \frac{(5) - (-15)}{2} = \frac{20}{2} = 10$$

The same result would be obtained if we averaged the operator effect for each machine. It indicates that there is, on the average, a 10-unit reversal in effect due to the specific operator-machine combination involved.

Specifically, in going from Dianne to Tom on the new machine we get on the average, a 10-unit increase; whereas in going from Dianne to Tom on the old machine we get on the average, a 10-unit decrease. For computational purposes in a 2^2 design, the interaction effect is measured as the difference of the averages down the diagonals of the table. Algebraically, this gives the same result as the above calculation since:

$$\begin{aligned}\text{Interaction} &= \frac{(X_{22} - X_{12}) - (X_{21} - X_{11})}{2} \\ &= \frac{(X_{22} + X_{11}) - (X_{21} + X_{12})}{2} \\ &= \frac{(\text{Southwest diagonal}) - (\text{Southeast diagonal})}{2} \\ &= \frac{15 + 20 - 5 - 10}{2} \\ &= 10\end{aligned}$$

There is a straightforward method to calculate the effects in more complicated experiments. It requires that the treatment combinations be properly identified. If we designate operator as factor A and machine as factor B, each with two levels ($-$ and $+$), the treatment combinations (cells) can be identified by simply showing the letter of a factor if it is at the $+$ level and not showing the letter if the factor is at the $-$ level. We show (1) if all factors are at the $-$ level. This is illustrated in Table 10.3.

The signs themselves indicate how to calculate an effect. Thus, to obtain the A (operator) effect we subtract all those observations under the $-$ level for A from those under the $+$ level and divide by the number of observations that go into either the $+$ or $-$ total to obtain an average. The signs that identify what to add and subtract in calculating an interaction can be found by multiplying the signs of its component factors together as in Table 10.4.

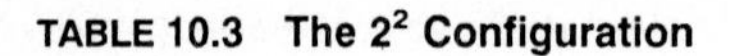

TABLE 10.3 The 2^2 Configuration

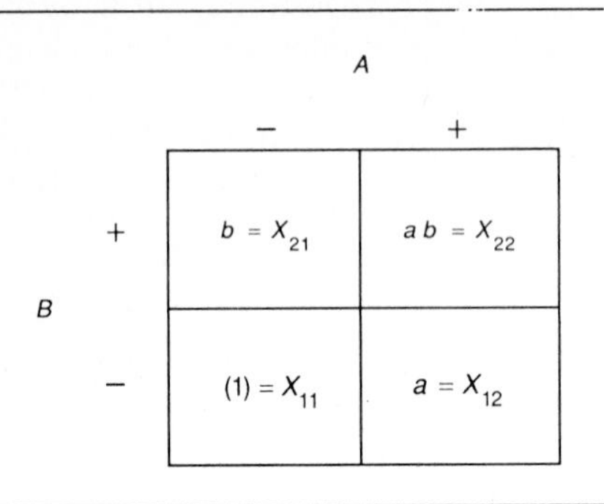

		A −	A +
B	+	$b = X_{21}$	$ab = X_{22}$
B	−	$(1) = X_{11}$	$a = X_{12}$

And we have

$$A \text{ effect (operators)} = (ab + a - b - 1)/2 = 0$$

$$B \text{ effect (machines)} = (b + ab - 1 - a)/2 = -5$$

$$\text{Interaction } (A \times B) = (ab + 1 - b - a)/2 = 10$$

10.3 Sums of Squares

Process control and troubleshooting attempt to reduce variability. It is possible to calculate how much each factor contributes to the total variation in the data by determining the sums of squares (SS)[2] for that factor. For an effect E associated with the factor the calculation is

$$SS(E) = r2^{p-2}E^2$$

TABLE 10.4 Signs of Interaction

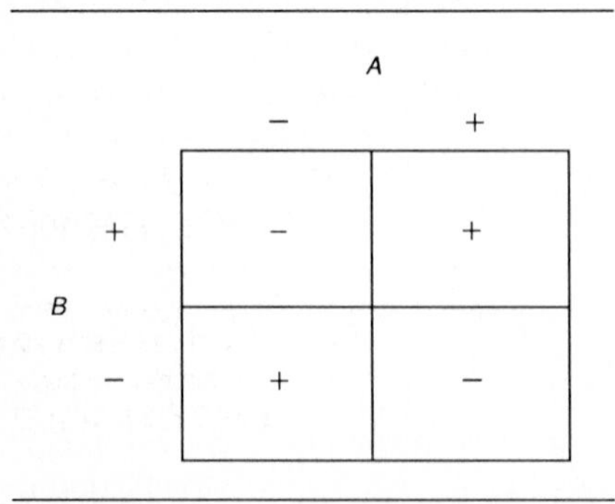

		A −	A +
B	+	−	+
B	−	+	−

[2]Sums of squares, SS, are simply the numerator in the calculation of the variance, the denominator being degrees of freedom, df. Thus, s^2 = SS/df and s^2 is called a mean square, MS, in analysis of variance.

where r is the number of observations per cell and p is the number of factors. Here, $r = 1$ and $p = 2$, so

$$SS(A) = (0)^2 = 0$$

$$SS(B) = (-5)^2 = 25$$

$$SS(A \times B) = (10)^2 = 100$$

To measure the variance $\hat{\sigma}^2$ associated with an effect, we must divide the sums of squares by the appropriate degrees of freedom to obtain mean squares (MS). The effects E (contrasts) each have one degree of freedom so that

$$\hat{\sigma}_E^2 = \text{Mean square } (E) = SS(E)/1 = SS(E)$$

The total variation in the data is measured by the sample variance from all the data taken together, regardless of where it came from. This is

$$\sigma_T^2 = s_T^2 = \frac{SS(T)}{df(T)} = \frac{\Sigma(X - \overline{X})^2}{r2^p - 1} = [(15 - 12.5)^2 + (5 - 12.5)^2 + (10 - 12.5)^2 + (20 - 12.5)^2]/[1(4) - 1] = \frac{125}{3} = 41.67$$

We can then make an analysis of variance table (Table 10.5) showing how the variation in the data is split up. We have no estimate of error since our estimation of the sums of squares for the three effects uses up all the information (degrees of freedom) in the experiment. If two observations per cell were taken, we would have been able to estimate the error variance in the experiment as well.

The F test[3] can be used to assess statistical significance of the mean squares when a measure of error is available. Since the sums of squares and degrees of freedom add to the total, the error sum of squares and degrees of freedom for error may be determined by difference. Alternatively, sometimes an error estimate is available from previous experimentation. Suppose, for example, an outside measure of error for this experiment was obtained and turned out to be $\hat{\sigma}_e^2 = 10$ with 20 degrees of freedom. Then for machines $F = 25/10 = 2.5$ and the F table for $\alpha = 0.05$ and 1 and 20 degrees of freedom shows $F^* = 4.35$ would be exceeded 5 percent of the time. So we are unable to declare that machines show a significant difference from chance variation. On the other hand, interaction produces $F = 100/10 = 10$ which clearly exceeds the critical value of 4.35, so we declare interaction significant at the $\alpha = 0.05$ level of risk. Note that this is a one-tailed test.

TABLE 10.5 Analysis of Variance

Effect	SS	df	MS
Operator (A)	0	1	0
Machine (B)	25	1	25
Interaction ($A \times B$)	100	1	100
Error	No estimate		
Total	125	3	

[3]See Sec. 4.5.

10.4 Yates Method

The calculation of the effects and the sums of squares in a 2^p experiment may be accomplished by an algorithm called Yates method which is easily incorporated in a computer spread sheet. To use that method, the observations have to be put in Yates standard order. This is obtained by starting at 1 and multiplying the previous results by the next letter available. We obtain: 1, *a*, *b*, *ab*, *c*, *ac*, *bc*, *abc*, *d*, *ad*, *bd*, *abd*, *cd*, *acd*, *bcd*, *abcd*, *e*, *ad*, *be*, *abe*, *ce*, *ace*, *bce*, *abce*, *de*, *ade*, *bde*, *abde*, *cde*, *acde*, *bcde*, *abcde*, etc. We select that portion of the sequence which matches the size of the experiment.

The Yates method consists of the following steps:

1. Write the treatment combinations in Yates order.
2. Write the response corresponding to each treatment combination. When there are r observations per cell, write the cell total.
3. Form column 1 and successive columns by adding the observations in pairs and then subtracting them in pairs, subtracting the top observation in the pair from the one below it.
4. Stop after forming p columns (from 2^p).
5. Estimate effects and sum of squares as

$$\text{Effect} = \frac{\text{Last column}}{r2^{p-1}}$$

$$\text{SS} = \frac{(\text{Last column})^2}{r2^p}$$

TABLE 10.6 Yates Method for 2^2 Experiment

		A −	*A* +
B	+	y_3	y_4
B	−	y_1	y_2

Yates order	Observation	Col. 1	Col. 2	Yates effect	Sum of squares
1	y_1	y_1+y_2	$y_1 + y_2 + y_3 + y_4 = T$		
a	y_2	y_3+y_4	$y_2 - y_1 + y_4 - y_3 = Q_a$	$Q_a/2$	$Q_a^2/4$
b	y_3	y_2-y_1	$y_3 + y_4 - y_1 - y_2 = Q_b$	$Q_b/2$	$Q_b^2/4$
ab	y_4	y_4-y_3	$y_4 - y_3 - y_2 - y_1 = Q_{ab}$	$Q_{ab}/2$	$Q_{ab}^2/4$

TABLE 10.7 Yates Analysis of Production Data

Yates order	Observation	Col. 1	Col. 2	Yates effect	Sum of squares
1	20	30	50		
a	10	20	0	0	0
b	5	−10	−10	−5	25
ab	15	10	20	10	100

For a 2^2 experiment, this requires 2 columns and is shown in Table 10.6.

Note that T is the total of all the observations.

So for the production data from Table 10.2, the Yates analysis is shown in Table 10.7.

Table 10.8 summarizes the Yates method of analysis.

These concepts are easily extended to higher 2^p factorials. For a 2^3 experiment we have a configuration as indicated in Table 10.9.

TABLE 10.8 Yates Method with *r* Replicates per Treatment Combination

Treatment	Observation total	Col. 1	Col. 2	...	Col. p (from 2^p)	Effect	SS
1	t_1	t_a+t_1	Repeat on Col. 1	Repeat, etc.	Final repetition		
a	t_a	Add				$\frac{\text{Col.}p}{r(2^{p-1})}$	$\frac{(\text{Col.}p)^2}{r(2^p)}$
b	t_b	pairs					
ab	t_{ab}						
...	...						
...	...	t_a-t_1					
...	...	Subtract pairs					
...	...	(upper from					
$(abc\ldots)$	...	lower)					

Compute:

$\sum(\text{Obsn. total}) = 1$

$\sum(\text{Obsn. total})^2 = 2$

$\sum(\text{Obsn.})^2 = 3$

$\sum(\text{SS}) = 4$

Then:

Check: $\frac{1}{r}\left[2 - \frac{(1)^2}{(2^p)}\right] = 4$

Residual error SS $= 3 - \frac{2}{r}$

Residual error MS $= \left(\frac{3 - \frac{2}{r}}{\text{Residual df}}\right)$

TABLE 10.9 2^3 Configuration

	B −		B +	
	A −	A +	A −	A +
C +	c y_5	ac y_6	bc y_7	abc y_8
C −	1 y_1	a y_2	b y_3	ab y_4

Effects may be calculated from first principles. For example, if the appropriate signs are appended to the observations, the effects may readily be calculated from the signs of Table 10.10.

So we have

$$\text{Main effects } \overline{A} = \frac{1}{4}(y_2 + y_4 + y_6 + y_8) - \frac{1}{4}(y_1 + y_3 + y_5 + y_7)$$

$$B = \frac{1}{4}(y_3 + y_4 + y_7 + y_8) - \frac{1}{4}(y_1 + y_2 + y_5 + y_6)$$

$$C = \frac{1}{4}(y_5 + y_6 + y_7 + y_8) - \frac{1}{4}(y_1 + y_2 + y_3 + y_4)$$

$$\text{Interaction } AB = \frac{1}{4}(y_1 + y_4 + y_5 + y_8) - \frac{1}{4}(y_2 + y_3 + y_6 + y_7)$$

$$AC = \frac{1}{4}(y_1 + y_3 + y_6 + y_8) - \frac{1}{4}(y_2 + y_4 + y_5 + y_7)$$

$$BC = \frac{1}{4}(y_1 + y_2 + y_7 + y_8) - \frac{1}{4}(y_3 + y_4 + y_5 + y_6)$$

$$ABC = \frac{1}{4}(y_2 + y_3 + y_5 + y_8) - \frac{1}{4}(y_1 + y_4 + y_6 + y_7)$$

And the Yates method becomes as shown in Table 10.11.

For example, consider the following data[4] from an experiment in which there were no replicates and the response was y = [yield (lb) − 80]. Note from the data display in Table 10.12 there is clearly a B effect.

The Yates analysis is shown in Table 10.13.

Suppose it is possible to assume no interactions exist and so to "pool" all but main effects as an estimate of error. The analysis of variance table is displayed in Table 10.14.

We see that when we change B from the low to the high level the response varies by an amount which exceeds chance at the $\alpha = 0.05$ level. Therefore

[4]O. L. Davies (ed.), "The Design and Analysis of Industrial Experiments," Oliver & Boyd Ltd., London, 1954, pp. 264–268.

TABLE 10.10 Signs for Effect Calculation

Observation	A	B	C	AB	AC	BC	ABC
	Effect						
y_1	−	−	−	+	+	+	−
y_2	+	−	−	−	−	+	+
y_3	−	+	−	−	+	−	+
y_4	+	+	−	+	−	−	−
y_5	−	−	+	+	−	−	+
y_6	+	−	+	−	+	−	−
y_7	−	+	+	−	−	+	−
y_8	+	+	+	+	+	+	+

TABLE 10.11 Yates Method for 2^3 Experiment

Yates order	Observation	Col. 1	Col. 2	Col. 3	Yates effect	Sum of squares
1	y_1	y_1+y_2	$y_1+y_2+y_3+y_4$	T*		
a	y_2	y_3+y_4	$y_5+y_6+y_7+y_8$	Q_a†	$Q_a/4$	$Q_a^2/8$
b	y_3	y_5+y_6	$y_2-y_1+y_4-y_3$	etc.		
ab	y_4	y_7+y_8	$y_6-y_5+y_8-y_7$			
c	y_5	y_2-y_1	$y_3+y_4-y_1-y_2$			
ac	y_6	y_4-y_3	$y_7+y_8-y_5-y_6$			
bc	y_7	y_6-y_5	$y_4-y_3-y_2+y_1$			
abc	y_8	y_8-y_7	$y_8-y_7-y_6+y_5$			

* $T = \Sigma y = 2^3\bar{y}$

† $Q_a = (y_2 - y_1 + y_4 - y_3) + (y_6 - y_5 + y_8 - y_7)$

TABLE 10.12 Illustrative Example of 2^3

	B −		B +	
	A −	A +	A −	A +
C +	6.7	9.2	3.4	3.7
C −	7.2	8.4	2.0	3.0

TABLE 10.13 Yates Analysis of Illustrative Example

Yates order	Observation	Col. 1	Col. 2	Col. 3	Yates effect	Sum of squares
1	7.2	15.6	20.6	43.6 = T	$2\bar{y}$ = 10.9	237.62
a	8.4	5.0	23.0	5.0 = 4A	A = 1.25	3.12
b	2.0	15.9	2.2	−19.4 = 4B	B = −4.85	47.04
ab	3.0	7.1	2.8	−2.4 = 4AB	AB = −0.6	0.72
c	6.7	1.2	−10.6	2.4 = 4C	C = 0.6	0.72
ac	9.2	1.0	−8.8	0.6 = 4AC	AC = 0.15	0.04
bc	3.4	2.5	−0.2	1.8 = 4BC	BC = 0.45	0.40
abc	3.7	0.3	−2.2	− 2.0 = 4ABC	ABC = 0.5	0.50
						290.16

TABLE 10.14 ANOVA of Illustrative Example

Source	SS	DF	MS	F	F^*0.05
A	3.12	1	3.12	7.52	7.71
B	47.04	1	47.04	113.35	7.71
C	0.72	1	0.72	1.73	7.71
Error	0.72+0.04 + 0.40 + 0.50 = 1.66	4	0.415		
Total	290.16−237.62 = 52.54	7			

factor B is statistically significant and the effect on yield in going from the low to the high levels of B is estimated to be −4.85 lb.

10.5 Blocking

Occasionally it is impossible to run all of the experiment under the same conditions. For example, the experiment must be run with 2 batches of raw material or on four different days or by two different operators. Under such circumstances it is possible to "block" out such changes in conditions by confounding, or irrevocably combining, them with selected effects. For example, if in the previous experiment units y_1, y_2, y_3, and y_4 were run by one operator and y_5, y_6, y_7, and y_8 were run by another operator, it would be impossible to distinguish the C effect from any difference that might exist between operators. This would be unfortunate, but it is possible to run the experiment in a way that an unwanted change in conditions, such as operators, will be confounded with a preselected higher order interaction in which there is no interest or which is not believed to exist. For example, looking at the signs associated with the ABC interaction, if y_2, y_3, y_5, and y_8 were run by the first operator and y_1, y_4, y_6, and y_7 by the second, the operator effect would be irrevocably combined, or confounded, with the ABC interaction. No other effect in the experiment would be changed by performing the experiment in this way. That is why the *structure* and randomization are so important.

Appendix Table A.17 gives blocking arrangements for various 2^p factorial designs. It will be seen that the pattern suggested will be found under design 1 for blocking a 2^p experiment.

10.6 Fractional Factorials

Sometimes it is possible to reduce the number of experimental runs by utilizing only a portion of structure of the full factorial experiment. Consider running only the cells of the previous 2^3 experiment which are not shaded. The result is shown in Table 10.15. The treatment combination notation has been given in each cell. Patterns such as this have been discovered which will allow running an experiment with some fraction of the units required for the full factorial. In this case only half of the units would be used so this is called a one-half replication of a 2^p factorial. The price of course is aliasing, that is irrevocably combining effects in the analysis. If only the runs indicated were performed and analyzed, the following effects would be aliased together:

$$A \text{ and } BC$$

$$B \text{ and } AC$$

$$C \text{ and } AB$$

If we were in a situation in which we did not expect any 2 factor or higher interactions to exist, we would be able to estimate the A, B, and C main effects by regarding the BC, AC, and AB effects to be zero. This is what is done in the analysis of fractional factorials.

Assume the observations obtained were as before, namely

Treatment	Response
(1)	9.2
ac	9.2
bc	3.4
ab	3.0

TABLE 10.15 Fraction of a 2^3

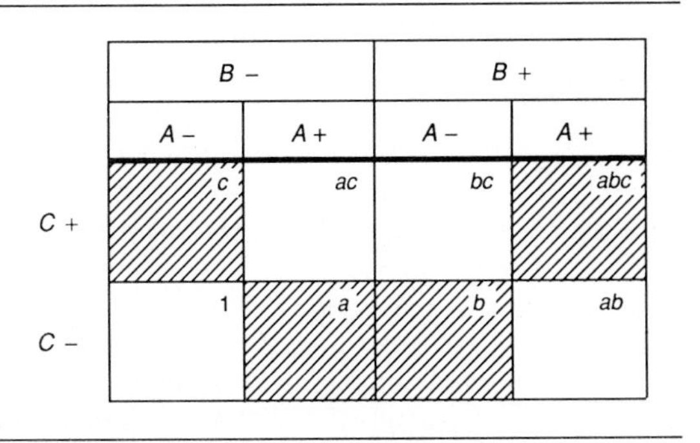

We could estimate the effects by simply subtracting the average response when the units are at the low level of a factor from those which are made at the high level. Thus

$$A = \frac{(ac + ab)}{2} - \frac{(1 + bc)}{2} = \frac{(9.2 + 3.0)}{2} - \frac{(7.2 + 3.4)}{2} = 6.1 - 5.3 = 0.8$$

$$B = \frac{(bc + ab)}{2} - \frac{(1 + ac)}{2} = 3.2 - 8.2 = -5.0$$

$$C = \frac{(ac + bc)}{2} - \frac{(1 + ab)}{2} = 6.3 - 5.1 = 1.2$$

Note that these are reasonably close to the estimates obtained from the full factorial.

In larger experiments, the Yates method may be used for the analysis. The procedure is as follows:

1. Write down Yates standard order for the size of the fractional factorial experiment run. That is, write as many terms as there are treatment combinations.
2. Match the treatment combinations run with the Yates order by writing the unused letter(s) in parenthesis after a treatment combination shown to match the units run.
3. Perform Yates analysis.
4. Identify the aliased effects represented in the effects and sum of squares column. This may be done by multiplying through the defining contrast given for the fraction used by the effect calculated by Yates, to obtain the aliased effects. Any squared terms are eliminated in this process.

Every fractional factorial is associated with a defining contrast. For the fraction used above it is

$$I = -ABC$$

If we treat I as if it were a 1 (one) and multiply both sides through by A, B, and C we get

$$A = -A^2BC = -BC$$

$$B = -AB^2C = -AC$$

$$C = -ABC^2 = -AB$$

Carrying out the Yates analysis on the yield data we get the results shown in Table 10.16.

Note that the effects calculated are not in parenthesis. The aliased effects are obtained from the defining contrast.

Fractional factorials have been tabulated and show the treatment combinations to be run and the defining contrast. Appendix Table A.18 is such a tab-

TABLE 10.16 Yates Analysis of Fraction of Illustrative Example

Yates order	Observation	Col. 1	Col. 2	Effect	Sum of squares	Aliased effects
1	7.2	16.4	22.8			
a(c)	9.2	6.4	1.6	0.8	0.64	$A-BC$
b(c)	3.4	2.0	−10.0	−5.0	25.00	$B-AC$
ab	3.0	−0.4	−2.4	−1.2	1.44	$AB-C$

ulation and shows the treatment combinations to be run, already in Yates order, with the corresponding effects that will be estimated by the row indicated by the treatment combination shown. Note that only two factor interactions are shown in the table, any higher interactions being ignored. Also when a ¼ fraction is run, the defining contrast contains 3 interactions.

If at some later time it is possible to run the rest of the fractional factorial the data from the fractions may be combined and analyzed. In that case only the interactions shown in the defining contrast will be lost. The other effects will be clear of confounding. Recombining the fractions acts like a blocked experiment with the interactions shown in the defining contrast confounded with any block effect that might come about from running at different times. This can be seen from appendix Table A.17 where the fraction run in the example is part of 2 blocks in the 2^p blocking arrangement shown. The confounded interaction is ABC.

10.7 Conclusion

This is a cursory discussion of design of experiments. It touches only on the most rudimentary aspects. It will serve, however, as an introduction to the concepts and content.

CASE HISTORY 10.1

2^5 Experiment on Fuses

J. H. Sheesley[2] has reported on an experiment in which the safe operation of a specialty lamp system depended on the safe and sure operation of a thermal fuse. Since this system was to be used in a new application, the behavior of the fuse was examined under various conditions. The data is shown here as a 2^3 experiment selected from the overall data to illustrate the procedure. The three factors were: line voltage (A), ambient temperature (B), and type of start (C). The response was temperature of the fuse after 10 min of operation as measured by a thermocouple on the fuse. The levels used are as follows:

Line voltage	$A+$	120 V
	$A-$	110 V

[5]J. H. Sheesley, "Use of Factorial Designs in the Development of Lighting Products," *ASQC Electron. Div. Newsletter—Tech. Supp.*, issue 4, Fall 1985, pp. 23–27.

Temperature	$B+$	110°
	$B-$	75°
Start	$C+$	Hot
	$C-$	Cold

The resulting data in terms of average temperature ($n = 10$) after 10 min is shown in Table 10.17.

TABLE 10.17 Average Temperature after 10 min (minus 200°C)

	A −		A +	
	B −	B +	B −	B +
C +	*c* 43.7	*bc* 47.3	*ac* 64.4	*abc* 65.8
C −	1 0.5	*b* 29.8	*a* 10.9	*ab* 48.2

The Yates analysis is as follows:

Yates order	Observation	Col. 1	Col. 2	Col. 3	Yates effect	Sum of squares
1	0.5	11.4	89.4	310.6	77.7 = $2\bar{y}$	12059.0
a	10.9	78.0	221.2	68.0	17.0 = A	578.0
b	29.8	108.1	28.8	71.1	17.8 = B	631.9
ab	48.2	113.1	39.2	5.8	1.4 = AB	4.2
c	43.7	10.4	66.6	131.8	33.0 = C	2171.4
ac	64.4	18.4	5.0	10.4	2.6 = AC	13.5
bc	47.3	20.7	8.0	−61.1	−15.3 = BC	466.7
abc	65.8	18.5	−2.2	−10.2	−2.6 = ABC	13.0

If the *ABC* interaction is assumed not to exist, its sum of squares can be used as a measure of error and we have:

Source		Sum of squares	df	Mean square	F	$F_{0.05}$
Line voltage	A	578.0	1	578.0	44.5	164.1
Temperature	B	631.9	1	631.9	48.6	164.1
Interaction	AB	4.2	1	4.2	0.3	164.1
Start	C	2171.4	1	2171.4	167.0	164.1
Interaction	AC	13.5	1	13.5	1.0	164.1
Interaction	BC	466.7	1	466.7	35.9	164.1
Error	(ABC)	13.0	1	13.0		
Total		3878.7	7			

We see that, even with this limited analysis of the fuse data, we are able to show that start (C) has a significant effect with a risk of $\alpha = 0.05$. The effect of start from cold to hot is 33.0°.

10.8 Practice Exercises

1. Consider the following data on height of Easter lilies taken from Table 14.2.

Storage Period

   a. Estimate the main effects and interaction effects from the basic formula.

		Storage period	
		Short	Long
Conditioning time	Long	28 26 30	48 37 38
	Short	31 35 31	37 37 29

   b. Perform Yates analysis.
   c. Estimate error.
   d. Set up an analysis of variance table.
   e. Test for significance of the effects at the $\alpha = 0.05$ level.

2. Given the following data on capacitances of batteries from Table 14.4:

Nitrate concentration (C)	Low				High				
Shim (B)	In		Out		In		Out		
Hydroxide (A)	New	Old	New	Old	New	Old	New	Old	
	−0.1	1.1	0.6	0.7	0.6	1.9	1.8	2.1	Day 1
	1.0	0.5	1.0	−0.1	0.8	0.7	2.1	2.3	
	0.6	0.1	0.8	1.7	0.7	2.3	2.2	1.9	
	−0.1	0.7	1.5	1.2	2.0	1.9	1.9	2.2	Day 2
	−1.4	1.3	1.3	1.1	0.7	1.0	2.6	1.8	
	0.5	1.0	1.1	−0.7	0.7	2.1	2.8	2.5	
Treatment combination	1	2	3	4	5	6	7	8	

a. Suppose the first three observations in each set of six had been run on one day and the last three observations on the next day. Estimate the block effect.
b. Perform Yates analysis.
c. Set up an analysis of variance table ignoring the fact that the experiment was run on different days and test at the $\alpha = 0.05$ level.
d. Set up an analysis of variance table as if the experiment were blocked as in (*a*) above and test at the $\alpha = 0.05$ level.
e. What are the advantages and disadvantages of blocking on days?

3. Suppose in Exercise 2 that treatment combinations 1, 4, 6, and 7 were tested on one piece of equipment and combinations 2, 3, 5, and 8 were tested on another. What would that do to the analysis and interpretation of the results? (*Hint:* Write out the treatment combinations and use Table 10.15.)
4. Consider the following data from Table 14.9 (recoded by adding 0.30) on contact potential of an electronic device with varying plate temperature (*A*), filament lighting schedule (*B*), and aging schedule (*C*).

Factor			
A	*B*	*C*	Response
−	−	−	0.16, 0.13, 0.15, 0.19, 0.11, 0.10
+	+	−	0.45, 0.48, 0.37, 0.38, 0.38, 0.41
−	+	+	0.26, 0.34, 0.41, 0.24, 0.25, 0.25
+	−	+	0.12, 0.18, 0.08, 0.09, 0.12, 0.09

a. What type of factorial experiment do these data represent?
b. What treatment combinations were run?
c. Place the treatment combinations in Yates order.
d. Perform Yates analysis.
e. Set up an analysis of variance table and test for significance at the $\alpha = 0.05$ level.

5. What is the defining contrast in Exercise 4? (*Hint:* Write out the treatment combinations and use Table A.18.)
6. Show what effects are aliased together in the analysis of variance for Exercise 4.
7. Write out the treatment combinations and defining contrast for a fractional factorial investigating five different factors, each at two levels, when it is possible to make only eight runs.
8. The defining contrast for a 2^{5-2} is $I = -BCE - ADE + ABCD$. What are all the effects aliased with *C*?
9. Linewidth, the width of the developed photoresist in critical areas, is of vital importance in photolithographic processes for semiconductors. In an attempt to optimize this response variable, Shewhart charts were run on the process, but even after identifying a number of assignable causes, the process remained out of control. In an attempt to improve the process and isolate other potential assignable causes several statistically designed exper-

iments were run. Among them was a 2^3 factorial experiment on the following factors, each, of course, at two levels as follows:

Factor	Levels
A: Print GAP spacing	Proximity print, soft contact print
B: Bake temperature	60°C, 70°C
C: Bake time	5 min, 6 min

The results of the experiment as given by Stuart Kukanaris,[6] a student of Dr. Ott, are as follows:

A:	Proximity print				Soft contact print			
B:	60°C		70°C		60°C		70°C	
C:	5 min	6 min	5 min	6 min	5 min	6 min	5 min	6 min
	373	368	356	356	416	397	391	407
	372	358	351	342	405	393	391	404
	361	361	350	349	401	404	396	403
	381	356	355	342	403	409	395	407
	370	372	355	339	397	402	403	406
Total	1857	1815	1767	1728	2022	2005	1976	2027

Once discovered, interactions play an important part in identifying assignable causes apart from naturally occurring process fluctuations. Often the process is so tightly controlled that naturally occurring slight changes in important factors do not indicate their potential impact. This designed experiment was useful in gaining further insight into the process.

a. Perform a Yates analysis and plot the analysis of means chart.

b. Confirm that *A* (spacing), *B* (temperature), *AB* (spacing-temperature interaction) and *AC* (spacing-time interaction) are significant. The physical importance of these effects is indicated by the effects column of the Yates analysis.

[6]Stuart Kukunaris, "Operating Manufacturing Processes Using Experimental Design," *ASQC Electron. Div. Newsletter—Tech. Supp.*, issue 3, Summer 1985, pp. 1–19.

Chapter

11

Troubleshooting with Attributes Data

11.1 Introduction

Perhaps the presence of an assignable cause has been signaled by a control chart. Or perhaps it is known that there are too many rejects, too much rework, or too many stoppages. These are important attributes problems. Perhaps organized studies are needed to determine which of several factors—materials, operators, machines, vendors, processings—have important effects upon quality characteristics. In this chapter, methods of analysis are discussed with respect to the quality characteristics of an attributes nature.

Not much has been written about process improvement and troubleshooting of quality characteristics of an attributes nature. Yet in almost every industrial process there are important problems where the economically important characteristics of the product are attributes: an electric light bulb will give light or it will not; an alarm clock will or will not ring; the life of a battery is or is not below standard. There are times when it is expedient to gauge a quality characteristic (go no-go) even though it is possible to measure its characteristic as a variable. This chapter discusses some effective *designed studies* using enumerative or attributes data and methods of analysis and interpretation of resulting data.

Explanations of *why* a process is in trouble are often based on subjective judgment. How can we proceed to get objective evidence in the face of all the plausible stories as to why this is not the time or place to get it? Data of the attributes type often imply the possibility of personal carelessness. Not everyone understands that perfection is unattainable; a certain onus usually attaches to imperfection. Thus it is important to find ways of enlisting the active support and participation of the department supervisors, the mechanics, and possibly some of the operators. This will require initiative and ingenuity.

In many plants, little is known about differences in machine performance. Just as two autos of the same design may perform differently, so do two or three machines of the same make. Or, a slight difference in a hand operation which is not noticed (or is considered to be inconsequential) may have an important effect on the final performance of a kitchen mixer or a nickel-cadmium battery. Experience indicates that there will be important differences in as few as two or three machines; or in a like number of operators, shifts, or days. Several case histories are presented in this chapter to illustrate important principles of investigation. In each, it is the intent to find *areas of differences.* Independent variables or factors are often chosen to be omnibus-type variables.[1] Once the presence and localized nature of important differences are identified, ways can usually be found by engineers or production personnel to improve operations.

Data from the case histories have been presented in graphical form for a variety of reasons. One compelling reason is that the experiment or study is valuable only when persons in a position to make use of the results are convinced that the conclusions are sensible. These persons have had long familiarity and understanding of graphical presentations; they respond favorably to them. Another reason is that the graphical form shows relationships and *suggests possibilities* of importance not otherwise recognized.

11.2 Ideas from Sequences of Observations over Time

The methods of Chap. 2 are applicable to sequences of attributes data as well as to variables data. Control charts with control limits, runs above and below the median—these procedures suggest ideas about the presence and nature of unusual performance. As each successive point is obtained and plotted, the chart is watched for evidence of economically important assignable causes in the process even while it is operating under conditions considered to be stable.[2]

If the process is stable (in statistical control), each new point is expected to fall within the control limits. Suppose the new point falls outside the established 3-sigma control limits. Since this is a very improbable event when the process is actually stable, such an occurrence is recognized as a signal that *some* change has occurred in the process. We investigate the process to establish the nature of the assignable cause. The risk of an unwarranted investigation from such a signal is very small—about three in a thousand.

In troubleshooting, it is often important to make an investigation of the process with a somewhat greater chance (*risk*) of an unwarranted investigation than three in a thousand; lines drawn at $\bar{p} \pm 2\hat{\sigma}_p$ will be more sensitive to the presence of assignable causes. A somewhat larger risk of making an unwarranted investigation of the process is associated with a point outside 2-sigma limits; it is about one chance in 20 (about a 5 percent risk). *However, there is now a smaller risk* (β) *of missing an important opportunity to investigate,* especially important in a process-improvement study.

In process control we set the control limits at $\pm 3\hat{\sigma}$ arbitrarily and com-

[1]See Chap. 9.
[2]See Chap. 5.

pute the resulting α. In troubleshooting, the decision limits use just the opposite approach.

11.3 Decision Lines[3] Applicable to *k* Points Simultaneously

Introduction

When each point on a Shewhart control chart is *not* appraised at the time it is plotted for a possible shift in process average, there is a conceptual difference in probabilities to consider. For example, consider decision lines drawn at $\bar{p} \pm 2\hat{\sigma}_p$. The risk associated with them is indeed about 5 percent if we apply them as criteria to a single point just observed. But if applied to an *accumulated* set of 20 points, about one out of twenty is expected to be outside of them even when there has been no change in the process. Evidently then, decision lines to study $k = 20$ points simultaneously, with a 5 percent risk of unnecessary investigation, must be at some distance beyond $\bar{p} \pm 2\hat{\sigma}_p$.

Troubleshooting is usually concerned with whether one or more sources—perhaps machines, operators, shifts, or days—can be identified as performing significantly differently from the average of the group of k sources. The analysis will be over the *k sources simultaneously*, with risk α.

In the examples and case histories considered here, the data to be analyzed will not usually relate to a previously established standard. For example, the data of Case History 11.1 represent the percent of rejects from 11 different spot-welding machine-operator combinations. In this *typical* troubleshooting case history, there is *no given standard* to use as a basis for comparison of the 11 machine-operator combinations. They will be compared to their own group average.

Data in Fig. 11.1, pertaining to the percent winners in horse racing, are basically a different type; *there is a given standard.* If track position is not important, then it is expected that one-eighth of all races will be won in each of the eight positions.

This graphical analysis of k groups of size n_g simultaneously is called the *analysis of means*[4] and is abbreviated ANOM. It uses the normal approximation to the binomial and therefore requires a fairly large sample size. It is recommended that $n_g \bar{p} > 5$.

---

[3]Even one point outside *decision lines* will be evidence of nonrandomness among a set of k points being considered simultaneously. Some persons prefer to use the term "control limits." Many practitioners feel strongly that only those lines drawn at ± 3 sigma about the average should be called control limits. At any rate, we shall use the term *decision lines* in the sense defined above.

[4]Ellis R. Ott and Sidney S. Lewis, "Analysis of Means Applied to Per-Cent Defective Data," Rutgers Stat. Cent. Tech. Rep. no. 2, Prepared for Army, Navy and Air Force under Contract NONR 404(11), (Task NP 042–21) with the Office of Naval Research, February 10, 1960. Ellis R. Ott, "Analysis of Means—A Graphical Procedure," *Ind. Qual. Control,* vol. 24, no. 2, pp. 101–109, August 1967. Also see Chaps. 13, 14, and 15.

Probabilities associated with k comparisons, standard given[5]

Values of a factor Z_α to provide proper limits are given in Table 11.1 (or A.7) for values of α = 10, 5, and 1 percent. *Upper and lower decision lines* to judge the extent of maximum expected random variation of points about a given group standard p' or percent defective P' of k samples are:

$$\begin{aligned} \text{UDL}(\alpha) &= p' + Z_\alpha \sigma_p \qquad & \text{UDL}(\alpha) &= P' + Z_\alpha \sigma_P \\ \text{LDL}(\alpha) &= p' - Z_\alpha \sigma_p \qquad & \text{LDL}(\alpha) &= P' - Z_\alpha \sigma_P \end{aligned} \tag{11.1}$$

If even one of the k points falls outside these decision lines, it indicates (statistically) different behavior from the overall group average.

The following derivation of entries in Tables A.7 and 11.1 may help the reader understand the problem involved in analyzing sets of data. The analysis assumes that samples of size n_g are drawn from a process whose known average is p' and n_g and p' are such that the distribution of p_i in samples of size n is essentially normal. (Approximately, $n_g p' > 4$ or 5; see Eq. (5.3) Chap. 5.) We now propose to select k independent random samples of n_g from the

TABLE 11.1 Nonrandom Variability

Standard given, df = ∞. See also Table A.7.

k	$Z_{.10}$	$Z_{.05}$	$Z_{.01}$
1	1.64	1.96	2.58
2	1.96	2.24	2.81
3	2.11	2.39	2.93
4	2.23	2.49	3.02
5	2.31	2.57	3.09
6	2.38	2.63	3.14
7	2.43	2.68	3.19
8	2.48	2.73	3.22
9	2.52	2.77	3.26
10	2.56	2.80	3.29
15	2.70	2.93	3.40
20	2.79	3.02	3.48
24	2.85	3.07	3.53
30	2.92	3.14	3.59
50	3.08	3.28	3.72
120	3.33	3.52	3.93

[5]The material in this section is very important; however, it can be omitted without seriously affecting the understanding of subsequent sections.

process and consider all k values p_i simultaneously. Within what interval

$$p' - Z_\alpha \sigma_p \quad \text{and} \quad p' + Z_\alpha \sigma_p$$

will all k sample fractions p_i lie, with risk α or confidence $(1 - \alpha)$?

Appropriate values of Z_α, corresponding to selected levels α and the above assumptions, can be derived as follows. Let Pr represent the unknown probability that any *one* sample p_i from the process will lie between the lines to be drawn from Eq. (11.1). Then the probability that *all k* of the sample p_i will lie within the interval in (11.1) is Pr^k. If at least one point lies outside these decision lines, this is evidence of *nonrandom variability* of the k samples; that is, some of the sample p_i are different, risk α. The value of Z_α can be computed as follows:

$$\text{Pr}^k = 1 - \alpha \tag{11.2}$$

Then, corresponding to the value of Pr found from this equation, Z_α is determined from Table A.1. Values of Z_α found via Eq. (11.2) are shown in Table 11.1 for $\alpha = 0.10, 0.05, 0.01$, and selected values of k.

Numerical example

Compute $Z_{0.05}$ in Table 11.1 for $k = 3$:

$$(\text{Pr})^3 = 0.95$$

$$\log \text{Pr} = (1/3) \log (0.95) = 9.99257 - 10$$

and

$$\text{Pr} = 0.98304$$

for a probability Pr of 0.00848 in each tail of a two-tail test. From appendix Table A.1, we find that corresponding decision lines drawn at $p' \pm Z_{0.05}\sigma_p$ require that $Z_{0.05} = 2.39$.

Note 1. When lines are drawn at $p' \pm 2\sigma_p$ about the central line, it is commonly believed that a point outside these limits is an indication of an assignable cause with risk about 5 percent. The risk on an established control chart of a stable process is indeed about 5 percent if we apply the criterion to a single point just observed; but if applied, for example, to 10 points simultaneously, the probability of at least 1 point of the 10 falling outside 2-sigma limits by chance is

$$1 - (0.954)^{10} = 1 - 0.624 = 0.376$$

That is, if many sets of 10 points from a stable process are plotted with the usual 2-sigma limits, over one-third of the sets (37.6 percent) are expected to have one or more points just outside those limits. This is seldom recognized.

Also, just for interest, what about 3-sigma limits? If many sets of 30 points from a stable process are plotted with usual 3-sigma limits, then

$$1 - (.9973)^{30} = 1 - 0.922 \text{ or } 7.8\%$$

of the sets are expected to have one or more points outside those limits.

Conversely, in order to provide a 5 percent risk for a set of 10 points considered as a group, limits must be drawn at

$$\pm Z_{0.05}\sigma_p = \pm 2.80\sigma_p$$

as shown in Table 11.1.

Note 2. Or consider an accumulated set of 20 means ($k = 20$). About one out of twenty is expected to be outside the lines drawn at $p' \pm 2\sigma_p$. Consequently, decision lines to study 20 groups simultaneously must be at some distance beyond $p' \pm 2\sigma_p$. Table 11.1 shows that the lines should be drawn at

$$p' \pm 3.02\sigma_p \qquad \text{for } \alpha = 0.05$$

and at

$$p' \pm 3.48\sigma_p \qquad \text{for } \alpha = 0.01$$

Example 11.1 Consider the following intriguing problem offered by Siegel[6]: "Does the post position on a circular track have any influence on the winner of a horserace?"

Data on post positions of the winners in 144 eight-horse fields were collected from the daily newspapers and are shown in Table 11.2. Position 1 is that nearest the inside rail.

The calculations for an analysis of means (ANOM) with *standard given* and $p' = 1/8 = 0.125$ follows

$$\sigma_p = \sqrt{\frac{(0.125)(0.875)}{144}} = 0.0275$$

For $k = 7$, we have $Z_{0.05} = 2.68$ and $Z_{0.01} = 3.19$.[7] The decision limits are:

Risk	LDL	UDL
0.05	0.051	0.199
0.01	0.037	0.213

TABLE 11.2 Winners at Different Post Positions

Post position:	1	2	3	4	5	6	7	8	Total
No. of winners:	29	19	18	25	17	10	15	11	144
Percent:	20.1	13.2	12.5	17.3	11.8	7.0	10.4	7.7	$\bar{P} = 12.5\%$

[6]S. Siegel, *Nonparametric Statistics for the Behavioral Sciences,* McGraw-Hill Book Company, New York, 1956, pp. 45–46.

[7]Although there are eight positions in this case, there are only seven *independent* positions. (When any seven of the p_i are known, the eighth is also known.) We enter Table 11.1 with $k = 7$. It is evident that the decision is not affected whether $k = 7$ or $k = 8$ is used. This situation seldom if ever arises in a production application.

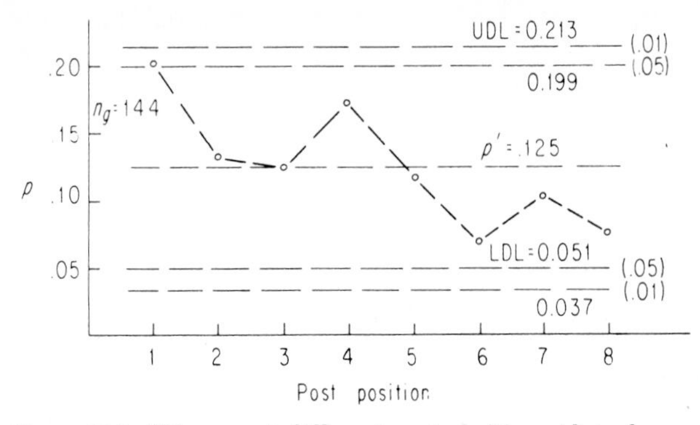

Figure 11.1 Winners at different post positions (data from Table 11.2).

These have been drawn in Fig. 11.1 following certain conventions:

1. The sample size, $n = 144$, is written in the upper-left corner of the chart.
2. The risks, 0.05 and 0.01, are shown at the end of the decision lines.
3. The points corresponding to the eight post positions are connected by a dotted line in order to recognize comparisons better.
4. The values of the decision lines are written adjacent to them.

Discussion The point corresponding to post position 1 is between the (0.05) and (0.01) upper lines[8]; this indicates that position 1 has a better than average chance of producing a winner ($\alpha < 0.05$). Figure 11.1 supports what might have been predicted: if positions have any effect, the best position would surely be that one nearest the rail, and the worst would be near the outside. Not only does the graph show position 1 in a favored light, it also indicates a general downward trend in the winners starting from the inside post positions. (There is not enough evidence to support, conclusively, the possibility that position 4 is superior to positions 2 and 3. There seems little choice among positions 6, 7, and 8.)

Factors to use in making *k* comparisons, no standard given

Factors for *standard given* were obtained easily in the preceding section. However, situations where they can be used in solving production problems seldom occur. In the great majority of troubleshooting situations, there is *no standard given;* but it is very useful to compare individual performances with the *overall average group performance.* The comparison procedure used here is called analysis of means, *no standard given.* It is similar to a p chart. The procedure is outlined in Table 11.3 and illustrated in several case histories. Factors, designated by H_α, provide *decision lines* for the important case of *no standard given*

$$\bar{p} \pm H_\alpha \hat{\sigma}_p \qquad \text{or} \qquad \bar{P} \pm H_\alpha \hat{\sigma}_P$$

[8]The authors' chi-square analysis of this data also indicates that there is a significant difference between positions with a risk between 0.05 and 0.01.

TABLE 11.3 Analysis of Means, Attributes Data, One Independent Variable

STEP 1: Obtain a sample of n_i items from each of k sources and inspect each sample. (It is preferable to have all n_i equal.) Let the number of defective or nonconforming units in the k samples be $d_1, d_2, \ldots, d_k$, respectively.

STEP 2: Compute the fraction or percent defective of each sample.

$$p_i = d_i/n_i \qquad P_i = 100d_i/n_i$$

STEP 3: Plot the points corresponding to the k values, p_i or P_i.

STEP 4: Compute the grand average $\bar{p}$ or $\bar{P}$ and plot it as a line:

$$\bar{p} = \sum d_i/\sum n_i \qquad \bar{P} = 100\sum d_i/\sum n_i$$

STEP 5: Compute a standard deviation, using average $\bar{n}$ initially if there is variation in sample size.

$$\text{if } n\bar{p} > 5 \text{ and } \bar{n}(-p) > 5 \qquad \hat{\sigma}_p = \sqrt{\frac{\bar{p}(1-\bar{p})}{\bar{n}}} \qquad \hat{\sigma}_P = \sqrt{\frac{\bar{P}(100-\bar{P})}{\bar{n}}}$$

STEP 6: From Table 11-4 or Appendix Table A-8, obtain the value of H corresponding to k and α. Draw decision lines:

$$\text{UDL}: \bar{p} + H_\alpha \hat{\sigma}_p \qquad \bar{P} + H_\alpha \hat{\sigma}_P$$

$$\text{LDL}: \bar{p} - H_\alpha \hat{\sigma}_p \qquad \bar{P} - H_\alpha \hat{\sigma}_P$$

STEP 7: Accept the presence of statistically significant differences (assignable causes) indicated by points above UDL and/or below LDL with risk α. Otherwise accept the hypothesis of randomness of the k means (i.e., no statistically significant differences).

STEP 8: Process improvement: consider ways of identifying the nature of and reasons for significant differences.

when analyzing attribute data, or

$$\bar{\bar{X}} \pm H_\alpha \sigma_{\bar{x}}$$

when analyzing variables data are needed. The computation of H_α is much more difficult than the earlier computation of Z_α. We use the normal Z test if σ is known and the t test if σ is unknown. Likewise, we use Z_α if μ is known and H_α if μ is not known. References are given to their derivation. Some factors, H_α, to use when analyzing attributes data which are reasonably normal are given in Table 11.4.[9]

When n and p do not permit the assumption of normality, a method of computing the exact probability of a percent defective to exceed a specified value is possible.[10]

[9]The binomial distribution is reasonably normal for those values of $n_g p$ and $n_g g$ greater than 5 or 6 and p and q are greater than say 0.05. See Eq. (5.3).

[10]Sidney S. Lewis, "Analysis of Means Applied to Percent Defective Data," *Proc. Rutgers All-Day Conf. Qual. Control,* 1958.

TABLE 11.4 Analysis of Means: No Standard Given; df = ∞
Comparing k groups with their group average (especially for use with attributes data): $\overline{P} \pm H_\alpha \hat{\sigma}_{\overline{P}}; \overline{\overline{X}} \pm H_\alpha \hat{\sigma}_{\bar{x}}$ (See also Table A.8 for df = ∞.)

k = no. of groups	$H_{0.10}$	$H_{0.05}$	$H_{0.01}$
2	1.16	1.39	1.82
3	1.67	1.91	2.38
4	1.90	2.14	2.61
5	2.05	2.29	2.75
6	2.15	2.39	2.87
7	2.24	2.48	2.94
8	2.31	2.54	3.01
9	2.38	2.60	3.07
10	2.42	2.66	3.12
15	2.60	2.83	3.28
20	2.72	2.94	3.39

The following notes pertain either to the use of ANOM or to an understanding of the procedures.

Note 1. The procedures for p_i and P_i can be combined easily as in Fig. 11.2. Simply compute UDL and LDL using percents, for example, and indicate the percent scale P on one of the vertical scales (the left one in Fig. 11.2). Then mark the fraction scale p on the other vertical scale.

Note 2. Values of H_α are given in Tables A.8 and 11.4 for three risks. We shall frequently draw both sets of decision lines expecting to bracket some of the points. (These three levels are to be considered as convenient reference values and not strict bases for decisions.) A risk of somewhat more than 5 percent is often a sensible procedure.

Note 3. When $k = 2$, and the assumption of normality is reasonable, the comparison of the two values of p is like "Student's t test."[11] Values of H_α corresponding to $k = 2$ are

$$H_\alpha = \frac{\sqrt{2}t_\alpha}{2}$$

where t_α is from a two-tailed t table corresponding to df = ∞. Thus the ANOM, for $k = 2$, is simply a graphical t test, with df = ∞, which amounts to a normal Z test.

In Case History 11.2, as in the previous Example 11.1, the prior author also used a chi-square analysis. His conclusion regarding the question of statistical significance of the data agrees with that of ANOM. It may be helpful to some readers to explain that ANOM is one alternative to a chi-square analysis. A chi-square analysis is not as sensitive as ANOM to the deviation of one or two

[11] See Sec. 13.4.

sources from average, or to trends and other order characteristics. A chi-square analysis is more sensitive to overall variation of k responses. The ANOM is very helpful to the scientist and engineer in identifying specific sources of differences, the magnitude of differences, and is a graphical presentation with all its benefits.

11.4 Introduction to Case Histories

The mechanics of ANOM using attributes data are used in the following case histories. Whatever analysis is used when analyzing data in a troubleshooting or process-improvement project is important only as it helps in finding avenues to improvement. Any analysis is incidental to the overall procedure of approaching a problem in production. The case histories have been chosen to represent applications to different types of processes and products. They have been classified below according to the number of independent variables employed, where independent variables are to be interpreted as discussed in Sec. 9.2.

The planned procedures for obtaining data in Secs. 11.5, 11.6, 11.7, and 11.8 are especially useful throughout industry; yet there is little organized information published on the subject.[12]

Many of the ideas presented here are applied also to variables data in Chaps. 13, 14, and 15.

Outline of case histories (CH) which follow in this chapter

Sec. 11.5 *One independent variable at k levels*

CH 11.1 Spot-welding electronic assemblies

CH 11.2 A corrosion problem with metal containers

CH 11.3 End breaks in spinning yarn

CH 11.4 Bottle-capping

Sec. 11.6 *Two independent variables*

CH 11.5 Alignment and spacing in a cathode-ray gun

CH 11.6 Phonograph pickup cartridges—a question of redesign

CH 11.7 Machine shutdowns (unequal r_i)

Sec. 11.7 *Three independent variables*

CH 11.8 Glass-bottle defectives; machines, shifts, days

CH 11.9 Broken and cracked plastic caps (three vendors)

Sec. 11.8 *Three independent variables; a special latin square*

CH 11.10 Grid-winding lathes

[12] For an excellent illustration of use of this procedure, see L. H. Tomlinson and R. J. Lavigna, "Silicon Crystal Termination—An Application of ANOM for Percent Defective Data," *J. Qual. Tech.*, vol. 15, no. 1, January 1983, pp. 26–32.

11.5 One Independent Variable with *k* Levels

CASE HISTORY 11.1

Spot-Welding Electronic Assemblies[13]

Excessive rejections were occurring in the mount assembly of a certain type of electronic device. Several hundreds of these mounts were being produced daily by operators, using their own spot-welding machine. The mount assemblies were inspected in a different area of the plant, and it was difficult to identify the source of welding trouble.[14]

The department supervisor believed that the trouble was caused by substandard components delivered to the department; an oxide coating was extending too far down on one component. In fact, there was evidence to support this view. The supervisor of the preceding operation agreed that the components were below standard. Even so, whenever any operation with as many as three or four operator-machine combinations is in trouble, a special short investigation is worthwhile. This is true even when the source of trouble is accepted as elsewhere.

This example discusses a straightforward approach to the type of problem described above. It is characterized by (1) the production of a product which can be classified only as "satisfactory" or "unsatisfactory," with (2) several different operators, machines, heads on a machine, or jigs and fixtures all doing the same operation. The procedure is to select in a carefully planned program small samples of the product for a *special study*, inspecting each one carefully, and recording these sample inspection data for careful analysis. Experience has shown that these small samples, obtained in a well-planned manner and examined carefully, usually provide more useful information for corrective action than information obtained from 100 percent inspection. It allows more effort to be allocated to fewer units.

Collecting data

An inspector was assigned by the supervisor to obtain five mounts (randomly) from each operator-welder combination at approximately hourly intervals for two days; then $n_g = (8)(2)(5) = 80$. Each weld was inspected immediately, and a record of each type of weld defect was recorded by operator-welder on a special record form. Over the two-day period of the study, records were obtained on 11 different operator welders as shown in Table 11.5: the percent defective from these eleven combinations, labeled *A, B, C,* . . . , *K*, have been plotted in Fig. 11.2. The average percent of weld rejects for the entire group for the two-day study was $\bar{p} = 66/880 = 0.075$ or $P = 7.5$ percent; this was just about the rate during recent production.

Discussion

Several different factors could have introduced trouble into this spot-welding operation. One factor was substandard components, as some believed. But were there also differences among spot welders, operators, or such factors as the time of day (fatigue), or the day of the week? Did some operators need training? Did some machines need mainte-

[13]Ellis R. Ott, "Trouble-Shooting," *Ind. Qual. Control,* vol. 11, no. 9, June 1955.

[14]In regular production, each operator was assigned to one specific welding machine. No attempt was made in this first study to separate the effects of the operators from those of the machines.

TABLE 11.5 Welding Rejects by Operator Machine

Samples of $n_g = 80$

Operator	No.	Percent
A	3	3.75
B	6	7.5
C	8	10.0
D	14	17.5
E	6	7.5
F	1	1.25
G	8	10.0
H	1	1.25
I	8	10.0
J	10	12.5
K	1	1.25

$\sum = 66$

$p = 66/880 = .075$

$P = 7.5\%$

$$\hat{\sigma} = \sqrt{\frac{(7.5)(92.5)}{80}} = 2.94\%$$

Decision line: $\alpha = .05,\ k = 11$

$\text{UDL} = 7.5\% + (2.70)(2.94)$
$= 15.4\%$

nance? We chose to study a combination of operators with their own regular machines; the supervisor decided to get data from 11 of them.

When we had the data, we plotted it and computed decision lines in Fig. 11.2. Combination D exceeded the upper limit ($\alpha = 0.01$); three combinations F, H, and K were "low." In discussing these four operators, the supervisor assured us without any hesitation that:

- Operator D was both "slow and careless."
- Operator F was very fast and also very careful, and it was the operator's frequent practice to *repeat* a weld.
- Operator H was slow, but careful.
- Operator K was one about whom little was known because the operator was not a regular.

Conclusion

Pooling the attributes information from small samples (of five per hour over a 2-day period) indicated the existence of important differences in operator-welder combinations. These differences were independent of the quality of components being delivered to the department. Efforts to improve the troublesome spraying oxide coating in the preceding department should be continued, of course.

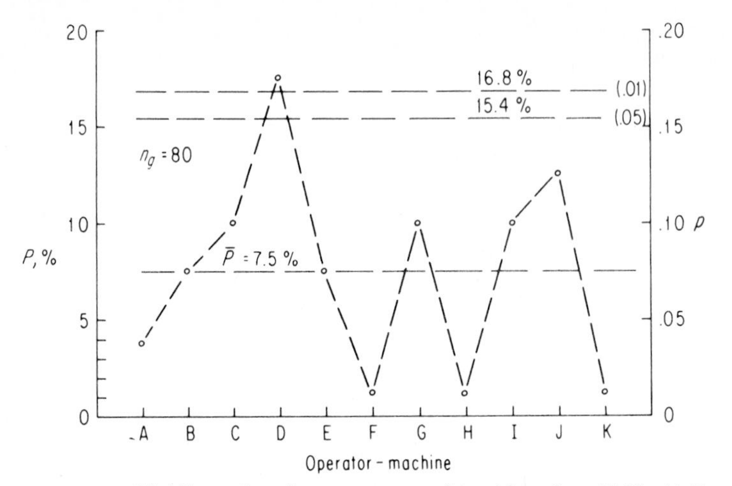

Figure 11.2 Welding rejects by operator machine (data from Table 11.5).

These observed differences in welding suggest also:

1. Combinations *F, H,* and *K* should be watched for clues to their successful techniques in the hope that they can then be taught to others.
2. Combination *D* should be watched to check the supervisor's unfavorable impression.
3. Also the desirability of studying the effect of repeat welding at subsequent stages in the manufacturing process should be studied. This may be an improvement at the welding stage; but its effect on through the assembly needs assessment.

CASE HISTORY 11.2

A Corrosion Problem with Metal Containers[15]

The effects of copper on the corrosion of metal containers was studied by adding copper in three concentrations. After being stored for a time, the containers were examined for failures of a certain type. The data are summarized in Table 11.6 and plotted in Fig.

TABLE 11.6 Effect of Copper on Corrosion

Level of copper, ppm	Containers examined, n_g	failures, d_i	Fraction failing	Percent failing
5	80	14	$p_1 = .175$	17.5
10	80	36	$p_2 = .450$	45.0
15	80	47	$p_3 = .588$	58.8
Totals	240		$\bar{p} = 97/240 = .404$	$\bar{P} = 40.4\%$

[15]H. C. Batson, "Applications of Factorial Chi-Square Analysis to Experiments in Chemistry," *Trans. Amer. Soc. Qual. Control,* 1956, pp. 9–23.

11.3. The large increase in defectives is very suggestive that an increase in parts per million (ppm) of copper produces a large increase in failures. The increase is significant both economically and statistically.

Formal analysis: ANOM

$$\hat{\sigma}_P = \sqrt{\frac{\overline{P}(100 - \overline{P})}{n_g}} = 5.5\% \qquad \text{for } \overline{P} = 40.4\%$$

$$n_g = 80$$

For $k = 3$ and $\alpha = 0.01$, Table A.8 gives $H_{0.01} = 2.38$. Then

$$\overline{P} \pm H_{0.01}\hat{\sigma}_P = 40.4 \pm (2.38)(5.5)$$

$$\text{UDL}(0.01) = 53.5\%$$

and

$$\text{LDL}(0.01) = 27.3\%$$

One point is below the LDL; one point is above the UDL. There is no advantage in computing decision lines for $\alpha = 0.05$.

Whether the suggested trend is actually linear will not be discussed here, but if we assume that it is linear, the increase in rejections per ppm from 5 to 15 ppm is

$$\text{Average increase} = \frac{58.8\% - 17.5\%}{10} = 4.13\% \text{ per ppm}$$

This was considered a very important change.

CASE HISTORY 11.3

End Breaks in Spinning Cotton Yarn[16]

The problem

An excessive number of breaks in spinning cotton yarn was being experienced in a textile mill. It was decided to make an initial study on a sample of eight frames to deter-

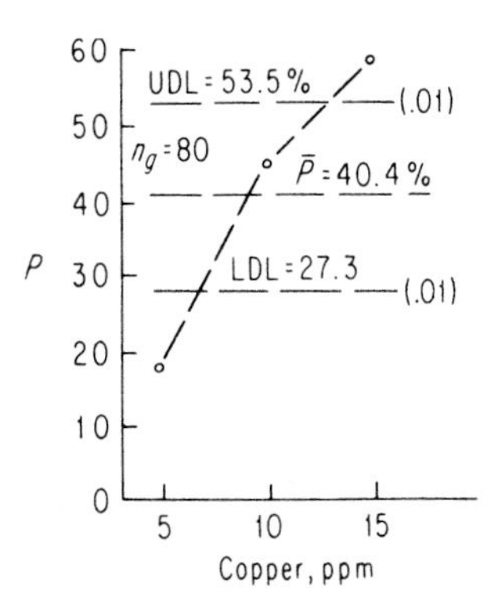

Figure 11.3 Effect of copper on corrosion (data from Table 11.6).

[16]Dr. Sundari Vaswani and Ellis R. Ott, "Statistical Aids in Locating Machine Differences," *Ind. Qual. Control,* vol. 11, no. 1, July, 1954.

mine whether there were any essential differences in their behavior. Rather than take all the observations at one time, it was decided to use random time intervals of 15 min until data were on hand for ten such intervals on each of the eight frames.

Each frame contained 176 spindles. As soon as a break occurred on a spindle, the broken ends were connected or "pieced" together and spinning resumed on that spindle. (The remaining 175 spindles continued to spin during the repair of the spindle.) Thus the number of end breaks during any 15-min interval is theoretically unlimited, but we know from experience that it is "small" during ordinary production.

The selection of a 15-min interval was an arbitrary decision for the initial study. Similarly, it was decided to include eight frames in the initial study.

The number of end breaks observed in 15 min per frame is shown in Table 11.7 and in Fig. 11.4.

Conclusions

It is apparent that there is an applicable difference between frames. Those with averages outside of the (0.01) decision lines are:

Excessive breaks: frames 5 and 8

Few breaks: frames 2, 3, and 7

The analysis using circles and triangles in Table 11.7 given in analysis 1 below, provides some insight into the performance of the frames.

TABLE 11.7 End Breaks during Spinning Cotton Yarn

Sample no.	Frame no. 1	2	3	4	5	6	7	8	Total
1	13	7	22	15	20	23	15	14	129
2	18	10	7	12	19	17	18	22	123
3	8	8	21	14	15	16	8	8	98
4	13	12	8	10	23	3	12	20	101
5	12	6	9	27	32	4	9	18	117
6	6	6	6	17	34	12	1	24	106
7	16	20	5	9	8	17	7	21	103
8	21	9	2	13	10	14	7	17	93
9	17	14	9	24	21	8	6	33	132
10	16	7	7	10	14	10	6	11	81
Frame Avg.	14.0	9.9	9.6	15.1	19.6	12.4	8.9	18.8	Grand Avg. = 13.54

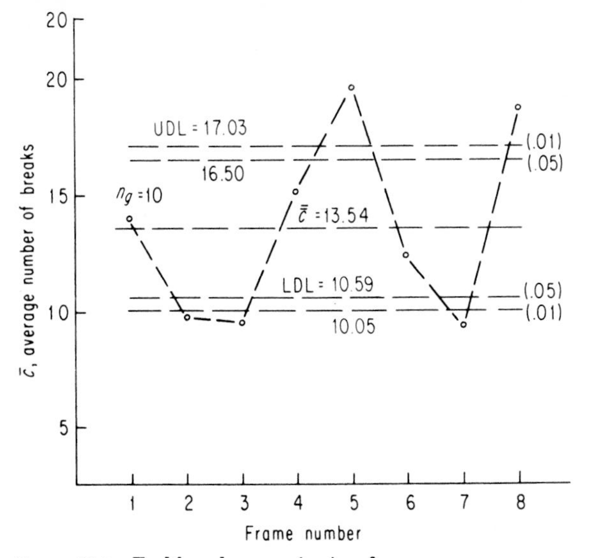

Figure 11.4 End breaks on spinning frames.

Analysis 1: a "quick analysis"

The average number of breaks per time interval is $\bar{c} = 13.54$. Then for a Poisson distribution (Sec. 3-4)

$$\hat{\sigma}_c = \sqrt{13.54} = 3.68$$

for individual entries in Table 11.7. Let us consider behavior at $\bar{c} \pm 2\sigma_c$ level since we are interested in detecting possible sources of trouble:

$$13.54 + 2(3.68) = 20.90 \text{ (figures in circles)}^{17}$$

$$13.54 - 2(3.68) = 6.18 \text{ (figures in triangles)}^{17}$$

Conclusions

A visual inspection of the individuals so marked suggests:

1. Frames 4, 5, and 8 are suspiciously bad since there are circles in each column and no triangles.
2. Frames 2 and 7 look good; there are at least two triangles in each and no circles. (Frame 3 shows excellent performance except for the two circled readings early in the study.)

Analysis 2: ANOM (the mechanics to obtain decision lines in Fig. 11.4)

Each frame average is of $n_g = 10$ individual observations. In order to compare them to their own group average, we compute

[17]In practice, we use two colored pencils. This analysis is a form of NL gauging; see Chap. 7.

$$\hat{\sigma}_{\bar{c}} = \hat{\sigma}/\sqrt{n_g} = 3.68/\sqrt{10} = 1.16$$

From Table A.8,[18] values of H_α for $k = 8$ are: $H_{0.05} = 2.54$ and $H_{0.01} = 3.01$.

Then for $\alpha = 0.05$

$$\text{UDL}(0.05) = 13.54 + 2.54(1.16) = 16.48$$

$$\text{LDL}(0.05) = 13.54 - 2.54(1.16) = 10.59$$

and for $\alpha = 0.01$

$$\text{UDL}(0.01) = 17.03$$

$$\text{LDL}(0.01) = 10.05$$

A further comment

Other proper methods of analyzing the data of Table 11.7 include chi-square and analysis of variance. Each of them indicates nonrandomness of frame performance; they need to be supplemented to indicate specific presses giving different behavior and the magnitude of that behavior difference.

Process action resulting from study

As a result of this initial study, an investigation was conducted on frames 5 and 8 which revealed, among other things, defective roller coverings and settings. Corrective action resulted in a reduction in their average breaks to 11.8 and 8, respectively; a reduction of about 50 percent. A study of reasons for better performance of frames 2, 3, and 7 was continued to find ways to make similar improvements in other frames in the factory.

CASE HISTORY 11.4

An Experience with a Bottle Capper

This capper has eight rotating heads. Each head has an automatic adjustable chuck designed to apply a designated torque. Too low a torque may produce a *leaker;* too high a torque may break the plastic cap or even the bottle.

It is always wise to talk to line operators and supervisors: "Any problem with broken caps?" "Yes, quality control has specified a high torque, and this is causing quite a lot of breakage." After watching the capper a few minutes, a simple tally of the number of broken caps from each head was made. (See Table 11.8 and Fig. 11.5.)

Head 8 is evidently breaking almost as many caps as all others combined (see formal analysis below). Too high a torque specification? Or inadequate adjustment on head 8? The answer is obviously the latter. In theory, broken caps may be a consequence of the capper, the caps, or the bottles. It is human to attribute the cause to "things beyond my responsibility."

[18]The individual observations are considered to be of Poisson type; this is somewhat skewed with a longer tail to the right. However, averages of as few as four such terms are essentially normally distributed. Consequently, it is proper to use Table A.8; see Theorem 3, Chap. 1.

TABLE 11.8 Plastic Caps Breaking at the Capper

Head no.	f (Number broken)
1	1
2	1
3	2
4	2
5	1
6	2
7	2
8	9

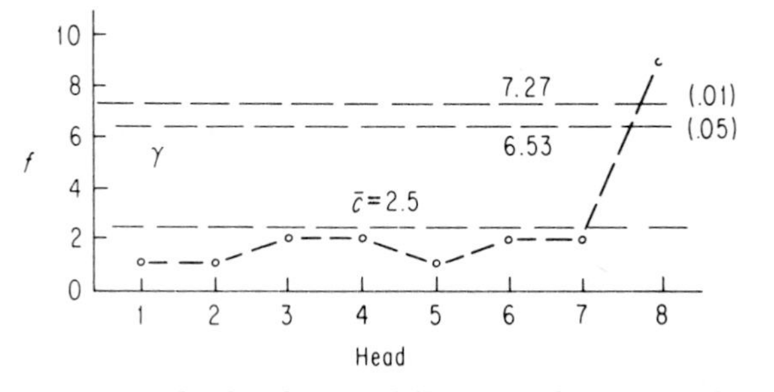

Figure 11.5 Cap breakage at different heads. Actual value of n is unknown, but 50 is a guess.

Discussion

If there had been no significant differences between heads, what then?

1. How many cavities are in the bottle mold? Probably four or eight. Let us hold out (collect) 20 or 30 broken-capped bottles and check the mold numbers of the bottles which are (or should be) printed during molding. Often, the great majority of defectives will be from one or two bottle molds.
2. How many cavities are producing caps? Probably 8 or 16. Let us take the same 20 or 30 broken-capped bottles and check the cavity numbers of the caps (also printed at molding). It is not unusual to find a few bottle-cap cavities responsible for a preponderance of broken caps.

Formal analysis

The *number* of breaks on each head is known for the period of observation. Opportunity for breaks was "large," but the incidence was "small" (see Sec. 5.4); a Poisson distribution is a reasonable assumption.

$$\bar{c} = 20/8 = 2.5; \hat{\sigma}_c = \sqrt{2.5} = 1.58$$

$$H_{0.05} = 2.54 \text{ for } k = 8$$

Then

$$\text{UDL}(.05) = \bar{c} + H_{0.05}\hat{\sigma} = 2.5 + 4.01 = 6.51$$

Also, UDL(.01) = 7.26

Conclusion

The point corresponding to head 8 is above UDL(0.01); this simply supports the intuitive visual analysis that head 8 is out of adjustment.

11.6 Two Independent Variables

Introduction

Our emphasis here and elsewhere will be upon planning and analyzing data from studies to identify *sources* of trouble; engineering and production personnel then use the resulting information to reduce that trouble. The ideas presented in Chap. 9 will be illustrated in these discussions. In particular, omnibus-type independent variables will be used frequently. Troubleshooting can usually be improved by *data collection plans* which employ more than one independent variable. Such plans speed up the process of finding the sources of trouble with little or no extra effort. This section will consider the important and versatile case of *two independent variables;* Sec. 11.7 the case of *three independent* variables.

When using temperatures of 100°, 120°, 140°, for example, it is said that the independent variable (temperature) has been used at three *levels*. Similarly, if a study considers three machines and two shifts, it is said that the study considers machines at three *levels* and shifts at two *levels*.

Consider a study planned to obtain data at a levels of variable A and b levels of variable B. When data from every one of the $(a \cdot b)$ possible combinations is obtained, the plan is called a *factorial design*. When A and B are each at *two* levels, there are $2^2 = 4$ possible combinations; the design is called a 2^2 factorial (two-squared factorial). Such designs are very effective and used frequently.

Two independent variables: a 2^2 factorial design

This procedure will be illustrated by Case History 11.5; then the analysis will be discussed.

CASE HISTORY 11.5

Comparing Effects of Operators and Jigs in a Glass-Beading Jig Assembly (Cathode Ray Guns)

Regular daily inspection records were being kept on twelve different hand-operated glass-beading jigs sometimes called machines. The records indicated that there were appreciable differences in the number of rejects from different jigs although parts in use came from a common source. It was not possible to determine whether the differences were attributable to jigs or operators without a special study. There was conflicting evidence, as usual. For example, one jig had just been overhauled and adjusted; yet it was producing more rejects than the departmental average. Consequently, its operator (Harry) was considered to be the problem.

The production supervisor, the production engineer, and the quality control engineer arranged an interchange of the operator of the recently overhauled jig with an operator from another jig to get some initial information. From the recent production, prior to interchange, 50 units from each operator were examined for two quality characteristics: (1) *alignment* of parts, and (2) *spacing* of parts. The results for alignment defects in the morning's sample are shown in Fig. 11.6 in combinations 1 and 4.

Then the two operators interchanged jigs. Again, a sample of 50 units of each operator's assembly was inspected; the results of the before and after interchange are shown in Fig. 11.6. The same inspector examined all samples.

Alignment defects

Totals for the 100 cathode ray guns assembled by each operator are shown at the bottom of Fig. 11.6; totals for the 100 guns assembled on each jig are shown at the right; and the numbers of rejects produced on the original and interchanged machines are shown at the two bottom corners.

Operators	Machines (jigs)	Interchange
Art: 16/100 = 16%	9: 11/100 = 11%	Original machines: 18/100 = 18%
Harry: 19/100 = 19%	10: 24/100 = 24%	Interchanged machines: 17/100 = 17%

Discussion

The difference in jig performance suggests a problem with jig 10. This difference is called a *jig main effect.* (Any significant difference between operators would be called an *operator main effect.*) This surprised everyone, but was accepted without any further analysis.

The performance of jig 10 is significantly worse than that of jig 9 (see Fig. 11.7) even though it had just been overhauled. The magnitude of the difference is large:

$$\bar{\Delta} = 24\% - 11\% = 13\%$$

Neither the small difference observed between the performances of the two operators nor between the performance of the operators *before* and *after* the interchange is statistically significant.

On the basis of the above information, all operators were called together and a program of extending the study in the department was discussed with them. The result of

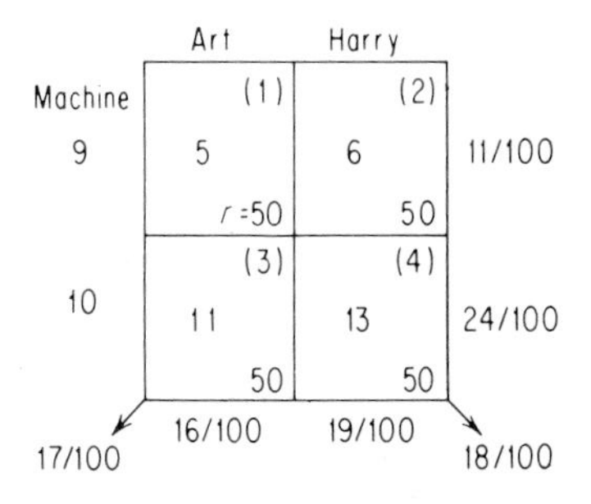

Figure 11.6 Alignment defects found in samples during an interchange of two operators on two machines. (The number of defects is shown in each square at the center and the sample size in the lower-right corner. See Case History 11.5.)

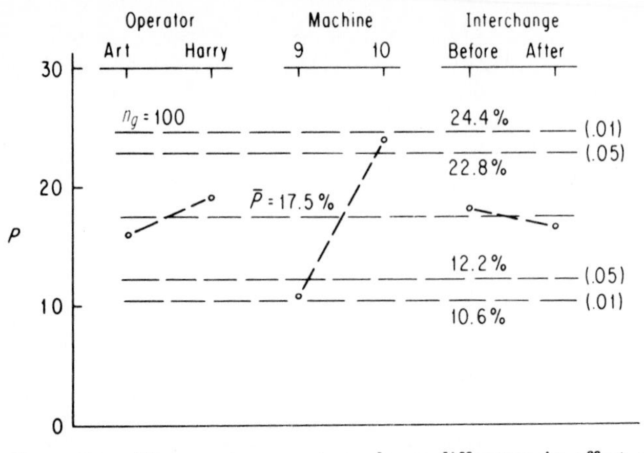

Figure 11.7 Alignment comparison shows difference in effect of machines, but not in operators or before-and-after effect (ANOM) (data from Fig. 11.6).

the interchange of the operators was explained; they were much interested and entirely agreeable to having the study extended.

An extension of the interpretation from the study

One interpretation of such an interchange would be that the two operators performed differently when using their own machine than when using a strange machine. Such preferential behavior is called an *interaction;* more specifically here, it is an operator-machine interaction.

A variation of this procedure and interpretation has useful implications. *If* the operators were not told of the proposed interchange until *after* the sample of production from their own machines had been assembled, then:

1. The number of defectives in combinations 2 and 3 made *after* the interchange (Fig. 11.6), *could well* be a consequence of more careful attention to the operation than was given *before* the interchange. Since the number of defects *after* the interchange is essentially the same as before, there is no suggestion from the data that "attention to the job" was a factor of any consequence in this study.
2. It would be possible in other 2^2 production studies that a proper interpretation of an observed difference between the two diagonals of the square (such as in Fig. 11.6) might be a mixture (combination) of "performance on their own machines" and "attention to detail." Such a mixture is said to represent a *confounding* of the two possible explanations.

Spacing defects

The same cathode ray guns were also inspected for *spacing* defects. The data are shown in Fig. 11.8.

Operators	*Jigs*
Art: 6/100 = 6%	9: 3/100 = 3%
Harry: 8/100 = 8%	10: 11/100 = 11%

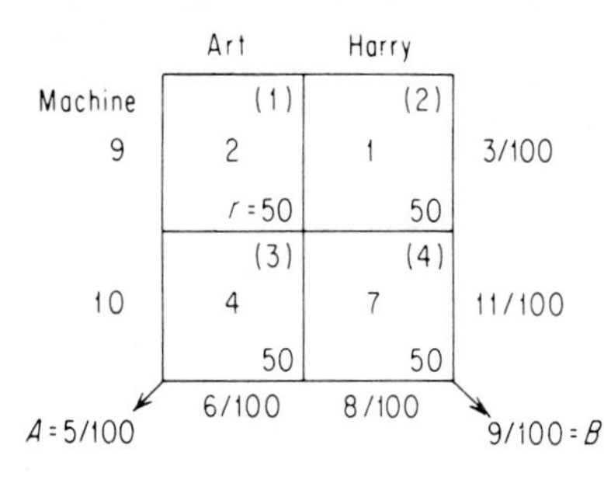

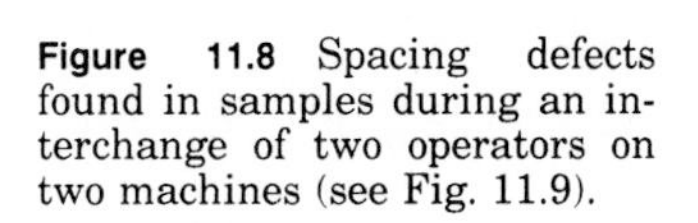
Figure 11.8 Spacing defects found in samples during an interchange of two operators on two machines (see Fig. 11.9).

Before-after interchange
Before: 9/100 = 9%
After: 5/100 = 5%

In Fig. 11.9, the difference between jigs is seen to be statistically significant for spacing defects, risk less than 5%. Since it seemed possible that the before-after interchange difference might be statistically significant for $\alpha = 0.10$, a third pair of decision lines have been included; the pair of points for *B*, *A*, lie inside them, however. There is the possibility that the interaction *might* prove to be statistically significant if a larger sample had been inspected.

Formal analysis

$\bar{P} = 7.0\%$; $k = 2$ in each of the three comparisons.

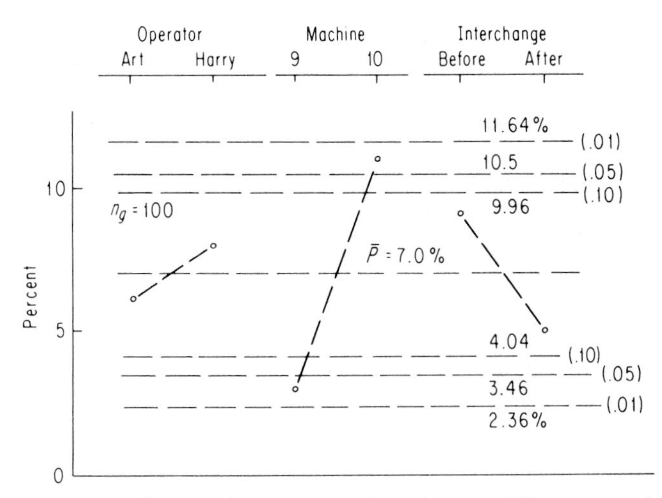

Figure 11.9 Spacing defects comparison showing difference in effect of machines, but not in operators or before-and-after interchange (data from Fig. 11.8).

$$\hat{\sigma}_P = \sqrt{\frac{7.0(93.0)}{100}} = 2.55$$

Decision lines (in Fig. 11.9)

For $\alpha = 0.05$, $H_{0.05} = 1.39$

$$\text{UDL} = 10.54$$
$$\text{LDL} = 3.46$$

For $\alpha = 0.10$, $H_{.10} = 1.16$

$$\text{UDL} = 9.96\%$$
$$\text{LDL} = 4.04\%$$

Two independent variables: a typical *axb* factorial design

The design of experiments has been described often with examples from agriculture and the chemical industries. It is at least equally important to use many of the same concepts in the electrical, electronic, and mechanical fields, but rather less complicated versions of them are recommended for troubleshooting studies in industry. The following example represents a small factorial experiment which was carried out three times in 3 days.

CASE HISTORY 11.6

A Multistage Assembly[19]

In many complicated assembly operations, we do not find the problems until the final inspection report made at the end of the assembly. Sometimes it is feasible to carry an identification system through assembly, and establish major sources of trouble on a regular, continuing basis. Many times, however, it is not feasible to maintain an identification system in routine production. In the study considered here, no one could determine who was responsible for a poor quality or inoperative unit found at final inspection. This operator says, "They are good when they leave me"; another says, "It's not my fault." No one accepts the possibility of being responsible.

We discuss an experience just like this in which a routing procedure through assembly was established on a sampling basis. The procedure was conceived in frustration; it is remarkably general and effective in application.

A particular new stereo pickup cartridge was well-designed in the sense that those cartridges which passed the final electrical test performed satisfactorily. However, too many acoustical rejects were being found at final testing, and the need of an engineering redesign had been considered and was being recommended.

There are, of course, many engineering ways of improving the design of almost any product; a redesign is often an appropriate approach to the solution of manufacturing quality problems. Perhaps a change in certain components or materials or other redesign is considered essential. But *there is an alternative method,* too frequently over-

[19]Ellis R. Ott, Achieving Quality Control, *Qual. Prog.*, May, 1969. (Figures 11.10, 11.12, 11.13, and 11.14 reproduced by permission of the editors.)

looked, and this is to determine whether the components, the assembly operators, and the jigs or fixtures—any aspect of the entire production process—are capable of a major improvement.

The following study was planned to compare assembly operators at two of the many stages of a complicated assembly process.

The two stages were chosen during a meeting called to discuss ways and means. Present were the production supervisor, the design engineer, and the specialist on planning investigations. During the meeting, different production stages thought to be possible contributors to the acoustical defects were suggested and discussed critically. Two omnibus-type variables were chosen for inclusion in this exploratory study: then at every other stage, each tray in the study was processed at the same machine, by the same operator, in the same way throughout as consistently as possible. No one was at fault; of course not. The purpose of the study was to determine whether substantial improvements might be possible within the engineering design of the cartridge by finding ways to adjust assembly procedures.

- Four *operators* at stage *A*, with their own particular machines, were included; the four operator-machine combinations[20] are designated as A_1, A_2, A_3, and A_4.
- Also included in the study were *three operators* performing a *hand operation* using only tweezers; these operators are designated as C_1, C_2, and C_3.
- A *three-by-four factorial* design with every one of the $3 \cdot 4 = 12$ combinations of *A* and *C* was planned. Twelve trays—or perhaps 16—can usually be organized and carried around a production floor without mishap. More than 15 or 16 is asking for trouble. Each standard production tray had spaces for 40 cartridges.
- A routing ticket as in Fig. 11.10 was put in each of the 12 trays to direct passage

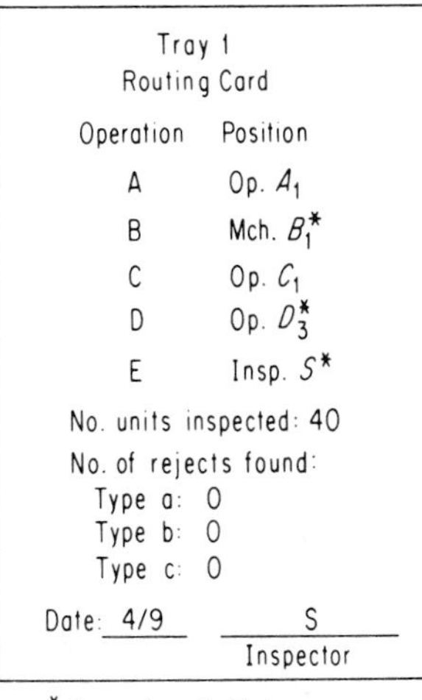

Tray 1
Routing Card

Operation	Position
A	Op. A_1
B	Mch. B_1^*
C	Op. C_1
D	Op. D_3^*
E	Insp. S^*

No. units inspected: 40
No. of rejects found:
Type a: 0
Type b: 0
Type c: 0
Date: 4/9 S Inspector

*Same for all 12 trays

Figure 11.10 Routing card used to obtain data on pickup cartridge assembly.

[20]The operators and machines were confounded deliberately. This was an exploratory study, and problems associated with interchanging operators and machines were considered excessive in comparison to possible advantages.

through the assembly line. The 12 trays were numbered 1, 2, 3,..., 12. All components were selected at random from a common source.

- Each cartridge (unit) was inspected for all three listed defects, (a), (b), and (c) and the entries in Fig. 11.11 indicate the number of each type of defect found at inspection. Since defect type c was found most frequently, Fig. 11.12 is shown for type c only. A mental statistical analysis, or "look test" at the data of Fig. 11.12, indicates clearly that C_2 is an operator producing many rejects. Also A_1 is substantially the best of the four operator-machine combinations.
- This was surprising and important information. Since not all results are as self-evident, an analysis is given in Fig. 11.13, first for columns A and then for rows C.
- This analysis shows that the differences were large and of practical significance as well as being statistically significant. Also, on the basis of differences indicated by this study, certain adjustments were recommended by engineering and manufacturing following a careful study and comparison of the operators at their benches. The study was repeated two days later. The very few type c defects found are shown in Fig. 11.14. The greatly improved process average of $\bar{P} = (6)(100)/480 = 1.2$ percent compares strikingly with the earlier $\bar{P} = 11.2$ percent which had led to considering an expensive redesign of components to reduce rejects. The improvement was directed to the same people who had decided that a redesign was the only solution to the problem.
- Formal analysis is shown for Figs. 11.12 and 11.13. The overall percent defective is

$$\bar{P} = 100(54)/(12)(40) = 5400/480 = 11.2\%$$

Then

$$\hat{\sigma}_P = \sqrt{\frac{(11.2)(88.8)}{n_g}}$$

where $n_g = (3)(40) = 120$ when comparing columns: $\hat{\sigma}_P = 2.9\%$
and $n_g = (4)(40) = 160$ when comparing rows: $\hat{\sigma}_P = 2.5\%$

Decision lines for the analysis of means are as follows:

Columns A: $k = 4$, $\hat{\sigma}_P = 2.9\%$, $H_{0.05} = 2.14$, $H_{0.01} = 2.61$

$r = 40$	Operator A_1	A_2	A_3	A_4
Operator C_1	a: 0 b: 0 c: 0	a: 1 b: 1 c: 1	a: 0 b: 0 c: 7	a: 0 b: 0 c: 3
C_2	a: 0 b: 0 c: 2	a: 0 b: 0 c: 14	a: 0 b: 0 c: 8	a: 2 b: 2 c: 5
C_3	a: 0 b: 0 c: 1	a: 0 b: 1 c: 5	a: 0 b: 2 c: 2	a: 0 b: 1 c: 6

Figure 11.11 Record of the defects of each type found in the first study, arranged according to the combination of operators from whom they originated for defect types a, b, and c.

$r = 40$	A_1	A_2	A_3	A_4	Σ	
C_1	0 ($r = 40$)	1	7	3	11	11/160 = 6.9%
C_2	2	14	8	5	29	29/160 = 18.1%
C_3	1	5	2	6	14	14/160 = 8.75%
Σ	3/120	20/120	17/120	14/120	54/480	
	2.5%	16.6%	14.2%	11.6%	11.2%	

Figure 11.12 Defects of type c only (data from Fig. 11.11). Total defects of this type shown at right and at bottom by operator.

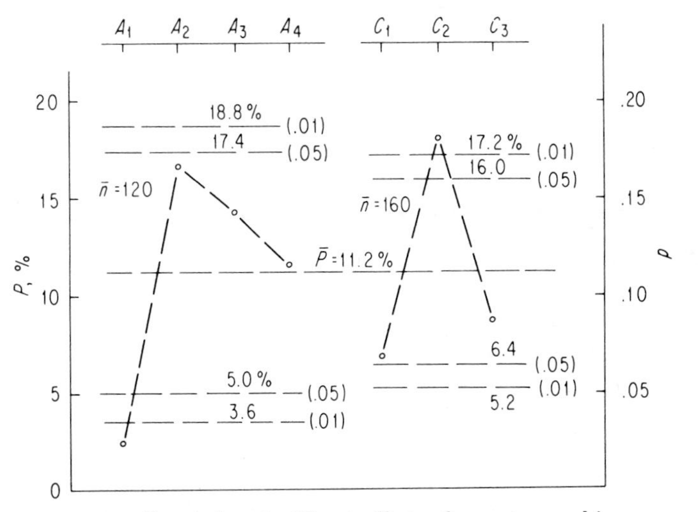

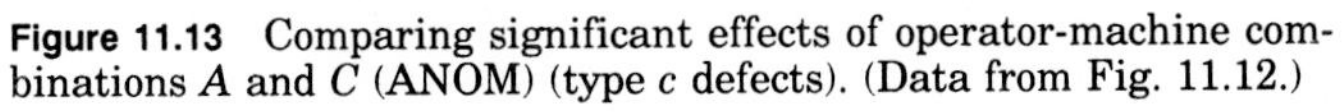

Figure 11.13 Comparing significant effects of operator-machine combinations A and C (ANOM) (type c defects). (Data from Fig. 11.12.)

	$\alpha = .05$	$\alpha = .01$
$\bar{P} \pm H_\alpha \hat{\sigma}_P$	UDL = 17.4	18.8%
	LDL = 5.0	3.6%

Rows C: $k = 3$, $\hat{\sigma}_P = 2.5$, $H_{0.05} = 1.91$, $H_{0.01} = 2.38$

$\alpha = .05$	$\alpha = .01$
UDL = 16.0	17.2%
LDL = 6.4	5.2%

$r = 40$	A_1	A_2	A_3	A_4	Σ
C_1	0 (40)	0	0	1	1
C_2	0	0	3	1	4
C_3	0	0	0	1	1
Σ	0	0	3	3	6/480 = 1.2%

Figure 11.14 Number of defects found in second study of pickup cartridge assemblies (type *c* defects).

CASE HISTORY 11.7

Machine Shutdowns (Unequal r_i)

This case history presents a variation in the type of data employed. It is different also in that the sample sizes are *not all equal.*

It is routine procedure in some plants to designate the source of data by the machine, shift, and/or operator. The data recorded[21] in Table 11.9 serve to illustrate a procedure of obtaining data for a process-improvement project and of providing an analysis when sample sizes are not all equal. The data relate to the performance of five different

TABLE 11.9 Talon's Press-Shift Performance Record

Press number	Shift	Number of times checked	Number of shutdowns
1	A	50	2
	B	55	7
	C	40	4
2	A	45	3
	B	55	3
	C	55	14
3	A	40	0
	B	35	3
	C	45	0
4	A	50	6
	B	55	9
	C	60	11
5	A	60	4
	B	45	3
	C	60	6

[21] J. Stuart Zahniser and D. Lehman, "Quality Control at Talon, Incorporated," *Ind. Qual. Control,* March 1951, pp. 32–36. (Reproduced by kind permission of the editor.)

presses (machines) over three shifts. Table 11.9 indicates both the number of times each press was checked and the number of times its performance was so unsatisfactory that the press was shut down for repairs. The quality of performance is thus indicated by *frequency of shutdowns*. We quote from the Zahniser-Lehman article:

> As the product comes from these presses, there are 57 separate characteristics that require an inspector's audit. It takes two or three minutes to complete the examination of a single piece. Yet the combined production from these presses reaches more than a million a day. To cover this job with a series of charts for variables would require a minimum of 15 inspectors, ... (We) use a ... "shutdown" chart. The shutdown chart is a p chart on which each point gives a percentage of checks resulting in shutdowns for a given press on a given shift during a two-week period. On this chart, r represents the number of times the press has been checked on this shift during the particular period....
>
> The percentage of checks resulting in shutdowns in the department is 10 percent. The average number of checks per press per shift is 750/15 = 50.
>
> The *speed* of the presses has been increased by 32 percent since control charts were first applied. Nevertheless, we find that the percentage of audits which result in shutdowns has been cut squarely in half. Obviously the quality of product coming from them is (also) vastly better than it was three years ago, even though our charts measure quality only indirectly.

The enhancement of records by charting a continuing history is demonstrated clearly.

The data of Table 11.9 have been rearranged in Table 11.10 in the form of an obvious three-by-five (3 × 5) factorial design. The numbers in the lower right-hand corner refer to the number of check inspections made. The percent of all shutdowns for each shift (across the five presses) has been indicated in the column at the right. The percent of shutdowns for each press across the three shifts is shown at the bottom.

TABLE 11.10 Table of Shutdowns

	Press					
Shift	1	2	3	4	5	Total, Percent
A	$d = 2$ $r = 50$	3 45	0 40	6 50	4 60	15/245 = 6.1
B	7 55	3 55	3 35	9 55	3 45	25/245 = 10.2
C	4 40	14 55	0 45	11 60	6 60	35/260 = 13.4
Total	13/145	20/155	3/120	26/165	13/165	$\bar{P}$ = 75/750 = 10%
Percent	9.0	12.9	2.5	15.7	7.9	

Press performance

The percent of shutdowns on presses 1 through 5 is

9.0%, 12.9%, 2.5%, 15.7%, and 7.9%

The excellent performance (2.5 percent) of press 3 should be of most value for improving the process.

We compute tentative decision lines as shown in Fig. 11.15*a* for risks (0.05) and (0.01). These will be adjusted subsequently.

Press performance (k = 5, $\bar{n}$ = 750/5 = 150)

Press 3

This is below the tentative (0.01) decision line. A look now at Table 11.10 shows that press 3 was checked only 120 times (compared to the average of $\bar{n}$ = 150). A recomputation of decision limits for n_g = 120 shows a slight increase in limits which does not affect materially the conclusion that press 3 performs significantly better than the overall average.

Press 4

This is just on the tentative (0.05) decision lines. A check shows that this press was checked n_g = 165 times—more than average. A recomputation of decision lines using n_g = 165 instead of 150 will shrink slightly the limits for press 4; see Fig. 11.16. There is reason to expect that some way can be found to improve its performance.

Shift performance (k = 3, $\bar{n}$ = 750/3 = 250)

Shift A

This is beyond (0.05) decision limit—and the shutdown rate is just about *one-third* that of shift C. Possible explanations include:

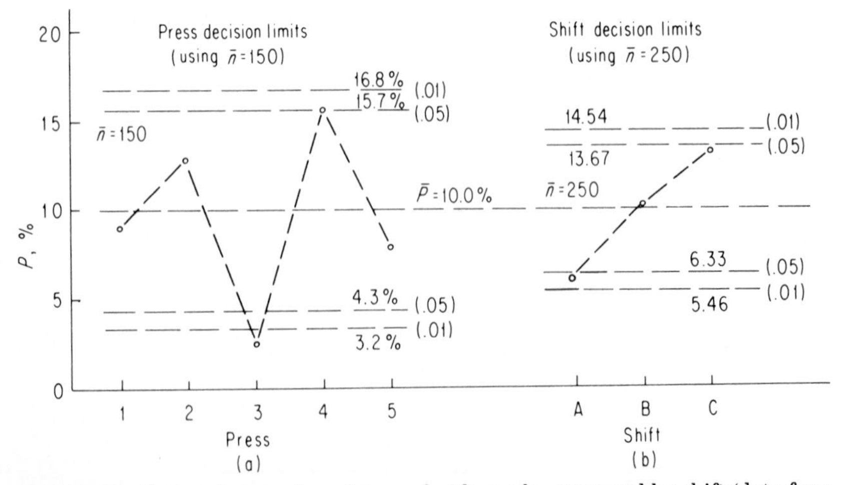

Figure 11.15 Comparing number of press shutdowns by press and by shift (data from Table 11.10).

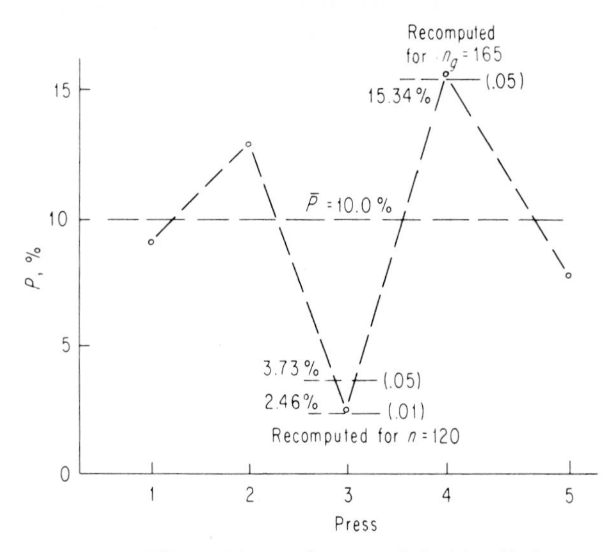

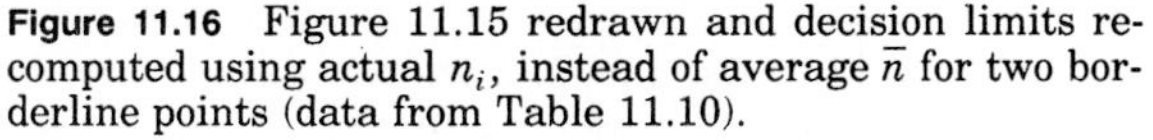

Figure 11.16 Figure 11.15 redrawn and decision limits recomputed using actual n_i, instead of average $\bar{n}$ for two borderline points (data from Table 11.10).

1. More experienced personnel on the shift A.
2. Better supervision on the shift A.
3. Less efficient checking on press performance.
4. Better conditions for press performance (temperature, lighting, humidity, as examples).

Adjusted decision lines

Now obtained for the actual values n_i, with $\alpha = 0.05$ for press 4 and $\alpha = 0.01$ for press 3. Decision lines in Fig. 11.15*a* were computed for average $\bar{n} = 150$. Only presses 3 and 4 are near the decision lines and are the only ones for which the decision might be affected when the actual values of n are used to recompute.

Press 3 (LDL)

$n_g = 120$ from Table 11.10

$$\hat{\sigma}_P = \sqrt{\frac{(10)(90)}{120}} = 2.74\%$$

$$\text{LDL}(0.01) = 10.0 - (2.75)(2.74) = 2.46\%$$

Press 4 (UDL)

$n_g = 165$

$$\hat{\sigma}_P = \sqrt{\frac{900}{165}} = 2.33\%$$

$$\text{UDL}(0.05) = 10.0 + (2.29)(2.33) = 15.34\%$$

These two changes in decision lines are shown in Fig. 11.16 by dotted lines. The slight changes would hardly affect the decision on whether to investigate. Unless the variation of an individual n_g from average $\bar{n}$ is as much as 40 percent it will probably not warrant recomputation.[22]

11.7 Three Independent Factors

Introduction

Some may have had actual experience using multifactor experimental designs. Some have not, and it is fairly common for new converts to propose initial experiments with four, five, and even more independent variables.

Anyone beginning the use of more than one independent variable (factor) is well-advised to limit the number of variables to two. There are several reasons for the recommendation. Unless someone industrially experienced in the methods is guiding the project, *something* is almost certain to go wrong in the implementation of the plan, or the resulting data will present difficulties of interpretation. These typical difficulties can easily lead to plant-wide rejection of the entire concept of multivariable studies. The advantage of just two independent variables over one is usually substantial; it is enough to warrant a limitation to two until confidence in them has been established. Reluctance to use more than one independent variable is widespread—it is the rule, not the exception. An initial failure can be serious, and the risk is seldom warranted.

After some successful experience with two independent variables—including an acceptance of the idea by key personnel, three variables often offer opportunities for getting even more advantages. The possibilities differ from case to case but they include economy of time, materials, and testing, and a better understanding of the process being studied. Some or all of the three variables should ordinarily be of the omnibus or chunky type, especially in troubleshooting expeditions.

There are two designs (plans) using three independent factors which are especially useful in exploratory projects in a plant. The 2^3 *factorial* design (read "two cubed")—i.e., three independent variables each at two levels—is least likely to lead to mixups in carrying out the program and will result in minimal confusion when interpreting results. Methods of analysis and interpretation are discussed in Case Histories 11.8 and 11.9.

The 2^3 factorial design provides for a comparison of the following effects:

1. *Main effects* of each independent variable (factor).
2. *Two-factor interactions* of the three independent variables.

These different comparisons are obtained with the same number of experi-

[22]Since the sample sizes were not very different, this analysis did not use the special H_α** factors available for use with unequal sample sizes. This kept the analysis as simple and straightforward as possible. See L. S. Nelson, "Exact Critical Values for Use with the Analysis of Means," *J. Qual. Tech.*, vol. 15, no. 1, January 1983, pp. 43–44. See also Table A.21.

mental units as would be required for a comparison of the main effects and at the same level of significance. An added advantage is that the effect of each factor can be observed under differing conditions of the other two factors.

3. *Three-factor interaction* of the three variables. Technically this is a possibility. Experience indicates, however, that this is of lesser importance in troubleshooting projects with attribute data. The *mechanics* of three-factor interaction analysis will be indicated in some of the case histories.

CASE HISTORY 11.8

Strains in Small Glass Components

The production of many industrial items involves combinations of hand operations with machine operations. This discussion will pertain to cathode ray tubes. Excessive failures had arisen on a reliable type of tube at the "boiling-water" test in which samples of the completed tubes were submerged in boiling water for 15 sec and then plunged immediately into ice water for 5 sec. Too many stem cracks were occurring where the stem was sealed to the wall tubing. A team was organized to study possible ways of reducing the percent of cracks. They decided that the following three questions related to the most probable factors affecting stem strength.

1. Should air be blown on the glass stem of the tube after sealing? If so, at what pounds per square inch (psi)?
2. Should the mount be molded to the stem by hand operation or by using a jig (fixture)?
3. The glass stems can be made to have different strain patterns. Can the trouble be remedied by specifying a particular pattern?

Factors selected

It was decided to include two levels of each of these three factors.

Air A_1: 2.5 psi air blown on stem after sealing
A_2: No air blown on stem after sealing
Jig B_1: Stem assembly using a jig
B_2: Stem assembly using only hand operations (no jig)
Stem tension C_1: Normal stem (neutral to slight compression)
C_2: Tension stem (very heavy)

Approximately 45 stems were prepared under each of the eight combinations of factors, and the number of stem failures on the boiling water test is indicated in the squares of Table 11.11*a* and *b*. These are two common ways of displaying the data. Manufacturing conditions in combination 1, for example, were at levels A_1, B_1, C_1, for $r = 42$ stems; there were two stem cracks. Stems in combination 2 were manufactured under conditions A_2, B_1, C_2, and so forth.

Main effects

Half the combinations were manufactured under air conditions A_1; those in the four groups (1), (3), (5), and (7). The other half (2), (4), (6), (8) were manufactured under air

TABLE 11.11 A Study of Stem Cracking: A 2^3 Production Design

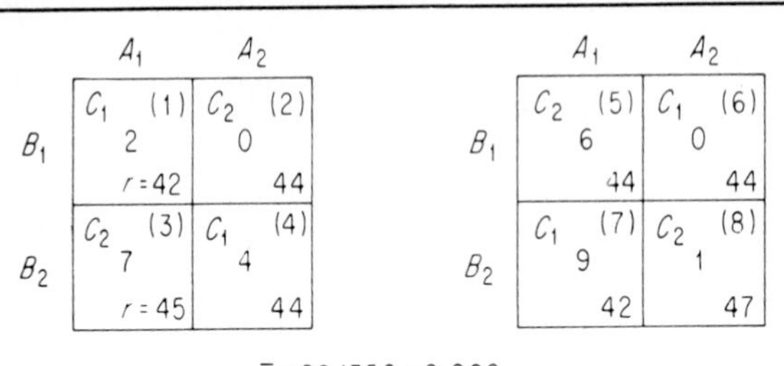

	A_1	A_2
B_1	C_1 (1) 2, $r = 42$	C_2 (2) 0, 44
B_2	C_2 (3) 7, $r = 45$	C_1 (4) 4, 44

	A_1	A_2
B_1	C_2 (5) 6, 44	C_1 (6) 0, 44
B_2	C_1 (7) 9, 42	C_2 (8) 1, 47

$\bar{p} = 29/352 = 0.082$

(a)

	A_1		A_2	
	B_1	B_2	B_1	B_2
C_1	(1) 2, $r = 42$	(7) 9, 42	(6) 0, 44	(4) 4, 44
C_2	(5) 6, 44	(3) 7, 45	(2) 0, 44	(8) 1, 47

(b)

conditions A_2. Then, in comparing the effect of A_1 versus A_2, data from these two groupings are pooled:

A_1: (1,3,5,7) had 24 stem cracks out of 173; 24/173 = 0.139.

A_2: (2,4,6,8) had 5 stem cracks out of 179; 5/179 = 0.028

Similarly, combinations for the Bs and Cs are displayed in Table 11.12.

Points corresponding to values shown in Table 11.12 have been plotted in Fig. 11.17. The logic and mechanics of computing decision lines are given below, following a discussion on decisions.

Decisions from Fig. 11.17

The most important single difference is the advantage of A_2 over A_1; 13.9 percent versus 2.8 percent. The magnitude of this difference is of tremendous importance and is statistically significant. Fortunately, it was a choice which could easily be made in production.

A second result was quite surprising; the preference of B_1 over B_2. The converse had been expected. The approximate level of risk in choosing B_1 over B_2 is about $\alpha = 0.01$;

TABLE 11.12 Computations for Analysis of Means
Data from Table 11.11

Summary		Interaction
		AC:
Air	A_1: (1,3,5,7) 24/173 = 0.139	Like: (1,2,7,8): 12/175 = 0.069
	A_2: (2,4,6,8) 5/179 = 0.028	Unlike: (3,4,5,6): 17/177 = 0.096
		AB:
Stem	B_1: (1,2,5,6) 8/174 = 0.046	Like: (1,4,5,8): 13/177 = 0.073
Assembly	B_2: (3,4,7,8) 21/178 = 0.118	Unlike: (2,3,6,7): 16/175 = 0.091
		BC:
Stem	C_1: (1,4,6,7) 15/172 = 0.087	Like: (1,3,6,8): 10/178 = 0.056
Tension	C_2: (2,3,5,8) 14/180 = 0.078	Unlike: (2,4,5,7): 19/174 = 0.109
		ABC:
		(1,2,3,4): 13/175 = 0.074
		(5,6,7,8): 16/177 = 0.090

Figure 11.17 Comparing effects of three factors on glass stem cracking: three main effects and their interactions (data from Table 11.12).

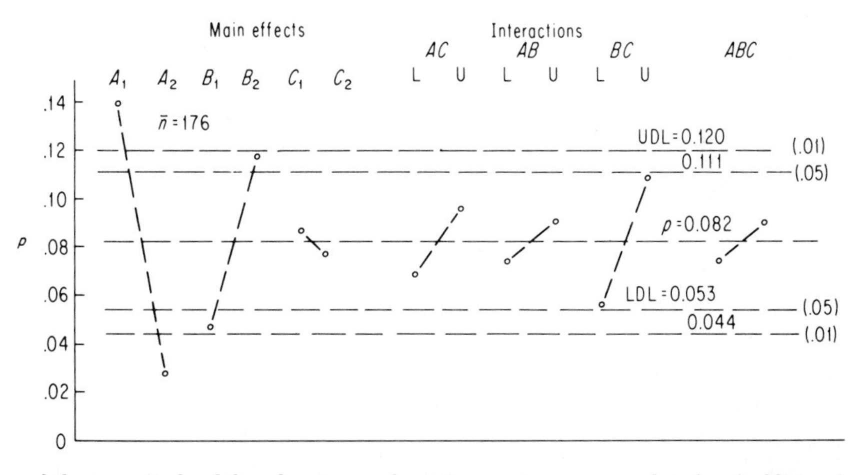

and the magnitude of the advantage—about 7 percent—was enough to be of additional economic interest.

No two-factor interaction is statistically significant at the 0.05 level and the BC interaction "just barely" at the 0.10 level. This latter decision can be checked by using the factor $H_{0.10}$ = 1.16 from Table A.14, using df = ∞, since this is for attributes. Then, the recommended operating conditions are A_2B_1 and two possibilities follow:

$$A_2B_1 \text{ and } C_1 \text{ or } A_2B_1 \text{ and } C_2$$

i.e., combinations 2 and 6 with 0 and 0 cracked stems.

Some *slight* support for a choice may be seen from the BC-interaction diagram (Fig. 11.17), which gives "the nod" to those B and C having *like* subscripts, i.e., to combination 6.

Mechanics of analysis—ANOM (Fig. 11.17)

For the initial analysis, we consider all eight samples to be of size $\bar{n} = 44$. Then each comparison for main effects and interactions will be between $k = 2$ groups of four samples each, with $\bar{n} = 4(44) = 176$ stems.

The total defectives in the eight combinations is 29

$$\bar{p} = 29/352 = 0.082 \text{ or } 8.2\%$$

and

$$\hat{\sigma}_P = \sqrt{\frac{(8.2)(91.8)}{176}} = 2.1\% \qquad \text{and} \qquad \hat{\sigma}_p = 0.021$$

Decision lines (for Fig. 11.17)

$\underline{\alpha = 0.05}$

$$\bar{p} \pm H_{0.05}\hat{\sigma}_p = 0.082 \pm (1.39)(0.021)$$

$$\text{UDL} = 0.111$$

$$\text{LDL} = 0.053$$

$\underline{\alpha = 0.01}$

$$\text{UDL} = 0.120$$

$$\text{LDL} = 0.044$$

These decision lines[23] are applicable not only when comparing main effects but also to the three two-factor interactions and to the three-factor interaction.

Two-factor interactions—discussion

In the study, it is possible that the main effect of variable A is significantly different under condition B_1 than it is under condition B_2; then there *is said to be an* "$A \times B$ ('A times B') two-factor interaction."

If the main effect of A is essentially (statistically) the same for B_1 as for B_2, then there is no $A \times B$ interaction. The mechanics of analysis for any one of the three possible two-factor interactions have been given in Table 11.12 without any discussion of the meaning.

In half of the eight combinations, A and B have the same or *like* (L) subscripts; namely (1,4,5,8). In the other half, they have different or *unlike* (U) subscripts; namely (2,3,6,7). The role of the third variable is always disregarded when considering a two-factor interaction:

AB: Like (1,4,5,8) has 13/177 = 0.073 or 7.3%

AB: Unlike (2,3,6,7) has 16/175 = 0.091 or 9.1%

The difference between 9.1 percent and 7.3 percent does not seem large; and it is seen in Fig. 11.17 that the two points corresponding to AB(L) and AB(U) are well within the decision lines. We have said this means there is *no significant* $A \times B$ *interaction.* This procedure deserves some explanation.

---

[23]Since the sample sizes were not very different, this analysis did not use special procedures available for the analysis of samples of different size. Use of $\bar{n}$ simplifies the analysis here with very little effect on statistical validity.

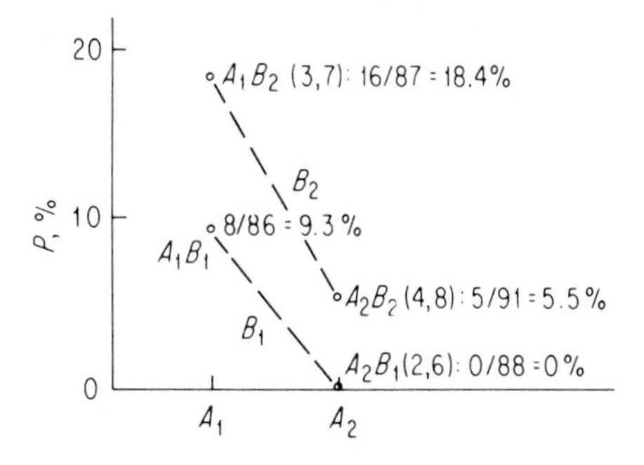

Figure 11.18 A graphical comparison of effect on stem cracks. The upper line shows the decrease in changing from A_1 to A_2 under condition B_2; the lower line shows the decrease in changing from A_1 to A_2 under condition B_1. Intuitively the effects seem quite comparable. (Data from Table 11.11.)

In Fig. 11.18, the decreases in stem failures in changing from A_1 to A_2 are as follows:

Under condition B_1: drops from 8/86 = 9.3% to 0/88 = 0%: a drop of 9.3%.
Under condition B_2: drops from 16/87 = 18.4% to 5/91 = 5.5%: a drop of 12.9%.

Are these changes significantly different (statistically)? If the answer is "yes," then there is an $A \times B$ interaction.

Consider the differences under conditions B_1 and B_2 and assume that they are not statistically equal.

$$A_1B_2 - A_1B_1 \neq A_2B_2 - A_2B_1 \tag{11.3}$$

i.e., is 9.1% ≠ 5.5% statistically?

This can be rewritten as

$$A_1B_2 + A_2B_1 \neq A_1B_1 + A_2B_2 \tag{11.4}$$

From Table 11.12, corresponding combinations are

Unlike (2,3,6,7) ≠ Like (1,4,5,8)

i.e., is 9.1% ≠ 7.3%?

That is, the two lines are *not* parallel if the points corresponding to (AB: Like) *and* (AB: Unlike) are statistically different i.e., fall outside the decision lines of Fig. 11.17. Relation in Eq. (11.4) is a simple mechanical comparison to compute when testing for AB interaction. The decision lines to be used in Fig. 11.17 are exactly those used in comparing main effects.

Similar combinations to be used when testing for $A \times C$ and $B \times C$ interactions are shown in Table 11.12.

Note

Also, the combinations used to test for a three-factor, $A \times B \times C$, interaction are given in Table 11.12. If an apparent three-factor interaction is observed, it should be recomputed for possible error in arithmetic, choice of combinations, or unauthorized changes in the experimental plan, such as failure to "hold constant" or randomize—all factors not included in the program. With very few exceptions, main effects will provide most opportunities for improvement in a process; but two-factor interactions will sometimes have enough effect to be of interest.

CASE HISTORY 11.9

A Problem in a High-Speed Assembly Operation (Broken Caps)

Cracked and broken caps (see Fig. 11.19)

Cracked and broken caps are a constant headache in pharmaceutical firms. One aspect of the problem was discussed in Case History 11.4. It was recognized that too many caps were being cracked or broken as the machine screwed them onto the bottles. Besides the obvious costs in production, there is the knowledge that some defective caps will reach the ultimate consumer; no large-scale 100 percent inspection can be perfect. When the department head was asked about the problem we were told that it had been investigated; "We need stronger caps." Being curious, we watched the operation awhile. It seemed that cracked caps were not coming equally from the four lines.

Production was several hundred per minute. The filling-capping machine had four adjacent lines to four capper positions. At each, beginning with an empty bottle, a soft plastic ring was squeezed on, a filler nozzle delivered the liquid content, a hard plastic marble was pressed into the ring, and a cap was applied and tightened to a designated torque.

Each of the four lines had a separate mechanical system, including a capper and torque mechanism. An operator-inspector stood at the end of the filling machine who would remove bottles with cracked caps whenever they were spotted. We talked to the department head about the possibilities of a quick study to record the number of defects from the four capper positions. There were two possibilities: a bucket for each position where defective caps could be thrown and then counted, or an ordinary sheet of paper where the operator-inspector could make tally marks for the four positions. The department head preferred the second method and gave necessary instructions.

Figure 11.19 Components in a toiletry assembly. The complete labeled assembly; a soft plastic ring; a hard plastic marble; a hard plastic cap; a glass bottle. (Case History 11.9.)

After some 10 or 15 min, the tally showed:

Capper head	Cracked caps
1	2
2	0
3	17
4	1

We copied the record and showed it to the department head who said, "I told you. We need stronger caps!" Frustrated a bit, we waited another hour or so for more data.

Capper head	Cracked caps
1	7
2	4
3	88
4	5

We now discussed the matter for almost 30 min and the department head was sure the solution was stronger caps. Sometimes we were almost convinced! But then it was agreed that capper head 3 could hardly be so *unlucky* as to get the great majority of weak caps. The chief equipment engineer was called and we had a three-way discussion. Yes, they had made recent adjustments on the individual torque heads. Perhaps the rubber cone in head 3 needed replacement.

Following our discussion and their adjustments on the machine, head 3 was brought into line with the other three. The improved process meant appreciably fewer cracked and broken caps in production *and* fewer shipped to customers. It is rarely possible to attain perfection. We dismissed the problem for other projects.

Stronger caps?

Then it occurred to us that we ought to give some thought to the department head's earlier suggestion: stronger caps. With a lot of experience in this business the department head might now possibly be a little chagrined; besides there might be a chance to make further improvement. So we reopened the matter. Exclusive of the capper heads, why should a cap break? Perhaps because of any of the following:

1. *Caps (C):* wall thickness, angle of threads, distances between threads, different cavities at molding; irregularities of many dimensions—the possibilities seemed endless.
2. *Bottles (B):* the angle of sharpness of threads on some or all molds, diameters, distances. This seemed more likely than caps.
3. *Rings (R):* perhaps they exert excessive pressures some way.
4. *Marbles (M):* these were ruled out as quite unlikely.

Selection of independent variables to study

The traditional approach would be to measure angles, thickness, distances, and so forth; this would require much manpower and time.

The question was asked: "How many vendors of caps? Of bottles? Of rings?" the answer was two or three in each case. Then how about choosing vendors as omnibus variables? Since a 2^3 study is a very useful design, we recommended the following vendors: two for caps, two for bottles, and two for rings. This was accepted as a reasonable procedure.

A 2^3 assembly experiment

Consequently, over the next two or three weeks, some 8000 glass bottles were acquired from each of two vendors B_1, B_2; also, 8000 caps from each of two vendors, C_1, C_2; and 8000 rings from each of two vendors, R_1, R_2.

It took some careful planning by production and quality control to organize the flow of components through the filling and capping operation and to identify cracked caps with the eight combinations. The assembly was completed in one morning. The data are shown in Tables 11.13*a*, *b*. From the table alone, it seemed pretty clear that the difference in cracked caps between vendors C_2 and C_1 must be more than just chance; 132 compared to 70. (Also see Fig. 11.20 and Table 11.14.)

It was a surprise to find that combination 5, $B_1R_1C_2$, which produced 51 rejects, was assembled from components *all from the same vendor!*

Formal analysis (see Fig. 11.20)

The percent of cracked caps is small and n_g is large; we shall simplify computations by assuming a Poisson distribution as close approximation to the more detailed analysis using the binomial $c = 202/8 = 25.25$.

$$\hat{\sigma}_c = \sqrt{25.25} = 5.02 \qquad \text{for individuals}$$

$$\hat{\sigma}_{\bar{c}} = \hat{\sigma}_c/\sqrt{4} = 2.51 \qquad \text{for averages of 4 cells}$$

Summary–Discussion (see Fig. 11.20)

Caps

The largest difference is between vendors of caps. The difference is of practical interest as well as being statistically significant ($\alpha < 0.01$); the overall breakage from C_2 is almost twice that of C_1. Also C_2 caps crack more often than C_1 caps for *each* of the four bottle-ring combinations. This is a likely reason that the department head had believed they needed "stronger caps," since any change from vendor C_1 to C_2 would increase cracking. Thus the reason for excessive cracking was from a joint origin; the adjustment of the capper had effected an improvement, and the selection of cap usage offers some opportunities. There were good reasons why it was not feasible for purchasing to provide manufacturing with only the best combination $B_1R_1C_1$. What should be done? The vendor of C_2 can study possible reasons for excessive weakness of the caps: Is the trouble in specific molding cavities? Is it in general design? Is it in plastic or molding temperatures? In certain critical dimensions?

TABLE 11.13 Data from a 2^3 Factorial Production Study of Reasons for Cracked Caps

Data from Case History 11.9, displayed on two useful forms

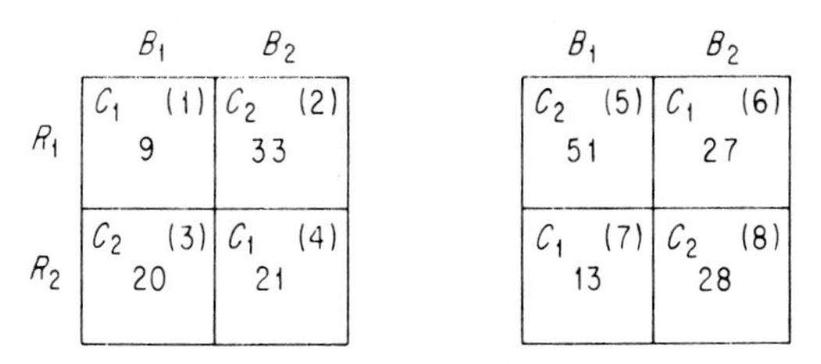

	B_1	B_2	B_1	B_2
R_1	C_1 (1) 9	C_2 (2) 33	C_2 (5) 51	C_1 (6) 27
R_2	C_2 (3) 20	C_1 (4) 21	C_1 (7) 13	C_2 (8) 28

Table 11-13a

	B_1		B_2	
	R_1	R_2	R_1	R_2
C_1	9 (1)	13 (7)	27 (6)	21 (4)
C_2	51 (5)	20 (3)	33 (2)	28 (8)

Table 11-13b

The number* of assemblies in each combination was 2,000; the number of cracked and broken caps is shown in the center.

$\bar{c} = 202/8 = 25.25$

$\hat{\sigma} = \sqrt{25.25}$ and $\hat{\sigma}_{\bar{c}} = \hat{\sigma}/\sqrt{4}$ when comparing averages of 4 groups

$\Sigma B_1(1,3,5,7) = 93;\ \bar{B}_1 = 23.25$ $\quad\Sigma C_1(1,4,6,7) = 70;\ \bar{C}_1 = 17.5$

$\Sigma B_2(2,4,6,8) = 109;\ \bar{B}_2 = 27.25$ $\quad\Sigma C_2(2,3,5,8) = 132;\ \bar{C}_2 = 33.0$

$\Sigma R_1(1,2,5,6) = 120;\ \bar{R}_1 = 30.0$

$\Sigma R_2(3,4,7,8) = 82;\ \bar{R}_2 = 20.5$

* We have chosen to analyze the data as being Poisson type.

In the meantime, *can purchasing favor* vendor C_1? Production itself should *avoid* using the very objectionable $B_1R_1C_2$ combination. (A rerun of the study a few days later showed the same advantage of C_1 over C_2; and that $B_1R_1C_2$ gave excessive rejects.)

Other main effects

Rings

The effect of rings is seen to be statistically significant (α about 5 percent); the magnitude of the effect is less than for caps. It can also be seen that the effect of rings when

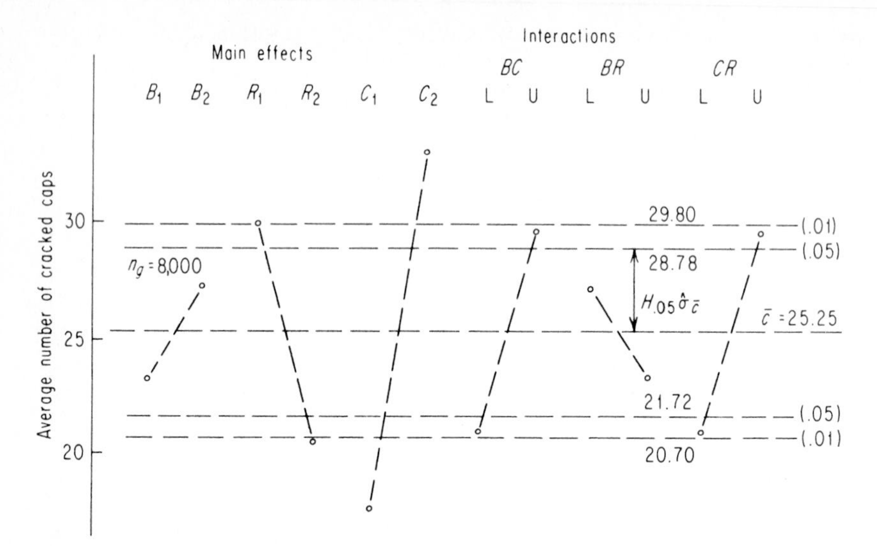

Figure 11.20 Comparing effects of bottles, rings, and caps from different vendors on cracked caps (main effects and two-factor interactions). (Data from Tables 11.13*b* and 11.14.)

TABLE 11.14 Computations for Two-Factor Interactions

Cracked cap: data from Table 11.13

Decision lines		*Computation for two-factor interactions*		
$\alpha = .05$				
$\text{UDL} = \bar{c} + H_{.05}\hat{\sigma}_{\bar{c}}$	BC:	Like (1,2,7,8) = 22 + 61 = 83	Avg. =	20.75
$= 25.25 + (1.39)(2.5)$		Unlike (3,4,5,6) = 48 + 71 = 119		29.75
$= 25.25 + 3.53 = 28.78$		202		
$\text{LDL} = 25.25 - 3.53 = 21.72$	BR:	Like (1,5,4,8) = 60 + 49 = 109		27.25
		Unlike (2,6,3,7) = 33 + 60 = 93		23.25
$\alpha = .01$		202		
$\text{UDL} = 25.25 + (1.82)(2.5)$				
$= 25.25 + 4.55 = 29.80$	CR:	Like (1,3,6,8) = 36 + 48 = 84		21.0
$\text{LDL} = 25.25 - 4.52 = 20.70$		Unlike (2,4,5,7) = 34 + 84 = 118		29.5
		202		

using C_2 is quite large (combining B_1 and B_2); the effect is negligible when using C_1. This interdependence is also indicated by the CR interaction.

Bottles

The effect of bottles as a main effect is the least of all three. (Most surprising to us!) However consider *that half* of the data in Table 11.15; these data are from Table 11.13*b* for only the better cap C_1.

TABLE 11.15 Effects of Rings and Bottles Using Only Caps C_1
Data from Table 11.13

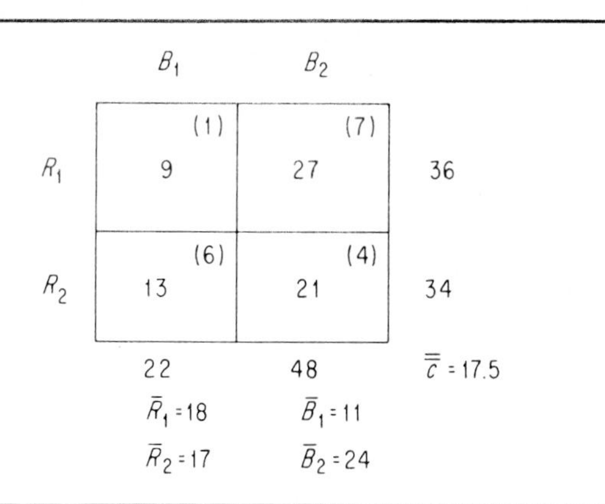

	B_1	B_2	
R_1	(1) 9	(7) 27	36
R_2	(6) 13	(4) 21	34
	22	48	$\bar{\bar{c}} = 17.5$
	$\bar{R}_1 = 18$	$\bar{B}_1 = 11$	
	$\bar{R}_2 = 17$	$\bar{B}_2 = 24$	

The data for this 2^2 design, using C_1 only, shows a definite advantage of using bottle B_1 in any combination with R. Also in Fig. 11.21, the advantage is seen to be statistically significant, $\alpha = 0.01$. (Note: this is a BC interaction.)

Formal analysis caps C_1 only (see Fig. 11.21)

$\bar{c} = 70/4 = 17.5$; $\hat{\sigma}_c = \sqrt{17.5} = 4.18$; $\hat{\sigma}_{\bar{c}} = 4.18/\sqrt{2} = 2.96$ for averages of two cells. Then

$$\begin{aligned} \text{UDL}(0.01) &= 17.50 + (1.82)(2.96) \\ &= 17.50 + 5.39 \\ &= 22.89 \\ \text{LDL}(0.01) &= 12.11 \end{aligned}$$

11.8 A Very Important Experimental Design[24]: ½ × 2^3

The 2^2 and 2^3 designs are useful strategies to use initially when investigating problems in many industrial processes. A third equally important strategy, discussed here, is more or less a combination of the 2^2 and 2^3 designs. It enables the experimenter to study effects of three different factors with only four combinations of them instead of the eight in a 2^3 design. This "half-rep of a two cubed" design is especially useful in exploratory studies in which the quality characteristics are attributes.

In Table 11.16*a* and *b*, two particular halves of a complete 2^3 factorial design are shown. Some data based on this design are shown in Fig. 11.22.

The reasons for choosing only a special half of a 2^3 factorial design are re-

[24]See also Sec. 14.6.

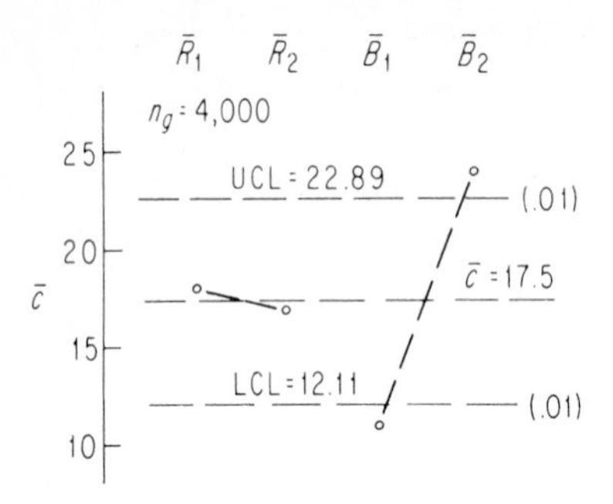

Figure 11.21 Comparing effects of bottles and rings from different vendors when using caps from the better vendor (data from Table 11.15).

ductions in time, effort, and confusion. Especially when it is expected that the effects of the three factors are probably independent, one should not hesitate to use such a design. In this design, the main effect of any variable may possibly be confounded with an interaction of the other two variables. This possible complication is often a fair price to pay for the advantage of doing only half the experimental combinations.

CASE HISTORY 11.10

Winding Grids

This case history presents two practical procedures:

1. A half-rep of a 2^3 design using attributes data.
2. A graphical presentation of data allowing simultaneous comparisons of four types of defects instead of the usual single one.

Introduction[25]

Grids are important components of electronic tubes. They go through the following manufacturing steps:

TABLE 11.16 Two Special Halves of a 2^3 Factorial Design

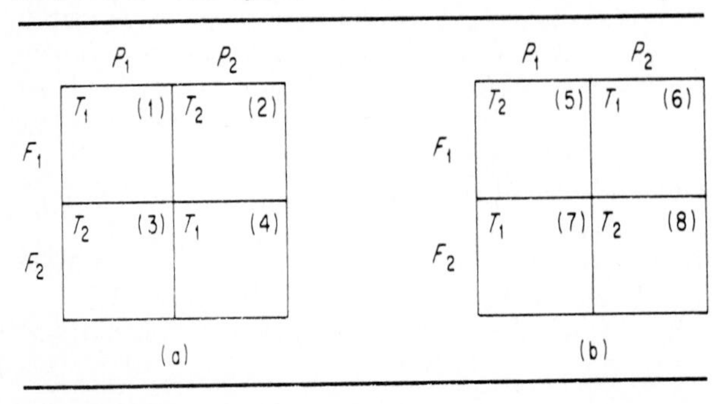

(a)

	P_1	P_2
F_1	T_1 (1)	T_2 (2)
F_2	T_2 (3)	T_1 (4)

(b)

	P_1	P_2
F_1	T_2 (5)	T_1 (6)
F_2	T_1 (7)	T_2 (8)

[25] Fred Ennerson, Ralph Fleischmann, and Doris Rosenberg, "A Production Experiment Using Attribute Data," *Ind. Qual. Control*, vol. 8, no. 5, March 1952, pp. 41–44.

1. They are wound on a grid lathe. There are several grid lathes in use and the tension T on the lateral wire can be varied on each. After winding, they are transported to an operator called a hand-puller.
2. The hand-puller P uses a pair of tweezers to remove loose wire from the ends. The loose wire is inherent in the manufacturing design. The grids are then transported to a forming machine.
3. Each forming machine has its forming operator F. After forming, the grids go to inspection.

Following these operations, inspectors examine the grids for the four characteristics of an attributes nature, go no-go; spaciness, taper, damage, slant.

Design of the experiment

In a multistage operation of this kind, it is difficult to estimate just how much effect the different steps may be contributing to the total rejections being found at the end of the line. Since the percent of rejections was economically serious, it was decided to set up a production experiment to be developed jointly by representatives of production, production engineering, and quality control. They proposed and discussed the following subjects.

- *Grid lathes:* It was decided to remove any effect of different grid lathes by using grids from only one lathe. The results of this experiment cannot then be transferred automatically to other lathes.

  However, it was thought that the tension of the lateral wire during winding on the grid lathe might be important. Consequently, it was decided to include *loose* tension T_1 and *tighter* tension T_2 in the experiment. Levels of T_1 and T_2 were established by production engineering.

- *Pullers:* The probable effect of these operators was expected to be important. Two of them were included in the experiment; both considered somewhat near average. They are designated by P_1 and P_2. Others could be studied later if initial data indicated an important difference between these two.
- *Forming operators:* It was judged that the machine operators had more effect than the machines. Two operators F_1 and F_2 were included, but both operated the same machine.
- *Inspection:* All inspection was done by the *same* inspector to remove any possible difference in standards.
- *Design of the experiment:* Actually the design of the experiment was being established during the discussion which led to the selection of three factors and two levels of each factor.

  It would now be possible to proceed in either of two ways:

  1. Perform all $2^3 = 8$ possible combinations, or
  2. Perform a special half of the combinations chosen according to an arrangement such as in Table 11.16*a*.

     Since it was thought that the effects of the different factors were probably *independent*, the half-rep design was chosen instead of the full 2^3 factorial.

- *Number of grids to include in the experiment:* The number was selected after the design of the experiment had been chosen. Since grids were moved by production in trays of 50, it was desirable to select multiples of 50 in each of the four squares. It was decided to include 100 grids in each for a total of 400 grids in the experiment. This had two advantages: it allowed the entire experiment to be completed in less than a day and had minimum interference with production, and it was expected that this many grids would detect economically important differences.

To obtain the data for this experimental design, 200 grids were wound with loose tension T_1 on the chosen grid lathe and 200 with tighter tension T_2. These 400 grids were then "hand-pulled" and "formed" according to the chosen schedule. The numbers of rejects for each characteristic, found at inspection, are shown in the center of squares in Fig. 11.22. The number of grids in each combination is shown in the lower-right corner.

Analysis of the data (ANOM)

Main effects

By combining the data within the indicated pairs of squares we have the percents of rejects for spaciness as follows:

$$P_1(1,3)\text{: } 45/200 = 22.5\%$$

$$P_2(2,4)\text{: } 41/200 = 20.5\%$$

$$F_1(1,2)\text{: } 32/200 = 16.0\%$$

$$F_2(3,4)\text{: } 54/200 = 27.0\%$$

$$T_1(1,4)\text{: } 47/200 = 23.5\%$$

$$T_2(2,3)\text{: } 39/200 = 19.5\%$$

These are shown in Fig. 11.22*a*.

Conclusions

By examining the four charts in Fig. 11.22, we see where to place our emphasis for improving the operation. It should be somewhat as follows:

Tension: A loose tension (T_1) on the grid lathe is preferable to a tight tension T_2 with respect both to taper and damage (Fig. 11.22*b*, *c*).

Forming operators: These have an effect on spaciness and slant (Charts *a*, *d*). Operator F_2 needs instruction with respect to spaciness; and operator F_1 with respect to slant. This is very interesting.

Hand pullers: Again, one operator is better on one characteristic and worse on another. Operator P_1 is a little more careful with respect to damage, but is not careful with respect to slant, ruining too many for slant (Fig. 11.22*c*, *d*).

It may be that there are other significant effects from these three factors; but if so, larger samples in each block would be needed to detect such significance. This experiment has provided more than enough information to suggest substantial improvements and extensions in the processing of the grids.

It is interesting and helpful that each operator is better on one quality characteristic and worse on another. Each operator can be watched for what is done right (or wrong), and whatever is learned can be taught to others. Sometimes it is argued that people are only human, so that while they are congratulated for good performance, they also can be taught how to improve their work.

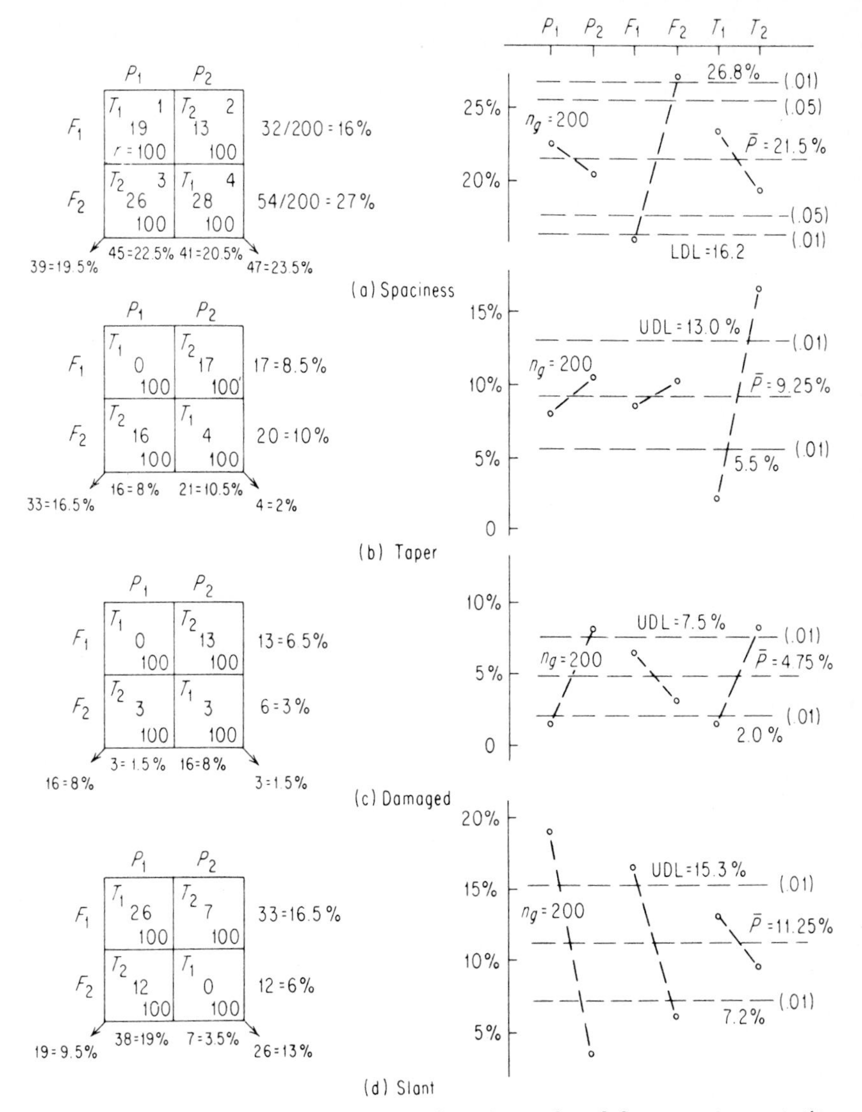

Figure 11.22 Effects of pullers, formers, and tension on four defect types (computations from Table 11.17).

Computation of decision lines (ANOM): half-rep of a 2^3

The computation in this half-rep design proceeds exactly as in a 2^2 factorial design (Table 11.17). First, the overall percent defective $\overline{P}$, is determined for each defect type. Then *decision lines are drawn* at: $\overline{P} \pm H_\alpha \hat{\sigma}_P$. Each comparison is between two totals of $n_g = 200$.

TABLE 11.17 Computations of Decision Lines (ANOM)
Data of Fig. 11.22

- *Spaciness:* Comparing $k = 2$ groups of 200 each, $\bar{P} = 86/400 = 21.5\%$

$$\hat{\sigma}_P = \sqrt{\frac{(21.5)(78.5)}{200}} = 2.90\%$$

For $\alpha = .05$:

$$\begin{aligned} \text{UDL} &= \bar{P} + H_\alpha \hat{\sigma}_P \\ &= 21.5 + (1.39)(2.90) \\ &= 25.5\% \\ \text{LDL} &= \bar{P} - H_\alpha \hat{\sigma}_P \\ &= 17.5\% \end{aligned}$$

For $\alpha = .01$:

$$\begin{aligned} \text{UDL} &= 21.5 + (1.82)(2.90) \\ &= 26.8\% \\ \text{LDL} &= 16.2\% \end{aligned}$$

- *Taper:* $k = 2$ groups of 200 each: $\bar{P} = 37/400 = 9.25\%$

$$\hat{\sigma}_P = \sqrt{\frac{(9.25)(90.75)}{200}} = 2.05\%$$

For $\alpha = .05$:
(Not needed)

For $\alpha = .01$:

$$\begin{aligned} \text{UDL} &= 9.25 + (1.82)(2.05) \\ &= 13.0\% \\ \text{LDL} &= 5.5\% \end{aligned}$$

- *Damaged:* $k = 2$ groups of 200 each: $\bar{P} = 19/400 = 4.75\%$

$$\hat{\sigma}_P = \sqrt{\frac{(4.75)(95.25)}{200}} = 1.50\%$$

For $\alpha = .05$:
(Not needed)

For $\alpha = .01$:

$$\begin{aligned} \text{UDL} &= 4.75 + (1.82)(1.50) \\ &= 7.50\% \\ \text{LDL} &= 2.0\% \end{aligned}$$

- *Slant:* $k = 2$, $\bar{P} = 45/400 = 11.25\%$

$$\hat{\sigma}_P = \sqrt{\frac{(11.25)(88.75)}{200}} = 2.23\%$$

For $\alpha = .05$:

$$\begin{aligned} \text{UDL} &= 11.25 + (1.39)(2.23) \\ &= 14.35 \\ \text{LDL} &= 8.15 \end{aligned}$$

For $\alpha = .01$:

$$\begin{aligned} \text{UDL} &= 11.25 + (1.82)(2.23) \\ &= 15.3 \\ \text{LDL} &= 7.2 \end{aligned}$$

11.9 Case History Problems

Problem 11.1—based on a case history

Defective glass bottles. The use of quite small attributes samples taken at regular time intervals during production can provide evidence of important differences in the production system and indicate sources to be investigated for improvements.

Background. A meeting was arranged by telephone with a quality control representative from a company whose only product was glass bottles. This was one of the few times Ellis Ott ever attempted to give specific advice to anyone in a meeting without visiting the plant or having had previous experience in the industry. The plant was a hundred miles away and they had a sensible discussion when they met.

The following points were established during the discussion:

1. There were too many rejects: the process was producing about 10 percent rejects of different kinds.
2. Knowledge of rejects was obtained from an acceptance sampling operation or a 100 percent inspection of the bottles. The inspection station was in a warehouse separate from the production areas; the usual purpose of inspection was to cull out the rejects before shipping the bottles to their customers. The information was not of much value for any process improvement effort; often obtained a week after bottles were made and too late to be considered representative of current production problems.
3. Large quantities of glass bottles were being produced. Several machines were operating continuously on three shifts, and for seven days per week. Each machine had many cavities producing bottles.

The visitor returned to the plant and made the following arrangements:

- A plant committee was organized representing production, inspection, and industrial engineering, to study causes and solutions to the problem of defects.
- An initial sampling procedure was planned for some quick information. From the most recent production, samp[illegible] per hour were to be chosen at random from: (1) each of three machines [illegible] end), and (2) each of three shifts, and (3) over seven days.
- The sample bottles were to be placed in slots in an egg-carton-type box marked to indicate the time of sampling as well as machine number, shift, and date.
- After the bottles were collected and inspected, the number and type of various defects were recorded. The data in Table 11.18 show only the total of all rejects. (A breakdown by type of defect was provided but is not now available.)

Major conclusions which you may reach

1. There were fundamental differences between the three machines, and differences between shifts.
2. There was a general deterioration of the machines, or possibly in raw materials, over the seven days. A comparison of this performance pattern with

TABLE 11.18 Defective Glass Bottles from Three Machines—Three Shifts and Seven Days

		Machine		
Date	Shift	1	2	3
8/12	A	1	4	4
	B	4	0	4
	C	12	6	9
8/13	A	3	6	30
	B	2	8	46
	C	2	7	27
8/14	A	2	1	1
	B	8	11	15
	C	8	7	17
8/15	A	4	11	10
	B	5	7	11
	C	4	6	11
8/16	A	10	8	9
	B	6	12	10
	C	7	15	19
8/17	A	7	11	15
	B	12	9	19
	C	24	8	18
8/18	A	8	6	16
	B	10	12	17
	C	8	19	15

Number of days: 7
Number of machines: 3
Number of shifts: 3
Number of hr/shift: 8
Number of items/hr: 15
Total n = (63)(120) = 7,560
r = (8)(15) = 120/shift/day/machine = cell size

the scheduled maintenance may suggest changes in the maintenance schedule.

3. Each machine showed a general uptrend in rejects; one machine is best and another is consistently the worst.
4. There was an unusual increase in rejects on all shifts on August 13 on one machine only. Manufacturing records should indicate whether there was any change in the raw material going to that one machine. If not, then something was temporarily wrong with the machine. The records should show what adjustment was made.

Suggested exercises. Discuss possible effects and *type* of reasons which might explain differences suggested below. Prepare tables and charts to support your discussion.

1. *Effect of days.* All machines and shifts combined. Is there a significant difference observed over the seven days?
2. *Effect of machines.* Each machine with three shifts combined. What is the behavior pattern of *each* machine over the seven days.
3. *Effect of shift.* Each shift with three machines combined. What is the behavior pattern of each shift over the seven days?

Problem 11.2—wire treatment

During processing of wire it was decided to investigate the effect of three factors on an electrical property of the wire.

- Three factors were chosen to be investigated:

  T: Temperature of firing

  D: Diameter of heater wire

  P: the pH of a coating

- It was agreed to study these three factors at two levels of each. The experimental design was a half-rep of a 2^3.
- The quality characteristic was first measured as a variable; the shape of the measurement distribution was highly skewed with a long tail to the right. Very low readings (measurements) were desired; values up to 25 units were acceptable but not desirable. (Upper specification was 25.) It was agreed, arbitrarily for this study, to call *very low* readings (< 5) *very good;* and high readings (> 14) *bad.* Then the analysis was to be made on *each* of these two *attribute* characteristics.
- After this initial planning, it was suggested and accepted that the fractional factorial would be carried out under each of three *firing* conditions:

  A: Fired in air

  S: Fired by the standard method already in use

  H: Fired in hydrogen

The data are shown in Table 11.19. Whether the *very good* or *bad* qualities are most important in production is a matter for others to decide.

Suggested exercises

1. Compare the effectiveness of the three firing conditions, *A, S,* and *H.* Then decide whether or not to pool the information from all three firing conditions when studying the effects of *T, D,* and *P.*
2. Assume now that it is sensible to pool the information of *A, S,* and *H;* make an analysis of *T, D,* and *P* effects.
3. Prepare a one- or two-page report of recommendations.

TABLE 11.19 Wire Samples from Spools Tested after Firing under Different Conditions of Temperature, Diameter, and pH

A. *Very Good Quality*

	T_1	T_2
D_1	P_1 — A: 8/12, S: 4/12, H: 9/12	P_2 — 10/12, 10/12, 8/12
D_2	P_2 — A: 8/12, S: 9/12, H: 7/12	P_1 — 7/12, 7/12, 5/12

B. *Bad Quality*

	T_1	T_2
D_1	P_1 — A: 0/12, S: 7/12, H: 1/12	P_2 — 1/12, 0/12, 4/12
D_2	P_2 — A: 3/12, S: 1/12, H: 3/12	P_1 — 5/12, 1/12, 5/12

11.10 Practice Exercises

1. Given three samples of 100 units each, the fraction defective in each sample is 0.05, 0.06, and 0.10. Compute $\bar{p}$, and ANOM control limits for alpha = 0.01 and alpha = 0.05. Do any of the three samples have significantly different fraction defective?
2. Recompute the ANOM decision chart, excluding workers *D, F, H,* and *K* in Table 11.5.
3. Based on the chart shown in Fig. 11.3, what action should be taken?
4. *a.* Explain why the author used the Poisson rather than the binomial model in Case History 11.3.
   b. Explain the meaning of the triangles and the circles in Table 11.7.
5. Assume, in Case History 11.6, that the data from operator A_4 had to be excluded from the analysis. Reanalyze the remaining data and state conclusions.
6. Explain the difference between the upper 5 percent decision limit of 15.7 percent shown on Fig. 11.15 and of 15.34 percent shown on Fig. 11.16.
7. Create an industrial example to illustrate the meaning of the terms "interaction" and "confounding." Prepare for oral presentation.
8. Work Case History Problem 11.1.
9. Work Case History Problem 11.2.
10. Make a histogram of the data of Table 11.7 to see if it appears to conform to the Poisson distribution.
11. Plot interaction response curves for the *AB* and *BC* interactions in Case History 11.8 using the data of Table 11.11.

Chapter

12

Special Strategies in Troubleshooting

The special methods of this chapter can be very effective in many situations:

- Disassembly and reassembly
- A special screening program for many treatments
- The relationship between two variables

They involve obtaining insight from the patterns in which data fall.

12.1 Ideas from Patterns of Data

Introduction

A set of numbers may be representative of one type of causal system when they arise in one pattern, or a different causal system when the same data appear in a different pattern. A control chart of data which represents a record of a process over a time period almost invariably carries much more meaning than the same data accumulated in a histogram which obscures any time effects.

There are times, however, when variations in the data appear to be irretrievably lost; sometimes, as discussed below, some semblance of order can be salvaged to advantage.

CASE HISTORY 12.1

Extruding Plastic Components

Hundreds of different plastic products are extruded from plastic pellets. Each product requires a mold which may have one cavity or as many as 16, 20, or 32 cavities producing items purported to be "exactly" alike. The cavities have been machined from two

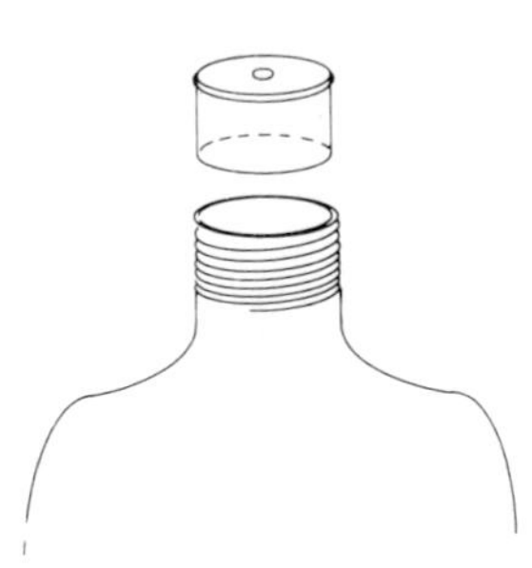

Figure 12.1 Plastic bottle and plug insert.

mating stainless steel blocks; plastic is supplied to all cavities from a common stream of semi-fluid plastic.

It is sometimes recognized that the cavities do *not* perform alike, and it is prudent foresight to require that a cavity number be cut into each cavity; this is then a means of identifying the cavity which has produced an item in the large bin of molded parts. The importance of these numbers in a feedback system is potentially tremendous.

In assemblies, there are two types of defects. Those which occur in a random fashion and those which occur in patterns but which are seldom recognized as such. A pattern, when recognized, can lead to corrective action; an ability to identify these patterns is of real value in troubleshooting.

A bottle for a well-known men's hair toiletry has a soft plastic plug (as in Fig. 12.1). It is inserted by machine into the neck of a plastic bottle. It controls excess application of toiletry during use.

Incoming inspection was finding epidemics of "short-shot plugs" in shipments from an out-of-state supplier. Several discussions were held, by telephone, with the vendor. The short-shot plugs were incompletely formed, as the term implies. When they got into production they often jammed the equipment and were also a source of dissatisfaction when they reached the consumer.

A knowledgeable supervisor got involved in the problem and decided to obtain some data at incoming materials inspection. The importance of determining whether these defects were occurring randomly from the many cavities or in some pattern was known. Several boxes of plugs were inspected, and the defective short-shot plugs kept separate. Just over 100 of them were obtained; by examining them, the cavities which produced them were identified (see Table 12.1). It is very evident [1] that the defective plugs do not come randomly from the 32 cavities as identified from their mold numbers. The supervisor then reasoned as follows:

- Since rejects were found from cavities 1 and 32, it must be a 32-cavity mold, even though some cavities produced no defective plugs.

- A 32-cavity mold must be constructed as a 4 × 8 mold rather than a 16 × 2 mold; and the obvious numbering of molds must be somewhat as in Table 12.2*a*. From Table 12.1, the supervisor filled in the number of short-shot plugs corresponding to their cavity of origin (see Table 12.2*b*).

- It is evident that almost all defective plugs were produced at the two ends of the mold,

[1]For those to whom it is not "evident," a formal analysis can be provided. It is possible and important for a troubleshooter to develop a sense of nonrandomness and resort to formal analysis when the evidence is borderline. Of course, when the use of data results in a major improvement, any question of "significance" is academic.

TABLE 12.1 Number of Defective Plastic Plugs (Short-Shot) from Each of 32 Cavities in a Mold

Cavity no.	No of defectives	Cavity no	No. of defectives
1	13	17	17
2	1	18	2
3	0	19	0
4	1	20	1
5	0	21	0
6	1	22	0
7	4	23	1
8	10	24	9
9	3	25	8
10	0	26	0
11	0	27	0
12	0	28	0
13	0	29	0
14	0	30	0
15	5	31	1
16	9	32	15
		N	101

TABLE 12.2a Numbering on Cavities in the Mold

1	2	3	4	5	6	7	8
9	10	11	12	13	14	15	16
17	18	19	20	21	22	23	24
25	26	27	28	29	30	31	32

TABLE 12.2b Pattern of Short-Shot Plugs

13	1	0	1	0	1	4	10
3	0	0	0	0	0	5	9
17	2	0	1	0	0	1	9
8	0	0	0	0	0	1	15

and essentially none were produced near the center. Certainly this is not random. What could produce such a pattern?

- In any molding process, plastic is introduced into the mold at a source and forced out to individual cavities through small channels in the mating blocks. Then, the supervisor reasoned, the source of plastic must be at the center and not enough was reaching the end cavities. This was an educated surmise, so the supervisor telephoned the vendor and asked, "Does your mold have 32 cavities in a 4 × 8 pattern?" "Yes," the vendor answered. "Does it have a center source of plastic?" Again, "Yes, what makes you think so?" Then the supervisor explained the data that had been obtained and the reasoning formulated from the data. During the telephone conversation, different ways were suggested as possible improvements for the physical extrusion problem:

1. Clean out or enlarge portions of the channels to the end cavities,
2. Increase the extrusion pressure, and/or
3. Reduce the viscosity of the plastic by increasing certain feed temperatures.

After some production trials, these suggestions resulted in a virtual elimination of short-shot plugs. Case dismissed.

CASE HISTORY 12.2

Automatic Labelers

Labels are applied to glass and plastic bottles of many kinds: beverage, food, pharmaceuticals. They may be applied at 400 or 500 a minute or at slower rates. It is fascinating to watch the intricate mechanism pick up the label, heat the adhesive backing of the label, and affix it to a stream of whirling bottles. But it can cause headaches from the many defect types: crooked labels (see Fig. 12.2), missing labels, greasy labels, wrinkled labels, and others.

Figure 12.2 Plastic bottle and crooked label.

There are many theories advanced by production supervisors to explain crooked label defects, for example. Crooked bottles was a common explanation. Production has even been known to keep one particular crooked bottle in a drawer; whenever a question was raised about crooked labels, the crooked bottle in the drawer was produced. Now crooked bottles (with nonvertical walls) will surely produce a crooked label. It takes a bit of courage and tenacity to insist that there may be other more important factors producing crooked labels than just crooked bottles. Such insistence will mean an effort to collect data in some form. Is it really justified in the face of that crooked bottle from the drawer? We thought so.[2]

There were two simple methods of getting data on this problem:

1. Collect some bottles with crooked labels, and measure the (minimum) angle of a wall. This procedure was not helpful in this case. In some other problems, defects have been found to come from only certain mold cavities.
2. Collect some 25 (or 50) bottles from each of the six label positions. This requires help from an experienced supervisor. Then inspect and record the extent of crooked labels from each head. One of our first studies showed crooked labels all coming from *one* particular position—*not* from the six positions randomly.

It is not easy to identify differences in the performance of six heads on a labeler (operating at several hundred per minute) or on any other high-speed multiple-head machine. And it is easy to attribute the entire source of trouble to crooked bottles, defective labels, or to the responsibility of other departments or vendors. No one wants to bear the onus: "It's not *my* fault," But someone in your organization must develop methods of

[2]One must frequently operate on the principle that "My spouse is independently wealthy," whether or not it is true.

getting sample data from high-speed processes—data which will permit meaningful comparisons of these heads as one phase of a problem-solving project or program. Then, controls must be established to prevent a relapse following the cure.

12.2 Disassembly and Reassembly

Any assembled product which can be readily disassembled and then reassembled can be studied by the general procedure of this chapter. The method with two components as in the first example below is standard procedure. But it is not standard procedure when three or four components are reassembled; the method has had many applications since the one in Case History 12.3.

Example 12.1 While walking through a pharmaceutical plant, we saw some bottles with crooked caps; the caps were crooked on some bottles but straight on others. The supervisor said, "Yes, some of the caps are defective." Out of curiosity, we picked up a bottle with a crooked cap and another with a straight cap and interchanged them. But the "crooked" cap was now straight and the "straight" cap was now crooked. Obviously, it was not this cap but the *bottle* which was defective. No additional analysis was needed, but a similar check on a few more crooked assemblies would be prudent.

CASE HISTORY 12.3

Noisy Kitchen Mixers[3]

During the production of an electric kitchen mixer, a company was finding rejects for noise at final inspection. Different production experiments had been run to determine causes of the problem, but little progress had been effected over several months of effort. It was agreed that new methods of experimentation should be tried.

Choice of factors to include in the experiment

There were different theories advanced to explain the trouble; each theory had some evidence to support it and some to contradict it. A small committee representing production, engineering, and quality control met to discuss possible reasons for the trouble. The usual causative variables had been tested without much success.

One gear component (henceforth called "gears") was suspected; it connected the top half and bottom half of the mixer. But there was no assurance that gears were a source of noise or the only source. A production program to inspect gears for "out of round" was being considered; then it was expected to use only the best in assembly.

Then the question was asked: "If it isn't the gears, is the trouble in the top half or the bottom half of the mixer?" There was no answer to the question. "Is it feasible to disassemble several units and reassemble them in different ways?" The answer was, "Yes." Since each half is a complicated assembly in itself, it was agreed to isolate the trouble in one-half of the mixer rather than look for the specific reason for trouble. There were now three logical factors to include in the study:

[3]Ellis R. Ott, "A Production Experiment with Mechanical Assemblies," *Ind. Qual. Control,* vol. 9, no. 6, 1953.

Tops	(T)
Bottoms	(B)
Gears	(G)

The trouble must certainly be caused by one or more of these separate factors (main effect) or perhaps by the interrelation of two (interaction). Further experiments might be required after seeing the results of this preliminary study.

Arrangement of factors in the experiment

The object of this troubleshooting project was to determine why some of the mixers were noisy. There was no test equipment available to measure the degree of noise of the mixer, but only to determine whether it was good G or noisy N.

It was agreed to select six mixers which were definitely noisy; and an equal number of good mixers. Then a program of interchange (reassemblies of tops, bottoms, and gears) was scheduled. During the interchanges, those mixer reassemblies which included a gear from a noisy mixer should test noisy *if* the gear was the cause; or if the top or bottom was the cause, then any reassembly which included a top or bottom from a noisy mixer should test noisy. Thus it might be expected that half of the reassemblies (containing a gear from a noisy mixer) would test noisy, and the other half test good.

The reassembly of components from the three different groups was scheduled according to the design of Table 12.3 which shows also the number of noisy mixers, out of a possible six, in each of the reassembled combinations. After reassembly (with parts selected at random), each mixer reassembly was rated by the original inspector as either noisy or good. In case of doubt, a rating of one-half was assigned. For example, group 7 indicates all six mixers from the reassembly of G gears, N tops, and G bottoms were noisy. No one type of component resulted in all *noisy* mixers nor all *good* mixers. But tops from noisy mixers were a major source of trouble. This was a *most unexpected* development and produced various explanations from those who had been intimately associated with the problem.

Two things are immediately evident from Table 12.3:

1. Noisy gears *do not* reassemble consistently into noisy mixers, and
2. Noisy tops *do* reassemble almost without exception into noisy mixers.

TABLE 12.3 Data on Reassemblies of Mixers

	N Gears		G Gears	
$n_g = 6$	N tops	G tops	N tops	G tops
N-bottoms	(1) $4\frac{1}{2}$	(2) 2	(3) 6	(4) 2
G-bottoms	(5) $4\frac{1}{2}$	(6) 3	(7) 6	(8) $1\frac{1}{2}$

Formal analysis

There were 6 × 8 = 48 reassemblies; the fractions of noisy ones are shown in Table 12.4 and graphically in Fig. 12.3.

Formal analysis

$$\hat{\sigma}_p = \sqrt{\frac{(0.615)(0.385)}{24}} = 0.099$$

Decision lines (Fig. 12.3)

$\alpha = 0.05$

$$p \pm H_{0.05}\sigma_p = 0.615 \pm (1.39)(0.099) = 0.615 \pm 0.138$$

$$\text{UDL} = 0.75$$

$$\text{LDL} = 0.48$$

$\alpha = 0.01$

$$\text{UDL} = 0.615 + 0.180 = 0.79$$

$$\text{LDL} = 0.615 - 0.180 = 0.43$$

Discussion

Before this study, gears had been the suspected source of trouble. Can we say, now, that gears have no effect? The interaction analysis in Fig. 12.3 indicates that the only effect which might be suspect is the top-gear relationship. Although no top-gear decision lines have been drawn for $\alpha = 0.10$, top-gear combinations *may* warrant a continuing suspicion that more extreme out-of-round gears result in noisy mixers with some tops.

But the most important result is that of the tops, and engineering studies can be designed to localize the source of trouble within them. This was done subsequently by

TABLE 12.4 Computations for Main Effects and Interactions (ANOM)

Number of noisy reassemblies *Main effects;*			*Two-factor interactions:*		
Gears:	N: (1,2,5,6):	14/24 = 0.583	TB:	(Like) (1,3,6,8):	15/24 = 0.625
	G: (3,4,7,8):	15.5/24 = 0.646		(Unlike) (2,4,5,7):	14.5/24 = 0.604
Tops:	N: (1,3,5,7):	21/24 = 0.875	TG:	(Like) (1,4,5,8):	12.5/24 = 0.520
	G: (2,4,6,8):	8.5/24 = 0.354		(Unlike) (2,3,6,7):	17/24 = 0.708
Bottoms:	N: (1,2,3,4):	14.5/24 = 0.604	BG:	(Like) (1,2,7,8):	14/24 = 0.583
	G: (5,6,7,8):	15/24 = 0.625		(Unlike) (3,4,5,6):	15.5/24 = 0.646
		$\bar{p} = 0.6145$	*Three-factor interactions:*		
			TBG:	(−) (1,4,6,7):	15.5/24 = 0.646
				(+) (2,3,5,8):	14/24 = 0.583

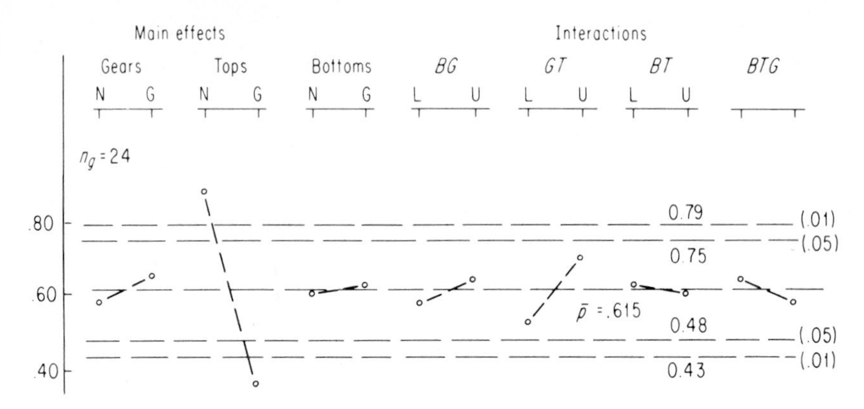

Figure 12.3 A formal comparison of mixer performance (analysis assumes independence) in reassemblies using subassemblies from six noisy and six good mixers (data from Table 12.4).

dividing tops into three areas and making reassemblies in a manner similar to that used in this study.

The decision to choose six noisy mixers and six good mixers was dictated primarily by expediency. With this number of mixers, it was estimated that the reassembling and testing could be completed in the remaining 2 hr of the workday. This was an important factor because some of the committee members were leaving town the next morning and it was hoped to have some information on what would happen. Also, it was thought that the cause of the trouble would be immediately apparent with even fewer reassembled mixers than six of each kind.

This type of 2^3 study with mechanical or electrical-mechanical assemblies is an effective and quick means of localizing the source of trouble. It is quite general, and we have used it many times with many variations since this first experience.

Note

See Case Histories 12.1 and 12.2 on a general strategy in troubleshooting in Sec. 12.1, "Ideas from patterns of data."

12.3 A Special Screening Program for Many Treatments[4]

Introduction

The periodic table contains many trace elements. Some will increase the light output from the phosphor coating on the face of a radar tube. This type of light output is called *cathodeluminescence.* The coating produces luminescence by an electron beam. How do we determine which trace element to add to a new type of coating?

[4]Ellis R. Ott and Frank W. Wehrfritz, "A Special Screening Program for Many Treatments," *Stat. Neerl.*, special issue in honor of Prof. H. C. Hamaker, July 1973, pp. 165–170.

There are many other examples of problems involving the screening of many possible treatments. Many chemicals have been suggested as catalysts in certain usages. Many substances are the possible irritants causing a particular allergy. Many substances are tested as possible clues for a cancer. How can the testing of many possibilities be expedited?

This chapter considers one general strategy to provide answers to problems of this type where there are many items to be investigated. How should they all be examined using only a few tests? The amount of testing can be reduced drastically under certain assumptions. Testing sequences involving many factors have been developed which can do the job. These sequences have their origins in the two-level factorial experiment. In Case History 11.10 it was discussed how the 2^3 factorial design can be split into two halves of four runs each; thus with a change in the assumptions, we can test the three factors with only four of the eight combinations and not perform the other four. As the number of factors increases, there are many other ways of dividing the 2^n runs into fractions; each strategy involves the extent to which the experimenter decides to relax the assumptions.

On the other hand, the basic 2^n factorial array can be augmented with additional data points, and the assumptions can then support quadratic estimates. Thus it is apparent from the developmental history of applied statistics that the number of tests or data points bear a direct relationship to the assumptions made at the planning stage for gathering data and to the testing of significance of effects seen when the data are finally at hand.

A screening program. In the case of screening many possible factors or treatments, it is possible to reduce drastically the amount of testing under certain assumptions.

Theorem. With n separate tests, it is possible to screen $(2^n - 1)$ treatments at the chosen levels as having a positive or negative effect, and to identify the effective treatment under assumptions given below. We then have:

Number of tests	n	2	3	4
Number of factors	$2^n - 1$	3	7	15
Number of treatments/test	2^{n-1}	2	4	8

Assumptions

1. It is feasible to combine more than one treatment in a single test. For example, several trace elements can be included in the same mix of coating.
2. No treatment is an inhibitor (depressant) for any other treatment. It is not unusual to observe a synergistic effect from two combined substances; such an effect would favor their joint inclusion. (The combining of two treatments which adversely affect each other would not be desirable.)
3. The effect of any treatment is either positive or negative (effective or not effective). For example, it is possible to determine whether the light output of one mix of phosphor is brighter than that of another.
4. The effect of a treatment is consistent (not intermittent).

5. No more than one effective treatment is among those being studied. (There may be none.)

These assumptions are the basis of this initial discussion. However, in application each assumption can be modified or disregarded as warranted by the conditions of a specific experiment. Modifications of the assumptions would require a reconsideration of the logic. For example, a numerical value of the yield of a process might be available on each run; this measure can then be used instead of considering only whether the run is a success or a failure. Replicates of each run would then provide a measure of the process variability. The following discussion is based on the listed assumptions, however.

Examples of screening programs

Screen 7 different treatments with three tests ($n = 3$).

Screen 15 different treatments with four tests ($n = 4$).

Screen 31 different treatments with five tests ($n = 5$).

Each of the n tests will include $2^n - 1$ treatments.

Example 12.2 As an example, when the experimenter is planning three tests ($n = 3$), seven treatments or factors ($k = 2^n - 1 = 7$) can be allocated to the three tests. Each test will include four treatments ($2^{n-1} = 4$). The treatments or factors are designated as $X_1, X_2, X_3, X_4, X_5, X_6, X_7$ and assigned to these numbers as desired. Table 12.5 shows which of the treatments to combine in each test. The array of plus and minus signs is the *design matrix* indicating the *presence* (+) or *absence* (−) of a treatment (or high and low concentrations of a treatment). The experimental responses are designated as E_i; each E is simply a (+) or a (−) when assumption 3 is being accepted.

The identification of the effective treatment is made by matching the pattern of experimental responses, E_1, E_2, E_3, with one of the columns of treatment identification. When the (+) and (−) results of the three tests match a column of (+) and (−) in Table 12.5, this *identifies* the single causative treatment. If, for example, the three treatments yield the vertical sequence (+ − +), then X_3 is identified as the cause.

Discussion A discussion of possible answer patterns for the three tests involving the seven treatments follows:

No positive results: If none of the three tests yields a positive result, then none of the seven treatments is effective (assumptions 2 and 4).

TABLE 12.5 A Screening Design for $2^3 - 1 = 7$ Factors

Treatments* included	Treatment identification							Experimental response
	X_1	X_2	X_3	X_4	X_5	X_6	X_7	
Test 1: X_1, X_2, X_3, X_5	+	+	+	−	+	−	−	E_1
Test 2: X_1, X_2, X_4, X_6	+	+	−	+	−	+	−	E_2
Test 3: X_1, X_3, X_4, X_7	+	−	+	+	−	−	+	E_3

*NOTE: Each test combines four of the treatments which are represented by +. A full factorial would require not 3 but $2^7 = 128$ tests.

One positive result: If, for example T_1 alone yields a positive effect, then treatment X_5 is effective. The reasoning is as follows: the only possibilities are the four treatments in T_1. It can't be either X_1 or X_2 because they are both in T_2 which gave a negative result: it can't be X_3 which is also in T_3 which gave a negative result. Then it must be X_5.

Two positive results: If T_1 and T_2 both yield positive results and T_3 yields negative (E_1 and E_2 are +; E_3 is −), then X_2 is the effective treatment. The reasoning is similar to that for one positive experimental result; the decision that it is X_2 can be made easily from Table 12.5. Simply look for the order (+ + −) in a column; it appears uniquely under X_2.

Three positive results: The (+ + +) arrangement is found under treatment X_1. It is easily reasoned that it cannot be X_2 since X_2 does not appear in test 3 which gave a positive response; and similar arguments for each of the other variables other than X_1.

Example 12.3 When $n = 4$, let the $2^4 - 1 = 15$ treatments be

$$X_1, X_2, X_3, \ldots, X_{15}$$

Then a representation of the screening tests and analysis is given in Table 12.6. Note that each test combines $8 = 2^{n-1}$ treatments.

There is a unique combination of the plus and minus signs corresponding to each of the 15 variables; this uniqueness identifies the single cause (assumption 5). No variable is effective if there is no positive response.

Fewer than $2^n - 1$ treatments to be screened Frequently, the number of treatments k to be screened is not a number of the form $2^n - 1$. Then designate the k treatments as

$$X_1, X_2, X_3, \ldots, X_k$$

When $n = 3$ and $k = 5$, for example, carry out the three tests T_1, T_2, T_3, disregarding entirely treatments X_6 and X_7 as in Table 12.7.

Note: We would not expect either T_2 alone or T_3 alone to give a positive result; that would be a puzzler under the assumptions. (Possible explanations: a gross experimental blunder or an intermittent effect or interaction. For example if X_1 and X_4 together were required to give a positive result.)

Again, if there are fewer than 15 treatments to be screened (but more than 7, which employs only three tests), assign numbers to as many treatments as there are and ignore the balance of the numbers. The analysis in Table 12.6 is still applicable.

Ambiguity if assumption 5 is not applicable One seldom can be positive, in advance, whether there *may possibly be two or more effective treatments* in the group to be screened. If during the experiment, some combination of the treatments gives a greatly superior performance, this is an occasion for rejoicing. But economics is a factor which may require or desire the specific identification of causes. Let us consider the case of seven treatments being screened (Table 12.5):

- *One positive test response:* No ambiguity possible; the indicated treatment is the only possibility.
- *Two positive test responses:* If both T_1 and T_2 give positive responses, then it may be X_2 alone or any two or three of the three treatments X_2, X_5, and X_6.

TABLE 12.6 A screening Design for $2^4 - 1 = 15$ Factors

	Treatment identification*															Experimental Response
	1	2	3	4	5	6	7	8	9	10	11	12	13	14	15	
$T_1: X_1 X_2 X_3 X_4 X_6 X_8 X_{10} X_{12}$	+	+	+	+	−	+	−	+	−	+	−	+	−	−	−	E_1
$T_2: X_1 X_2 X_3 X_5 X_6 X_9 X_{11} X_{13}$	+	+	+	−	+	+	−	−	+	−	+	−	+	−	−	E_2
$T_3: X_1 X_2 X_4 X_5 X_7 X_8 X_{11} X_{14}$	+	+	−	+	+	−	+	+	−	−	+	−	−	+	−	E_3
$T_4: X_1 X_3 X_4 X_5 X_7 X_9 X_{10} X_{15}$	+	−	+	+	+	−	+	−	+	+	−	−	−	−	+	E_4

* The treatments may also be designated by the letter $A, B, C, \ldots, O$: (Case History 12.4)

Test 1: A B C D F H J L
Test 2: A B C E F I K M
Test 3: A B D E G H K N
Test 4: A C D E G I J O

TABLE 12.7 A Screening Design for Five Factors

	Treatment identification					Experimental response
	X_1	X_2	X_3	X_4	X_5	
$T_1: X_1, X_2, X_3, X_5$	+	+	+	−	+	E_1
$T_2: X_1, X_2, X_4$	+	+	−	+	−	E_2
$T_3: X_1, X_3, X_4$	+	−	+	+	−	E_3

The possible ambiguity can be resolved in a few tests; an obvious procedure would be to run tests with X_2, X_5, and X_6 individually. However, when there seems to be little possibility of having found more than a single positive treatment, run a single test with X_5 and X_6 combined. If the result is negative, then the single treatment is X_2.

CASE HISTORY 12.4

Screening Some Trace Elements

Some of the earth's trace elements have important effects upon some quality characteristics of cathode ray tubes. Something is known about these effects, but not enough. Hoping to identify any outstanding effect upon the light-output quality characteristic, a screening study was planned. Fifteen trace elements (designated *A*, *B*, *C*,..., *O*) were mixed in four phosphor slurries in the combination shown in Table 12.8 (eight different trace elements were included in each slurry run). A (+) indicates those elements which were included in a test and a (−) indicates those which were not included. The measured output responses of the four slurries are shown in the column at the right.

- Previous experience with this characteristic had shown that the process variability was small; consequently, the large response from run 2 is "statistically significant."
- Also, the response from test 2 was a definite improvement over ordinary slurries which had been averaging about 45 to 50. The other three output responses, 46, 44, and 51, were typical of usual production. The four test responses can be considered to be: (− + − −). Under the original assumptions of this study plan, the identity of the responsible trace element must be *M*.

TABLE 12.8 Variables Data in a Screening Design for 15 Factors (Trace Elements)

	C_4^4	C_3^4				C_2^4						C_1^4				Experimental response
	A	*B*	*C*	*D*	*E*	*F*	*G*	*H*	*I*	*J*	*K*	*L*	*M*	*N*	*O*	
Test 1	+	+	+	+	−	+	−	+	−	+	−	+	−	−	−	46
Test 2	+	+	+	−	+	+	−	−	+	−	+	−	+	−	−	84
Test 3	+	+	−	+	+	−	+	+	−	−	+	−	−	+	−	44
Test 4	+	−	+	+	+	−	+	−	+	+	−	−	−	−	+	51

- One cannot be sure, of course, that the uniqueness assumption is applicable. Regardless, the eight trace elements of run 2 combine to produce an excellent output response. At this point in experimentation, the original 15 trace elements have been reduced to the eight of run 2 with strong evidence that the improvement is actually a consequence of *M* alone. Whether it is actually a consequence of *M* alone or some combination of two or more of the eight elements of run 2 can now be determined in a sequence of runs. Such a sequence might be as follows:

  The first run might be with *M* as the only one, the second with the seven other than *M*. If there is still uncertainty, experiments could continue eliminating one trace element at a time.

12.4 Other Screening Strategies

One can find other screening strategies discussed in the technical literature. Anyone who has reason to do many screening tests should consider the preceding strategy and consult the published literature for others.

The foregoing allocation of experimental factors to the tests is based on combinatorial arrays. As such, this strategy serves to reduce the initial large number of factors to more manageable size. This is indicated in the readings of Table 12.8.

12.5 Relationship of One Variable to Another

Engineers often use scatter diagrams to study possible relationships of one variable to another. They are equally useful in studying the relationship of data from two sources—two sources which are presumed to produce sets of data where either set should predict the other. Scatter diagrams are helpful in studying an expected relationship between two sets of data by displaying the actual relationship which does exist under these specific conditions. They often show surprising behavior patterns which give an insight into the process which produced them. In other words, certain relationships which are expected by scientific knowledge may be found to exist for the majority of the data but not for all. Every type of nonrandomness (lack of control) of a single variable mentioned earlier—outliers, gradual and abrupt shifts, bimodal patterns—is a possibility in a scatter diagram which displays the relationship of one set of data to another. These evidences of nonrandomness may lead an engineer to investigate these unexplained behavior patterns and discover important facts about the process.

For example, one step in many physical or chemical processes is the physical treatment of a product for the purpose of producing a certain effect. How well does the treatment actually accomplish its expected function on the items? Does it perform adequately on the bulk of them but differently enough on some items to be economically important? Any study of a quality-characteristic relationship between *before* and *after* a treatment of importance can begin with a display in the form of a scatter diagram. This is especially important when problems of unknown nature and sources exist. The method is illustrated below.

CASE HISTORY 12.5

Geometry of an Electronic Tube

Changes in the internal geometry in a certain type of electronic tube during manufacture can be estimated indirectly by measuring changes in the *electrical capacitance.* This capacitance can be measured on a tube while it is still a "mount," i.e., before the mount structure has been sealed into a glass bulb (at high temperature).

During successive stages of a tube assembly, the internal geometry of many (or all) was being deformed at some unknown stage of the process. It was decided to investigate the behavior of a mount before being sealed into a bulb and then the tube after sealing—a *before-and-after* study. Also, it was decided to seal some tubes into bulbs at 800° and some at 900° and to observe the effect of sealing upon these few tubes.

Data are shown in Table 12.9 for 12 tubes sealed at 900° and another 12 sealed at 800°; two of the latter were damaged and readings after sealing were not obtainable. In preparing the scatter diagram, we use principles of grouping for histograms, Sec. 1.4, Chap. 1. The range of the *R*-1 readings is from a low of 0.56 to high of 0.88, a difference of 0.32. Consequently, a cell width of 0.02 gives us 16 cells, which is reasonable for such a scattergram. Similarly, a cell width of 0.02 has been used for the *R*-2 data. In practice, tally marks are made with two different colored pencils; in Fig.

TABLE 12.9 12 Tubes at 900° and 12 Tubes at 800°

At 900°			At 800°		
Tube no.	R-1 before stage A	R-2 after stage A and before stage B	Tube no.	R-1 before stage A	R-2 after stage A and before stage B
1.	0.66	0.60	13.	0.75	0.68
2.	0.72	0.57	14.	0.56	0.49
3.	0.68	0.55	15.	0.72	0.59
4.	0.70	0.60	16.	0.66	0.56
5.	0.64	0.64	17.	0.62	0.52
6.	0.70	0.58	18.	0.56	—
7.	0.72	0.56	19.	0.65	—
8.	0.73	0.62	20.	0.88	0.87
9.	0.82	0.62	21.	0.56	0.46
10.	0.66	0.60	22.	0.76	0.77
11.	0.72	0.62	23.	0.72	0.69
12.	0.84	0.87	24.	0.74	0.70

SOURCE: Ellis R. Ott, A Scatterdiagram Used to Compare "Before" and "After" Measurements, *Ind. Qual. Control,* vol. 8, no. 12, June, 1957. (Reproduced by permission of the editor.)

12.4, an x has been used for tubes processed at 900° and an o for tubes processed at 800°. Tube no. 1, at 900°, for example, is represented by an x at the intersection of column 66–67 and row 60–61.

Analysis and interpretation: stage A

There is an obvious difference in the patterns corresponding to the 12 tubes sealed at 900° and the 10 sealed at 800°; all 22 tubes had had the same manufacturing assembly and treatment in all other respects. The 900° sealing temperature is deforming the internal geometry in an unpredictable manner; tubes processed at 800° in stage A show a definite tendency to line up. If the quality characteristics of the tubes can be attained at the 800°, well and good. If not, then perhaps the mount structure can be strengthened to withstand the 900°, or perhaps an 850° sealing temperature could be tried.

In any case, the stage in the process which was deforming the geometry has been identified, and the problem can be tackled.

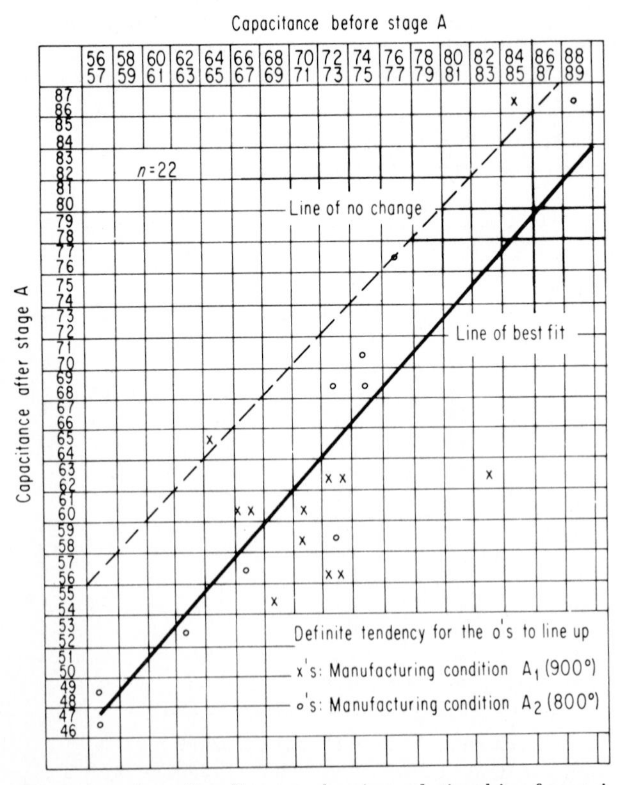

Figure 12.4 A scatter diagram showing relationship of capacitance on same individual tubes before and after stage A (data from Table 12.9).

Discussion

A small number of tubes was sufficient to show a difference in behavior between the two groups. No analytical measure[5] of the relationship would give additional insight into the problem.

The *line of no change* has been drawn in Fig. 12.4 by joining the point at the maximum corner of the (86–87) cell on both R-1 and R-2 with the similar corner point of the (56–57) cell. R-1 = R-2 at each point of the line. It is evident that the electrical capacitance was unchanged on only three or four tubes. For all others, it was appreciably reduced during stage A, especially for tubes with lower capacitances. This represents an unsolved problem.

Summary: Scatter diagram, before and after

In Chap. 2, we discussed certain patterns of data involving one *independent* variable which indicate assignable causes which are of interest to an engineer. We also discussed certain tests available to detect different data patterns. When we study possible relationships between two variables, we have an interest in the pattern of points with respect to a diagonal line (such as a line of no change). The data may show variations of the same or different types as in Chap. 2: gross errors, mavericks, trends, abrupt shifts, short cycles, and long cycles. The tests suggested for the existence of these patterns with respect to a horizontal line *can be extended or adjusted to diagonal lines.* A simple scattergram gives valuable clues which are often entirely adequate[6] for engineering improvements.

12.6 Practice Exercises

1. State the three factors of Case History 12.3.
2. Since there were 6 × 8 = 48 reassemblies in this experiment why is n = 24 used in computing sigma-hat?
3. What was the advantage of conducting the reassembly experiment instead of inspecting for out-of-round gears?
4. What was the main conclusion of the experiment?
5. What are the general conditions that lead to screening programs of the type described in Sec. 12.3?
6. How many different treatments can be screened with six tests? What can be learned about interaction between factors in such a screening program?
7. Make up a set of data for Table 12.5 to illustrate a significant effect for treatment 5 but not for any others.

[5]*Note:* There will be occasions when an explicit measure of the before and after relationship is useful to the troubleshooter, but not in this case. However, the correlation coefficient $r = 0.84$ and the least-squares line of best fit $Y = -1.66 + 1.119X$ for the benefit of those who commonly use such measures. Note that a simple nonparametric test for the existence of a relationship is presented in Paul S. Olmstead and John W. Tukey, "A Corner Test for Association," *Annal. Math. Stat.*, vol. 18, December 1947, pp. 495–513.

[6]They often provide more information of value than the more traditional computation of the correlation coefficient r. The computation of r from data containing assignable causes may give a deceptive measure of the underlying relationship between the two variables—sometimes too large and sometimes too small. Any computation of r should be postponed at least until the scattergram has been examined.

8. *a*. Would you expect a scatter diagram to be of help in presenting the comparison of the chemical analysis and the materials balance computations of the data in Table 3.1?
   b. Prepare a scatter diagram of the data
9. *a*. Would you expect a scatter diagram of machine 49 versus 50, for example, in Table 4.3 to be helpful?
   b. Prepare a scatter diagram.

Chapter

13

Comparing Two Process Averages

13.1 Introduction

This section provides a transition to analysis of means for measurement data in the special case where two treatments are being compared and sample sizes are the same.

The discussion begins with a comparison at two levels of just one independent variable. When data from two experimental conditions are compared, how can we judge objectively whether they justify our initial expectations of a difference between the two conditions? Three statistical methods are presented to judge whether there is objective evidence of a difference greater than expected only from chance. This is a typical decision to be made, with *no standard given.*

13.2 Tukey's Two-Sample Test to Duckworth's Specifications[1]

There are important reasons for becoming familiar with the Tukey procedure: no hand calculator is needed; such a procedure may well be used more often and "compensate for (any) loss of mathematical power. Its use is to indicate the weight of the evidence roughly. If a delicate and critical decision is to be made, we may expect to replace it or augment it with some other procedure."

Tukey procedure

Given two groups of r_1 and r_2 measurements taken under two conditions are the criteria for the Tukey–Duckworth procedure. The requirement for comparing the experimental conditions by this criteria is that

[1]John W. Tukey, "A Quick Compact, Two-Sample Test to Duckworth's Specifications," *Technometrics,* vol. 1, no. 1, February 1959, pp. 31–48.

the largest observation of the two be in one sample (A_2) and the smallest in the other (A_1). Let the number of observations in A_2 which are larger than the largest in A_1 be a, and let the number in A_1 smaller than the smallest in A_2 be b where neither a nor b is zero. (Count a tie between A_1 and A_2 as 0.5.) Critical values of the sum of the two counts, $a + b$, for a two-sided test are given in Table 13.1. The test is essentially independent of sample sizes if they are not too unequal, i.e., the ratio of the larger to the smaller is less than 4:3.

Note, the Tukey-Duckworth test may be one- or two-sided.

TABLE 13.1 Critical Values of the Tukey–Duckworth Sum (Two-Sided)
Also see Table A.13

Approximate risk	Two-sided critical values of the sum $a + b$	One-sided* critical value of the sum $a + b$
0.09	6	5
0.05	7	6
0.01	10	9
0.001	13	12

*Kindly provided by Dr. Larry Rabinowitz who also studied under Dr. Ott. Note that, for the two-sided test $\alpha \cong (a + b)/2^n$ and for the one-sided test $\alpha \cong (a + b)/2^{n-1}$

CASE HISTORY 13.1

Nickel-Cadmium Batteries

In the development of a nickel-cadmium battery, a project was organized to find[2] some important factors affecting capacitance.

The data in Table 13.2 were obtained at stations C_1 and C_2 (other known independent variables believed to have been held constant). Is the difference in process averages from the two stations statistically significant?

Some form of graphical representation is always recommended and the credibility of the data considered. The individual observations have been plotted in Fig. 13.1. There is no obvious indication of an outlier or other lack of stability in either set. Also, the criteria for the Tukey–Duckworth procedure are satisfied, and the sum of the two counts is

$$a + b = 6 + 6 = 12$$

This exceeds the critical sum of 10 required for risk $\alpha \cong 0.01$ (Table 13.1).

13.3 Analysis of Means, $k = 2$, $n_g = r_1 = r_2 = r$

There is hardly need of any additional evidence than the Tukey two-sample analysis to decide that changing from C_1 to C_2 (Table 13.2) will increase ca-

[2]See Case History 14.2 for additional discussion.

TABLE 13.2 Data: Capacitance of Nickel-Cadmium Batteries Measured at Two Stations

C_1	C_2
0.6	1.8
1.0	2.1
0.8	2.2
1.5	1.9
1.3	2.6
1.1	2.8
$\bar{C}_1 = 1.05$	$\bar{C}_2 = 2.23$
$R_1 = 0.9$	$R_2 = 1.0$

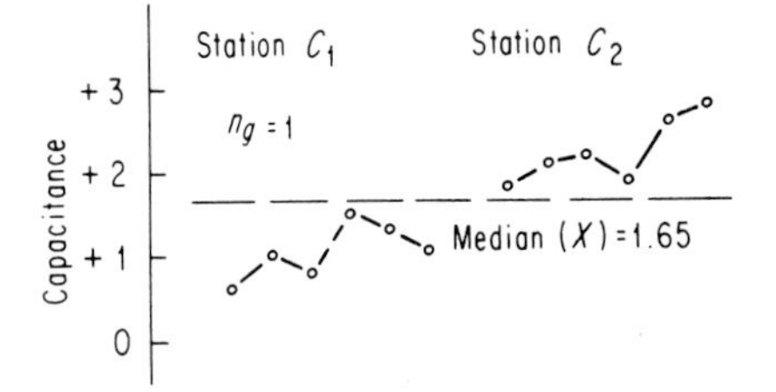

Figure 13.1 Comparing level of a battery characteristic manufactured at two different stations (data from Table 13.2).

pacitance. However, analysis of means (ANOM) for variables data, discussed previously in Chap. 11 for attributes data, will be presented here and used later with many sets of variables data.

ANOM will be used first to compare two processes represented by samples, then applied in Chap. 14 to 2^2 and 2^3 experimental designs. The importance of 2^2 and 2^3 designs in troubleshooting, pilot-plant studies, and initial studies warrants discussion separate from the more general approach in Chap. 15 where the number of variables and levels of each is not restricted to two.

Just as with attributes data it is often good strategy to identify possible problem sources quickly and leave a definitive study till later. The choice of some independent variables to be omnibus-type variables is usually an important short cut in that direction.

Formal analysis

- From Table 13.2 ($k = 2$, $r = 6$), values of the two averages and ranges are known. They are shown in Fig. 13.2. The two range points are inside the control limits.

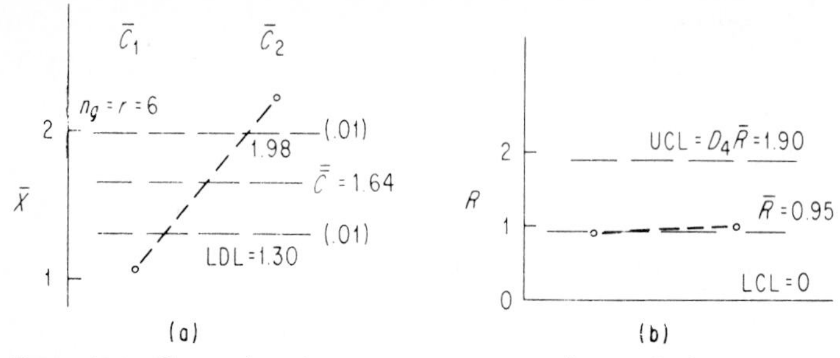

Figure 13.2 Comparing two process averages by analysis of means (variables) (data from Table 13.2).

- From Table A.11,

$$\hat{\sigma} = \bar{R}/d_2^* = 0.95/2.60 = 0.365$$

and

$$\hat{\sigma}_{\bar{X}} = \hat{\sigma}/\sqrt{r} = (0.365)/\sqrt{6} = 0.149$$

$$\text{df} \cong (0.9)2(6 - 1) = 9$$

- From Table A.14 for $k = 2$ and df = 9; $H_{0.05} = 1.60$, $H_{0.01} = 2.30$.
- *Decision lines*

$$\bar{\bar{X}} \pm H_{0.01}\hat{\sigma}_{\bar{X}} = 1.64 \pm (2.30)(0.149)$$

For $\alpha = 0.01$

$$\text{UDL} = 1.64 + 0.34 = 1.98$$

$$\text{LDL} = 1.64 - 0.34 = 1.30$$

TABLE 13.3 Summary: Mechanics of Analysis of Means, ANOM, for Two Small Samples (*A* and *B*) with $r_1 = r_2 = r$

STEP 1. Obtain and plot the two sample ranges. Find $\bar{R}$ and $D_4\bar{R}$. If both points fall below $D_4\bar{R}$, compute†

$$\hat{\sigma} = \bar{R}/d_2^* \quad \text{and} \quad \hat{\sigma}_{\bar{X}} = \hat{\sigma}/\sqrt{r}$$

Also df $\cong (0.9)k(r - 1) = 1.80(r - 1)$ for $k = 2$ (or see Table A.11).

STEP 2. Plot points corresponding to $\bar{A}$, $\bar{B}$, and their average $\bar{\bar{X}}$.

STEP 3. Compute $H_\alpha\hat{\sigma}_{\bar{X}}$, and draw decision lines

$$\text{UDL} = \bar{\bar{X}} + H_\alpha\hat{\sigma}_{\bar{X}}$$

$$\text{LDL} = \bar{\bar{X}} - H_\alpha\hat{\sigma}_{\bar{X}}$$

usually choosing values of $\alpha = 0.10$, 0.05, and 0.01 to bracket the two sample averages. H_α is from Table A.14.

STEP 4. When the pair of points falls outside a pair of decision lines, their difference is statistically significant, risk α.

NOTE: Points $\bar{A}$ and $\bar{B}$ will be symmetrical with $\bar{\bar{X}}$ when $r_1 = r_2$.
†Find d_2^* in Table A.11 for $k = 2$ and r.

■ These two decision lines are shown in Fig. 13.2*a*; the two C points are outside them. We conclude that there is a statistically significant difference in capacitance resulting in a change from station C_1 to C_2. This is in agreement with the Tukey procedure above. A summary of this procedure is given in Table 13.3.

13.4 Student's *t* and *F* Test Comparison of Two Stable Processes

Note. This section may be omitted without affecting understanding of subsequent sections.

Example 13.1 Again use data of Table 13.2; $k = 2$, $r_1 = r_2 = r = 6$.)

The *t statistic* to compute is:

$$t = \frac{\bar{C}_2 - \bar{C}_1}{s_p\sqrt{\dfrac{1}{r_1} + \dfrac{1}{r_2}}} \tag{13.1}$$

STEP 1: Compute

$$s_1{}^2 = \hat{\sigma}_1{}^2 = \frac{r\Sigma x^2 - (\Sigma x)^2}{r(r-1)} = \frac{6(7.15) - (6.3)^2}{30} = 0.107$$

$$s_2{}^2 = \hat{\sigma}_2{}^2 = \frac{6(30.70) - (13.4)^2}{30} = 0.155$$

STEP 2: Check for evidence of possible inequality of variances with the F test, Eq. (4.6).

$$F = \frac{s_2^2}{s_1^2} = \frac{0.155}{0.107} = 1.45 \qquad \text{with df} = F(5,5)$$

In Table A.12, we find critical values: $F_{0.05} = 5.05$ and $F_{0.10} = 3.45$ for risk $\alpha = 0.10$ and $\alpha = 0.20$.

STEP 3: Since $F = 1.45$ is less than even the critical value $F_{0.10}$, we accept equality of variances of the two processes and proceed to estimate their common pooled variance. From Eq. (4.5*b*)

$$s_p{}^2 = \frac{s_1{}^2 + s_2{}^2}{2} = \frac{(0.107) + (0.155)}{2} = 0.131$$

and

$$s_p = \sqrt{0.131} = 0.362$$

Since $r_1 = r_2$ in Eq. (13.1), the denominator becomes

$$\hat{\sigma}_{\bar{\Delta}} = \frac{\sqrt{2}s_p}{\sqrt{r}} = \sqrt{2}s_{\bar{X}} = 0.209 \tag{13.2}$$

Note that when $r_1 \neq r_2$, then

$$s_p^2 = \frac{(r_1 - 1)s_1^2 + (r_2 - 1)s_2^2}{r_1 + r_2 - 2} \tag{13.3}$$

See Eq. (4.5a).

STEP 4: Then finally compute from Eq. (13.1)

$$t = \frac{\overline{C}_2 - \overline{C}_1}{0.209} = \frac{1.18}{0.209} = 5.64 \qquad \text{df} = 10$$

The critical value found in Table A.15 for df = 10 is: $t_{0.01} = 3.17$.

STEP 5: *Decision:* Since our $t = 5.64$ is larger than $t_{0.01} = 3.17$, we decide that the process represented by the sample C_2 is operating at a higher average than the process represented by the sample C_1 with risk less than $\alpha = 0.01$.

STEP 6: *Inequality of variances:* It may happen that the F test of step 2 rejects the hypothesis of equality of variances. When this happens, it is inappropriate to calculate the pooled variance s_p^2 because there is no one variance that describes the variability in the data. Under such circumstances, we may appeal to the Welch–Aspin test,[3] which can be regarded as a modification of the t test in which

$$t = \frac{\overline{X}_1 - \overline{X}_2}{\sqrt{\dfrac{s_1^2}{r_1} + \dfrac{s_2^2}{r_2}}} \qquad \text{df} \cong \frac{(r_1 - 1)(r_2 - 1)\left(\dfrac{s_1^2}{r_1} + \dfrac{s_2^2}{r_2}\right)^2}{(r_2 - 1)\left(\dfrac{s_1^2}{r_1}\right)^2 + (r_1 - 1)\left(\dfrac{s_2^2}{r_2}\right)^2}$$

the test proceeds in the manner described above for the standard t test. Note, a little algebra will show that when $r_1 = r_2 = r$,

$$\text{df} \cong \frac{(r - 1)(1 + F)^2}{(1 + F^2)}$$

where F is the ratio of the two variances $F = s_2^2 / s_1^2$.

Some Comparisons of *t* Test and ANOM in Analyzing Data of Table 13.2 In Fig. 13.2b, both range points fall below UCL(R), and we accept homogeneity of variability in the two processes. This agrees with the results of the F test above.

Then $\hat{\sigma} = \overline{R}/d_2^* = 0.95/2.60 = 0.365$. This estimate $\hat{\sigma}$ agrees closely with the pooled estimate $s_p = 0.362$ in step 3.

The decision lines in Fig. 13.2a are drawn about C at a distance $\pm H_\alpha \hat{\sigma}_{\overline{X}}$. It can be shown that

$$\pm H_\alpha \hat{\sigma}_{\overline{X}} = \pm \frac{1}{2} t_\alpha \frac{s_p\sqrt{2}}{\sqrt{r}} \qquad \text{i.e.} \qquad H_\alpha = \frac{t_\alpha}{\sqrt{2}}$$

Thus the decision between the two process averages is made from looking at Fig. 13.2a instead of looking in the t table. The ANOM is just a graphical t test *when* $k = 2$. It becomes an extension of the t test when $k > 2$.

When $r_1 \neq r_2$ or when r *is not small,* we use Eq. (4.5a) in estimating $\hat{\sigma}$ for ANOM.

[3]B. L. Welch, "The Generalization of Student's Problem when Several Different Population Variances Are Involved," *Biometrika,* vol. 34, 1947, pp. 28–35.

When $r_1 = r_2 = r$ *is small*—say less than 6 or 7, the efficiency of the range in estimating $\hat{\sigma}$ is very high (see Table 4.2); the loss in degrees of freedom (df) is only about 10 percent as we have seen. It is, of course, possible to increase the sample size to compensate for this loss of efficiency.

13.5 Magnitude of the Difference between Two Means

At least as important as the question of statistical significance is the question of *practical or economic significance.* The observed *sample difference* in capacitance in Table 13.2 is

$$\bar{\Delta} = \bar{C}_2 - \bar{C}_1 = 2.23 - 1.05 = 1.18$$

This was found to be statistically significant. It is now the scientist or engineer who must decide whether the observed difference is large enough to be of practical interest. If the data were not coded, it would be possible to represent the change as a percent of the average, $\bar{C} = 1.640$. In many applications, a difference of 1 or 2 percent is not of practical significance; a difference of 10 percent or so would often be of great interest. The decision must be made for each case, usually by design or quality engineers.

If the study in Table 13.2 were repeated with another pair of samples for stations C_1 and C_2, we would not expect to observe exactly the same average difference $\bar{\Delta}$ as observed this first time. (However, we *would* expect the average difference for $r = 6$ to be statistically significant, risk $\alpha \cong 0.01$.) The confidence limits on the difference are given (for any risk α) by the two extremes

$$\left.\begin{aligned} \bar{\Delta}_1 &= (\bar{C}_2 - \bar{C}_1) + 2H_\alpha \hat{\sigma}_{\bar{X}} \\ \bar{\Delta}_2 &= (\bar{C}_2 - \bar{C}_1) - 2H_\alpha \hat{\sigma}_{\bar{X}} \end{aligned}\right\} \text{risk } \alpha \qquad (13.4)$$

We have, with risk $\alpha = 0.01$ for example:

$$\bar{\Delta}_1 = 1.18 + 2(2.30)\left(\frac{0.362}{\sqrt{6}}\right) = 1.86$$

and

$$\bar{\Delta}_2 = 1.18 - 2(2.30)\left(\frac{0.362}{\sqrt{6}}\right) = 0.50$$

Or, we found the effects of C_1 and C_2 to differ by 1.18 units; the two processes which $\bar{C}_1$ and $\bar{C}_2$ represent may actually differ by as much as 1.87 units or as little as 0.49 units. Thus, in Eq. (13.4), the experimenter has a measure of the extreme differences which can actually be found in a process as a result of shifting between levels C_1 and C_2, risk α.

Sometimes, the observed difference may not be of practical interest in itself but may suggest the *possibility* that a larger change in the independent variable might produce a larger effect which *would* then be of interest. These are matters to discuss with the engineer or scientist.

CASE HISTORY 13.2

Height of Easter Lilies on Date of First Bloom[4]

Botanists have learned that many characteristics of plants can be modified by man. For example, "Easter lilies" grown normally in the garden, in many states, bloom in July or August—not at Easter time. You would probably not give a second thought to such characteristics as the range of heights you would favor when buying an Easter lily or the number of buds and blooms you would prefer, but they are important factors to the horticulturist. The referenced study employed a more complex design than either the one presented here or the 2^2 design in Table 14.2. Botanists and agriculturalists usually have to wait through one or more growing seasons to acquire data. Their experimental designs often need to be quite complicated to get useful information in a reasonable time. Industrial troubleshooting and process improvement can often move much faster; additional information can often be obtained within a few hours or days. Several less complicated experiments are usually the best strategy here. This is one reason for our emphasis on three designs: the 2^2, the 2^3, and the fractional factorial. A study was made of the height of Easter lilies (on the date of first bloom). Under two different conditioning times after storage, T_1 and T_2, of Easter lily bulbs (all other factors believed to have been held constant), the measured heights of plants were

Condition:	T_1	T_2	
	28	31	
	26	35	
	30	31	
$\overline{T}_1 =$	28.0	32.3	$= \overline{T}_2$
$R_1 =$	4	4	$= R_2$

These heights are plotted in Fig. 13.3.

Analysis

- The *Tukey–Duckworth count* is $a + b = 6$, which is significant at $\alpha \cong 0.10$.
- *A second analysis (ANOM).* In Fig. 13.4, points corresponding to $\overline{T}_1$ and $\overline{T}_2$ fall outside the $\alpha = 0.10$ lines and inside the 0.05 lines.

Conclusion

From this analysis, there is some evidence (risk less than 0.10 and greater than 0.05) that a change from condition T_1 to T_2 may produce an increase in the height. The

[4]Richard H. Merritt, "Vegetative and Floral Development of Plants Resulting from Differential Precooling of Planted Croft Lily Bulbs," *Proc. Amer. Soc. Hortic. Sci.*, vol. 82, 1963, pp. 517–525.

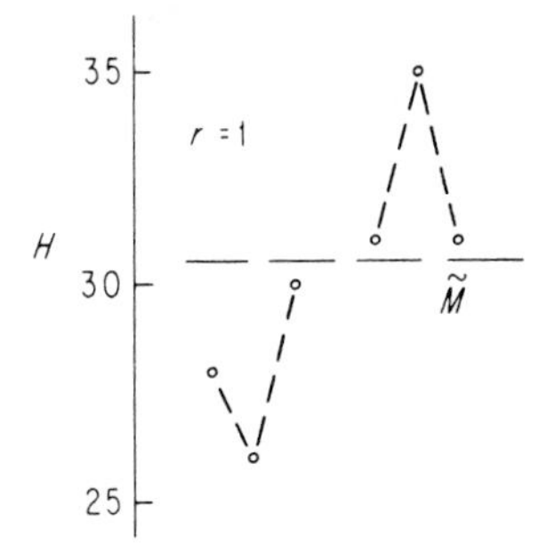

Figure 13.3 Heights of lilies under two different storage conditions.

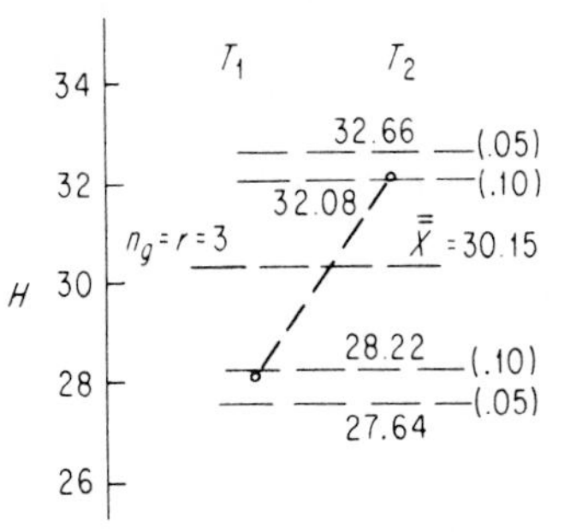

Figure 13.4 Comparing average heights of lilies under two different conditions (ANOM).

amount of increase is discussed in the following section. The choice of conditions to use in raising Easter lilies and/or whether to study greater differences in levels of T must be made by the scientist. (Also see Case History 14.1.)

Mechanics of computing decision lines (Fig. 13.4)

$$\hat{\sigma} = \bar{R}/d_2^* = 4.0/1.81 = 2.21$$

$$\hat{\sigma}_{\bar{X}} = \hat{\sigma}/\sqrt{r} = 2.21/1.73 = 1.28$$

$$\text{df} \cong (0.9)k(r - 1) = (0.9)(2)(2) = 3.6$$

Or from Table A.11, df = 3.8 ≅ 4.

Decision lines: $\bar{\bar{X}} \pm H_\alpha \hat{\sigma}_{\bar{x}}$

$\alpha = 0.05$

$$\text{UDL} = 30.15 + 2.51 = 32.66$$

$$\text{LDL} = 30.15 - 2.51 = 27.64$$

$\alpha = 0.10$

$$\text{UDL} = 30.15 + 1.93 = 32.08$$

$$\text{LDL} = 30.15 - 1.93 = 28.22$$

- Magnitude of difference

For $\alpha = 0.10$

$$\bar{\Delta}_1 = (\bar{T}_2 - \bar{T}_1) + 2H_{10}\hat{\sigma}_{\bar{X}} = 4.30 + 2(1.93) = 4.30 + 3.86 = 8.16 \textit{ in}$$

$$\bar{\Delta}_2 = (\bar{T}_2 - \bar{T}_1) - 2H_{10}\hat{\sigma}_{\bar{X}} = 4.30 - 3.86 = 0.44 \textit{ in}$$

Thus the expected average difference may actually be as small as 0.44 in or as large as 8.16 in, risk $\alpha = 0.10$.

For $\alpha = 0.05$

$$\bar{\Delta}_1 = 4.30 + 2H_{0.05}\hat{\sigma}_{\bar{X}} = 4.30 + 2(2.51) = +9.32$$

$$\bar{\Delta}_2 = 4.30 - 2(2.51) = -0.72$$

A negative sign on $\bar{\Delta}_2$ means that there is actually a small chance that condition T_1 might produce taller plants than T_2; it is a small chance but a possibility when considering confidence limits of $\alpha = 0.05$.

CASE HISTORY 13.3

Vials from Two Manufacturing Firms

The weights in grams of a sample of 15 vials manufactured by firm A and 12 vials by firm B are given in Table 13.4. Are vials manufactured by firm A expected to weigh significantly more than those manufactured by firm B?

We shall discuss the problem from different aspects.

Informal analysis 1

We begin by plotting the data in a single graph (Fig. 13.5). We note that all observations from firm B lie below the average $\bar{A}$ of firm A. Little additional formal analysis is necessary to establish that the process average from firm A exceeds the process average of firm B.

Analysis 2

The required conditions for the Tukey–Duckworth test are satisfied, and the counts are: $a + b = 8 + 4.5 = 12.5$. This count exceeds the critical count of 9 for risk $\alpha \cong 0.01$ for a one-sided test. This is in agreement with analysis 1.

Analysis 3

Clearly, the Student t test is inappropriate here because the variances of the vials from firm A and firm B are unequal. This is indicated by an F test as shown in Case History 4.12. In this case the Welch–Aspin test is in order. We calculate

TABLE 13.4 Data: Vials from Two Manufacturing Firms

Firm	Weight, grams
A:	7.6, 8.3, 13.6, 14.9, 12.7, 15.6, 9.1, 9.3, 11.7, 9.6, 10.7, 8.0, 9.4, 11.2, 12.8 ($r_1 = 15$)
B:	7.1, 7.6, 10.1, 10.1, 8.7, 7.2, 9.5, 10.2, 9.5, 9.0, 7.3, 7.4 ($r_2 = 12$)

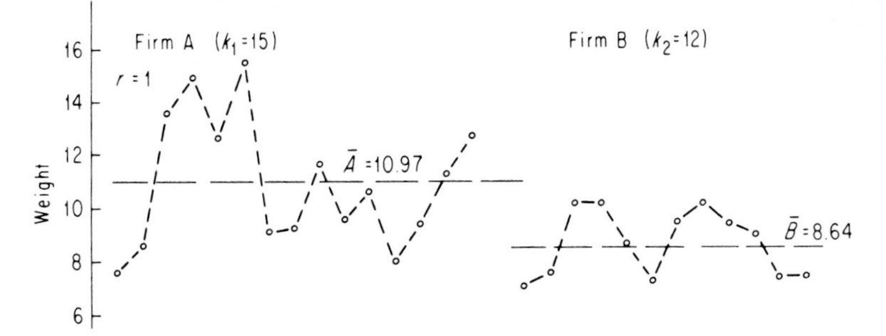

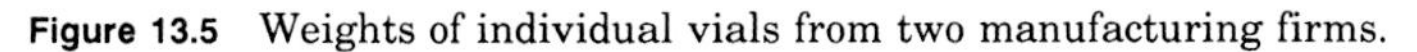

Figure 13.5 Weights of individual vials from two manufacturing firms.

$$t = \frac{\bar{A} - \bar{B}}{\sqrt{\dfrac{s_A^2}{r_1} + \dfrac{s_B^2}{r_2}}}$$

$$\bar{A} = 10.97 \qquad \bar{B} = 8.64$$

$$s_A^2 = 6.34 \qquad s_B^2 = 1.55$$

$$r_1 = 15 \qquad r_2 = 12$$

$$t = \frac{10.97 - 8.64}{\sqrt{\dfrac{6.34}{15} + \dfrac{1.55}{12}}}$$

$$= \frac{2.33}{0.74}$$

$$= 3.15$$

with degrees of freedom

$$\text{df} \cong \frac{(r_1 - 1)(r_2 - 1)\left(\dfrac{s_1^2}{r_1} + \dfrac{s_2^2}{r_2}\right)^2}{(r_2 - 1)\left(\dfrac{s_1^2}{n_1}\right)^2 + (r_1 - 1)\left(\dfrac{s_2^2}{n_2}\right)^2}$$

$$\cong \frac{14(11)\left(\dfrac{6.34}{15} + \dfrac{8.64}{12}\right)^2}{11\left(\dfrac{6.34}{15}\right)^2 + 14\left(\dfrac{8.64}{12}\right)^2}$$

$$\cong \frac{201.06}{9.22}$$

$$\cong 21.8 \sim 22$$

The t table shows $t_{0.01} = 2.83$ with 22 df and so the critical value is clearly exceeded with a one-sided risk less than 0.005.

Discussion

Thus the three analyses are in agreement that the average product expected from firm A should exceed that of firm B, *provided* the basic assumptions are satisfied.

However, consider the patterns of the data in Fig. 13.5. There are four *consecutive* vials from firm A which are appreciably higher than the others. There is enough evidence to raise certain important questions:

1. Is the product from firm A of only one kind, or is it of two or more kinds?
2. Do the four high values from firm A represent higher weights, or is the test equipment unstable (in view of the consecutive peculiarities)?
3. *Summary:* Is it possible to predict future weights?

Comments

Two basic assumptions in comparing samples from two populations are that *two stable and normally distributed populations are being sampled.* In this example, these assumptions are surely not justified. Of course, if a decision must now be made to choose the firm producing "larger" vials, then firm A would be the choice. But it will then be prudent to sample succeeding lots of product to ensure that the noncontrolled process of firm A does not begin producing vials of a low weight.

We seldom find two sets of data where it is adequate to be satisfied with a single routine test (such as a t test or a Tukey test). Additional worthwhile information comes from "looking at" the data in two or more different ways. Many patterns of nonrandomness occur. Some statistical tests are robust in detecting one pattern, some in detecting others.

CASE HISTORY 13.4

Average of Electronic Devices

It is the *usual* experience to find one set of data (or both) originating from a nonstable source instead of just one stable source, as often assumed. Let us consider the data of Table 13.5 pertaining to two batches of nickel cathode sleeves (see Example 4.2). Using cathode sleeves made from one batch of nickel (melt A), a group of 10 electronic devices was processed; then an electrical characteristic (transconductance, G_m) was read on a bridge. Using nickel-cathode sleeves from a new batch of nickel (melt B), a second group of 10 devices was processed and G_m read. Is there evidence that devices processed from melt B will average a higher G_m than devices from melt A, as is hoped? (See Fig. 4.2.)

Analysis: a routine procedure (not recommended)

If we were to proceed directly with a formal t test, we would first average each group and compute each variance

$$\bar{A} = 4184.0 \qquad \bar{B} = 4861.0$$

$$s_A^2 = 1{,}319{,}604.4 \qquad s_B^2 = 1{,}886{,}387.8$$

TABLE 13.5 Data: Measurements on Electronic Devices Made from Two Batches of Nickel Cathode Sleeves

Melt A	Melt B
4,760	6,050
5,330	4,950
2,640	3,770
5,380	5,290
5,380	6,050
2,760	5,120
4,140	1,420
3,120	5,630
3,210	5,370
5,120	4,960
$\bar{A} = 4{,}184.0$	$\bar{B} = 4{,}861.0$
$(n_1 = 10)$	$(n_2 = 10)$
	$\bar{B}' = 5{,}243.3$
	$(n_2' = 9)$

Since $r_1 = r_2$, we use the simplified form of the t test from Sec. 13.4, step 3,

$$s_p^2 = \tfrac{1}{2}(s_A^2 + s_B^2) = 1{,}602{,}996.1$$

Then

$$s_p = 1266.1 \quad \text{and} \quad \sqrt{2}s_{\bar{X}} = \frac{\sqrt{2}(1266.1)}{\sqrt{10}} = 566.2$$

and

$$t = 677/566.2 = 1.20 \qquad \text{df} = n_1 + n_2 - 2 = 18$$

Critical values of t corresponding to 18 df are

$$t_{0.20} = 1.330$$

and

$$t_{0.10} = 1.734$$

Thus, we do *not* have evidence of statistical significance even with risk $\alpha \cong 0.20$ (or 80 percent confidence).

This routine application of the t test is not recommended. The suspected outlier of 1420 in melt B has major effects on the result of a t test, as well as on the average of the process using melt B.

Further analysis

Consider again the data in Table 13.5 (Fig. 4.2). After excluding 1420 (the seventh point) in melt B as a suspected maverick, the Tukey counts are easily seen from Fig. 4.2 to be

$a = 3$ (number in melt B larger than any in melt A)

$b = 4$ (number in melt A smaller than any in melt B)

and $(a + b) = 7$

This test on the modified data indicates that the G_m of devices made from nickel-cathode sleeves from melt A will average lower than those made from melt B (with risk of about $\alpha = 0.05$).

Analysis: Student's *t* test applied to the modified data

We may now recompute the t test after excluding the 1420 observation.

$$\bar{A} = 4184.0 \qquad (r_1 = 10) \qquad \bar{B}' = 5243.3 \qquad (r_2 = 9)$$

$$\hat{\sigma}_A^2 = 1{,}319{,}604 \qquad \hat{\sigma}_{B'}^2 = 487{,}675$$

Then from Eq. (13.3)

$$s_p = \sqrt{\frac{9(1{,}319{,}604) + 8(487{,}675)}{17}} = \sqrt{928{,}108} = 963.4$$

From Eq. (13.1)

$$s_p\sqrt{\frac{1}{10} + \frac{1}{9}} = (963.4)\sqrt{\frac{19}{90}} = 442.2$$

Then $$t = \frac{\bar{B}' - \bar{A}}{442.0} = \frac{1059.3}{442.0} = 2.40 \qquad \text{df} = 17$$

From appendix Table A.15, critical values of t, for 17 df are

$$t_{0.02} = 2.567$$

$$t_{0.05} = 2.110$$

Consequently, this t test (on the modified data) indicates a significant difference between population means with risk about $\alpha \cong 0.03$ or 0.04. This result is consistent with analysis 2 but not with analysis 1.

Conclusion

Clearly, the one suspected observation in melt B has a critical influence on conclusions about the two process averages. Now it is time to discuss the situation with the engineer. The following points are pertinent:

- The melt B average is an increase of about 25 percent over the melt A average. Is the increase enough to be of practical interest? If not, whether 1420 is left in or deleted is of no concern.
- There is a strong suspicion that the data of melt A comes from two sources (see Fig. 4.2 and the discussion of Example 4.2). Are there two possible sources (in the process) which may be contributing two-level values to G_m from nickel of melt A?
- It appears that a serious study should be made of the stability of the manufacturing process when using any single melt. The question of "statistical significance" between the two melt-sample averages may be less important than the question of process stability.

13.6 Practice Exercises

1. Use moving ranges ($n_g = 2$) on the following:
   a. The data of melt A, $n = 10$, to obtain $\hat{\sigma}_A$, and compare this with $s_A = 1149$
   b. The data for melt B′, $n = 9$, to obtain $\hat{\sigma}_B$, and compare with $s_{B'} = 691.1$
2. Plot moving average and range charts, $n_g = 2$, for the data of firm A in Case History 13.3. What evidence does this present, if any, regarding the randomness of the data from firm A?
3. Consider all 15 + 12 = 27 points from the two firms in Case History 13.3, and repeat the procedure of Exercise 2. Does the number and/or length of runs provide evidence of interest?
4. What do we conclude by applying the Tukey test to the ranges of Fig. 4.3? Are the conclusions using the range-square ratio test F_r and Tukey's test in reasonable agreement?
5. Compare the process averages represented by the samples from two machines, Table 4.3. Possibilities include:
   a. Dividing each group into subgroups ($n_g = 5$, for example)
   b. Using a *t* test on all 25 observations in each group; or perhaps the first (or last, or both) five or ten of each.
6. Construct a data set and pick a significance level such that Tukey's test and the *t* test give opposite results. Which is more likely to be correct? Why would you prefer one of these tests over the other?
7. Use the data set from Exercise 6 to show that, for $k = 2$, the *t* test and ANOM are roughly equivalent. (The only difference is in the use of *s* or $\bar{R}$ to estimate sigma.)
8. The following data were acquired from two parallel sets of processing equipment. Examine them with the Tukey–Duckworth, the *t* test and the *F* test to determine whether H_0 can be rejected or accepted.

Equipment A	Equipment B
40	53
22	61
25	30
37	45
20	39
26	32
27	32
28	42
47	38
52	45
	59
	50

a. In Exercise 6 of Chap. 4, above, formulate H_0 for all 25 units of the samples from both processes and use both the Tukey–Duckworth and the t test to attempt to reject H_0, the hypothesis of no difference.

b. Graph the $\overline{X}$ and R values for each of the five samples on $\overline{X}$ and R control charts. Do both process centers appear to be operating in good control? Why or why not?

c. Conduct an ANOM according to the process results. Explain the difference in philosophy between the $\overline{X}$ control chart and the ANOM.

9. Set up a Shewhart chart with $n = 2$ using the data for firm A from Table 13.4. Plot firm B on this chart. What does this tell you?
10. Perform analysis of means on the data from Table 13.4. Use $\alpha = 0.05, 0.01$ limits.

Chapter

14

Troubleshooting with Variables Data

14.1 Introduction

This chapter covers 2^2, 2^3, and fractional factorial designs which are the fundamental structures for troubleshooting experiments in industry.

The ideas on troubleshooting with attributes data, discussed in Chap. 11, are equally applicable when using variables data. Identifying economically important problems, enlisting the cooperation of plant and technical personnel, deciding what independent variables and factors to include—these are usually more important than the analysis of resulting data. These remarks are repeated here to emphasize the importance of reasonable planning prior to the collection of data. The reader may want to review the ideas of Chaps. 9 and 11 before continuing.

This chapter will discuss three very important designs: two factorial[1] designs, 2^2 and 2^3, and the "half-replicate" design, $\frac{1}{2} \times 2^3$. These are very effective designs, especially when making exploratory studies and in troubleshooting. In Chap. 15, some case histories employing more than two levels of some independent variables are discussed.

Results from a study involving the simultaneous adjustment of two or more independent variables are often not readily accepted by those outside the planning group. For many years, engineers and production supervisors were taught to make studies with just one independent variable at a time. Many are skeptical of the propriety of experiments which vary even two independent variables simultaneously. Yet they are the ones who must accept the analysis and conclusions if the findings in a study are to be implemented. If they are not implemented, the experiment is usually wasted effort. It is critical that an analysis of the data be presented in a form which is familiar to engineers and which suggests answers to important questions. For these reasons the graphical analysis of means is recommended and emphasized here.

[1]Chap. 11, Secs. 11.6, 11.7, and 11.8.

14.2 Suggestions in Planning Investigations—Primarily Reminders

The two "levels" of the independent variable may be of a continuous variable which is recognized as a possible causative variable. Then we may use a common notation: A_- to represent the lower level of the variable and A_+ to represent the *higher* level. Frequently, however, the two "levels" should be omnibus-type variables as discussed in Chap. 9: two machine-operator combinations, two shifts, two test sets, two vendors. Then we may use a more representative notation such as A_1 and A_2 to represent the two levels.

Some amount of replication is recommended, that is, $r > 1$ and perhaps as large as 5 or 6. In many investigations, there is little difficulty in getting replicates at each combination of the independent variables. A single replicate may possibly represent the process adequately *if* the process is actually stable during the investigation, which is an assumption seldom satisfied. It should certainly be checked beforehand. Outliers and other evidences of an unstable process are common even when all known variables are held constant (Chap. 3). It is even more likely that a single observation would be inadequate when two or three variables are being studied at different levels in a designed experiment.

There may be exceptional occasions where it is practical and feasible to use a design with four or five variables, each at two levels (a 2^4 or a 2^5 design or a fraction thereof). But leave such complexities to those experienced in such matters.

Any study that requires more than 15 or 20 trays of components to be carried about manually in the plant will require extreme caution in planning and handling to prevent errors and confusion in identification. It is very difficult to maintain reasonable surveillance when only 8 or 10 trays must be routed and identified through production stages.

Scientists often use appreciably more data than a statistician might recommend for the following reasons:

1. It has been traditional in many sciences to use very large sample sizes and to have little confidence in results from smaller samples.
2. Any analysis assumes that the sample chosen for the study is *representative* of a larger population; a larger sample may be required to satisfy this important condition. Usually, however, replicates of 3, 4, or 5 are adequate.

Evolutionary operation[2]. The 2^2 design with its center point is the basis of the well-known evolutionary operation (EVOP). It has appealed especially to the chemical industry in efforts to increase process yields. During planning of an EVOP program, two independent process variables, A and B, are selected to study. They may be temperature, throughput rate, percent catalyst, and so on. Two levels of each variable are chosen within manufacturing specifications, one toward the upper specification limit and one toward the lower spec-

[2]G. E. P. Box, "Evolutionary Operation: A Method for Increasing Industrial Productivity," *Appl. Stat.*, vol. 6, 1957, pp. 81–101.

ification limit. Since the difference in levels is usually not large, several replicates may be needed to establish significant differences.

14.3 Analysis of Means: A 2^2 Factorial Design

The method of Chap. 13 is now extended to this very important case of two independent variables (two levels of each).

Main effects and two-factor interaction

In Fig. 14.1, the $\overline{X}_{ij}$ represent the average quality characteristic or response under the indicated conditions. If the average response $\overline{A}_1$ under conditions B_1 and B_2 is statistically different from the average response $\overline{A}_2$ under the same two conditions, then *factor A is said to be a significant main effect.*

Consider the $\overline{X}_{ij}$ in Fig. 14.1 to be averages of r replicates. For simplicity we shall abbreviate the notation by writing, for example,

$$\overline{A}_1 = ½(1,3) \text{ for } ½[(1) + (3)] = ½(\overline{X}_{11} + \overline{X}_{12})$$

$$\overline{A}_1 = ½(1,3) \qquad \overline{B}_1 = ½(1,2) \qquad \overline{L} = ½(1,4)$$

$$\overline{A}_2 = ½(2,4) \qquad \overline{B}_2 = ½(3,4) \qquad \overline{U} = ½(2,3)$$

where $\overline{L}$ again represents the average of the two combinations 1 and 4 of A and B having *like* subscripts; $\overline{U}$ is the average of the two *unlike* combinations.

As before, the mechanics of testing for a main effect with the analysis of means is to compare $\overline{A}_1$ with $\overline{A}_2$ and $\overline{B}_1$ with $\overline{B}_2$ using *decision lines*

$$\overline{\overline{X}} \pm H_\alpha \hat{\sigma}_{\bar{X}} \tag{14.1}$$

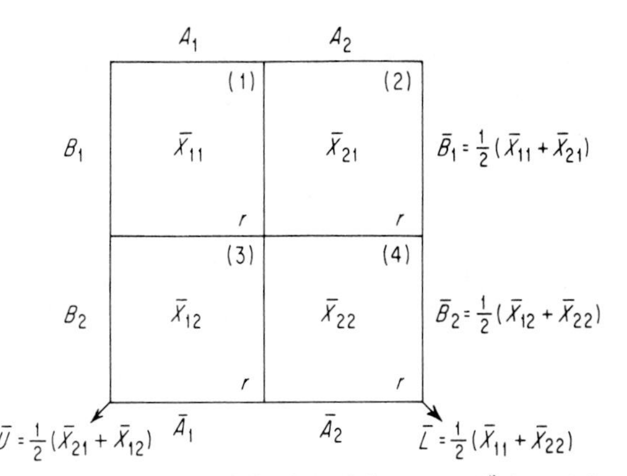

Figure 14.1 A general display of data in a 2^2 factorial design, r replicates in each average, X_{ij}.

Actually,there are three comparisons which can be tested against the decision lines in Eq. (14.1). The mechanics of testing for main effects with ANOM can be extended as follows to test for the interaction of A and B:

$\bar{A}_1 = ½[(1)+(3)]$ versus $\bar{A}_2 = ½[(2)+(4)]$ to test for an A main effect

$\bar{B}_1 = ½[(1) + (2)]$ versus $\bar{B}_2 = ½[(3) + (4)]$ to test for a B main effect

$\bar{L} = ½[(1) + (4)]$ versus $\bar{U} = ½[(2) + (3)]$ to test for an AB interaction

Discussion

I. Change in response to A under condition B_1 (A_1 to A_2) = (2) − (1) = $\bar{X}_{21} - \bar{X}_{11}$

II. Change in response to A under condition B_2 (A_1 to A_2) = (4) − (3) = $\bar{X}_{22} - \bar{X}_{12}$

$$\text{Total change in } A = [(2) + (4)] - [(1) + (3)]$$

$$\text{Average change in } A = 0.5([(2) + (4)] - [(1) + (3)])$$

This average change is called the *main effect of A*.

Definition. If the A effect I is different under B_1 than II under B_2, there is said to be a *two-factor* or *AB interaction*.

When the changes I and II are unequal,

$$(2) - (1) \neq (4) - (3)$$

then

$$(2) + (3) \neq (4) + (1)$$

But [(2) + (3)] is the sum of the two cross combinations of A and B with *unlike* U subscripts while [(1) + (4)] is the sum of *like* L subscripts in Fig. 14.1. Briefly, when there is an AB interaction, the sum (and average) of the *unlike* combinations are not equal to the sum (and average) of the *like* combinations.

Conversely, if the sum (and average) of the two terms with *like* subscripts equals statistically[3] the sum (and average) of the two with *unlike* subscripts

$$[(2) + (3)] = [(4) + (1)]$$

then

$$[(2) - (1)] = [(4) - (3)]$$

that is,

$$\text{I} = \text{II}$$

[3]Being *equal statistically* means that their *difference* is *not* statistically significant, risk α.

THEOREM 14.1 *To test for a two-factor interaction AB, obtain the cross-sums, like [(1) + (4)] and unlike [(2) + (3)]. There is an AB interaction if, and only if, the like sum is not equal to the unlike sum, i.e., their averages are not equal, statistically.*

It is both interesting and instructive to plot the four combination averages as shown in Fig. 14.2. It always helps in understanding and interpreting the meaning of the interaction. When [(2) + (3)] = [(1) + (4)], the two lines are essentially *parallel,* and there is *no* interaction. Also, when they are *not* essentially parallel, there *is* an $A \times B$ interaction.
This procedure is summarized in Table 14.1 and will be illustrated in Case History 14.1.

CASE HISTORY 14.1

Height of Easter Lilies[4]

Introduction

Consider data from *two independent variables* in a study of Easter lilies raised in the Rutgers University greenhouse. The two independent factors considered in this analysis are:

Storage period (*S*)

The length of time bulbs were stored in a controlled dormant state, (S_1 and S_2).

Time (*T*)

The length of time the bulbs were conditioned after the storage period, (T_1 and T_2).

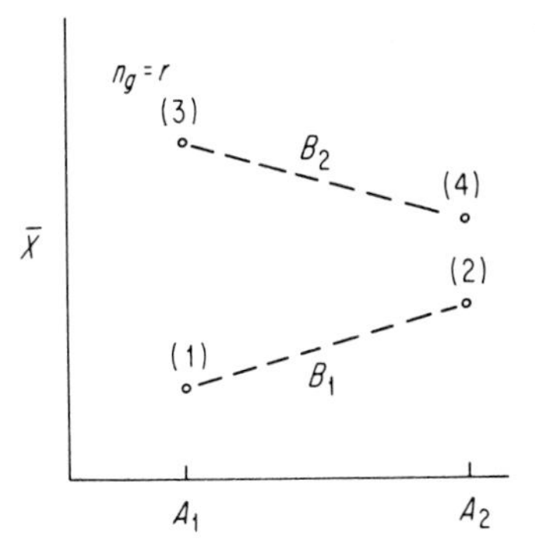

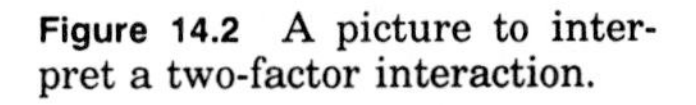
Figure 14.2 A picture to interpret a two-factor interaction.

[4]Other important independent variables and quality characteristics were reported in the research publication. See Richard H. Merritt, "Vegetative and Floral Development of Plants Resulting from Differential Precooling of Planted Croft Lily Bulbs," *Proc. Amer. Soc. Hortic. Sci.*, vol. 82, 1963, pp. 517–525.

TABLE 14.1 Analysis of Means in a 2^2 Factorial Design, *r* Replicates

STEP 1. Obtain and plot the four ranges. Find $\bar{R}$ and $D_4\bar{R}$ and use the range chart as a check on possible outliers. Obtain d_2^* from Table A.11; compute $\hat{\sigma} = \bar{R}/d_2^*$ and $\hat{\sigma}_{\bar{X}} = \hat{\sigma}/\sqrt{n_g} = \hat{\sigma}/\sqrt{2r}$.

$$\text{df} \cong (0.9)k(r-1) = 3.6(r-1) \qquad \text{for } k = 4 \text{ ranges}$$

(Or see Table A.11)

STEP 2. Plot points corresponding to the two main effects and interaction:

$$\bar{A}_1 = \tfrac{1}{2}[(1)+(3)] \quad \bar{B}_1 = \tfrac{1}{2}[(1)+(2)] \quad \bar{L} = \tfrac{1}{2}[(1)+(4)] \quad \bar{\bar{X}} = \tfrac{1}{4}[(1)+(2)+(3)+(4)]$$

$$\bar{A}_2 = \tfrac{1}{2}[(2)+(4)] \quad \bar{B}_2 = \tfrac{1}{2}[(3)+(4)] \quad \bar{U} = \tfrac{1}{2}[(2)+(3)]$$

STEP 3. Obtain H_α from Table A.8 for $k = 2$ and df as calculated. Compute and draw lines at

$$\text{UDL} = \bar{\bar{X}} + H_\alpha \hat{\sigma}_{\bar{X}}$$

$$\text{LDL} = \bar{\bar{X}} - H_\alpha \hat{\sigma}_{\bar{X}}$$

usually choosing $\alpha = 0.05$ and then 0.10 or 0.01 to bracket the extreme sample averages.

STEP 4. When a pair of points falls outside decision lines, their difference is statistically significant, risk α. If points corresponding to $\bar{L}$ and $\bar{U}$ fall outside (or near) the decision lines, graph the interaction as in Fig. 14.2.

NOTE: In step 1, the value $\hat{\sigma} = \bar{R}/d_2^*$ is a measure of the *within-group* variation, i.e., an estimate of inherent variability even when all factors are thought to be held constant. This *within-group* variation is used as a yardstick to compare *between-factor* variation by the decision lines of step 3.

The well-known analysis of variance (ANOVA) measures *within* group variation by a *residual sum of squares*, $\hat{\sigma}_E^2$, whose square root $\hat{\sigma}_E$ will be found to approximate $\hat{\sigma} = \bar{R}/d_2^*$ quite closely in most sets of data. ANOVA compares *between-group* variation by a series of variance ratio tests (F tests) instead of decision lines.

It is important to stress that a replicate is the result of a repeat of the experiment as a whole and is not just another unit run at a single setting.

The researchers specified levels of S and of T from their background of experience. The quality characteristic (dependent variable) considered here is a continuous variable, the height H in inches of a plant on the date of first bloom.

Table 14.2 represents data from the four combinations of T and S, with $r = 3$ replicate plants in each.

Formal analysis (Fig. 14.3)

Main effects

$$\bar{S}_1 = 30.15 \qquad \bar{T}_1 = 34.65$$

$$S_2 = 37.8 \qquad \bar{T}_2 = 33.3$$

Interaction

$$\overline{L} = 31.15 \qquad \overline{\overline{X}} = 33.98$$

$$\overline{U} = 36.8$$

All four R points fall below $D_4\overline{R} = 17.99$ in Fig. 14.3*b*. Then

$$\hat{\sigma} = \overline{R}/d_2^* = 7.0/1.75 = 4.0$$

$$\hat{\sigma}_{\overline{X}} = 4.0/\sqrt{6} = 1.63$$

$$\text{df} \cong (0.9)k(r - 1) \cong 7$$

Decision lines: Fig. 14.3*a*

$\overline{\overline{X}} \pm H_\alpha \hat{\sigma}_{\overline{X}}$:

$\alpha = 0.05$

$$\text{UDL} = 33.98 + (1.67)(1.63) = 36.70$$

$$\text{LDL} = 33.98 - (1.67)(1.63) = 31.26$$

$\alpha = 0.01$

$$\text{UDL} = 33.98 + (2.47)(1.63) = 38.01$$

$$\text{LDL} = 33.98 - (2.47)(1.63) = 29.95$$

The risks have been drawn in parentheses at the right end of the decision lines.

- We decide that the major effect is storage period, with a risk between 0.05 and 0.01. There is also a two-factor interaction, risk slightly less than 0.05.
- If customers prefer heights averaging about 28 in, then the combination T_1S_1 is indicated. If the preference is for heights of about 38 in, additional evidence is needed since the 49-in plant in (2) and the 29-in plant in (4) represent possible outliers.

TABLE 14.2 Height of Easter Lilies (inches)

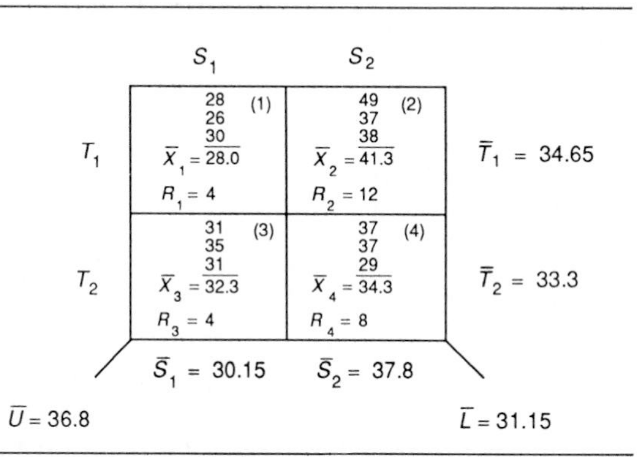

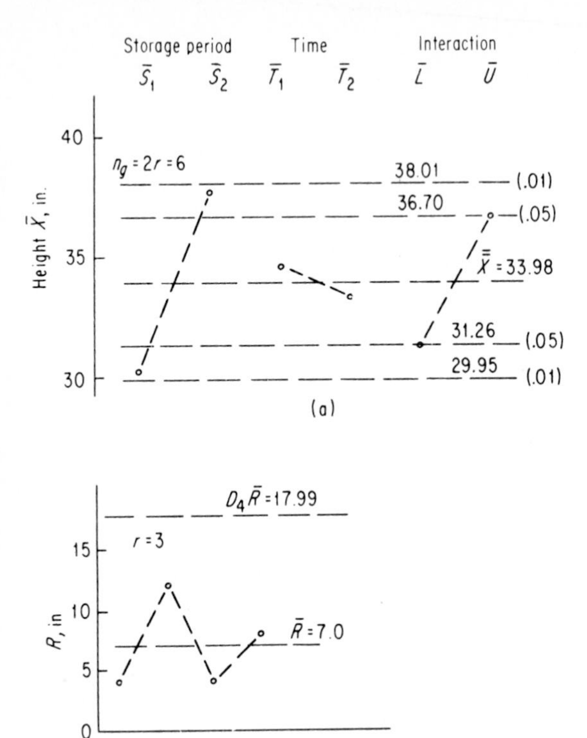

Figure 14.3 ANOM data from Table 14.2. (*a*) Height of Easter lilies; (*b*) ranges of heights.

But the effect of storage period is certainly substantial and the effect of time probably negligible.

Magnitude of the difference

From Sec. 13.5, Eq. (13.4), the magnitude of a difference may vary by $\pm 2H_\alpha \hat{\sigma}_{\bar{X}}$ from the observed difference. Then the expected magnitude of the S main effect in a large production of Easter lily bulbs is

$$(\bar{S}_2 - \bar{S}_1) \pm 2H_\alpha \hat{\sigma}_{\bar{X}} = 7.65 \pm 2(2.72)$$

$$= 7.65 \pm 5.44 \qquad (\alpha \cong 0.05)$$

that is by as much as 13.09 in and as little as 2.21 in.

Discussion of the time-by-storage interaction

Since the $T \times S$ was shown to be statistically significant, risk about 5 percent, there is a preferential pairing of the two variables. The change in height corresponding to a change from T_1 to T_2 is different when using S_1 than when using S_2. To interpret the interaction, we plot the averages of the four subgroups from Table 14.2 as shown in Fig.

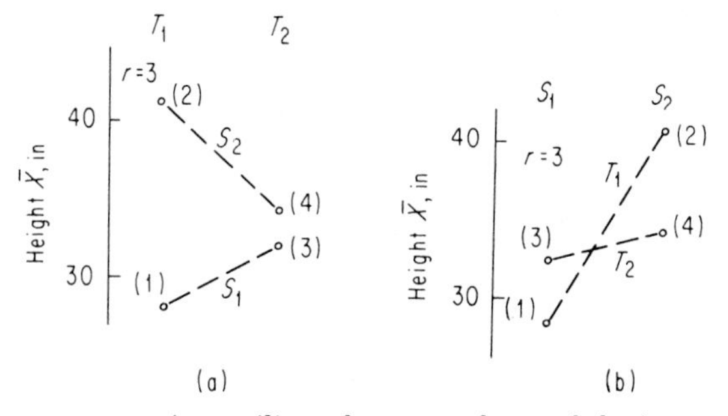

Figure 14.4 An auxiliary chart to understand the interaction of S with time (data of Table 14.2).

14.4*a* and *b*. Height *increases* when changing from condition T_1 to T_2 at level S_1 but *decreases* when changing at level S_2. The lines in Fig. 14.4*b* are *not essentially parallel;* there is a $T \times S$ interaction. It was shown in Fig. 14.3*a* that the interaction effect is statistically significant, $\alpha \cong 0.05$.

Pedigree of some data

The three heights in combination 2 warrant checking for presence of an outlier. From Table A.9,

$$\mathbf{r}_{10} = \frac{49 - 38}{49 - 37} = 0.915$$

Critical values of $\mathbf{r}_{10}$ are

$$0.941 \text{ for } \alpha = 0.05$$
$$0.925 \text{ for } \alpha = 0.10$$

The observed ratio is roughly equal to the critical value for $\alpha = 0.10$. The record books of the scientist should be checked to see whether an error has been made in transcribing the 49-in observation or whether there are (were) possible experimental clues to explain such a tall plant. In the formal analysis above, we have included the suspected observation but should now keep a suspicious mind about the analysis and interpretation.

The 29-in height in combination 4 is also suspect.

If no blunder in recording is found, some other important and possibly unknown factor in the experiment may not be maintained at a constant level.

14.4 Three Independent Variables: A 2^3 Factorial Design

This discussion is about quality characteristics measured on a continuous scale; it parallels that of Sec. 11.7 which considers quality characteristics of an attribute nature. The mechanics of analysis as summarized in Table 14.3 are slight variations of those in Sec. 14.3 and will be presented here in connection with actual data in the following Case History 14.2.

TABLE 14.3 General Analysis of a 2^3 Factorial Design, $r > 1$

STEP 1: Plot an $\bar{R}$ chart (Fig. 14.5*b*). Check on possible outliers; when all range points fall below $D_4\bar{R}$, compute

$$\hat{\sigma} = \bar{R}/d_2^*$$

$$\hat{\sigma}_{\bar{X}} = \hat{\sigma}/\sqrt{n_g} = \hat{\sigma}/\sqrt{4r}$$

$$\text{df} \cong (0.9)k(r - 1)$$

STEP 2. Obtain averages as shown in Table 14.6 ($n_g = 4r$)

a. Main effects as in Table 14.6.

b. Two-factor interactions as in Tables 14.6 and 14.7.

Plot averages as in Fig. 14.5*a*.

STEP 3. Compute decision lines for $k = 2$, $n_g = 4r$, and $\alpha = 0.05$ and then for 0.10 or 0.01 as appropriate:

$$\text{UDL} = \bar{\bar{X}} + H_\alpha\hat{\sigma}_{\bar{X}}$$

$$\text{LDL} = \bar{\bar{X}} - H_\alpha\hat{\sigma}_{\bar{X}}$$

Draw the decision lines as in Fig. 14.5*a*.

STEP 4. Any pair of points outside the decision lines indicates a statistically significant difference, risk about α. Differences which are of practical significance indicate areas to investigate or action to be taken.

STEP 5. It is sometimes helpful to compute confidence limits* on the magnitude of the observed differences, whether a main effect or a two-factor interaction:

$$(\bar{X}_1 - \bar{X}_2) \pm 2H_\alpha\hat{\sigma}_{\bar{X}}$$

*Recall $H_\alpha = \dfrac{t}{\sqrt{2}}$ so $2H_\alpha\hat{\sigma}_{\bar{x}} = t\sqrt{2}\hat{\sigma}_{\bar{x}}$ which is the usual construction for a confidence interval on the difference of two means.

CASE HISTORY 14.2[5]

Assembly of Nickel-Cadmium Batteries

A great deal of difficulty had developed during production of a nickel-cadmium battery. As a consequence, a unified team project was organized to find methods of improving certain quality problems.

In an exploratory experiment, three omnibus-type variables were included:

A_1: Processing on production line 1—using one concentration of nitrate

A_2: Processing on production line 2—a different nitrate concentration

(A difference between A_1 and A_2 might be a consequence of lines or concentrations.)

B_1: Assembly line B-1—using a shim in the battery cells

B_2: Assembly line B-2—not using a shim

(A difference between B_1 and B_2 might be a consequence of lines or shims.)

C_1: Processing on station C-1—using fresh hydroxide

[5]From a term paper prepared for a graduate course at Rutgers University by Alexander Sternberg. The data have been coded by subtracting a constant from each of the observations; this does not affect the relative comparison of effects.

C_2: Processing on station C-2—using reused hydroxide

(A difference between C_1 and C_2 might be a consequence of stations or hydroxide.)

All batteries ($r = 6$) were assembled from a *common supply* of components in each of the eight combinations. The 48 batteries were processed according to a randomized plan, and capacitance was measured at a single test station. The measurements are shown in Table 14.4 and the combination averages in Table 14.5. (Large capacitances are desired.)

The variation within any subgroup of six batteries can be attributed to three possible sources in some initially unknown way:

1. Variation attributable to components and materials
2. Variation attributable to manufacturing assembly and processing
3. Variation of testings

A measure of *within-subgroup* variability is

$$\hat{\sigma} = \bar{R}/d_2^* = 1.45/2.55 = 0.57$$

If none of the independent variables is found to be statistically significant or scientifically important, the variation from components and materials is more important than the effect of changes of processing and assembly which have been included in the study.

The following analysis is a direct extension of the analysis of means for a 2^2 investigation, Table 14.1 and Sec. 11.7. Of the eight combinations in Table 14.4, half were produced at level A_1—those in columns 1, 3, 5, and 7. Half were produced at level A_2—those in columns 2, 4, 6 and 8. We shall designate the *average* of these four column averages as

$$\bar{A}_1 = 0.64 \qquad \text{with } n_g = 4r = 24$$

Also

$$\bar{A}_2 = 1.73$$

TABLE 14.4 Capacitance of Individual Nickel-Cadmium Batteries in a 2^3 Factorial Design (Data Coded)

The numbering of the eight columns is consistent with that in Table 14.5

	A_1				A_2			
	B_1		B_2		B_1		B_2	
	C_1	C_2	C_1	C_2	C_1	C_2	C_1	C_2
	(1)	(5)	(7)	(3)	(6)	(2)	(4)	(8)
	−0.1	1.1	0.6	0.7	0.6	1.9	1.8	2.1
	1.0	0.5	1.0	−0.1	0.8	0.7	2.1	2.3
	0.6	0.1	0.8	1.7	0.7	2.3	2.2	1.9
	−0.1	0.7	1.5	1.2	2.0	1.9	1.9	2.2
	−1.4	1.3	1.3	1.1	0.7	1.0	2.6	1.8
	0.5	1.0	1.1	−0.7	0.7	2.1	2.8	2.5
$\bar{X}$	0.08	0.78	1.05	0.65	0.91	1.65	2.23	2.13
R	2.4	1.2	0.9	2.4	1.4	1.6	1.0	0.7

TABLE 14.5 Averages of Battery Capacitances (r = 6) in a 2^3 Factorial Design; Displayed As Two 2 × 2 Tables

Data from Table 14.4

	A_1	A_2
B_1	C_1 (1) $\bar{X}$ = 0.08 r = 6	C_2 (2) 1.65
B_2	C_2 (3) 0.65	C_1 (4) 2.23

	A_1	A_2
B_1	C_2 (5) 0.78	C_1 (6) 0.91
B_2	C_1 (7) 1.05	C_2 (8) 2.13

An outline of the mechanics of computation for main effects and two-factor interactions is given in Table 14.3; further details follow:

- $\bar{R} = 1.45$ and $D_4\bar{R} = (2.00)(1.45) = 2.90$
  All range points fall below $D_4\bar{R}$ (Fig. 14.5b), and we accept homogeneity of ranges, and compute

$$\hat{\sigma} = \bar{R}/d_2^* = 0.57$$

$$\hat{\sigma}_{\bar{X}} = \frac{\hat{\sigma}}{\sqrt{24}} = \frac{0.57}{4.90} = 0.116 \qquad \text{for } n_g = 4r = 24$$

$$\text{df} \cong 36$$

- Averages of the 3 *main* effects and 3 two-factor interaction effects are shown in Table 14.6. Each average is of $n_g = 4r = 24$ observations.
  The decision lines for each comparison are: ($k = 2$, df $\cong$ 36)

$$\bar{\bar{X}} \pm H_\alpha \hat{\sigma}_{\bar{X}}$$

For $\alpha = 0.05$: $H_\alpha = 1.43$

$$\text{UDL}(0.05) = 1.185 + (1.43)(0.12)$$
$$= 1.185 + 0.17 = 1.35$$
$$\text{LDL}(0.05) = 1.185 - 0.17 = 1.02$$

- Since two sets of points are outside these decision lines, we compute another pair.

For $\alpha = 0.01$, $H_\alpha = 1.93$

$$\text{UDL}(0.01) = 1.185 + 0.232 = 1.42$$
$$\text{LDL}(0.01) = 1.185 - 0.232 = 0.95$$

- Figure 14.5a indicates large A and B main effects, and a BC interaction all with risk $\alpha < 0.01$. The large A main effect was produced by using one specific concentration of nitrate which resulted in a much higher quality battery. Combinations (4) and (8) with A_2 and B_2 evidently are the best. Besides the demonstrated advantages of A_2

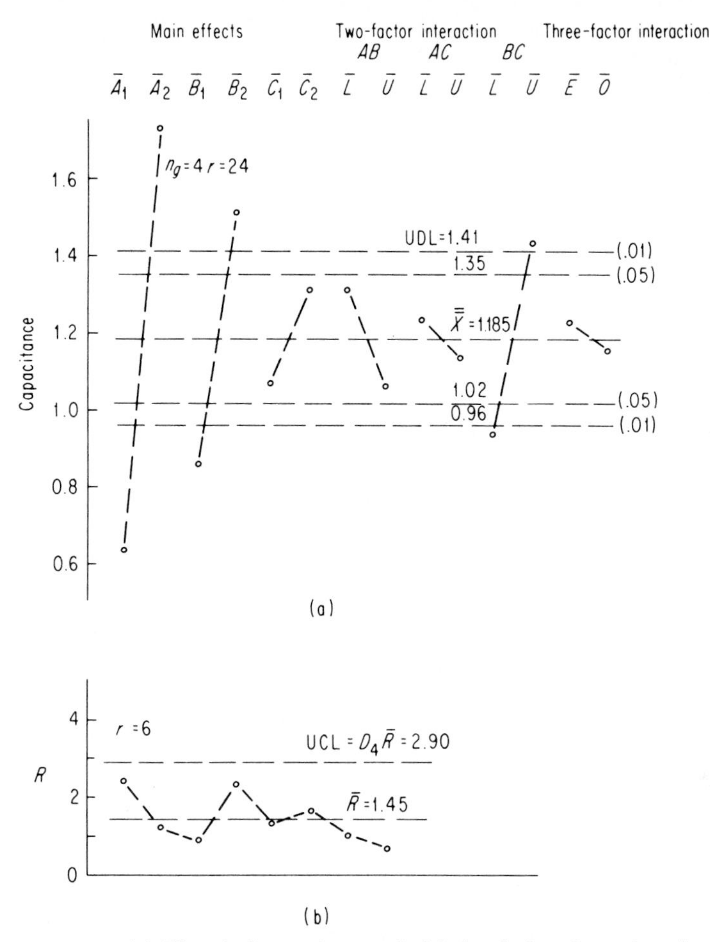

Figure 14.5 (*a*) Electrical capacitance of nickel-cadmium batteries: the ANOM comparisons; (*b*) range chart, nickel-cadmium batteries (data from Table 14.4).

over A_1 and of B_2 over B_1, a bonus benefit resulted from no significant difference due to C. This result indicated that a certain expensive compound could be reused in manufacturing and permitted a very substantial reduction in cost.

The presence of a BC interaction caused the group to investigate. It was puzzling at first to realize that the four groups of batteries assembled on line B_1 and processed at station C_2 and those assembled on line B_2 and processed at station C_1 average significantly better than combinations B_1C_1 and B_2C_2. This resulted in the detection of specific differences in the processing and assembling of the batteries. The evaluation led to improved and standardized process procedures to be followed and a resulting improvement in battery quality.

When all parties concerned were brought together to review the results of the production study, they decided to manufacture a pilot run to check the results. This

TABLE 14.6 Averages to Test for Main Effects and Two-Factor Interactions (data of Table 14.4); $n_g = 4r = 24$

Main effects	Two-factor interactions	
$\bar{A}_1 = 0.64$	AB:	$\bar{L} = 1.305$
$\bar{A}_2 = 1.73$		$\bar{U} = 1.065$
$\bar{B}_1 = 0.855$	AC:	$\bar{L} = 1.23$
$\bar{B}_2 = 1.515$		$\bar{U} = 1.14$
$\bar{C}_1 = 1.07$	BC:	$\bar{L} = 0.94$
$\bar{C}_2 = 1.30$		$\bar{U} = 1.43$

retest agreed with the first production study; changes were subsequently made in the manufacturing process which were instrumental in improving the performance of the battery and in reducing manufacturing costs.

From Eq. (13.4), limits on the magnitude of the main effects (for $\alpha = 0.01$) are:

$$\bar{\Delta}_A = (\bar{A}_2 - \bar{A}_1) \pm 2H_\alpha \hat{\sigma}_{\bar{X}}$$
$$= (1.73 - 0.64) \pm 2(0.232)$$
$$= 1.09 \pm 0.46$$
$$\bar{\Delta}_B = (\bar{B}_2 - \bar{B}_1) \pm 2H_\alpha \hat{\sigma}_{\bar{X}}$$
$$= (1.515 - 0.8555) \pm 2(0.232), \alpha = 0.01$$
$$= 0.66 \pm 0.46$$

14.5 Computational Details for Two-Factor Interactions in a 2^3 Factorial Design

There are three possible two-factor interactions: *AB, AC,* and *BC*. As discussed previously, Theorem 14.1, a test for a two-factor interaction compares those combinations having like subscripts with those having unlike subscripts ignoring the third variable (see Table 14.7).

TABLE 14.7 Diagram to Display a Combination Selection Procedure to Compute $\bar{L}$ and $\bar{U}$ in Testing *AB* Interaction

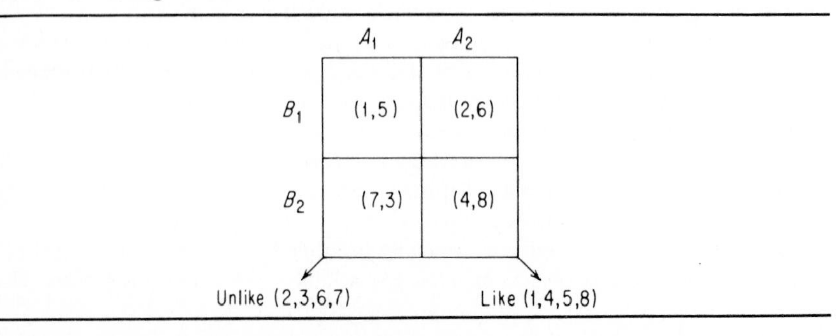

The four AB combinations having *like subscripts* in Table 14.5 are:

Like: A_1B_1: combinations (1), (5)
A_2B_2: combinations (4), (8)
Unlike: A_1B_2: combinations (7), (3)
A_2B_1: combinations (2), (6)

Then the two factor AB interaction can be tested by ignoring the third variable C and comparing the averages

$$\bar{L}_{AB} = (0.08 + 2.23 + 0.78 + 2.13)/4 = 1.305$$

$$\bar{U}_{AB} = (1.65 + 0.65 + 0.91 + 1.05)/4 = 1.065$$

The comparison above is between the two diagonals of Table 14.7. Similar comparisons provide tests for the two other two-factor interactions.

These averages are plotted in Fig. 14.5*a*. Each of these averages is (again) of $n_g = 4r = 24$ observations just as in testing main effects; the two-factor interactions can be compared to the same decision lines as for main effects. Since the pair of points for BC is outside the decision lines for $\alpha = 0.01$, there is a BC interaction.

Three-factor interaction in a 2^3 factorial design

Consider the subscripts, each of value 1 or 2. When all three subscripts are added together for the treatment combinations, half the resulting sums are even E and half are odd O. A comparison of those whose sums are even with those which are odd provides a test for what is called a *three-factor interaction.*

$$\overline{ABC}_E = (0.78 + 1.05 + 0.91 + 2.13)/4 = 1.22 \text{ (columns 5, 6, 7, 8)}$$

$$\overline{ABC}_O = 1.15 \text{ (columns 1, 2, 3, 4)}$$

The difference between the three-factor (ABC) averages is quite small (Fig 14.5*a*), and the effect is not statistically significant; it is quite unusual for it ever to appear significant.

Many remarks could be made about the practical scientific uses obtained from three-factor interactions which "appear to be significant." The following suggestions are offered to those of you who find an apparently significant three-factor interaction:

1. Recheck each set of data for outliers.
2. Recheck for errors in computation and grouping.
3. Recheck the method by which the experiment was planned. Is it possible that the execution of the plan was not followed?

4. Is the average of one subgroup "large" in comparison to all others? Then discuss possible explanations with the scientist.
5. Discuss the situation with a professional applied statistician.

14.6 A Very Important Experimental Design: ½ × 2^3

In this chapter we have just discussed 2^2 and 2^3 factorial designs. They were also discussed in Chap. 11 for data of an attributes nature. We shall conclude this chapter, as we did Chap. 11, with an example of a *half replicate* of a 2^3. Some reasons why the design is a very important one were listed in Sec. 11.8; the reasons are just as applicable when studying response characteristics which are continuous variables.

Example 14.1 Consider only that portion of the data for a nickel-cadmium battery in Table 14.5 corresponding to the four combinations shown here in Table 14.8. The computations for the analysis are the same as for an ordinary 2^2 factorial design. The averages and ranges are shown in Table 14.8 and plotted in Fig. 14.6.

Mechanics of Analysis From Fig 14.6*b* the four ranges fall below

$$D_4\overline{R} = (2.00)(1.85) = 3.70 \qquad \text{for } k = 4 \qquad r = 6$$

$$\hat{\sigma} = \overline{R}/d_2{}^* = 1.85/2.57 = 0.72$$

$$\hat{\sigma}_{\overline{X}} = 0.71/\sqrt{12} = 0.207 \qquad n_g = 2r = 12$$

$$\text{df} \cong (0.9)(4)(5) = 18$$

TABLE 14.8 Battery Capacitances: A Special Half of a 2^3 Design

Data from Table 14.5

	A_1	A_2	
B_1	C_1 (1) $\overline{X} = 0.08$ $R = 2.4$ $r = 6$	C_2 (2) $\overline{X} = 1.65$ $R = 1.6$	$\overline{B}_1 = 0.865$
B_2	C_2 (3) $\overline{X} = 0.65$ $R = 2.4$	C_1 (4) $\overline{X} = 2.23$ $R = 1.0$	$\overline{B}_2 = 1.44$
$\overline{C}_2 = 1.15$	$\overline{A}_1 = 0.365$	$\overline{A}_2 = 1.94$	$\overline{C}_1 = 1.155$ $\overline{\overline{X}} = 1.152$

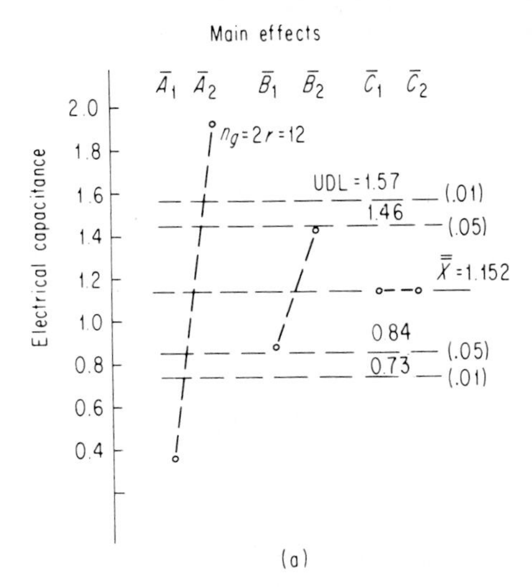

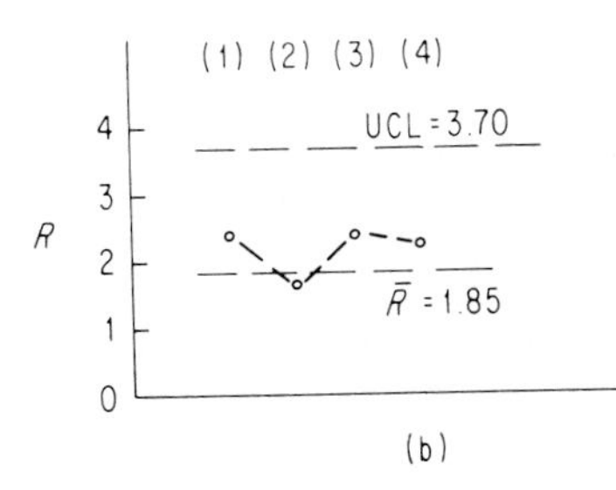

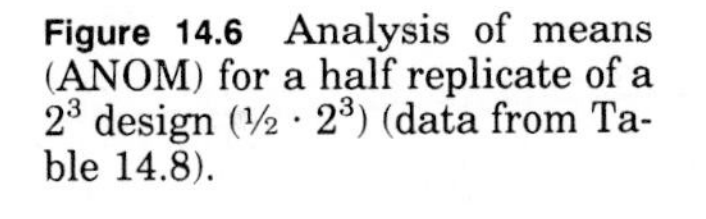
Figure 14.6 Analysis of means (ANOM) for a half replicate of a 2^3 design ($\frac{1}{2} \cdot 2^3$) (data from Table 14.8).

For $\alpha = 0.05$:

$$\text{UDL} = 1.152 + (1.49)(0.207)$$
$$= 1.46$$
$$\text{LDL} = 0.84$$

For $\alpha = 0.01$:

$$\text{UDL} = 1.152 + (2.04)(0.207)$$
$$= 1.57$$
$$\text{LDL} = 0.73$$

For $\alpha = 0.10$:

Although not computed, the (0.10) with the (0.05) lines would clearly bracket the B points.

Some comments about the half-replicate and full factorial designs.

The two points corresponding to A fall outside the 0.01 decision lines; the two points corresponding to B are just inside the 0.05 decision lines. We note that the magnitude

$$\bar{B}_2 - \bar{B}_1 = 0.575$$

in this half-replicate is almost the same as

$$\bar{B}_2 - \bar{B}_1 = 0.660$$

in the complete 2^3 design (data of Table 14.6).

We see that the principal reason the B effect shows significance more strongly in the 2^3 study than in the half replicate is the wider decision lines in Fig. 14.6*a*. These decision lines are based on only half as many batteries (and df) as those in Fig. 14.5*a*, namely $n_g = 2r = 12$ compared to $n_g = 4r = 24$. This reduction in sample size results in a larger $\hat{\sigma}_{\bar{X}}$ and requires a slightly larger H_α. There is a possible ambiguity as to whether the diagonal averages represent a comparison of a *C main effect, or* an *AB interaction.* Similarly each apparent main effect factor may be confounded with an interaction of the other two factors.

When the magnitude of the difference is of technical interest, there are two possible alternatives to consider: (1) Decide on the basis of scientific knowledge—from previous experience or an extra test comparing C_1 with C_2—whether a main effect is more or less plausible than an $A \times B$ interaction. (2) It is very unlikely that there is a genuine $A \times B$ interaction unless either one or both of A and B is a significant main effect. Since A and B both have large main effects in this case history, an interaction of these two is not precluded. The ambiguity can also be resolved by completing the other half of the 2^3 design; this effort will sometimes be justified. The recommended strategy is to proceed on the basis that main effects are dominant and effect all possible improvements. The advantages of this design are impressive, especially in troubleshooting projects.

CASE HISTORY 14.3

An Electronic Characteristic

Important manufacturing conditions are frequently difficult to identify in the manufacture of electronic products. The factors which control different quality characteristics of a particular product type often seem difficult to adjust to meet specifications. Materials in the device are not always critical provided compensating steps can be specified for subsequent processing stages. Designed production studies with two, three, and sometimes more factors are now indispensable in this and other competitive industries. It was decided to attempt improvements in contact potential quality by varying three manufacturing conditions considered to affect it in manufacture.

The three production conditions (factors) recommended by the production engineer for this experiment were the following:

- Plate temperature, designated as P
- Filament lighting schedule F
- Electrical aging schedule A

On the basis of his experience, the production engineer specified two levels of each of the three factors; levels which he thought should produce substantial differences yet which were expected to produce usable product. At the time of this production study, these three factors were being held at levels designated by P_1, F_1 and A_1. It was agreed that these levels would be continued in the production study; second levels are designated by P_2, F_2, and A_2. A half-replicate design was chosen for this production study in preference to a full 2^3 factorial.

Twelve devices were sent through production in each of the four combinations of P, F, and A shown in Table 14.9. All units in the study were made from the same lot of components, assembled by the same production line, and processed randomly through the same exhaust machines at approximately the same time. After they were sealed and exhausted, each group was processed according to the plan shown in Table 14.9. Then, electronic readings on contact potential were recorded on a sample of six of each combination. (All 12 readings are shown in Table 14.10.)

Some conclusions about Case History 14.3

1. From Fig. 14.7*a* the *change in aging* to A_2 produced a very large improvement. Also the change from F_1 to F_2 had the undesirable significant effect of lowering contact potential; the change from P_1 to P_2 had a statistically significant improvement (at the 0.05 level) but of lesser magnitude than the A effect.

The production engineer considered combination 2 to be a welcome improvement and immediately instituted a change to it in production (A_2, P_2 and F_1). The reduction in rejected items was immediately evident.

TABLE 14.9 Coded Contact Potential Readings in a Half Replicate of a 2^3

	P_1	P_2	
F_1	A_1 (1) -0.14 -0.17 -0.15 -0.11 -0.19 -0.20 $\bar{X}$ = -0.160 R = 0.09	A_2 (2) +0.15 +0.18 +0.07 +0.08 +0.08 +0.11 +0.112 0.11	$\bar{F}_1$ = -0.024
F_2	A_2 (3) -0.04 +0.04 +0.11 -0.06 -0.05 -0.05 $\bar{X}$ = -0.008 R = 0.17	A_1 (4) -0.18 -0.12 -0.22 -0.21 -0.18 -0.21 -0.187 0.10	$\bar{F}_2$ = -0.098
$\bar{A}_2$ = +0.052	$\bar{P}_1$ = -0.084	$\bar{P}_2$ = -0.038	$\bar{A}_1$ = -0.174

SOURCE: Doris Rosenberg and Fred Ennerson, Production Research in the Manufacture of Hearing Aid Tubes, *Ind. Qual. Control*, vol. 8, no. 6, May, 1952, *Practical Aids*, pp. 94–97. Data reproduced by permission.

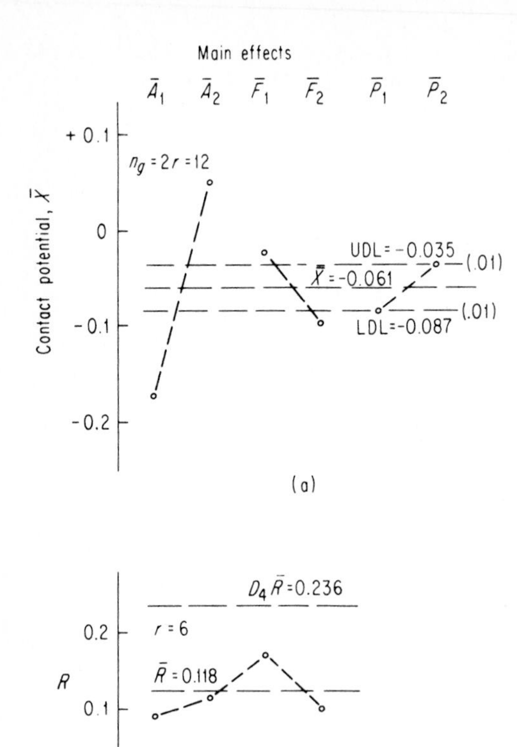

Figure 14.7 Analysis of three factors in their effects on contact potential. (*a*) ANOM in a ½ × 2^3 experiment; (*b*) ranges (data from Table 14.9).

2. A *control chart* on different characteristics, including contact potential, had been kept before the change was made from combination 1 to 2. Figure 14.8 shows the sustained improvement in $\bar{X}$ after the change.

3. *Further studies* were carried out to determine whether there were two-factor interaction effects and how additional changes in A and P (in the same direction) and in F (in the opposite direction) could increase contact potential still further. Some of these designs were full factorial; some were half replicates.

Formal analysis (Fig. 14.7)

$$\text{For } k = 4,\ r = 6;\ \hat{\sigma} = \bar{R}/d_2^* = 0.118/2.57 = 0.046:$$

$$\hat{\sigma}_{\bar{X}} = 0.046/\sqrt{12} = 0.0133 \qquad \text{for } n_g = 2r = 12$$

Decision lines: $k = 2$, df = 18, $H_{0.05} = 1.49$ and $H_{0.01} = 2.04$

$\alpha = 0.05$

$$\text{UDL} = -0.061 + (1.49)(0.0133) = -0.04$$

$$\text{LDL} = -0.061 - 0.020 = -0.08$$

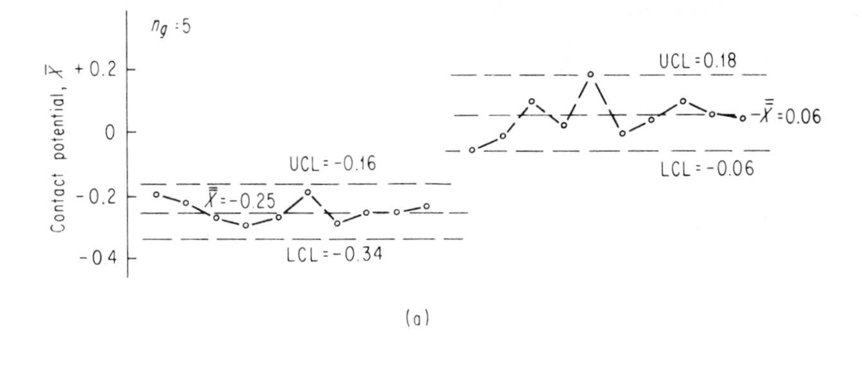

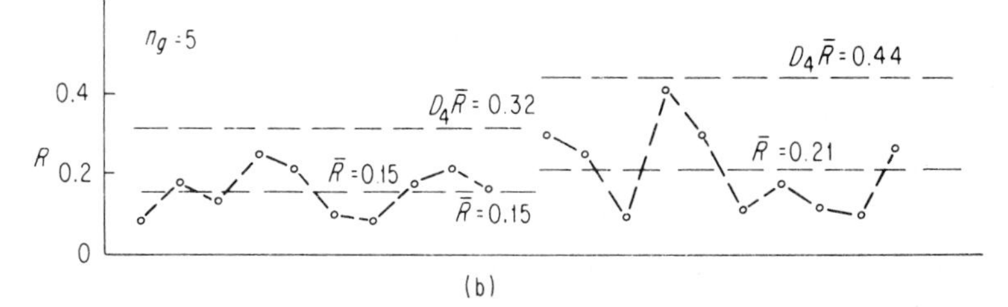

Figure 14.8 $\overline{X}$, R control charts from production before and after changes made as a consequence of the study discussed in Case History 14.3.

$\alpha = 0.01$

$$\text{UDL} = -0.061 + (2.04)(.0133) = -0.03$$

$$\text{LDL} = -0.061 - 0.027 = -0.09$$

14.7 Practice Exercises

Possible (useful) things to do with the data of Table 14.10 (use $r = 4$):

1. Check each group for outliers and other types of nonhomogeneity by whatever methods you choose. Does the manufacturing process within each group appear reasonably stable (excepting the effects of *P, F, A*)?
2. Form subgroups, vertically in each column, of 4 each.
   a. Obtain ranges, $r = 4$ and make an R chart, for all groups combined.
   b. If the chart shows control, compute $\hat{\sigma} = \overline{R}/d_2^*$ and $\hat{\sigma} = \overline{R}/d_2$ and compare.
3. Complete an ANOM.
4. Select at random 6 of each group; do a formal analysis.
5. How well do the conclusions from the data of Table 14.10 agree with the data, $r = 6$, of Table 14.9?

TABLE 14.10 Contact Potential in a Half Replicate of a 2^3 Design, r = 12; P = Plate Temperature; F = Filament Lighting; A = Aging

See Case History 14.3

	(1) $P_1F_1A_1$	(2) $P_2F_1A_2$	(3) $P_1F_2A_2$	(4) $P_2F_2A_1$
	−0.20	+0.28	−0.08	−0.22
	−0.17	+0.07	+0.11	−0.15
	−0.18	+0.17	−0.05	−0.10
	−0.20	+0.15	−0.01	−0.20
	−0.17	+0.16	−0.07	−0.17
	−0.25	+0.05	−0.15	−0.14
	−0.14	+0.15	+0.04	−0.12
	−0.17	+0.18	+0.11	−0.22
	−0.15	+0.07	−0.06	−0.21
	−0.11	+0.08	−0.04	−0.18
	−0.19	+0.11	−0.05	−0.21
	−0.20	+0.08	−0.05	−0.18
$\bar{X}$ =	−0.1775	+0.129	−0.025	−0.175

6. In this chapter, the authors present step-by-step procedures for analyzing the 2^2 factorial (Table 14.1) and the 2^3 factorial (Table 14.3) designs. Following this format, write a procedure for analyzing a ½ × 2^3 design. Pay close attention to the definition and values of r, k, and n_g. Note the difference in meaning of r and k depending on whether we are preparing the R chart or the ANOM chart. (This is good to note for experimental designs in general.)
7. In Sec. 12.1 the author offers practical advice concerning the planning of experiments for industrial process improvements. Similar suggestions are scattered throughout the text. Go through the entire text and prepare an index to these suggestions.

   Note: This is a good exercise to conduct in small groups. Later, in your professional practice, when you are thinking about running a process-improvement study, you may find such an index very helpful.
8. Suppose an experiment is run on 56 units with eight subgroups of seven units each. The average range for these eight groups is 5.5 and the overall data average is 85.2. Furthermore, these eight subgroups are aggregated into a 2^3 factorial design.
   a. Compute the control limits on the R chart.
   b. Compute the decision limits on the ANOM chart, alpha = 0.05 and alpha = 0.01.
9. Consider Table 14.4.
   a. What columns would you use to perform a test on factor A with B held constant at level 1, and C held constant at level 2?

b. Calculate $\overline{S}$ to estimate sigma and compare with the text estimate of sigma. If this approach had been used what would have been the degrees of freedom?

10. Analyze the data in Table 14.9 as a 2^3 factorial on aging schedule and filament lighting schedule. Compute or estimate the p value for interaction as an indication of significance.
11. Regard the data of Table 14.2 as a half-rep of a 2^3 with factors S, T, and (L,U). Perform an analysis of means. Use $\alpha = 0.05$.
12. Make a table of A_5 factors which when multiplied by $\overline{R}$ and added and subtracted from the grand mean will give analysis of means limits.

Chapter

15

More Than Two Levels of an Independent Variable

15.1 Introduction

A vital point was introduced in Chap. 11 that is seldom appreciated even by the most experienced plant personnel: *Within a group of no more than four of five units*—molds, machines, operators, inspectors, shifts, production lines, etc.—*there will be at least one that performs in a significantly different manner.* The performance may be better or worse, but the difference will be there. Because even experienced production or management personnel do not expect such differences, demonstrating their existence is an important hurdle to overcome on starting a process improvement study. However, it does not require much investigation to convince the skeptics that there actually are many sources of potential improvement. The case histories in this book were not rare events; they are typical of those in our experience.

Some simple design strategies aid immeasurably in finding such differences. This too is contrary to the notion that an experienced factory hand can perceive differences of any consequence. Once production personnel learn to expect differences, they can begin to use effectively the various strategies that help identify the specific units that show exceptional behavior. Some very effective strategies for this purpose were discussed in Chap. 11.

The importance of designs using two and three variables at two levels of each (2^2 and 2^3 designs) was discussed in Chap. 14. They are especially effective when looking for clues to the sources of problems. Some industrial problems warrant the use of more than two levels of an independent variable (factor), even in troubleshooting. This chapter extends the graphical methods of analysis to three and more levels. Just as in Chap. 11, there are two basic procedures to consider: (1) a *standard given* and (2) *no standard given.*

15.2 An Analysis of k Independent Samples—Standard Given—One Independent Variable

Given a stable process (i.e., one in statistical control) with known average μ and standard deviation σ, we obtain k independent random samples of r each from the given process, and consider all k means simultaneously. Within what interval

$$\mu \pm Z_\alpha \sigma_{\bar{X}}$$

will all k means lie with probability $(1 - \alpha)$?

Under the generally applicable assumption that averages from the process or population are normally distributed, values of Z_α corresponding to $\alpha = 0.10$, 0.05, and 0.01 were derived in Sec. 11.3 for attributes data and are given in Table A.7. They are equally applicable for variables data.[1]

The limits for various cases of standards given are shown in Table 15.1 where the symbol $H_{\alpha,\nu}$ indicates the risk α used and the degrees of freedom ν to be employed.

TABLE 15.1 Limits for Standards Given

Standard (s) given	Limits
None	$\bar{\bar{X}} \pm H_{\alpha,\nu}\hat{\sigma}/\sqrt{n_g}$
σ	$\bar{\bar{X}} \pm H_{\alpha,\infty}\,\sigma/\sqrt{n_g}$
μ	$\mu \pm H_{\alpha,\nu}\left(\hat{\sigma}/\sqrt{n_g}\right)\sqrt{k/(k-1)}$
μ, σ	$\mu \pm H_{\alpha,\infty}\left(\sigma/\sqrt{n_g}\right)\sqrt{k/(k-1)}$ or $\mu \pm Z_\alpha\,\sigma/\sqrt{n_g}$

NOTE: $Z_\alpha = H_{\alpha,\infty}\sqrt{k/(k-1)}$ for comparison with the results of Sec. 11.3.

Suppose an individual enters a casino where dice are being thrown at eight tables. Data is taken on 25 throws at each of the tables to check on the honesty of the house. It can be shown that for the sum of two fair dice on one toss $\mu = 7$ and $\sigma = 2.42$. Therefore, if the resulting sample averages were 7.23, 6.46, 7.01, 6.38, 6.68, 7.35, 8.12, and 7.99 we do not have evidence of dishonesty at the $\alpha = 0.05$ level of risk since the decision limits are as follows:

$\alpha = 0.05$

$$7 \pm 2.54\left(\frac{2.42}{\sqrt{25}}\right)\sqrt{\frac{8}{8-1}}$$

$$7 \pm 1.31$$

[1]When r is as large as 4, this assumption of normality of means is adequate even if the population of individuals is rectangular, right-triangular, or "almost" any other shape. (See Theorem 3, Sec. 1.8.)

$$\text{LDL}_{0.05} = 5.69 \qquad \text{UDL}_{0.05} = 8.31$$

$\alpha = 0.01$

$$7 \pm 3.01\left(\frac{2.42}{\sqrt{25}}\right)\sqrt{\frac{8}{8-1}}$$

$$7 \pm 1.56$$

$$\text{LDL}_{0.01} = 5.44 \qquad \text{UDL}_{0.01} = 8.56$$

And the plot shows no evidence of a departure from the hypothesized distribution as shown in Fig. 15.1.

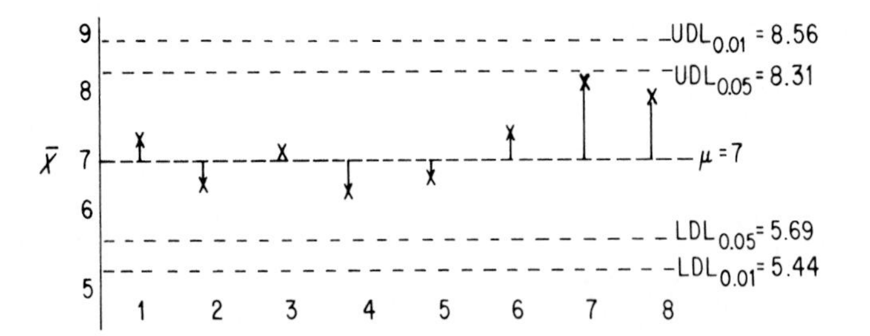

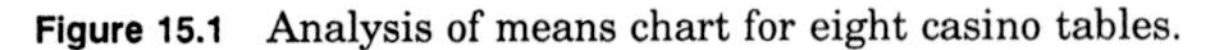

Figure 15.1 Analysis of means chart for eight casino tables.

One other example, standard given, was discussed in Sec. 11.3. The very important area of no standard given with variables data follows.

15.3 An Analysis of *k* Independent Samples—No Standard Given—One Independent Variable

This analysis is a generalization of the analysis of means in Sec. 14.3. It compares $k > 2$ means (averages) with respect to their own grand mean $\bar{\bar{X}}$ instead of being restricted to $k = 2$. More formally, the procedure is the following.

Given k sets of r observations each, but no known process average or standard deviation, the k means will be analyzed simultaneously for evidence of nonrandomness (significant differences). Decision lines, UDL and LDL, will be drawn at

$$\bar{\bar{X}} \pm H_\alpha \hat{\sigma}_{\bar{X}}$$

Thus the k means are compared to their own group mean $\bar{\bar{X}}$. If any mean lies outside the decision lines, this is evidence of nonrandomness, risk α.

The factors H_α are functions of both k and the degrees of freedom,[2] df, in estimating σ. The computation of H_α, no standard given, is much more com-

[2]See Table A.11 for df corresponding to the number of samples k of r each when using ranges. Otherwise use df of estimate employed.

$$\text{df} \cong (0.9)k(r - 1)$$

plicated than the computation of Z_α for the case of standard given, Sec. 5.3. Dr. L. S. Nelson[3] has succeeded in deriving the exact values of $h_\alpha = \sqrt{k/(k-1)}H_\alpha$. These were published in 1983. We give values of H_α in Table A.8 for α = 0.10, 0.05, and 0.01, without indicating the method of computation.

Table A.8 gives percentage points of the Studentized extreme deviate.[4]

$$\text{UDL} = (\overline{X}_{(k)} - \overline{\overline{X}})/\hat{\sigma}_{\overline{X}}$$

$$\text{LDL} = (\overline{\overline{X}} - \overline{X}_{(1)})/\hat{\sigma}_{\overline{X}}$$

for selected values of k from $k = 2$ to $k = 60$ and selected degrees of freedom.

CASE HISTORY 15.1

Possible Advantage of Using a Selection Procedure for Ceramic Sheets

During the assembly of electronic units, a certain electrical characteristic was too variable. In an effort to improve the uniformity, attention was directed toward an important ceramic component of the assembly. Ceramic sheets were purchased from an outside vendor. In production these ceramic sheets were cut into many individual component strips. How does the overall variability of assemblies using strips cut from many different sheets compare with variability corresponding to strips within single sheets? Could we decrease the overall variability by rejecting some sheets on the basis of different averages of small samples from them?

There was no record of the order of manufacture of the sheets, but it was decided to cut seven strips from each of six different ceramic sheets. The six sets were assembled into electronic units through the regular production process. The electrical characteristics of the final 42 electronic units are shown in Table 15.2 (also see Table 1.7).

The troubleshooter should ask the question: "Is there evidence from the sample data that some of the ceramic sheets are significantly different from their own group average?" If the answer is "no, the data simply represent random or chance variation about their own average," there is no reason to expect improvement by using selected ceramic sheets.

Analysis of means

Analysis of means (ANOM) applied to the data of Table 15.2 (one independent variable at k levels)

STEP 1: Plot a range chart (Fig. 15.2*b*). All points are between $D_3\overline{R}$ and $D_4\overline{R}$. Then we compute

$$\hat{\sigma} = \overline{R}/d_2{}^* = 2.15/2.73 = 0.788$$

and

$$\hat{\sigma}_{\overline{X}} = 0.788/\sqrt{7} = 0.30$$

with df $\cong (0.9)k(r-1) \cong 32$.

STEP 2: Obtain the six averages from Table 15.2 and the grand average, $\overline{\overline{X}} = 15.97$. Plot the averages as in (Fig. 15.2*a*).

STEP 3: Compute the decision lines $\overline{\overline{X}} \pm H_\alpha\hat{\sigma}_{\overline{X}}$ for $k = 6$, df = 32.

---

[3]L. S. Nelson, "Exact Critical Values for Use with the Analysis of Means," *J. Qual. Tech.*, vol. 15, no. 1, January 1983, pp. 40–44.

[4]The symbols $\overline{X}_{(k)}$ and $\overline{X}_{(1)}$ represent the largest and smallest, respectively, of the k ordered means.

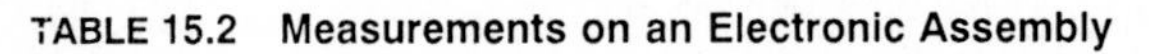

TABLE 15.2 Measurements on an Electronic Assembly

Ceramic sheet	1	2	3	4	5	6
	16.5	15.7	17.3	16.9	15.5	13.5
	17.2	17.6	15.8	15.8	16.6	14.5
	16.6	16.3	16.8	16.9	15.9	16.0
	15.0	14.6	17.2	16.8	16.5	15.9
	14.4	14.9	16.2	16.6	16.1	13.7
	16.5	15.2	16.9	16.0	16.2	15.2
	15.5	16.1	14.9	16.6	15.7	15.9
$\bar{X}$	16.0	15.8	16.4	16.5	16.1	15.0
R	2.8	3.0	2.4	1.1	1.1	2.5

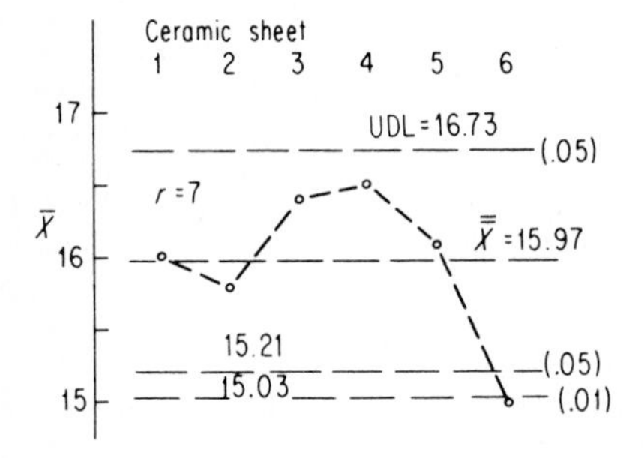

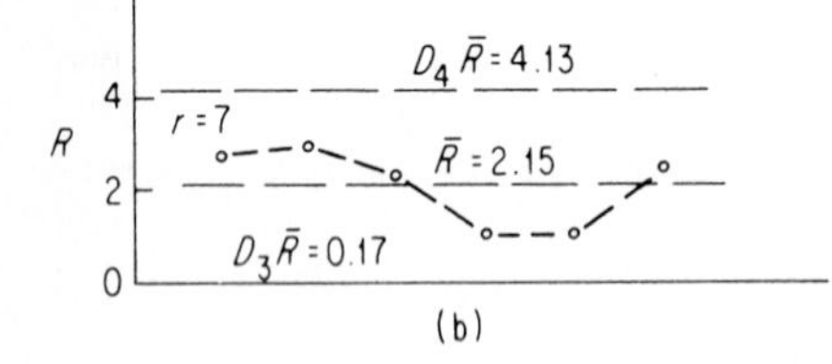

Figure 15.2 Analysis of means charts (averages and ranges) (ceramic sheet data from Table 15.2).

$\alpha = 0.05$

$$\text{UDL} = 15.97 + (2.54)(0.30) = 16.73$$

$$\text{LDL} = 15.21$$

$\alpha = 0.01$

$$\text{UDL} = 15.97 + (3.13)(0.30)$$

$$= 16.91$$

$$\text{LDL} = 15.03$$

The decision lines are drawn in Fig. 15.2*a*; the risk α is indicated in parentheses at the end of each decision line.

STEP 4: The point corresponding to sample 6, $\bar{X} = 15.0$, is below LDL(0.05) and very near LDL (0.01) = 15.03. Whether such a point is *just outside* or *just inside* the decision lines will not impress many troubleshooters as representing different bases for action. If they would reject or accept one, they would similarly reject or accept the other.

STEP 5: *Interpretation.* Sample 6 represents a ceramic sheet whose average is significantly different (statistically) from the grand average (risk $\alpha \cong 0.01$). This evidence supports the proposal to reject ceramic sheets such as 6.

Discussion

After removing sample 6, consider the grand average of the remaining five samples. Has the removal improved the average of the remaining ceramic sheets enough that they now represent a process average at the given specification of 16.5? To answer, we shall compare their average to decision lines (standard given) drawn around $\mu = 16.5$. The average of the combined 35 observations from the remaining five samples is $\bar{X} = 16.16$; this is shown as a circled point in Fig. 15.3.

Decision lines, using our previous $\hat{\sigma} = 0.788$ are simply $\mu \pm t\,\hat{\sigma}_{\bar{X}}$. Using Table A.15: $k = 1$, $n_g = 5r = 35$, df $\cong 32$.

For $\alpha = 0.05$

$$\text{LDL} = 16.5 - (2.04)\left(\frac{0.788}{\sqrt{35}}\right) = 16.23$$

For $\alpha = 0.01$

$$\text{LDL} = 16.5 - (2.75)\left(\frac{0.788}{\sqrt{35}}\right) = 16.13$$

Decision

The grand average of the 35 electronic units made from the 35 pieces of ceramic is below the LDL for $\alpha = 0.05$ and is therefore significantly lower (statistically) than $\mu = 16.5$, risk $\alpha \cong 0.05$. Thus, no plan of rejecting individual ceramic sheets by sam-

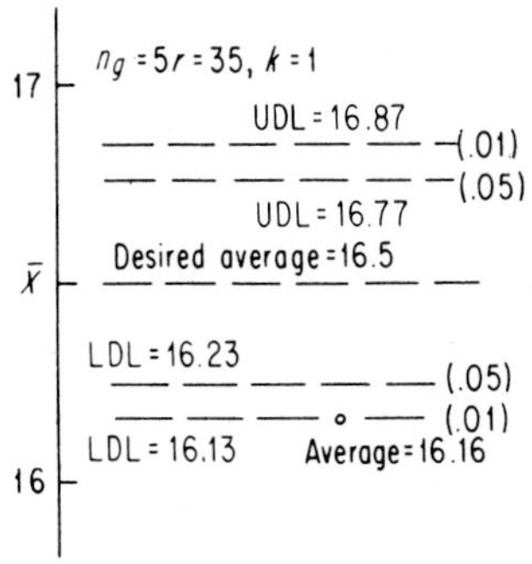

Figure 15.3 Comparing a group average with a given specification or a desired average (average of first five ceramic sheets compared to desired average).

pling can be expected to raise the grand average of the remaining to 16.5, risk < 0.05 and about 0.01.

Technical personnel need to consider three matters based on the previous analyses:

1. What can be done in processing ceramic sheets by the vendor to increase the average electrical characteristic to about 16.5? It may take much technical time and effort to get an answer.
2. Will it be temporarily satisfactory to assemble samples ($r = 7$) from each ceramic sheet, and either reject or rework any ceramic sheet averaging below 15.20 or 15.03? (See Fig. 15.2) This would be expected to improve the average somewhat.
3. Perhaps there are important factors other than ceramic sheets which offer opportunities. What can be done in the assembly or processing of the electronic assemblies to increase the electrical characteristic?

CASE HISTORY 15.2

Adjustments on a Lathe

A certain grid (for electronic tubes) was wound under five different grid-lathe tensions to study the possible effect on diameter.

Do the dimensions in Table 15.3 give evidence that tension (of the magnitude included in this experiment) affects the diameter? It was the opinion in the department that increased tension would reduce the diameter.

Interpretation

All of the points plotted in Fig. 15.4 lie within the decision lines; also there is no suggestion of a downward trend in the five averages, as had been predicted. We *do not* have evidence that the changes in grid-lathe tension affect the grid diameter.

15.4 Analysis of Means—No Standard Given—More Than One Independent Variable

Analysis of means of experiments involving multiple factors becomes more complicated. An extension of the methods of Ott[5] to the analysis of main effects and interactions in a designed experiment was developed by Schilling.[6,7,8] This approach is based on the experiment model and utilizes the departures, or differentials, from the grand mean that are associated with the levels of the treatments run in the experiment. These differentials, or treatment effects, are adjusted to remove any lower-order effects and plotted against decision limits using the analysis of means procedure. The following

[5] E. R. Ott, "Analysis of Means—A Graphical Procedure," *Ind. Qual. Control,* vol. 24, no. 2, August 1967.

[6] E. G. Schilling, "A Systematic Approach to the Analysis of Means," part 1, "Analysis of Treatment Effects," *J. Qual. Tech.,* vol. 5, no. 3, July 1973, pp. 93–108.

[7] E. G. Schilling, "A Systematic Approach to the Analysis of Means," part 2, "Analysis of Contrasts," *J. Qual. Tech.,* vol. 5, no. 4, October 1973, pp. 147–155.

[8] E. G. Schilling, "A Systematic Approach to the Analysis of Means," part 3, "Analysis of Non-Normal Data," *J. Qual. Tech.,* vol. 5, no. 4, October 1973, pp. 156–159.

TABLE 15.3 Grid Diameters under Tensions
See Fig. 15.4

	T_{20}	T_{40}	T_{60}	T_{80}	T_{120}
	42	48	46	48	50
	46	48	42	46	45
	46	46	42	42	49
	44	47	46	45	46
	45	48	48	46	48
$\bar{X}_i$:	44.6	47.4	44.8	45.4	47.6
R_i:	4	2	6	6	5

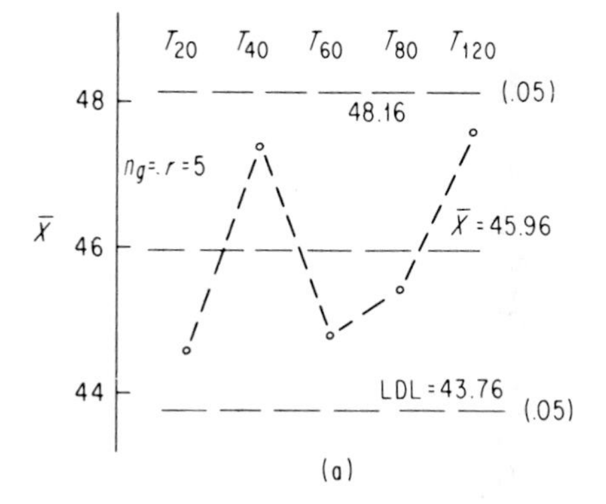

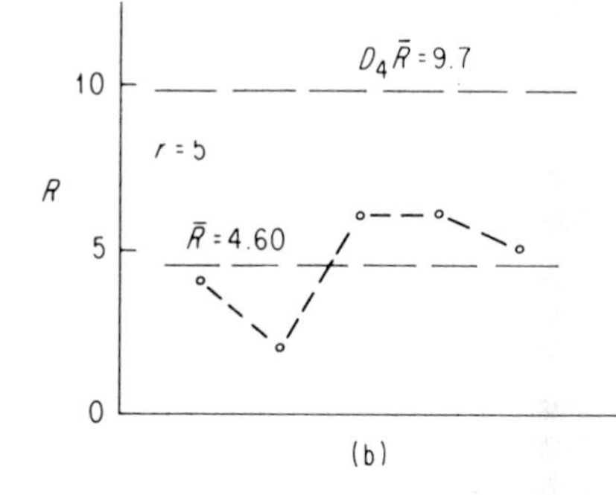

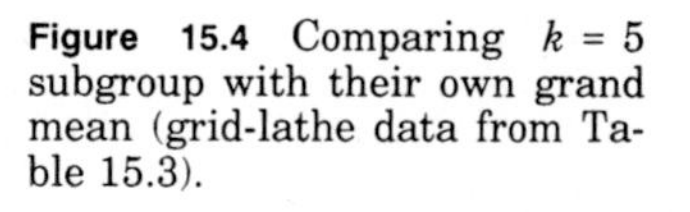
Figure 15.4 Comparing $k = 5$ subgroup with their own grand mean (grid-lathe data from Table 15.3).

is a simplification of this procedure which may be used for crossed and nested experiments.

15.5 Analysis of Two-Factor Crossed Designs

For a two-factor factorial experiment having a levels of factor A and b levels of factor B, with r observations per cell, as illustrated in Fig. 15.5, the procedure

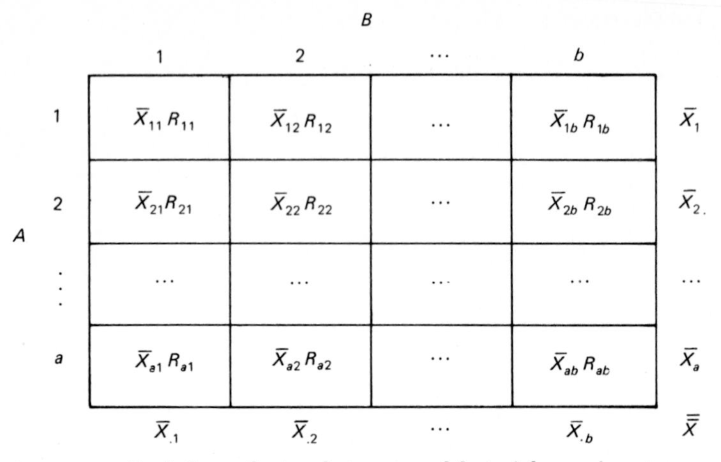

Figure 15.5 Basic form of a two-factor crossed factorial experiment.

is outlined as follows:

STEP 1: Calculate means and ranges as shown in Fig. 15.5.

STEP 2: Calculate treatment effects for the main effects as the difference between the level mean and the grand mean.

Main effects for factor A

$$A_1 = (\bar{X}_1 - \bar{\bar{X}})$$

$$A_2 = (\bar{X}_2 - \bar{\bar{X}}$$

$$\cdots\cdots\cdots\cdots\cdots\cdots$$

$$A_a = (\bar{X}_a - \bar{\bar{X}}$$

In general $$A_i = (\bar{X}_i - \bar{\bar{X}})$$

Main effects for Factor B

$$B_1 = (\bar{X}_1 - \bar{\bar{X}})$$

$$B_2 = (\bar{X}_2 - \bar{\bar{X}})$$

$$\cdots\cdots\cdots\cdots\cdots\cdots$$

$$B_b = (\bar{X}_b - \bar{\bar{X}})$$

In general $$B_j = (\bar{X}_j - \bar{\bar{X}})$$

STEP 3: Calculate treatment effects for interaction as the difference between the cell means and the grand mean less all previously estimated lower order effects which would be contained in the cell means. This gives the following:

Interaction effects (AB)

$$AB_{11} = (\overline{X}_{11} - \overline{\overline{X}}) - A_1 - B_1$$

$$AB_{12} = (\overline{X}_{12} - \overline{\overline{X}}) - A_1 - B_2$$

. . . .

$$AB_{1b} = (\overline{X}_{1b} - \overline{\overline{X}}) - A_1 - B_b$$

$$AB_{21} = (\overline{X}_{21} - \overline{\overline{X}}) - A_2 - B_1$$

$$AB_{22} = (\overline{X}_{22} - \overline{\overline{X}}) - A_2 - B_2$$

. . . .

$$AB_{2b} = (\overline{X}_{2b} - \overline{\overline{X}}) - A_2 - B_b$$

$$AB_{a1} = (\overline{X}_{a1} - \overline{\overline{X}}) - A_a - B_1$$

$$AB_{a2} = (\overline{X}_{a2} - \overline{\overline{X}}) - A_a - B_2$$

. . . .

$$AB_{ab} = (\overline{X}_{ab} - \overline{\overline{X}}) - A_a - B_b$$

In general

$$AB_{ij} = (\overline{X}_{ij} - \overline{\overline{X}}) - A_i - B_j$$

Refer to Fig. 15.5 to see how any given cell mean $\overline{X}_{ij}$ would be affected by not only the interaction effect AB_{ij} but also would be affected by the main effect A_i and the main effect B_j. Hence A_i and B_j must be subtracted out to give a legitimate estimate of the interaction effect.

STEP 4: Estimate experimental error.

$$\hat{\sigma}_e = \frac{\overline{R}}{d_2^*}$$

with

$$\text{df} = 0.9ab(r - 1)$$

Alternatively, experimental error can be estimated more precisely using the standard deviation as

$$\hat{\sigma}_e = \sqrt{\frac{\sum X^2 - \sum_{i=1}^{t} \sum_{j=1}^{ki} n_{ij} T_{ij}^2 - \frac{(\Sigma X)^2}{n}}{n - \sum_{i=1}^{t} (q_i - 1)}} \qquad (15.1)$$

with

$$\text{df} = n - \sum_{i-1}^{t} (q_i - 1)$$

where X = individual observation
t = number of effects tested (main effects, interactions, blocks, etc.)
k_i = number of individual treatment effects (means) for an effect tested
n = total number of observations in experiment
n_{ij} = number of observations in an individual treatment effect (mean)
T_{ij}^2 = treatment effect squared
q_i = degrees of freedom for an effect tested

STEP 5: Compute limits for the treatment effect differentials as

$$0 \pm \hat{\sigma}_e h_\alpha \sqrt{\frac{q}{n}}$$

where n = total number of observations in the experiment
q = degrees of freedom for effect tested
k = number of points plotted

A main effect $\quad q = a - 1 \quad k = a$

B main effect $\quad q = b - 1 \quad k = b$

AB interaction $\quad q = ab - a - b + 1 \quad k = ab$

Where h_α is obtained as follows

Main effects $\quad h_\alpha = H_\alpha \sqrt{k/(k - 1)}$ from Table A.8

Interactions $\quad h_\alpha = h_\alpha^*$ from Table A.19

Two different factors[9] are necessary since H_α is exact for main effects only. For interactions and nested factors, h_α^* is used because of the nature of the correlation among the points plotted.

STEP 6: Plot the chart as in Fig. 15.6.

The following example studies the effects of developer strength (A) and development time (B) on the density of a photographic film plate and will illustrate the method. Figure 15.7 presents the data.

[9]The factors h_α^* in Table A.19 follow from the approach to ANOM limits suggested in P. F. Ramig, "Applications of Analysis of Means," *J. Qual. Tech.*, vol 15, no. 1, January 1983, pp. 19–25 and are incorporated in the computer program for ANOM by P. R. Nelson, "The Analysis of Means for Balanced Experimental Designs," *J. Qual. Tech.*, vol. 15, no. 1, January 1983, pp. 45–84. Note that in the special case in which one or more factors in an interaction has two levels, the above interaction limits are somewhat conservative. A complete discussion with appropriate critical values is given in P. R. Nelson, "Testing for Interactions Using the Analysis of Means," *Technometrics,* vol. 30, no. 1, February 1988, pp. 53–61. It is pointed out that when one factor has two levels, k may be reduced by one-half. This fact is used in the above computer program. The approach used in the text is for consistency and ease of application and will be found to be adequate in most cases.

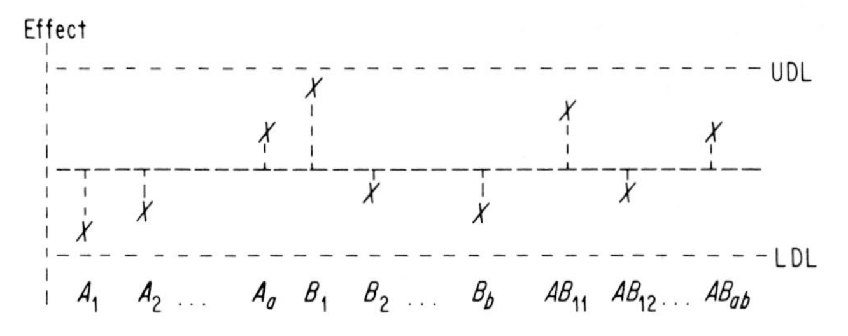

Figure 15.6 Analysis of means chart for two-factor experiment.

Development time (B)

Developer strength (A)	10	15	18	
1	0 $\bar{X}_{11} = 2.75$ 5 2 $R_{11} = 5$ 4	1 $\bar{X}_{12} = 2.50$ 4 3 $R_{12} = 3$ 2	2 $\bar{X}_{13} = 4.25$ 4 5 $R_{13} = 5$ 4	$\bar{X}_{1.} = 3.17$
2	4 $\bar{X}_{21} = 5.50$ 7 6 $R_{21} = 3$ 5	6 $\bar{X}_{22} = 7.00$ 7 8 $R_{22} = 2$ 7	9 $\bar{X}_{23} = 8.00$ 8 10 $R_{23} = 5$ 5	$\bar{X}_{2.} = 6.83$
3	7 $\bar{X}_{31} = 8.00$ 8 10 $R_{31} = 3$ 7	10 $\bar{X}_{32} = 8.75$ 8 10 $R_{32} = 3$ 7	12 $\bar{X}_{33} = 9.75$ 9 10 $R_{33} = 4$ 8	$\bar{X}_{3.} = 8.83$
	$\bar{X}_{.1} = 5.42$	$\bar{X}_{.2} = 6.08$	$\bar{X}_{.3} = 7.33$	$\bar{\bar{X}} = 6.28$

Figure 15.7 Density of photographic film plate.

STEP 1: See Fig. 15.7 for means and ranges.
STEP 2: Main effects are as follows:

$$A_1 = (3.17 - 6.28) = -3.11$$

$$A_2 = (6.83 - 6.28) = 0.55$$

$$A_3 = (8.83 - 6.28) = 2.55$$

$$B_1 = (5.42 - 6.28) = -0.86$$

$$B_2 = (6.08 - 6.28) = -0.20$$

$$B_3 = (7.33 - 6.28) = 1.05$$

STEP 3: Interaction effects are as follows:

$$AB_{11} = (2.75 - 6.28) - (-3.11) - (-0.86) = 0.44$$

$$AB_{12} = (2.50 - 6.28) - (-3.11) - (-0.20) = -0.47$$

$$AB_{13} = (4.25 - 6.28) - (-3.11) - (1.05) = 0.03$$

$$AB_{21} = (5.50 - 6.28) - (0.55) - (-0.86) = -0.47$$

$$AB_{22} = (7.00 - 6.28) - (0.55) - (-0.20) = 0.37$$

$$AB_{23} = (8.00 - 6.28) - (0.55) - (1.05) = 0.12$$

$$AB_{31} = (8.00 - 6.28) - (2.55) - (-0.86) = 0.03$$

$$AB_{32} = (8.75 - 6.28) - (2.55) - (-0.20) = 0.12$$

$$AB_{33} = (9.75 - 6.28) - (2.55) - (1.05) = -0.13$$

STEP 4: Experimental error is estimated here using the range. (See Secs. 15.6 and 15.7 for examples of the standard deviation method.)

$$\overline{R} = \frac{32}{9} = 3.56$$

$$\hat{\sigma}_e = \frac{3.56}{2.08} = 1.71 \qquad \text{df} = 0.9(3)(3)(4 - 1) = 24.3 \cong 25$$

STEP 5: Limits are

- Main effects: df = 25, $n = 36$, $k = 3$, $q = 2$, alpha = 0.05

$$h_\alpha = H_\alpha \sqrt{\frac{k}{k - 1}} = 2.04 \sqrt{\frac{3}{2}} = 2.50$$

$$0 \pm 1.71\,(2.50) \sqrt{\frac{2}{36}}$$

$$0 \pm 1.01$$

- Interaction: df = 25, $n = 36$, $k = 9$, $q = 4$, alpha = 0.05

$$h_\alpha = h_\alpha^* = 3.03$$

$$0 \pm 1.71\,(3.03) \sqrt{\frac{4}{36}}$$

$$0 \pm 1.73$$

STEP 6: Plot as in Fig. 15.8.

We see from Fig. 15.8 that developer strength (A) and development time (B) are both significant while interaction (AB) is not significant. Note that anal-

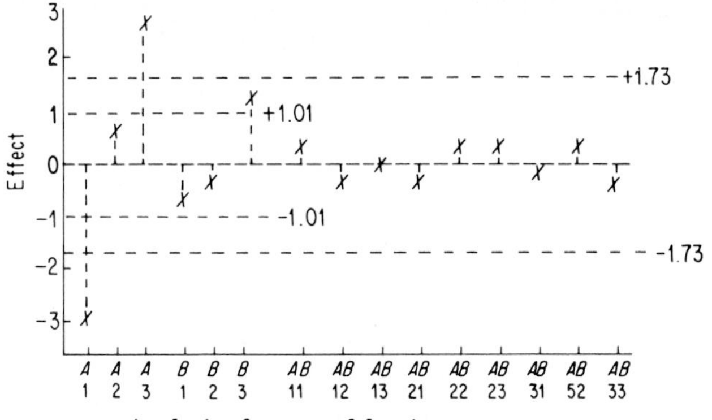

Figure 15.8 Analysis of means of density.

ysis of means indicates which levels are contributing to the significant results.

The analysis of variance for these results is given in Fig. 15.9. Since the results of analysis of variance and analysis of means are almost always consistent with each other, it is not surprising that ANOVA shows only main effects to be significant. In fact, the ANOVA table can be constructed directly from the treatment effects.

15.6 Analysis of Fully Nested Designs (Optional)

It is sometimes the case that levels of one factor are nested within another higher order factor, rather than applicable across all levels of the other. For example, within a plant, machines and operators may be interchanged, but if the machines were in different plants, or even countries, it is unlikely that all operators would be allowed to run all machines. The operators would only run the machine in their plant. They would then be nested within their own machine. Note that this is different from the crossed experiments discussed previously in that the average for an operator would only apply to one machine.

Nested experiments can be analyzed using the same steps as are shown for crossed experiments in Sec. 15.5 with the exception of step 3 and step 5. These should be modified as follows:

Analysis of Variance

Source	df	SS	MS	F	$F_{0.05}$
Strength	2	198.22	99.11	37.684	3.362
Time	2	22.72	11.36	4.319	3.362
Interaction	4	3.28	0.82	0.312	2.732
Error	27	71.00	2.63		
Total	35	295.22			

Figure 15.9 Analysis of variance of density.

STEP 3: Calculate the treatment effects as the difference between the level mean and the grand mean minus the treatment effects for the factors within which the factor is nested.

Treatment effects for factor A *(from* STEP *2)*

$$A_1 = (\overline{X}_1 - \overline{\overline{X}})$$

$$A_2 = (\overline{X}_2 - \overline{\overline{X}})$$

In general $A_i = (\overline{X}_i - \overline{\overline{X}})$

$$\cdots\cdots\cdots\cdots$$

$$A_a = (\overline{X}_a - \overline{\overline{X}})$$

Treatment effects for factor B *nested within* A

$$B_{1(1)} = (\overline{X}_{1(1)} - \overline{\overline{X}}) - A_1$$

$$B_{2(1)} = (\overline{X}_{2(1)} - \overline{\overline{X}}) - A_1$$

$$\cdots\cdots\cdots\cdots$$

$$B_{1(2)} = (\overline{X}_{1(2)} - \overline{\overline{X}}) - A_2$$

$$\cdots\cdots\cdots\cdots$$

$$B_{b(a)} = (\overline{X}_{b(a)} - \overline{\overline{X}}) - A_a$$

In general $B_{j(i)} = (\overline{X}_{j(i)} - \overline{\overline{X}}) - A_i$

Treatment effects for factor C *nested within* A *and* B

$$C_{1(1,1)} = (\overline{X}_{1(1,1)} - \overline{\overline{X}}) - A_1 - B_{1(1)}$$

$$\cdots\cdots\cdots\cdots$$

$$C_{1(2,2)} = (\overline{X}_{1(2,2)} - \overline{\overline{X}}) - A_2 - B_{2(2)}$$

$$\cdots\cdots\cdots\cdots$$

$$C_{c(a,b)} = (\overline{X}_{c(a,b)} - \overline{\overline{X}}) - A_a - B_{b(a)}$$

In general $C_{k(i,j)} = (\overline{X}_{k(i,j)} - \overline{\overline{X}}) - A_i - B_{j(i)}$

The pattern is continued for all subsequent nested factors.

Note. No interaction effects can be obtained from a fully nested design.

STEP 5: Calculate limits for the highest order factor using the main-effect limit formula. Calculate limits for the nested factors using the formula given for interactions as follows:

Highest order factor

$$h_\alpha = H_\alpha \sqrt{\frac{k}{k-1}} \text{ from Table A.8}$$

Nested factors

$$h_\alpha = h_2^* \text{ from Table A.19}$$

Consider the following fully nested experiment[10] in Fig. 15.10 showing the copper content (coded by subtracting 84) of two samples from each of 11 castings:

STEP 1: Means are shown below in Fig. 15-10.

STEP 2: Differentials ($\bar{X} - \bar{\bar{X}}$) are shown as part of step 3.

STEP 3: Treatment effects for the fully nested experiment are now computed.

Casting treatment effect	Sample treatment effect
$C_1 = 1.54 - 1.57 = -0.03$	$S_{1(1)} = (1.55-1.57) - (-0.03) = 0.01$
$C_2 = 1.41 - 1.57 = -0.16$	$S_{2(1)} = (1.52-1.57) - (-0.03) = -0.02$
$C_3 = 1.34 - 1.57 = 0.23$	$S_{1(2)} = (1.57-1.57) - (-0.16) = 0.16$
$C_4 = 1.24 - 1.57 = -0.33$	$S_{2(2)} = (1.25-1.57) - (-0.16) = -0.16$
$C_5 = 1.70 - 1.57 = 0.13$	$S_{1(3)} = (1.74-1.57) - (-0.23) = 0.40$
$C_6 = 1.82 - 1.57 = 0.25$	$S_{2(3)} = (0.94-1.57) - (-0.23) = -0.40$
$C_7 = 1.78 - 1.57 = 0.21$	$S_{1(4)} = (1.49-1.57) - (-0.33) = 0.25$
$C_8 = 2.14 - 1.57 = 0.57$	$S_{2(4)} = (1.00-1.57) - (-0.33) = -0.24$
$C_9 = 1.62 - 1.57 = 0.05$	$S_{1(5)} = (1.56-1.57) - (0.13) = -0.14$
$C_{10} = 1.47 - 1.57 = -0.10$	$S_{2(5)} = (1.84-1.57) - (0.13) = 0.14$
$C_{11} = 1.18 - 1.57 = -0.39$	$S_{1(6)} = (1.79-1.57) - (0.25) = -0.03$
$\Sigma C_i^2 = 0.8013$	$S_{2(6)} = (1.86-1.57) - (0.25) = 0.04$
$4\Sigma C_i^2 = 3.2052$	$S_{1(7)} = (1.74-1.57) - (0.21) = -0.04$
	$S_{2(7)} = (1.82-1.57) - (0.21) = 0.04$
	$S_{1(8)} = (2.12-1.57) - (0.57) = -0.02$
	$S_{2(8)} = (2.16-1.57) - (0.57) = 0.02$
	$S_{1(9)} = (1.48-1.57) - (0.05) = -0.14$
	$S_{2(9)} = (1.76-1.57) - (0.05) = 0.14$
	$S_{1(10)} = (1.04-1.57) - (-0.10) = -0.43$
	$S_{2(10)} = (1.90-1.57) - (-0.10) = 0.43$
	$S_{1(11)} = (1.14-1.57) - (-0.39) = -0.04$
	$S_{2(11)} = (1.21-1.57) - (-0.39) = 0.03$
	$\Sigma(S_{j(i)})^2 = 0.949$
	$2\Sigma(S_{j(i)})^2 = 1.898$

STEP 4: Error may be estimated using Eq. (15.1).

$$\sigma_e = \sqrt{\frac{\sum X^2 - \sum_{i=1}^{t}\sum_{j=1}^{ki} n_{ij}T_{ij}^2 - \frac{(\Sigma X)^2}{n}}{n - \sum_{i=1}^{t}(q_i - 1)}}$$

[10]C. A. Bennett and N. L. Franklin, *Statistical Analysis in Chemistry and the Chemical Industry,* John Wiley & Sons, Inc., New York, 1954, p. 364.

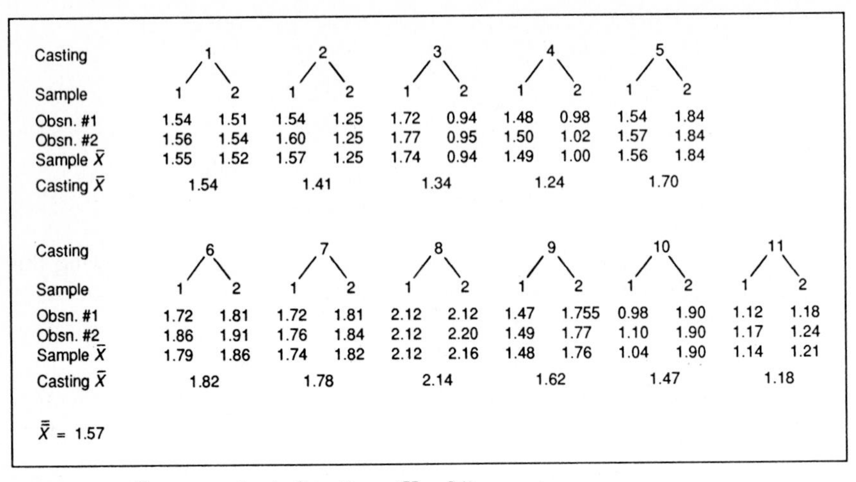

Casting	1		2		3		4		5	
Sample	1	2	1	2	1	2	1	2	1	2
Obsn. #1	1.54	1.51	1.54	1.25	1.72	0.94	1.48	0.98	1.54	1.84
Obsn. #2	1.56	1.54	1.60	1.25	1.77	0.95	1.50	1.02	1.57	1.84
Sample $\bar{X}$	1.55	1.52	1.57	1.25	1.74	0.94	1.49	1.00	1.56	1.84
Casting $\bar{X}$	1.54		1.41		1.34		1.24		1.70	

Casting	6		7		8		9		10		11	
Sample	1	2	1	2	1	2	1	2	1	2	1	2
Obsn. #1	1.72	1.81	1.72	1.81	2.12	2.12	1.47	1.755	0.98	1.90	1.12	1.18
Obsn. #2	1.86	1.91	1.76	1.84	2.12	2.20	1.49	1.77	1.10	1.90	1.17	1.24
Sample $\bar{X}$	1.79	1.86	1.74	1.82	2.12	2.16	1.48	1.76	1.04	1.90	1.14	1.21
Casting $\bar{X}$	1.82		1.78		2.14		1.62		1.47		1.18	

$\bar{\bar{X}} = 1.57$

Figure 15.10 Copper content of castings $(X - 84)$.

$$= \sqrt{\frac{113.340 - 4(0.8013) - 2(0.949) - \dfrac{(69)^2}{44}}{44 - 10 - 11 - 1}}$$

$$= \sqrt{\frac{0.0353}{22}} = 0.04$$

STEP 5: Limits using $\alpha = 0.05$ are as follows:

Castings C

$$h_\alpha = H_\alpha \sqrt{\frac{k}{k - 1}}$$

$$= 2.98 \sqrt{\frac{11}{10}}$$

$$= 3.13$$

$$0 \pm .04\,(3.13)\sqrt{\frac{10}{44}}$$

$k = 11$, df $= 22$, $q = 10$ $\qquad 0 \pm 0.0597$

Samples S

$$h_\alpha = h_\alpha^* = 3.42$$

$$0 \pm .04\,(3.42)\sqrt{\frac{11}{44}}$$

$k = 22$, df $= 22$, $q = 11$ $\qquad 0 \pm .0684$

STEP 6: The chart for this nested experiment (Fig. 15.11) shows a scale for

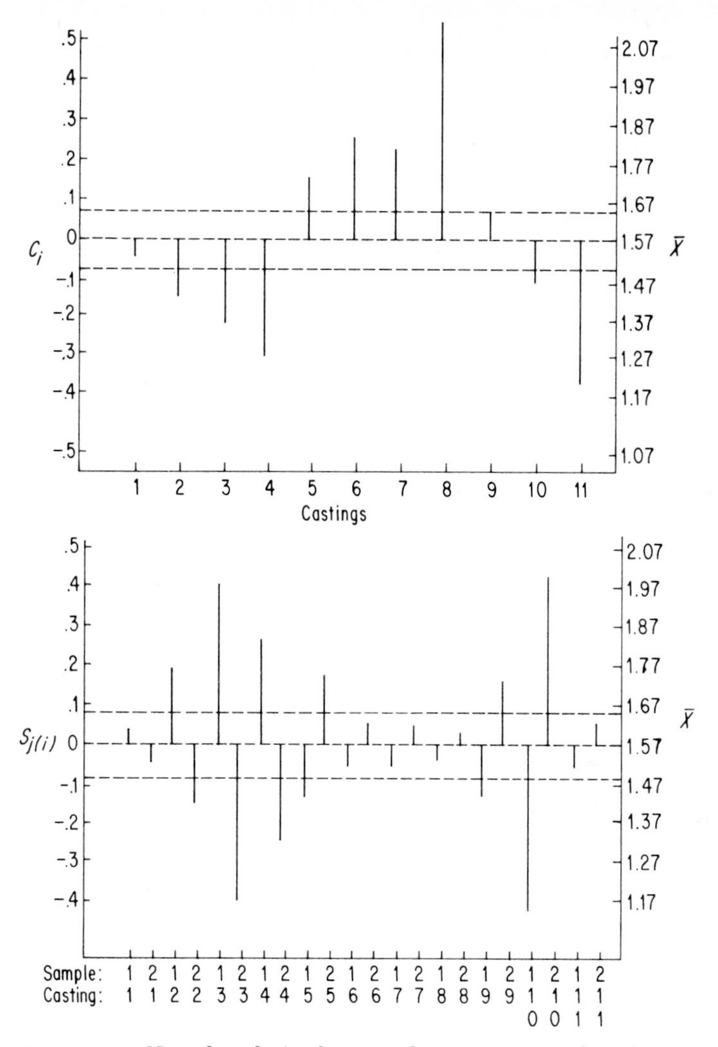

Figure 15.11 Nested analysis of means of copper content of castings.

the mean as well as for treatment effects since such a scale is meaningful in this case.

Thus, the result is the same as the analysis of variance, shown in Fig. 15.12, but in this case the plot reveals the nature of the considerable variation.

15.7 Analysis of Means for Crossed Experiments—Multiple Factors

The two-factor crossed analysis of means is easily extended to any number of factors or levels. The procedure remains essentially the same as that for a two-factor experiment with an extension to higher order interactions.

Source	SS	df	MS	F	$F_{0.05}$
Castings	3.2031	10	0.3202	200.1	2.30
Samples (within castings)	1.9003	11	0.1728	108.0	2.26
Residual	0.0351	22	0.0016		
Total	5.1385	43			

Figure 15.12 Analysis of variance of copper content of castings.

To calculate the differentials (or treatment effects) for a higher order interaction, calculate the appropriate cell means by summing over all factors not included in the interaction. Then, obtain the difference of these cell means from the grand mean. Subtract all main effects and lower order interactions contained in the cell means used. This implies that it is best to work from main effects to successive higher order interactions. For a three factor experiment, this gives:

$$A_i = (\overline{X}_i - \overline{\overline{X}})$$

$$B_j = (\overline{X}_j - \overline{\overline{X}}$$

$$C_k = (\overline{X}_k - \overline{\overline{X}})$$

$$AB_{ij} = (\overline{X}_{ij} - \overline{\overline{X}}) - A_i - B_j$$

$$AC_{ik} = (\overline{X}_{ik} - \overline{\overline{X}}) - A_i - C_k$$

$$BC_{jk} = (\overline{X}_{jk} - \overline{\overline{X}}) - B_j - C_k$$

$$ABC_{ijk} = (\overline{X}_{ijk} - \overline{\overline{X}}) - A_i - B_j - C_k - AB_{ij} - AC_{ik} - BC_{jk}$$

Suppose there are a levels of A, b levels of B, and c levels of C. Then the degrees of freedom q for each effect tested are:

$$q_A = a - 1$$

$$q_B = b - 1$$

$$q_C = c - 1$$

$$q_{AB} = (ab - 1) - q_A - q_B$$

$$q_{AC} = (ac - 1) - q_A - q_C$$

$$q_{BC} = (bc - 1) - q_B - q_C$$

$$q_{ABC} = (abc - 1) - q_A - q_B - q_C - q_{AB} - q_{AC} - q_{BC}$$

In other words, the differentials are replaced by the number of cell means minus one, and from that is subtracted the degrees of freedom of all the lower order effects contained therein.

We see also that k, the number of cell means or points to be plotted on the chart, is equal to the product of the number of levels of the factors included in the treatment effect. So

$$k_A = a \qquad k_{AB} = ab \qquad k_{ABC} = abc$$

$$k_B = b \qquad k_{AC} = ac$$

$$k_C = c \qquad k_{BC} = bc$$

If the experiment were expanded to include another factor D with d levels, we would also have

$$D_l = (\overline{X} - \overline{\overline{X}})$$

$$AD_{il} = (\overline{X}_{il} - \overline{\overline{X}}) - A_i - D_l$$

$$BD_{jl} = (\overline{X}_{jl} - \overline{\overline{X}}) - B_j - D_l$$

$$CD_{kl} = (\overline{X}_{kl} - \overline{\overline{X}}) - C_k - D_l$$

$$ABD_{ijl} = (\overline{X}_{ijl} - \overline{\overline{X}}) - A_i - B_j - D_l - AB_{ij} - AD_{il} - BD_{jl}$$

$$ACD_{ikl} = (\overline{X}_{ikl} - \overline{\overline{X}}) - A_i - C_k - D_l - AC_{ik} - AD_{il} - CD_{kl}$$

$$BCD_{jkl} = (\overline{X}_{jkl} - \overline{\overline{X}}) - B_j - C_k - D_l - BC_{jk} - BD_{jl} - CD_{kl}$$

$$ABCD_{ijkl} = (\overline{X}_{ijkl} - \overline{\overline{X}}) - A_i - B_j - C_k - D_l - AB_{ij} - AC_{ik} - AD_{il} - BC_{jk} - BD_{jl} - CD_{kl} - ABC_{ijk} - ABD_{ijl} - ACD_{ikl} - BCD_{jkl}$$

And clearly

$$q_D = d - 1$$

$$q_{AD} = (ad - 1) - q_A - q_D$$

$$q_{BD} = (bd - 1) - q_B - q_D$$

$$q_{CD} = (cd - 1) - q_C - q_D$$

$$q_{ABD} = (abd - 1) - q_A - q_B - q_D - q_{AB} - q_{AD} - q_{BD}$$

$$q_{ACD} = (acd - 1) - q_A - q_C - q_D - q_{AC} - q_{AD} - q_{CD}$$

$$q_{BCD} = (bcd - 1) - q_B - q_C - q_D - q_{BC} - q_{BD} - q_{CD}$$

$$q_{ABCD} = (abcd - 1) - q_A - q_B - q_C - q_D - q_{AB} - q_{AC} - q_{AD} - q_{BC} - q_{BD} - q_{CD} - q_{ABC} - q_{ABD} - q_{ACD} - q_{BCD}$$

with

$$k_D = d \qquad k_{ABD} = abd$$

$$k_{AD} = ad \qquad k_{ACD} = acd$$

$$k_{BD} = bd \qquad k_{BCD} = bcd$$

$$k_{CD} = cd \qquad k_{ABCD} = abcd$$

It should be noted that, regardless of the number of factors or levels, or the number of replicates r per cell, n is the total number of observations in the experiment. So for a four-factor experiment with r observations per cell:

$$n = abcdr$$

CASE HISTORY 15.3

2 × 3 × 4 Factorial Experiment—Lengths of Steel Bars

The example in Table 15.4 is given by Ott[11] and will illustrate the approach given here. Steel bars were made from two heat treatments (W and L) and cut on four screw machines (A,B,C,D) at three times (1, 2, 3—at 8:00 A.M., 11:00 A.M., and 3:00 P.M., all on the same day), with four replicates. The time element involved the possibility of fatigue on the part of the operator which may have included improper machine adjustment. The results are shown in Table 15.4 with averages summarized in Table 15.4*a*.

Suppose the main effects of time, machine, and heat were analyzed each separately as an analysis of k independent samples. We would proceed as follows:

The first step is to prepare a range chart, as in Fig. 15.13, with $\bar{R} = 5.29$ and

$$D_4\bar{R} = (2.28)(5.29) = 12.1$$

$$\hat{\sigma} = \frac{\bar{R}}{d_2^*} = \frac{5.29}{2.07} = 2.56$$

$$\text{df} = 0.9(24)(4 - 1) = 64.8 \sim 65$$

All the points lie below the control limit, and this is accepted as evidence of homogeneity of ranges. However, it may be noted that seven of the eight

[11] E. R. Ott, "Analysis of Means—A Graphical Procedure," *Ind. Qual Control,* vol. 24, no. 2, August 1967, pp. 101–109.

TABLE 15.4 A 2 × 3 × 4 Factorial Experiment (Data Coded)

Data: Lengths of Steel Bars*

	Heat Treatment W					Heat Treatment L			
	Machine					Machine			
	A	B	C	D		A	B	C	D
Time 1	6	7	1	6		4	6	−1	4
	9	9	2	6		6	5	0	5
	1	5	0	7		0	3	0	5
	3	5	4	3		1	4	1	4
$\bar{X}$	4.75	6.50	1.75	5.50		2.75	4.50	0.00	4.50
$\bar{\bar{X}}$		4.63					2.94		
R	8	4	4	4	$\bar{T}_1 = 3.78$	6	3	2	1
Time 2	6	8	3	7		3	6	2	9
	3	7	2	9		1	4	0	4
	1	4	1	11		1	1	−1	6
	−1	8	0	6		−2	3	1	3
$\bar{X}$	2.25	6.75	1.50	8.25		0.75	3.50	0.50	5.50
$\bar{\bar{X}}$		4.69					2.56		
R	7	4	3	5	$\bar{T}_2 = 3.63$	5	5	3	6
Time 3	5	10	−1	10		6	8	0	4
	4	11	2	5		0	7	−2	3
	9	6	6	4		3	10	4	7
	6	4	1	8		7	0	−4	0
$\bar{X}$	6.00	7.75	2.00	6.75		4.00	6.25	−0.5	3.50
$\bar{\bar{X}}$		5.63					3.31		
R	5	7	7	6	$\bar{T}_3 = 4.47$	7	10	8	7
Column $\bar{X}$	4.33	7.00	1.75	6.83		2.50	4.75	0.00	4.50
	$\bar{W} = 4.98$					$\bar{L} = 2.94$			
$\bar{\bar{X}} = 3.96$									$\bar{R} = 5.29$

*Prof. W. D. Baten, "An Analysis of Variance Applied to Screw Machines," *Ind. Qual. Control,* vol. 7, no. 100, April 1956.

TABLE 15.4*a* Summary of Averages (Main Effects)

Time	Machine	Heat
$\bar{T}_1 = 3.78$	$\bar{A} = 3.42$	$\bar{W} = 4.98$
$\bar{T}_1 = 3.63$	$\bar{B} = 5.88$	$\bar{L} = 2.94$
$\bar{T}_3 = 4.47$	$\bar{C} = 0.88$	
	$\bar{D} = 5.67$	
$n_g = 32$	$n_g = 24$	$n_g = 48$

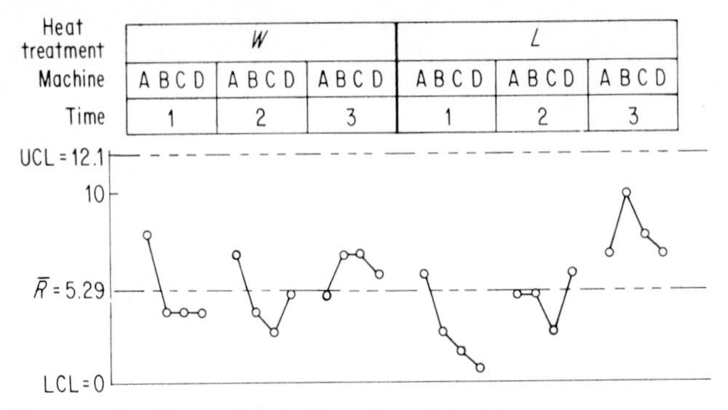

Figure 15.13 Range chart of lengths of steel bars.

points for time 3 are above the average range, R—which suggests increased variability at time 3.

The second step is to compute the averages—these are shown in Table 15.4*a*. It is immediately evident that the largest differences are between machines, and the least with time.

Next, decision limits are determined as in Fig. 15.14.

Then, the computed decision lines are drawn and the main effects are plotted on the analysis of means (ANOM) chart (Fig. 15.15).

The differences in *machine* settings contribute most to the variability in the length of the steel bars; this can probably be reduced substantially by the appropriate factory personnel. Just which machines should be adjusted, and to what levels, can be determined by reference to the specifications.

The effect of *heat treatments* is also significant (at the 0.01 level). Perhaps the machines can be adjusted to compensate for differences in the effect of heat treatment; perhaps the variability of heat treatment can be reduced in that area of processing. The magnitude of the machine differences is greater than the magnitude of the heat treatment differences.

Time did not show a statistically significant effect at either the 0.01 or 0.05 level. However, it may be worthwhile to consider the behavior of the individual machines with respect to time.

Whether the magnitudes of the various effects found in this study are enough to explain differences which were responsible for the study must be

Time	Machine	Heat
$\hat{\sigma}_{\bar{t}} = \dfrac{\hat{\sigma}}{\sqrt{32}}$	$\hat{\sigma}_{\bar{m}} = \dfrac{\hat{\sigma}}{\sqrt{24}}$	$\hat{\sigma}_{\bar{h}} = \dfrac{\hat{\sigma}}{\sqrt{48}}$
$= 0.454$	$= 0.524$	$= 0.371$
$k_t = 3$	$k_m = 4$	$k_h = 2$
$H_{0.05} = 1.96$	$H_{0.01} = 2.72$	$H_{0.01} = 1.88$
UDL = 4.84	UDL = 5.38	UDL = 4.69
LDL = 3.06	LDL = 2.52	LDL = 3.21

Figure 15.14 Decision limits for main effects for length of steel bars.

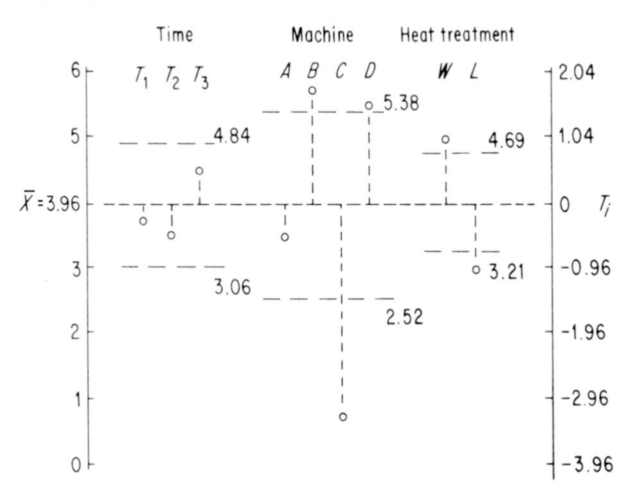

Figure 15.15 Analysis of means of length of steel bars—main effects.

discussed with the responsible factory personnel: statistical significance has been found. If they are not of practical significance, then additional possible causative factors need to be considered.

Certain combinations of these three factors (heat, machines, and time) may produce an effect not explained by the factors considered separately; such effects are called *interactions*. An answer to the general question of whether a two-factor interaction exists—and whether it is of such a magnitude to be of actual importance—can be presented using the analysis of means approach.[12] Averages are found by ignoring all factors except those being considered when there are more factors than those included in the interaction.

In troubleshooting projects, main effects will usually provide larger opportunities for improvement than interactions—but not always.

Notice the scale on the right side of the ANOM chart. It shows values of the means plotted minus the constant 3.96, which is the grand mean. In plotting the means for each treatment, we have constructed the chart for the main effect differentials or treatment effects, for time, machines, and heat treatment. Thus, by this simple transformation, the mean effect chart may be thought of in terms of the means themselves or in terms of the treatment effects or differentials from the grand mean brought about by the levels at which the experiment was run. Whether these differences are substantial enough to be beyond chance is indicated by their position relative to the decision lines.

For interactions, the differentials are interpreted as departures from the grand mean caused by the treatment effect plotted, that is, how much of a difference a particular set of conditions made. Its significance is again determined by the decision lines.

The steel bar data will now be analyzed in an analysis of means for treat-

---

[12]See Case Histories 11.5 and 14.1.

ment effects or ANOME. Underlying this analysis is the assumption of a mathematical model for the experiment, whereby the magnitude of an individual observation would be composed of the true mean of the data, μ, plus treatment effect differentials, up or down, depending on the particular set of treatments applied. That is,

$$X_{ijkl} = \mu + M_i + T_j + H_k + MT_{ij} + MH_{ik} + TH_{jk} + MTH_{ijk} + e_{l(ijk)}$$

The following is the complete analysis of means for treatment effects (ANOME) on the steel-bar data.

STEP 1: Means and ranges are calculated and summarized as shown in Table 15.5.

STEP 2: Compute the differentials to estimate the main effects. The sum of the squared treatment effects is also shown (to be used in estimating experimental error).

Machines (M)

$$M_i = \overline{X}_i - \overline{\overline{X}}$$

$$M_1 = 3.42 - 3.96 = -0.54$$

$$M_2 = 5.88 - 3.96 = 1.92$$

$$M_3 = 0.88 - 3.96 = -3.08$$

$$M_4 = 5.67 - 3.96 = 1.71 \qquad \Sigma(M_i)^2 = 16.3885$$

Times (T)

$$T_j = \overline{X}_j - \overline{\overline{X}}$$

$$T_1 = 3.78 - 3.96 = -0.18$$

$$T_2 = 3.63 - 3.96 = -0.33$$

$$T_3 = 4.47 - 3.96 = 0.51 \qquad \Sigma(T_j)^2 = 0.4014$$

Heats (H)

$$H_k = \overline{X}_k - \overline{\overline{X}}$$

$$H_1 = 4.98 - 3.96 = 1.02$$

$$H_2 = 2.94 - 3.96 = -1.02 \qquad \Sigma(H_k)^2 = 2.0808$$

STEP 3: Treatment effects for interactions are now computed.

Machine × time (MT)

$$MT_{ij} = (\overline{X}_{ij} - \overline{\overline{X}} - M_i - T_j$$

$$MT_{11} = (3.75 - 3.96) - (-0.54) - (-0.18) = 0.51$$

$$\Sigma(MT_{ij})^2 = 8.8803$$

	Averages, $\overline{X}_{ij}$				Treatment effects, MT_{ij}		
	Time				Time		
Machine	T_1	T_2	T_3		T_1	T_2	T_3
A	3.75	1.50	5.00	M_1	0.51	-1.59	1.07
B	5.50	5.13	7.00	M_2	-0.20	- 0.42	0.61
C	0.88	1.00	0.75	M_3	0.18	0.45	-0.64
D	5.00	6.88	5.13	M_4	-0.49	1.54	-1.05

Machine × heat (MH)

$$MH_{ik} = (\overline{X}_{ik} - \overline{\overline{X}}) - M_i - H_k$$

$$MH_{11} = (4.33 - 3.96) - (-0.54) - (+1.02) = -0.11$$

$$\Sigma(MH_{ik})^2 = 0.1284$$

	Averages, $\overline{X}_{ik}$			Treatment effects, MH_{ik}	
	Heat			Heat	
Machine	W	L		H_1	H_2
A	4.33	2.50	M_1	-0.11	0.10
B	7.00	4.75	M_2	0.10	-0.11
C	1.75	0.00	M_3	-0.15	0.14
D	6.83	4.50	M_4	0.14	-0.15

Time × heat (TH)

$$TH_{jk} = (\overline{X}_{jk} - \overline{\overline{X}}) - T_j - H_k$$

$$TH_{11} = (4.63 - 3.96) - (-0.18) - (+1.02) = -0.17$$

$$\Sigma(TH_{jk})^2 = 0.1046$$

	Averages, $\overline{X}_{jk}$				Treatment effects, TH_{jk}		
	Time				Time		
Heat	T_1	T_2	T_3		T_1	T_2	T_3
W	4.63	4.69	5.63	H_1	-0.17	0.04	0.14
L	2.94	2.56	3.31	H_2	0.18	-0.05	-0.14

Machine × time × heat (MTH)

$$MTH_{ijk} = (\overline{X}_{ijk} - \overline{\overline{X}}) - M_i - T_j - H_k - MT_{ij} - MH_{ik} - TH_{jk}$$

$$MTH_{111} = (4.75 - 3.96) - (-0.54) - (-0.18) - (+1.02) - (+0.51) - (-0.11) - (-0.17) = 0.26$$

$$\Sigma(MTH_{ijk})^2 = 2.4436$$

	Averages, $\overline{X}_{ijk}$							
	Heat, W				Heat, L			
	Machine				Machine			
Time	A	B	C	D	A	B	C	D
T_1	4.75	6.50	1.75	5.50	2.75	4.50	0.00	4.50
T_2	2.25	6.75	1.50	8.25	0.75	3.50	0.50	5.50
T_3	6.00	7.75	2.00	6.75	4.00	6.25	−0.50	3.50

	Treatment Effects MTH_{ijk}							
	Heat, H_1				Heat, H_2			
	Machine				Machine			
Time	M_1	M_2	M_3	M_4	M_1	M_2	M_3	M_4
T_1	0.26	0.05	0.17	−0.49	−0.26	−0.05	−0.18	0.49
T_2	−0.20	0.46	−0.41	0.17	0.22	−0.45	0.43	−0.16
T_3	−0.05	−0.51	0.24	0.32	0.06	0.52	−0.23	−0.32

STEP 4: Experimental error may be estimated using the range as above. We obtained

$$\hat{\sigma}_e = \frac{\overline{R}}{d_2^*} = \frac{5.29}{2.07} = 2.56$$

Alternatively, the treatment effects themselves may be used to estimate error based on the standard deviation. This will give us more degrees of freedom for error. The formula is that of Eq. (15.1).

$$\hat{\sigma}_e = \sqrt{\left[\Sigma X^2 - \sum_{i=1}^{t}\sum_{j=1}^{k_i} n_{ij}T_{ij}^2 - \frac{(\Sigma X)^2}{n}\right] \Big/ \left[n - \sum_{i=1}^{t}(q_i - 1)\right]}$$

where X = individual observation
t = number of effects tested (main effects, interactions, blocks, etc.)
k_i = number of individual treatment effects (means) for an effect tested
n = total number of observations in experiment
n_{ij} = number of observations in an individual treatment effect
T_{ij}^2 = treatment effect squared
q_i = degrees of freedom for an effect tested

For this experiment, we obtain

$$\sum_{i=1}^{t}\sum_{j=1}^{k_i} n_{ij}T_{ij}^2 = 24(16.3885) + 32(0.4014) + 48(2.0808) + 8(8.8803)$$
$$+ 12(0.1284) + 16(0.1046) + 4(2.4436) = 590.0784$$

$$n - \sum_{i=1}^{t}(q_i - 1) = 96 - (3 + 2 + 1 + 6 + 3 + 2 + 6) - 1 = 72$$

$$\hat{\sigma}_e = \sqrt{\frac{2542 - 590.0784 - \left(\frac{380^2}{96}\right)}{72}} = 2.49$$

Here we have 72 degrees of freedom versus 65 (using the range as an estimate of error). This estimate of error is the same as we would have obtained if we used the square root of the error mean square from an analysis of variance.

STEP 5: The decision limits for the treatment effect differentials (using the standard deviation as the estimate of experimental error) are as follows:

$$0 \pm \sigma_e h_a \sqrt{\frac{q}{n}}$$

Machines M:	$0 \pm 2.49(2.53)\sqrt{\frac{3}{96}}$	$h_a = 2.19\sqrt{\frac{4}{3}} = 2.53$
$k = 4$, df $= 72$	0 ± 1.11	
Times T:	$0 \pm 2.49(2.40)\sqrt{\frac{2}{96}}$	$h_a = 1.96\sqrt{\frac{3}{2}} = 2.40$
$k = 3$, df $= 72$	0 ± 0.86	
Heat H:	$0 \pm 2.49(2.00)\sqrt{\frac{1}{96}}$	$h_a = 1.41\sqrt{\frac{2}{1}} = 2.00$
$k = 2$, df $= 72$	0 ± 0.51	
MT:	$0 \pm 2.49(2.96)\sqrt{\frac{6}{96}}$	
$k = 12$, df $= 72$	0 ± 1.84	
MH:	$0 \pm 2.49(2.82)\sqrt{\frac{3}{96}}$	
$k = 8$, df $= 72$	0 ± 1.24	
TH:	$0 \pm 2.49(2.71)\sqrt{\frac{2}{96}}$	
$k = 6$, df $= 72$	0 ± 0.97	
MTH:	$0 \pm 2.49(3.20)\sqrt{\frac{6}{96}}$	
$k = 24$, df $= 72$	0 ± 1.99	

STEP 6: Plot the decision limits and the treatment effect differentials on the analysis of means chart. The chart appears as Fig. 15.16.

Machine and heat main effects are the only effects to show significance. The interactions are not significant; however, we will examine the *MH* inter-

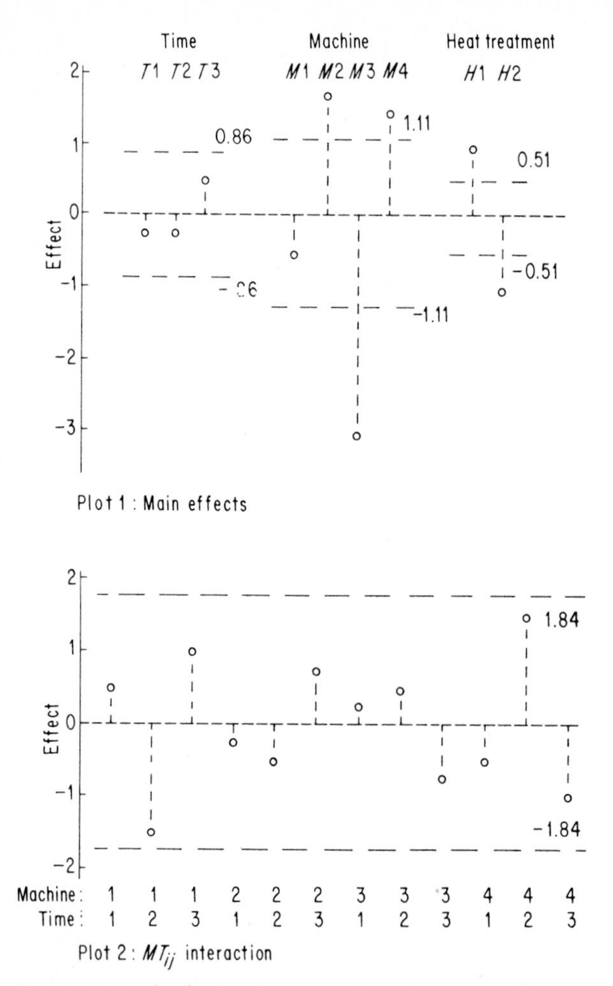

Figure 15.16 Analysis of means for treatment effects—length of steel bars.

action to illustrate a procedure developed by Ott[13] to analyze a $2 \times k$ interaction. Consider the interaction diagram shown in Fig. 15.17.

If the average difference $\overline{\Delta}_i$ between the heats are plotted for the four machines we obtain a plot as given in Fig. 15.18.

Here the differences plotted are 1.83, 2.25, 1.75, and 2.33, respectively, as can be seen from the table used to calculate treatment effects for the *MH* interaction. The differences are treated as if they were main effects with a standard deviation

[13]E. R. Ott, "Analysis of Means—A Graphical Procedure," *Ind. Qual Control,* vol. 24, no. 2, August 1967, pp. 101–109.

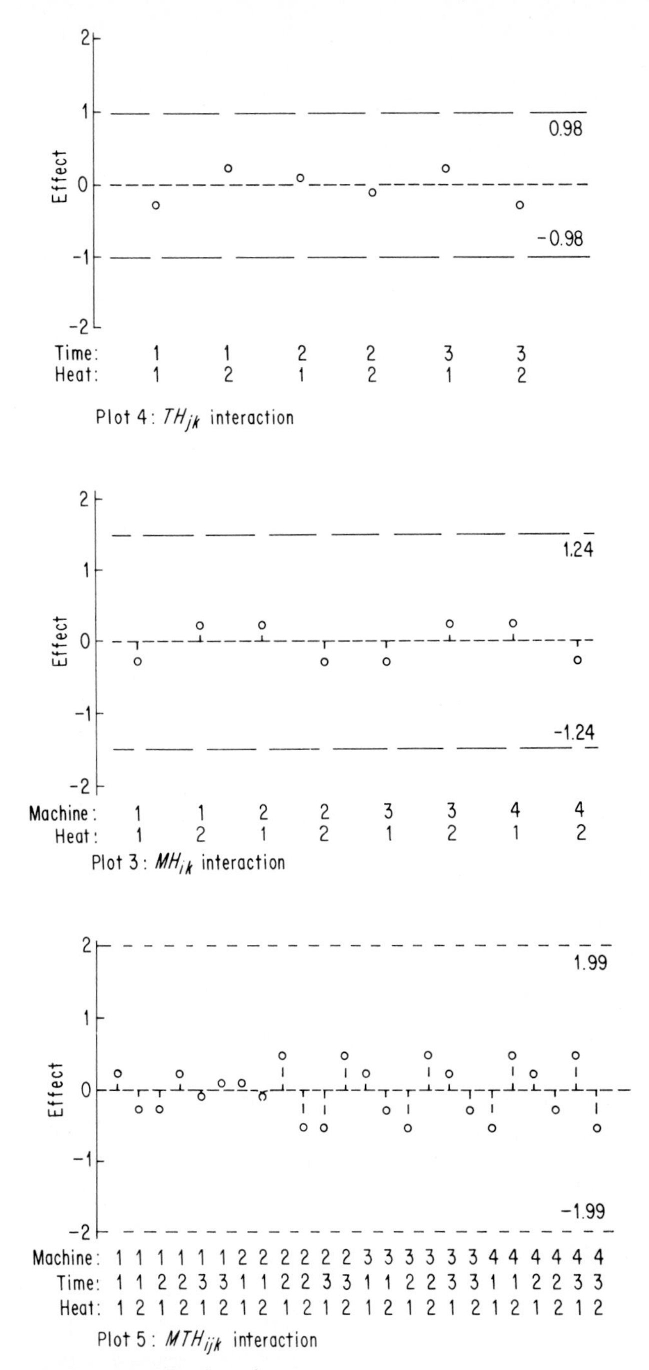

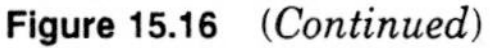
Figure 15.16 *(Continued)*

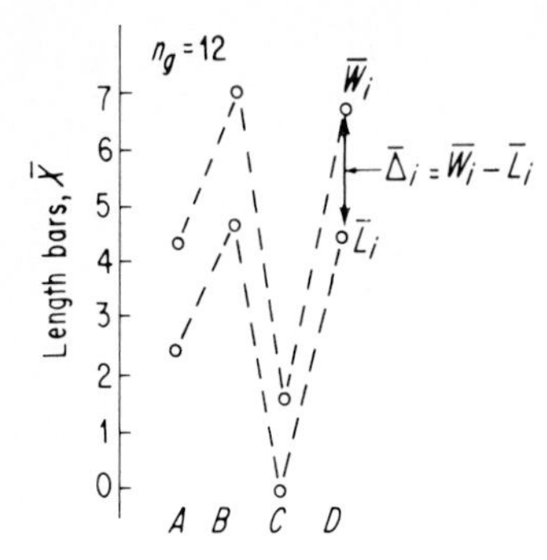

Figure 15.17 Interaction comparison of patterns $\overline{W}$ and $\overline{L}$.

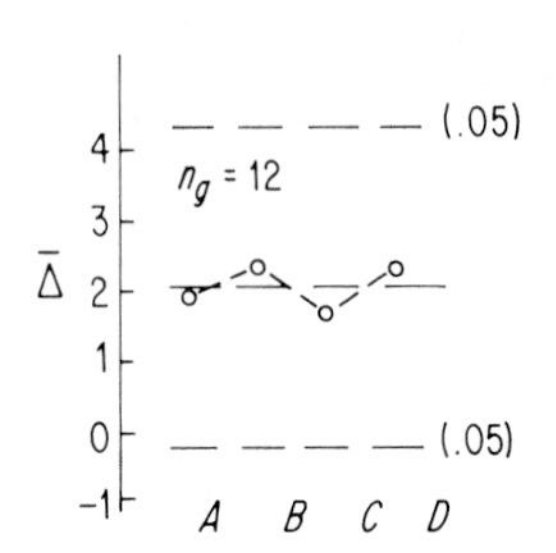

Figure 15.18 Interaction analysis, $\overline{W} \times \overline{L}$: ANOM.

$$\hat{\sigma}_{\Delta_i} = \hat{\sigma}_e \sqrt{\frac{1}{n_g} + \frac{1}{n_g}}$$

$$= \sqrt{2}\,\frac{\hat{\sigma}_e}{\sqrt{n_g}}$$

and the limits become

$$\overline{\overline{\Delta}} \pm H_\alpha \sqrt{2}\,\frac{\hat{\sigma}_e}{\sqrt{n_g}}$$

where n_g is the number of observations used to calculate each of the means constituting the difference. In this case $n_g = 12$ and $\overline{\overline{\Delta}} = 2.04$ so we have

$$2.04 \pm 2.20\sqrt{2}\,\frac{(2.49)}{\sqrt{12}}$$

$$2.04 \pm 2.24$$

giving UDL = 4.28 and LDL = − 0.20 as shown in Fig. 15.18. Peter R. Nelson[14] has extended this procedure by providing tables to facilitate the calculation of exact limits for this case, and also limits for the case in which both factors are at more than two and up to five levels. The reader should refer to his paper for this extension.

It is interesting to compare Fig. 15.18 with plot 3 for the *MH* interaction. The former shows that the difference between the heats does not vary from machine to machine. The latter shows estimates of the magnitude of the differences of the cell means from the grand mean in the *MH* interaction table if there were no other effects by making use of the experiment model. Each provides insight into the data patterns resulting from the experiment.

15.8 Nested Factorial Experiments (Optional)

When an experiment includes both crossed and nested factors it can be dealt with using the same approach as with fully nested or fully crossed experi-

[14]Peter R. Nelson, "Testing for Interactions Using the Analysis of Means," *Technometrics*, vol. 30, no. 1, February 1988, pp. 53–61.

ments, respectively. The analysis is essentially as if the experiment were crossed; however, any interactions between nested factors and those factors within which they are nested are eliminated from the computations. Thus, if factor C is nested within the levels of factor B, while B is crossed with factor A, the treatment effect calculations for A, B, and AB would be as crossed, while those for factor C would be:

$$C_{k(j)} = (\overline{X}_{kj} = \overline{\overline{X}}) - B_j$$

$$AC_{ik(j)} = (\overline{X}_{ijk} - \overline{\overline{X}}) - A_i - B_j - C_{k(j)} - AB_{ij}$$

Note that, for this experiment, there could be no BC or ABC interactions. Degrees of freedom for the effect may be calculated by substituting degrees of freedom for each of the terms in the treatment effect computation, with the term $(\overline{X} - \overline{\overline{X}})$ having degrees of freedom one less than the number of treatment effects for the effect being plotted.

Analysis of means for fully crossed or nested experiments is considerably simplified using the method presented. To apply analysis of means to more complicated factorial models, split-plots, or to incomplete block designs, see Schilling.[15]

15.9 Multifactor Experiments with Attributes Data

The methods presented for multiple factors are applicable also to attributes data: percent, proportion, or count. As discussed earlier, analysis of means for attributes data is usually done through limits set using the normal approximation to the binomial or the Poisson distribution. This implies that the sample size must be large enough for the approximation to apply. Sometimes transformations are useful, but experience has shown the results to be much the same in most cases with or without the use of such devices. Treatment effects may be calculated using the estimated proportions in place of the treatment means and the overall proportion or count provides an estimate of error. Thus, for proportions the standard deviation of a single observation is

$$\hat{\sigma}_e = \sqrt{\bar{p}(1 - \bar{p})}$$

with analogous results for percent or count data. Naturally, the factors for the decision limits are found using

$$\text{df} = \infty$$

as in a one-way 2^p experiment.

Consider, for example, some data supplied by Richard D. Zwickl[16] showing

[15] E. G. Schilling, "A Systematic Approach to the Analysis of Means," part 1, "Analysis of Treatment Effects," *J. Qual. Tech.*, vol. 5, no. 3, July 1973, pp. 93–108.

[16] Richard D. Zwickl, "Applications of Analysis of Means," Ellis R. Ott Conference on Quality Management and Applied Statistics in Industry, New Brunswick, N. J., April 7, 1987.

the proportion defective on three semiconductor wire-bonding machines over three shifts for a one-month period given in Table 15.5.

TABLE 15.5 Proportion Defective on Bonders (n = 1800)

Bonder	Shift 1	Shift 2	Shift 3	Bonder avg.
Number 600	0.028	0.042	0.017	0.029
Number 611	0.037	0.052	0.029	0.039
Number 613	0.023	0.045	0.015	0.028
Shift average	0.029	0.046	0.020	$\bar{p}$ = 0.032

The treatment effects are calculated as

Bonder

$$B_1\text{: } 0.029 - 0.032 = -0.003$$

$$B_2\text{: } 0.039 - 0.032 = 0.007$$

$$B_3\text{: } 0.028 - 0.032 = -0.004$$

Shift

$$S_1\text{: } 0.029 - 0.032 = -0.003$$

$$S_2\text{: } 0.046 - 0.032 = 0.014$$

$$S_3\text{: } 0.020 - 0.032 = -0.012$$

Interaction

$$BS_{11}\text{: } (0.028 - 0.032) - (-0.003) - (-0.003) = 0.002$$

$$BS_{12}\text{: } (0.042 - 0.032) - (-0.003) - (0.014) = -0.001$$

$$BS_{13}\text{: } (0.017 - 0.032) - (-0.003) - (-0.012) = 0$$

$$BS_{21}\text{: } (0.037 - 0.032) - (0.007) - (-0.003) = 0.001$$

$$BS_{22}\text{: } (0.052 - 0.032) - (0.007) - (0.014) = -0.001$$

$$BS_{23}\text{: } (0.029 - 0.032) - (0.007) - (-0.012) = 0.002$$

$$BS_{31}\text{: } (0.023 - 0.032) - (-0.004) - (-0.003) = -0.002$$

$$BS_{32}\text{: } (0.045 - 0.032) - (-0.004) - (0.014) = 0.003$$

$$BS_{33}\text{: } (0.015 - 0.032) - (-0.004) - (-0.012) = -0.001$$

Note the disparity in the interaction effects due to rounding.

The limits for $\alpha = 0.05$ are

$$\hat{\sigma}_e = \sqrt{0.032(0.968)} = 0.176$$

Main effects

$$0 \pm 0.176(1.91)\sqrt{\frac{2}{16{,}200}}$$

$$0 \pm 0.0037$$

Interaction

$$0 \pm 0.176(2.766)\sqrt{\frac{4}{16{,}200}}$$

$$0 \pm 0.0076$$

and the plot is as shown in Fig. 15.19.

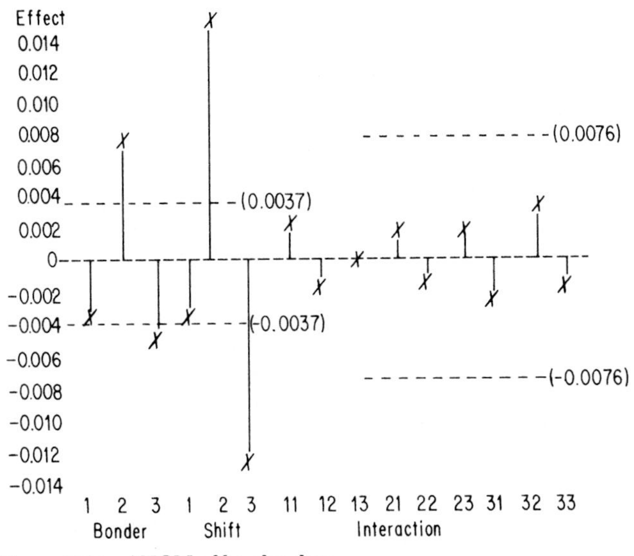

Figure 15.19 ANOM of bonder data.

Clearly, the significance of main effects is indicated. It should be noted that this approach to the analysis of proportions is approximate but, since so much industrial data is of this type, it provides an extension of the control-chart approach to the analysis of such data as a vehicle of communication and understanding.

An example of use of analysis of means count data in a multifactor experiment is also provided by Richard Zwickl.[17] An experiment was performed to find the best rinse conditions to minimize particulates on semiconductor wafers. The number of particles greater than 0.5μ was counted using a unit size of 10 wafers for various rinse times and temperatures. The results are shown in Table 15.6.

TABLE 15.6 Particle Count on Wafers

Temp. in degrees, D	Elapsed rinse time in minutes, M			Temperature average
	2 min	5 min	8 min	
25°C	205	111	48	364/3 = 121.3
85°C	138	97	23	258/3 = 86.0
Time avg.	343/2 = 171.5	208/2 = 104.0	71/2 = 35.5	$\hat{\mu}$ = 622/6 = 103.7

The treatment effects become

Time (M)

$$M_1 = 171.5 - 103.7 = 67.8$$

$$M_2 = 104.0 - 103.7 = 0.3$$

$$M_3 = 35.5 - 103.7 = -68.2$$

Temperature (D)

$$D_1 = 121.3 - 103.7 = 17.6$$

$$D_2 = 86.0 - 103.7 = -17.7$$

Interaction (MD)

$$MD_{11} = (205 - 103.7) - (67.8) - (17.6) = 15.9$$

$$MD_{12} = (138 - 103.7) - (67.8) - (-17.7) = -15.8$$

$$MD_{21} = (111 - 103.7) - (0.3) - (17.6) = -10.6$$

$$MD_{22} = (97 - 103.7) - (0.3) - (-17.7) = 10.7$$

$$MD_{31} = (48 - 103.7) - (-68.2) - (17.6) = -5.1$$

$$MD_{32} = (23 - 103.7) - (-68.2) - (-17.7) = 5.2$$

The limits for $\alpha = 0.05$ are as follows:

$$\sigma_e = \sqrt{103.7}$$

$$= 10.18$$

[17]Ibid.

Time

$$0 \pm 10.18(1.91)\sqrt{\frac{2}{6}}$$

$$0 \pm 11.23$$

Temperature

$$0 \pm 10.18(1.386)\sqrt{\frac{1}{6}}$$

$$0 \pm 5.76$$

Interaction

$$0 \pm 10.18(2.631)\sqrt{\frac{2}{6}}$$

$$0 \pm 15.46$$

and the plot is shown in Fig. 15.20.

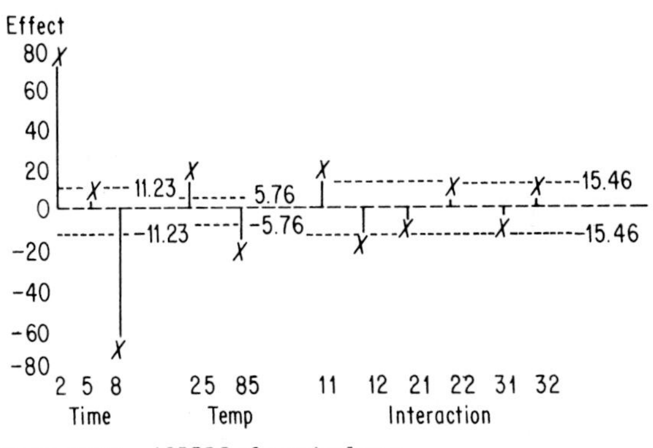

Figure 15.20 ANOM of particulates.

The main effects of time and temperature are clearly significant. Note the downward trend with increasing levels of both. Interaction is also barely significant at the 5 percent level. An interaction diagram is shown in Fig. 15.21.

Again, the analysis of means of count data such as this involves the approach to normality of the Poisson distribution (the mean of each cell should be greater than five) and is, of course, approximate. Experience has shown, however, that like the c chart, it is indeed adequate for most industrial applications. More detail will be found in the papers by Lewis and Ott[18] and also by Schilling.[19]

[18]S. Lewis and E. R. Ott, "Analysis of Means Applied to Percent Defective Data," *Tech. Rep. no. 2,* Rutgers University Statistics Center, February 10, 1960.

[19]E. G. Schilling, "A Systematic Approach to the Analysis of Means," part 3, "Analysis of Non-Normal Data," *J. Qual. Tech.,* vol. 5, no. 4, October 1973, pp. 156–159.

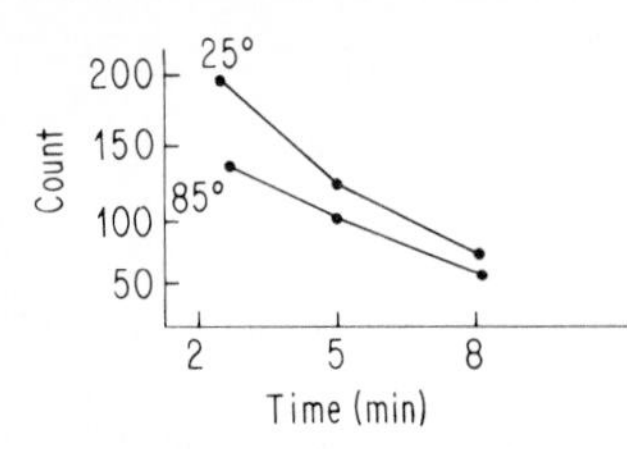

Figure 15.21 Interaction of particulates.

15.10 Comparing Variabilities

Introduction

The steel-rod lengths from the four machines, three times, and two heat treatments were being studied because of excessive variability in the finished rods. The comparison of average lengths (Fig. 15.15) shows two major special causes for variability; differences between machines and between heat treatments.

Now let us look at the inherent variability, or common causes. Some machine(s) may be innately more *variable* than others, independent of their average settings. We can compare variabilities from exactly two processes by using either a range-square-ratio test (F_R) or an F test.[20]

We can apply the method here to compare variabilities from the two heats W and L. There are 12 subgroup ranges in W and another 12 in L; in each subgroup, $r = 4$. Their averages are

$$\overline{R}_W = \frac{64}{12} = 5.33 \qquad \overline{R}_L = \frac{63}{12} = 5.25$$

These two values are surprisingly close; no further statistical comparison is necessary. A procedure, if needed, would be to compute $(\overline{R}_W/d_2^*)^2$ and $(\overline{R}_L/d_2^*)^2$ and form their ratio

$$F_R = (5.33/5.25)^2 \qquad \text{with df} \cong F(32,32)$$

and compare with values in Table A.12.

The range-square-ratio test in this form is applicable only to two levels of a factor. The following procedure *is applicable* to the four machines and the three times.

Analysis of means to analyze variability

1. Internal or within machine variability. Figure 15.22 is a rearrangement of the R chart, Fig. 15.13; it allows a ready, visual comparison of *machine* variabilities. A casual study of Fig. 15.22 suggests the possibility that machine A may be most variable and machines C or D the least; but the evidence is not

[20]See. Chap. 4. Also, for a more extensive discussion of use of analysis of means in comparing variabilities, see N. R. Ullman, "The Analysis of Means (ANOM) for Signal and Noise," *J. Qual. Tech.*, vol. 21, no. 2, April 1989, pp. 111–127.

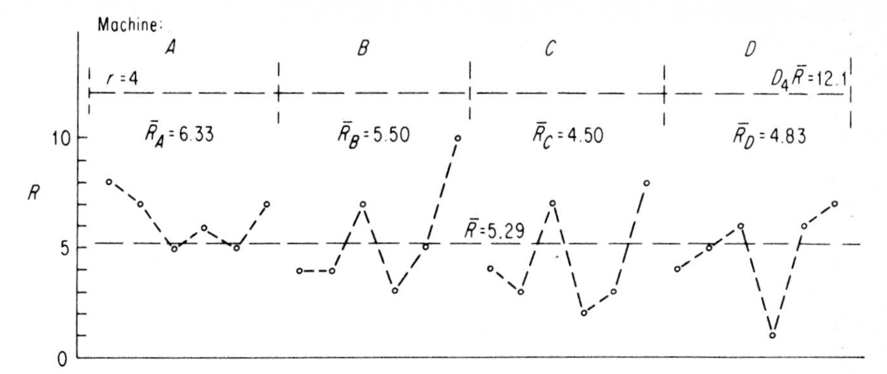

Figure 15.22 Subgroup ranges ($r = 4$) arranged by machines (data from Table 15.7).

very persuasive. An objective comparison of their variabilities is the following. The average machine ranges from Table 15.7 have been plotted in Fig. 15.23.

The computation of decision lines requires a measure $\hat{\sigma}_R$ of expected variation of the range R. Although ranges of individual subgroups are not normally distributed, *average* ranges of four (or more) subgroups are essentially normal (Chap. 1, Theorem 3). The standard deviation of ranges can be estimated as follows:

From Table A.4, the upper 3-sigma limit on R is $D_4\bar{R}$, where D_4 has been computed to give an upper control limit at $\bar{R} + 3\hat{\sigma}_R$

$$D_4\bar{R} = \bar{R} + 3\hat{\sigma}_R$$

TABLE 15.7 Subgroup Ranges

Data from Table 15.4; $n_g = r = 4$

	Heat Treatment								
	W				L				
	Machines								
Time	A	B	C	D	A	B	C	D	
1	8	4	4	4	6	3	2	1	$\bar{R}_1 = 32/8 = 4.00$
2	7	4	3	5	5	5	3	6	$\bar{R}_2 = 38/8 = 4.75$
3	5	7	7	6	7	10	8	7	$\bar{R}_3 = 57/8 = 7.12$

$\bar{R}_A = 38/6 = 6.33$
$\bar{R}_B = 33/6 = 5.50$
$\bar{R}_C = 27/6 = 4.50$
$\bar{R}_D = 29/6 = 4.83$

$\bar{R}_W = 5.33$
$\bar{R}_L = 5.25$

$\bar{\bar{R}} = 5.29$

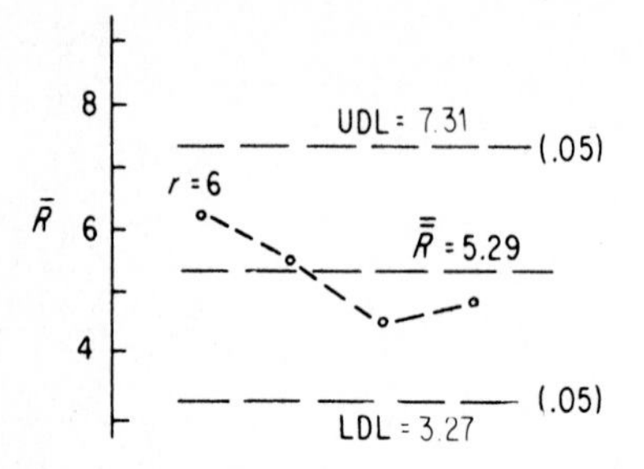

Figure 15.23 Comparing average machine variabilities (data from Fig. 15.22; each point is an average of $r = 6$ ranges).

Then $$\hat{\sigma}_R = \bar{R}(D_4 - 1)/3 = d_R\bar{R}$$

and $$\hat{\sigma}_R = \frac{d_3}{d_2}\bar{R}$$

Values of the factor d_R are given in Table 15.8 to simplify computation. When comparing any averages of these ranges,

$$\hat{\sigma}_R = d_R\bar{\bar{R}} = (0.43)(5.29) = 2.27$$

with degrees of freedom df $\cong (0.9)k(n - 1) = (0.9)24(3) = 65$

When comparing machine average ranges ($n_{\bar{R}} = 6$) of Fig. 15.23

$$\hat{\sigma}_{\bar{R}} = 2.27/\sqrt{6} = 0.92$$

Decision lines to compare averages of machine ranges are determined with: df = 65, $k = 4$, $H_{0.05} = 2.21$.

TABLE 15.8 Values of d_R Where $\hat{\sigma}_R = d_R\bar{R}$ and $d_R = (D_4 - 1)/3 = d_3/d_2$

r	d_R	D_4
2	0.76	3.27
3	0.52	2.57
4	0.43	2.28
5	0.37	2.11
6	0.33	2.00
7	0.31	1.92

$$\text{UDL}(0.05) = \overline{\overline{R}} + H_{0.05}\hat{\sigma}_{\overline{R}}$$

$$= 5.29 + (2.20)(0.92) = 7.31$$

$$\text{LDL}(0.05) = 3.27$$

All four points fall within the decision lines ($\alpha = 0.05$), and there does not appear to be a difference in variabilities of the four machines.

2. Variability at different times. The range data, Table 15.7, has been rearranged by time in Fig. 15.24.

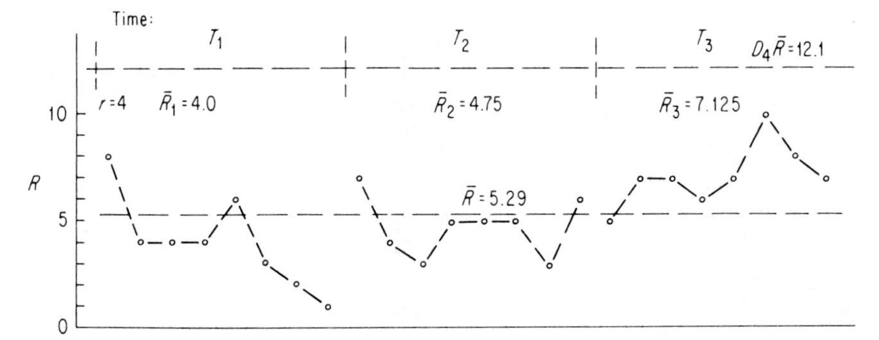

Figure 15.24 Subgroup ranges ($r = 4$) arranged by time periods (data from Table 15.7).

Analysis 1. Data for the third period T_3 appears to be significantly large. A comparison[21] of T_3 with a pooling of groups T_1 and T_2 shows an $(a + b)$ count of $(1 + 9) = 10$ which shows significance, $\alpha \cong 0.01$ by the Tukey–Duckworth test.

Analysis 2. Analysis of means (Fig. 15.25):

$$\overline{\overline{R}} = 5.29;\ k = 3,\ \text{df} \cong 65;\ n_g = 8$$

$$\hat{\sigma}_R = d_R\overline{\overline{R}} = 2.27 \text{ (each } R \text{ is of } r = 4)$$

$$\hat{\sigma}_{\overline{R}} = 2.27/\sqrt{8} = 0.80$$

$\alpha = 0.05$

$$\text{UDL} = 5.29 + (1.96)(0.80)$$

$$= 6.86$$

$$\text{LDL} = 3.72$$

[21]See Sec. 13.2.

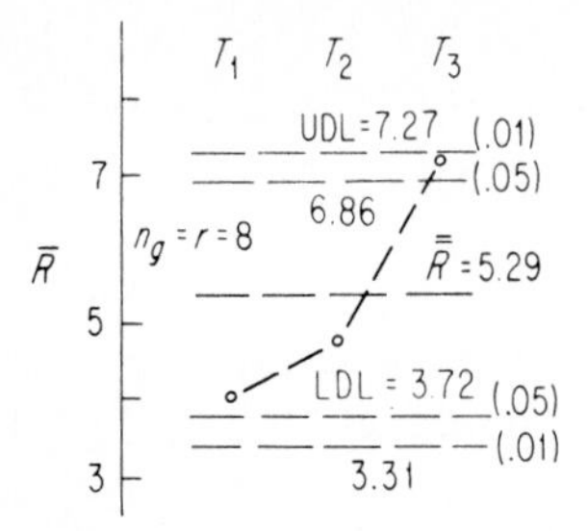

Figure 15.25 Comparing average time variabilities (data from Fig. 15.24; each point is an average of $r = 8$ ranges).

$\alpha = 0.01$

$$\text{UDL} = 5.29 + (2.47)(0.80)$$

$$= 7.27$$

$$\text{LDL} = 3.31$$

Interpretation. There is supporting evidence of a time effect on variability, risk $\alpha \cong 0.05$; with a definite suggestion that it became progressively more variable. The average at time T_1 is close to the lower (0.05) limit and at T_3 is outside the (0.05) and close to the (0.01) limit. Then we can consider the behavior of the different individual machines with respect to time (Table 15.9). A plot of the data is shown in Fig. 15.26. Surprisingly, this indicates that machine A appears affected altogether differently than the other three machines. This may be a consequence of operator fatigue or of machine maintenance, but it requires technical attention.

The biggest factor in variability is machine average—proper adjustments on individual machines should quickly minimize this. Secondly, the difference in heat treatment averages may possibly warrant adjustments for each new

TABLE 15.9 A Two-Way Table (Machine by Time) Ignoring Heat Treatment

Data from Table 15.7; each entry below is the average of two ranges

	T_1	T_2	T_3
A	7.0	6.0	6.0
B	3.5	4.5	8.5
C	3.0	3.0	7.5
D	2.5	5.5	6.5

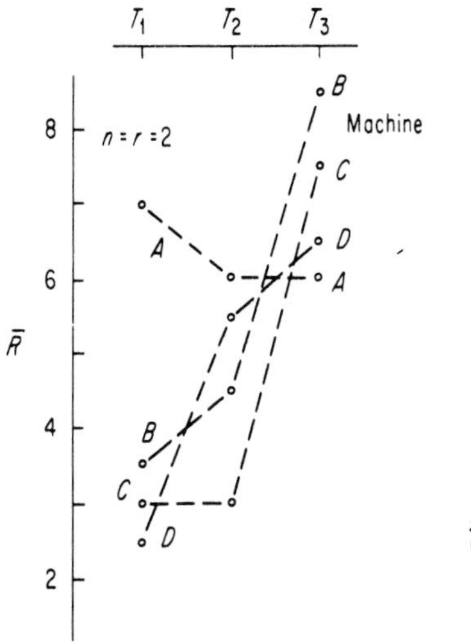

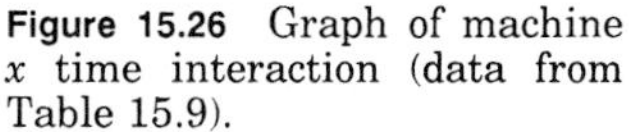

Figure 15.26 Graph of machine x time interaction (data from Table 15.9).

batch of rods, at least until the heat treatment process is stabilized. Probably third in importance is to establish reasons for the effect of time on within-machine variation; in fact, this may be of more importance than heat treatment.

15.11 Nonrandom Uniformity

Suppose we were to measure n consecutive steel bars all made on machine A from the same treatment. We would expect variation, not too much and *not too little.* If the n measurements were made and recorded in the order of manufacture, we could count the number of runs above and below the median, and compare them with the expected number (Table A.2). Note that there is a *minimum* number of runs expected (risk α) just as there is a maximum. Either *too few* or *too many* runs is evidence of an *assignable cause* in the process.

A variables control chart $(\bar{X},R)$ can signal an assignable cause by *too little* variation; we call this *nonrandom uniformity.*

Many articles have been written about evidence indicating the presence of assignable causes of nonrandomness and some about the identification[22] of the indicated assignable causes. These discussions have usually been concerned with the concept of nonrandom "excessive" variability. The literature has not emphasized that it is sometimes of scientific importance to discuss statistical evidence of *nonrandom uniformity* and to identify types of process behavior which may produce such patterns. Sources of data displaying

[22]Paul S. Olmstead, "How to Detect the Type of an Assignable Cause," part 1, "Clues for Particular Types of Trouble," part 2, "Procedure When Probable Cause is Unknown," *Ind. Qual Control,* vol. 9, no. 3, p. 32; vol. 9, no. 4, p. 22.

nonrandom uniformity include: differences in precision between analytical laboratories and sampling from a bimodal population or other sources of nonrational subgroups which produce exaggerated estimates of σ.

Nonrandom uniformity—standard given

As in Sec. 15.2, consider k samples of n_g each from a process in statistical control with average μ and standard deviation σ. If *all* k sample means lie *within narrow decision lines* drawn at

$$\mu \pm z_\alpha \hat{\sigma}_{\bar{x}}$$

this shall be considered evidence (with risk α) of *nonrandom uniformity.* Let Pr be the probability that a single point falls by chance between these lines. What must be the value of z_α in order that the probability of all k points falling within such a narrow band shall be only $\mathrm{Pr}^k = \alpha$?

Values of z_α are obtained from $\mathrm{Pr}^k = \alpha$ in the same manner as Z_α was obtained in Sec. 11.3. When $k = 3$, this becomes

$$\mathrm{Pr}^3 = 0.05$$

and

$$\mathrm{Pr} = 0.368$$

Then the corresponding $z_{0.05} = 0.48$ is found from a table of areas under the normal curve (Table A.1). Other selected values of z_α have been computed and are shown in Table 15.10.

For example, if in the casino example for standards given in Sec. 15.2, it was desired to check for nonrandom uniformity, the limits would be

$$k = 8, \mu = 7, \sigma = 2.42, n_g = 25$$

$$\alpha = 0.05 \qquad \mu \pm z_\alpha \frac{\sigma}{\sqrt{n_g}}$$

$$7 \pm 1.01 \frac{(2.42)}{\sqrt{25}}$$

$$7 \pm 0.49$$

$$\mathrm{LDL}_{0.05} = 6.51 \qquad \mathrm{UDL}_{0.05} = 7.49$$

and we have the following plot in Fig. 15.27.

Since all the points are not contained within the limits (in fact half the points are outside), there is no evidence to impugn the honesty of the casino on this basis.

TABLE 15.10 Factors to Judge Presence of Nonrandom Uniformity, Standard Given

k	$z_{.05}(k)$	$z_{.01}(k)$
2	.28	.13
3	.48	.27
4	.63	.41
5	.75	.52
6	.85	.62
7	.94	.70
8	1.01	.78
9	1.07	.84
10	1.13	.90
15	1.34	1.12
20	1.48	1.27
24	1.57	1.36
30	1.67	1.47
50	1.89	1.71
120	2.25	2.08

Nonrandom uniformity—no standard given

Some very interesting techniques of analysis are possible in this category. The critical values of N_α have been computed,[23] and selected values are given in Table A.16. It happens rather frequently that points on a control chart all lie very near the process average, and the erroneous conclusion is frequently made that the process is "in excellent control." The technique of this section provides an objective test of nonrandom uniformity. The computation of these entries in Table A.16 is much more complicated than for those in Table 15.10; the method is not given here. Decision lines to use in deciding whether our data indicate nonrandom uniformity are drawn at

$$\bar{\bar{X}} \pm N_\alpha \hat{\sigma}_{\bar{X}}$$

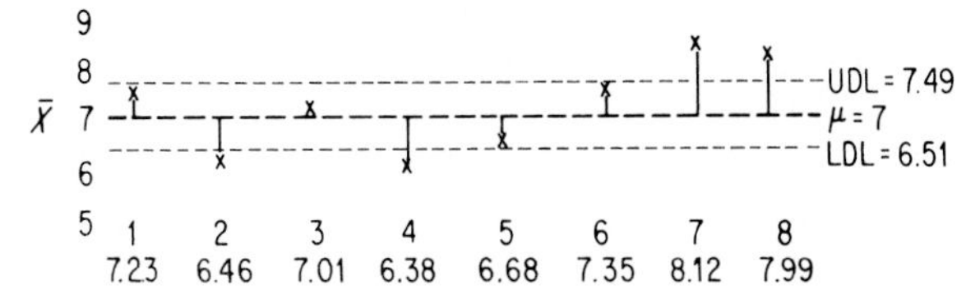

Figure 15.27 Nonrandom uniformity chart for eight casino tables.

[23]K. R. Nair, "The Distribution of the Extreme Deviate from the Sample Mean and Its Studentized Form," *Biom.*, vol. 35, 1948, pp. 118–144.

15.12 Development of Analysis of Means

The analysis of means was originally developed by Dr. E. R. Ott[24] and first reported in 1958. Subsequently, Sidney S. Lewis and Ellis R. Ott extended the analysis of means procedure to binomially distributed data when the normal approximation to the binomial distribution applies. Their results were reported[25] in 1960. In 1967, Dr. Ott published his Brumbaugh award winning paper, "Analysis of Means—A Graphical Procedure"[26] in *Industrial Quality Control*. Significantly, it was the Shewhart Memorial Issue.

The basic Ott procedure is intended for use with the means resulting from main effects in analysis of variance and in similar applications. Schilling[27] extended the analysis of means to the analysis of interactions and to a variety of experiment designs by providing a systematic method for the analysis of means, derived from the experiment model. This procedure used a modified factor h_α for computation of the limits where

$$h_\alpha = \sqrt{k/(k-1)}\, H_\alpha$$

from the Bonferroni inequality. Based on Ott's original analysis of 2^n experiments, Schilling[28] extended the procedure to the analysis of contrasts in various forms. He also provided a procedure for use with nonnormal distributions such as attributes data or the Weibull.[29]

L. S. Nelson[30] computed an extensive table of h_α factors using the Bonferroni inequality and later produced tables of exact h_α factors based on the theoretical development of P. R. Nelson.[31] P. R. Nelson[32] has also provided tables of sample size for analysis of means as well as power curves for the procedure.[33] The exact values of L. S. Nelson were modified by D. C. Smialek[34] to provide exact H_α factors equivalent to those used by Ott in the original procedure. Professor Smialek produced the table of h^* factors using

[24] E. R. Ott, "Analysis of Means," *Tech. Rep. no. 1*, Rutgers University Statistics Center, August 10, 1958.

[25] S. S. Lewis and E. R. Ott, "Analysis of Means Applied to Percent Defective Data," *Tech. Rep. no. 2*, Rutgers University Statistics Center, February 10, 1960.

[26] E. R. Ott, "Analysis of Means—A Graphical Procedure," *Ind. Qual. Control*, vol. 24, no. 2, August 1967, pp. 101–109.

[27] E. G. Schilling, "A Systematic Approach to the Analysis of Means," part 1, "Analysis of Treatment Effects," *J. Qual. Tech.*, vol. 5, no. 3, July 1973, pp. 93–108.

[28] E. G. Schilling, "A Systematic Approach to the Analysis of Means," part 2, "Analysis of Contrasts," *J. Qual Tech.*, vol. 5, no. 4, October 1973, pp. 147–155.

[29] E. G. Schilling, "A Systematic Approach to the Analysis of Means," part 3, "Analysis of Non-Normal Data," *J. Qual. Tech.*, vol. 5, no. 4, October 1973, pp. 156–159.

[30] L. S. Nelson, "Exact Critical Values for Use with the Analysis of Means," *J. Qual. Tech.*, vol. 15, no. 1, January 1983, pp. 40–44.

[31] P. R. Nelson, "Exact Critical Points for the Analysis of Means," *Comm. in Stat., Theo. and Meth.*, II, pp. 699–709.

[32] P. R. Nelson, "A Comparison of Sample Size for the Analysis of Means and the Analysis of Variance," *J. Qual. Tech.*, vol. 15, no. 1, January 1983, pp. 33–39.

[33] P. R. Nelson, "Power Curves for the Analysis of Means," *Technometrics*, vol. 27, no. 1, February 1985, pp. 65–73.

[34] E. G. Schilling and D. C. Smialek, "Simplified Analysis of Means for Crossed and Nested Experiments," *43d Ann. Qual. Control Conf.*, Rochester Section ASQC, Rochester, N. Y., March 10, 1987.

the Sidak approximation which appears here. P. R. Nelson[35] examined analysis of means for interactions when at least one factor is at two levels and provided critical values for other special cases.

A computer program for analysis of factorial experiments simultaneously by analysis of means and analysis of variance was developed by Schilling, Schlotzer, Schultz and Sheesley[36] and subsequently modified by P. R. Nelson[37] to include exact values. Sheesley[38] has provided a computer program to do control charts and single factor experiments for measurements or attributes data using the Bonferroni values, which allow for lack of independence among the points plotted.

Sheesley[39] has also provided tables of simplified factors for analysis of means, similar to control chart factors for use with the range. One of Ott's last papers on the topic was an insightful analysis of multiple-head machines coauthored with Dr. R. D. Snee.[40] Neil Ullman[41] has expanded the area of application by providing factors for analysis of means on ranges suitable for use in analysis of the Taguchi signal-to-noise ratio.

Analysis of means provides a vehicle for the simultaneous display of both statistical and engineering significance. The procedure brings to bear the intuitive appeal and serendipity of the control chart to the analysis of designed experiments. It is appropriate that the American Society for Quality Control's Shewhart Medal should bear an inscription of a control chart, and equally fitting that the Shewhart Medal should have been awarded to Ellis R. Ott in 1960, the year "Analysis of Means for Percent Defective Data" was published.

15.13 Practice Exercises

1. Recompute the decision lines of Exercise 14.7 assuming that the eight subgroups are from eight levels of a single factor.

2. Analyze the following data on an experiment comparable in nature to that presented in Case Histories 15.1 and 15.2.

---

[35] P. R. Nelson, "Testing for Interactions Using the Analysis of Means," *Technometrics,* vol. 30, no. 1, February 1988, pp. 53–61.

[36] E. G. Schilling, G. Schlotzer, H. E. Schultz, and J. H. Sheesley, "A Fortran Computer Program for Analysis of Variance and Analysis of Means," *J. Qual. Tech.,* vol. 12, no. 2, April 1980, pp. 106–113.

[37] P. R. Nelson, "The Analysis of Means for Balanced Experimental Designs," *J. Qual. Tech.,* vol. 15, no. 1, January 1983, pp. 45–56.

[38] J. H. Sheesley, "Comparison of *k* Samples Involving Variables or Attributes Data Using the Analysis of Means," *J. Qual. Tech.,* vol. 12, no. 1, January 1980, pp. 47–52.

[39] J. H. Sheesley, "Simplified Factors for Analysis of Means When the Standard Deviation is Estimated from the Range," *J. Qual. Tech.,* vol. 13, no. 3, July 1981, pp. 184–185.

[40] Ellis R. Ott and R. D. Snee, "Identifying Useful Differences in a Multiple-Head Machine," *J. Qual. Tech.,* vol 5, no. 2, April 1973, pp. 47–57.

[41] Neil R. Ullman, "Analysis of Means (ANOM) for Signal to Noise," *J. Qual. Tech.*, vol. 21, no. 2, April 1989, pp. 111–127.

Machine setting	1	2	3
Response values	876	1050	850
	933	895	748
	664	777	862
	938	929	675
	938	1005	837
	676	912	921
	614	542	681
	712	963	797
	721	937	752
	812	896	646

Note: this problem has a hidden "catch."

3. Calculate $z_{0.05}(8) = 1.01$ and $z_{0.01}(8) = 0.78$ as in Table 15.10.
4. Is there evidence of nonrandom uniformity in Case History 15.3 at alpha = 0.05 for any of the machines, days, or heat treatments?
5. Delete machine D and rework problem of Case History 15.3.
6. Assume that a possible assignable cause exists for the four vials appearing to be high in the data of Table 13.4 as analyzed in the plot of Fig. 13.5. Reanalyze using analysis of means for three levels.
7. The following data came out of an experiment to determine the effects of package plating on bond strength. Two levels of nickel thickness and two levels of current density were used. Data is supplied by Richard D. Zwickl.[42]

Breaking Strength of Wire Bonds

	Nickel thickness				
Current Density	Thin		Thick		
2 amp/ft^2	8.68		10.82		
	9.25	$\bar{X} = 9.272$	10.50	$\bar{X} = 10.326$	
	9.41		10.36		9.660
	9.77		9.26		
	9.25		10.64		
4 amp/ft^2	9.07		9.68		
	9.41	$\bar{X} = 9.910$	10.42	$\bar{X} = 10.182$	
	10.38		10.21		10.113
	10.69		10.00		
	10.00		9.93		
Average	9.794		9.979		

Analyze this experiment using analysis of means with $\alpha = 0.05$. Note that $\bar{\bar{X}} = 9.8865$ and $\bar{R} = 0.539$. Perform the analysis using (*a*) treatment effects and (*b*) Yates method as a 2^2 factorial.

[42]R. D. Zwickl, "Applications of Analysis of Means," The Ellis R. Ott Conference on Quality Management and Applied Statistics in Industry, New Brunswick, N. J., April 7, 1987.

8. The percentage of wire bonds with evidence of ceramic pullout (CPO) is given below for various combinations of metal-film thickness, ceramic surface, prebond clean and annealing time.

Percentage of Wire Bond with CPO

Metal-film thickness, ceramic surface		Normal		1.5 × normal	
Prebond clean	Annealing time	Unglazed	Glazed	Unglazed	Glazed
Normal	Normal	9/96	70/96	8/96	42/96
clean	4 × normal	13/64	55/96	7/96	19/96
No	Normal	3/96	6/96	1/64	7/96
clean	4 × normal	5/96	28/96	3/96	6/96

Note that the sample size is not maintained over all the cells. This is typical of real industrial data.[43] Analyze the experiment using analysis of means with $\alpha = 0.05$.

9. A 3^3 experiment was run on wire bonding to determine the effect of capillary, temperature, and force on wire bonding in semiconductor manufacture. Data adapted from that supplied by Richard D. Zwickl[44] is as follows:

Average Pull Strength

	Capillary								
	New			Worn			Squashed		
	Temperature, °C			Temperature, °C			Temperature, °C		
Force, psi	25	100	150	25	100	150	25	100	150
25	4.08	4.67	6.22	3.27	5.91	9.35	2.70	5.04	7.43
	4.77	2.96	7.67	4.18	5.60	8.49	3.28	4.66	8.97
40	2.50	4.83	8.62	3.32	5.81	8.53	4.01	5.82	8.57
	2.30	6.13	8.12	5.06	6.62	6.78	4.61	5.73	9.13
55	4.34	4.85	8.31	4.18	6.61	9.38	3.97	6.03	7.62
	3.32	4.15	8.38	5.71	7.32	9.21	4.02	6.24	10.44

The values shown are averages of 18 wirebonds. Analyze the experiment using analysis of means with $\alpha = 0.05$. What can be said about the nature of the temperature effect from the analysis of means plot?

[43] Richard D. Zwickl, "An Example of Analysis of Means for Attribute Data Applied to a 2^4 Factorial Design," *ASQC Electronics Div. Tech. Suppl.*, issue 4, Fall 1985, pp. 1–22.

[44] Richard D. Zwickl, "Applications of Analysis of Means," The Ellis R. Ott Conference on Quality Management and Applied Statistics in Industry, New Brunswick, N. J., April 7, 1987.

10. An experiment was conducted to determine the effect of developers and linewidth measuring equipment (aligned) on linewidth of a photolithographic process. The following data was obtained.[45]

	Linewidth				
	Wafer location				
	1	2	3	4	
	202	211	211	186	
	208	217	200	174	
Aligner 1	215	220	206	198	
	231	226	211	208	
	208	212	205	189	
$\overline{X}$	212.8	217.2	206.6	191.0	206.9
R	29	15	11	34	
	219	207	211	199	
	231	222	225	206	
Aligner 2	225	216	210	211	
	222	216	218	213	
	224	215	216	207	
$\overline{X}$	224.2	215.2	216.0	207.2	215.6
R	12	15	15	14	
	255	250	254	246	
	253	250	252	246	
Aligner 3	254	250	254	246	
	253	249	253	245	
	253	250	254	249	
$\overline{X}$	253.6	249.8	253.4	246.4	250.8
R	2	1	2	4	
	223	226	223	222	
	220	221	216	211	
Aligner 13	228	235	231	212	
	221	221	222	215	
	229	231	219	219	
$\overline{X}$	224.2	226.8	222.2	215.8	222.2
R	9	14	15	11	
	228.7	227.2	224.6	215.1	223.9

Analyze by analysis of means using $\alpha = 0.05$. Draw the interaction diagram.

[45]Stuart Kukunaris, "Operating Manufacturing Processes Using Experimental Design," *ASQC Electronics Div. Tech. Suppl.*, issue 3, Summer 1985, pp. 1–19.

11. The following results were obtained in an experiment to determine the effect of developers and linewidth measuring equipment on linewidth.[46]

	Linewidth			
	Aligner 1	Aligner 2	Aligner 3	Average
	215	223	226	
	225	225	240	
	211	230	237	
Developer 1	212	236	234	
	212	232	235	
	211	238	225	
	206	234	236	
$\bar{X}$	213.1	231.1	233.3	225.8
R	19	15	15	
	213	220	231	
	206	223	228	
	206	221	231	
Developer 2	211	220	223	
	215	222	228	
	227	228	245	
	217	228	229	
$\bar{X}$	213.6	223.1	230.7	222.5
R	21	8	22	
	213	218	238	
	219	225	228	
	211	225	231	
Developer 3	212	230	244	
	207	232	234	
	207	228	237	
	216	231	236	
$\bar{X}$	212.1	227.0	235.4	224.8
R	12	14	16	
Average	212.9	227.1	233.1	$\bar{\bar{X}} = 224.4$

Analyze the experiment by analysis of means using $\alpha = 0.05$. Draw the interaction diagram.

[46]Ibid.

12. Certain questions have arisen regarding the error-making propensity of four randomly selected work stations from two departments (C and D) on shifts A or B, Monday through Friday. Records from one week are shown below. Determine which department, shift and days of the week are to be preferred using a 95 percent confidence level.

	C		D	
	A	B	A	B
Mon.	10	15	16	19
	11	22	12	24
	16	12	13	15
	15	20	11	16
R	6	10	5	9
Tues.	9	11	15	21
	6	15	10	16
	10	19	9	17
	7	10	10	21
R	4	9	6	5
Wed.	14	18	15	14
	13	16	12	18
	10	14	14	13
	14	13	15	21
R	4	5	3	8
Thurs.	7	8	10	14
	13	15	12	12
	6	15	13	11
	11	17	9	13
R	7	9	4	3
Fri.	10	18	14	15
	13	16	17	22
	9	15	10	15
	7	21	17	15
R	6	6	7	7

13. A designed experiment involving two treatments, A and B, in four departments, w, x, y, and z, and five time periods, I, II, III, IV, and V, gave the results indicated below. Complete an analysis of means to determine which departments, treatments, or time periods are to be preferred at a 95 percent confidence level. High results are desired.

	w		x		y		z		
	A	B	A	B	A	B	A	B	
I	7.1	6.0	7.6	6.5	7.5	6.9	6.2	7.9	
	7.5	7.3	7.4	7.2	8.5	7.2	7.1	7.6	
	8.0	7.8	6.4	7.7	6.6	6.6	6.1	6.0	$\bar{I} = 7.11$
R	0.9	1.8	1.2	1.2	1.9	0.6	1.0	1.9	
II	6.6	7.6	7.5	8.0	7.4	6.7	6.7	7.7	
	7.3	6.2	7.3	6.5	7.2	6.8	6.3	6.7	
	6.3	6.3	6.5	7.5	7.1	7.0	6.9	7.7	$\overline{II} = 6.99$
R	1.0	1.4	1.0	1.5	0.3	0.3	0.6	1.0	
III	6.4	6.5	7.0	6.8	7.8	7.3	5.5	6.4	
	7.3	7.1	6.9	6.8	7.2	6.4	6.9	5.9	
	5.7	6.2	5.9	6.5	7.5	7.5	7.5	5.7	$\overline{III} = 6.70$
R	1.6	0.9	1.1	0.3	0.6	1.1	2.0	0.7	
IV	6.7	6.2	6.1	6.3	8.4	7.1	7.7	7.9	
	7.4	6.0	6.7	6.2	7.2	7.3	6.3	6.4	
	7.3	7.1	7.0	6.9	8.1	7.0	6.3	6.5	$\overline{IV} = 6.92$
R	0.7	1.1	0.9	0.7	1.2	0.3	1.4	1.5	
V	7.4	6.3	8.0	7.3	7.2	7.5	7.6	6.2	
	6.1	7.9	6.5	6.1	8.1	8.1	7.4	7.9	
	7.2	6.4	6.9	7.1	6.7	7.1	6.1	6.2	$\overline{V} = 7.05$
R	1.3	1.6	1.5	1.2	1.4	1.0	1.5	1.7	

$\bar{w} = 6.84$ $\bar{x} = 6.90$ $\bar{y} = 7.30$ $\bar{z} = 6.78$
$\bar{A} = 7.02$ $\bar{R} = 1.1225$
$\bar{B} = 6.89$ Mean of all sample data points $= 6.955$

Chapter

16

Epilogue

Every process and every product is maintained and improved by those who combine some underlying theory with some practical experience. More than that, they call upon an amazing backlog of ingenuity and know-how to amplify and support that theory. New-product ramrods are real "pioneers"; they also recognize the importance of their initiative and intuition and enjoy the dependence resting on their know-how. However, as scientific theory and background knowledge increase, dependence on native skill and initiative often decreases. An expert can determine just by listening that an automobile engine is in need of repair. Similarly, an experienced production man can often recognize a recurring malfunction by characteristic physical manifestations. However, problems become more complicated. Although familiarity with scientific advances will sometimes be all that is needed to solve even complicated problems—whether for maintenance or for improvement, many important changes and problems cannot be recognized by simple observation and initiative no matter how competent the scientist. It should be understood that no process is so simple that data from it will not give added insight into its behavior. The typical standard production process has unrecognized complex behaviors which can be thoroughly understood only by studying data from the product it produces. The "pioneer" who accepts and learns methods of scientific investigation to support technical advances in knowledge can be an exceptionally able citizen in an area of expertise. Methods in this book can be a boon to such pioneers in their old age.

This book has presented different direct procedures for acquiring data to suggest the character of a malfunction or to give evidence of improvement opportunities. Different types of data and different methods of analysis have been illustrated, which is no more unusual than a medical doctor's use of various skills and techniques in diagnosing the ailments of a patient. It cannot be stressed too much that the value and importance of the procedure or method are only in its applicability and usefulness to the particular problem at hand. The situation and the desired end frequently indicate the means.

Discussing the situation with appropriate personnel, both technical and supervisory, at a very early date, before any procedures are planned, will often prevent a waste of time and even avoid possible embarrassment. It will also often ensure their subsequent support in implementing the results of the study; but you may expect them to assure you that any difficulty "isn't my fault." Often a study should be planned, expecting that it will support a diagnosis made by one or more of them. Sometimes it does; sometimes it does not. But the results should pinpoint the area of difficulty, suggest the way toward the solution of a problem, or even sometimes give evidence of unsuspected problems of economic importance. Properly executed, the study will always provide some insight into the process. A simple remedy for a difficulty may be suggested where the consensus, after careful engineering consideration, had been that only a complete redesign or major change in specifications would effect the desired improvements.

An industrial consultant often has the right and authority to study *any* process or project. But this is not exactly a divine right. It is usually no more than a "hunting or fishing" license; you may hunt, but no game is guaranteed. So find some sympathetic cooperative souls to talk to. They may be able to clear the path to the best hunting ground. Some of the most likely areas are:

1. A spot on the line where rejects are piling up
2. On-line or final inspection stations
3. A process using selective assembly. It is fairly common practice in production to separate components A and B each into three categories; low, medium, and high, and then assemble low A with high B, etc. This process is sometimes a short-term necessary evil, but there are inevitable typical problems which result.

Many things need to be said about the use of data to assist in troubleshooting. We may as well begin with the following one, which differs from what we often hear.

Industry is a mass-production operation; it differs radically from most agricultural and biological phenomena which require a generation or more to develop data. If you do not get enough data from a production study today, more data can be had tomorrow with little or no added expense. Simple studies are usually preferred to elaborate nonreplicated designs which are so common in agriculture, biology, and some industry research and/or development problems.

Throughout this book much use has been made in a great variety of situations of a simple yet effective method of studying and presenting data, the graphical analysis of means. This method makes use of developments in applying control charts to data, and a similar development in designing and analyzing experiments. Let us look at some of its special advantages:

1. Computations are simple and easy. Often no calculator is necessary, but it is possible to program the ANOM for graphical printout on a computer.
2. Errors in calculation may be shown up, often apparent in a graphical presentation, even to the untrained.
3. The graphical comparison of effects presents the results in a way which

will be accepted by many as the basis for decision and action, encouraging the translation of conclusions into scientific action.

4. Dealing directly with means (averages), the method provides an immediate study of possible effects of the factors involved.
5. Not only is nonrandomness of data indicated, but (in contrast to the results from other analyses) the sources of such nonrandomness are immediately pinpointed.
6. This analysis frequently, as a bonus, suggests the unexpected presence of certain types of nonrandomness which can be included in subsequent studies for checking.
7. The graphical presentation of data is almost a necessity when interpreting the meaning of any interaction.

Troubleshooters and others involved in process-improvement studies who are familiar with analysis of variance will find the graphical analysis of means a logical interpretative followup procedure. Others, faced with studying multiple independent variables, will find that the graphical procedure provides a simple immediate and effective analysis and interpretation of data. It is difficult to repeat too often the importance, to the business of troubleshooting, of a well-planned *but simple design.*

Frequently, in setting up or extending a quality control program some sort of organized teaching program is necessary. Whenever possible, an outside consultant should be the instructor. It is important that the instructor play a key role in troubleshooting projects in the plant. The use of current in-plant data suggested by class members for study will not only provide pertinent and stimulating material as a basis for discussion of the basic techniques of analysis but may actually lead to a discussion of ways of improving some major production problem. However, not many internal consultants can keep sensitive issues often raised by such discussion in check without serious scars.

Quality control requires consciousness from top management to operator and throughout all departments. Therefore representatives from purchasing, design, manufacturing, quality, sales, and related departments should be included in the class for at least selected pertinent aspects of the program.

And what should the course include? Well, that is what this book is all about. But to start, keep it simple and basic, encouraging the application of the students' ingenuity and know-how to the use of whatever analytical techniques they learn, in the study of data already available.

Friends and associates of many years and untold experiences sometimes come to the rescue. Not too long ago, one responded when asked, "Bill, what shall I tell them?" Slightly paraphrased, here is what he scribbled on a note:

- Come right out and tell them to plot the data.
- The important thing is to get moving on the problem quickly; hence, use quick, graphical methods of analysis. Try to learn something quickly—not everything. Production is rolling. Quick, partial help *now* is preferable to

somewhat better advice postponed. Get moving. Your prompt response will trigger ideas from them too.

- Emphasize techniques of drawing out the choice of variables to be considered, asking "dumb, leading questions." (How does one play dumb?)
- Develop techniques of making the operators think it was all their idea.
- Make them realize the importance of designing little production experiments and the usefulness of a control chart in pointing up areas where experimentation is needed. The chart does not solve the problem, but it tells you where and when to look for a solution.
- Say something like "you don't need an $\overline{X}$, R control chart on every machine at first; a p chart may show you the areas where $\overline{X}$, R charts will be helpful."
- Introduce the outgoing product quality rating philosophy of looking at the finished product and noting where the *big* problems are.
- After the data are analyzed you have to tell someone about the solution—like the boss—to get action. You cannot demand that the supervisor follow directions to improve the process, but the *boss* can find a way. For one thing, the boss' remarks about how the supervisor worked out a problem with you can have a salutary effect—on the supervisor himself *and* on other supervisors.

Now Bill would not consider this little outline a panacea for all ailments, but these were the ideas which popped into his head and they warrant some introspection.

If you have read this far, there are two remaining suggestions:

1. Skim through the case histories in the book. If they do not trigger some ideas about your own plant problems, then at least one of us has failed.
2. If you did get an idea, then get out on the line and get some data! (Not too much, now.)

16.1 Practice Exercises

1 to ∞. Practice does not end here. Find some data. Get involved. Use the methods you have learned. The answers are not in an answer book, but you will know when you are correct and the rewards will be great.

Remember—Plot The Data!

Appendix

TABLE A.1 Areas under the Normal Curve*

Proportion of total area under the curve to the left of a vertical line drawn at $\mu + Z_\alpha$, where Z represents any desired value from $Z = 0$ to $Z = \pm 3.09$

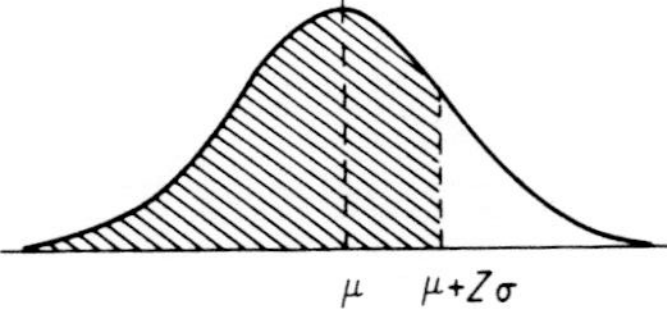

Z	−0.00	−0.01	−0.02	−0.03	−0.04	−0.05	−0.06	−0.07	−0.08	−0.09
− 3.0	0.00135	0.00131	0.00126	0.00122	0.00118	0.00114	0.00111	0.00107	0.00104	0.00100
− 2.9	0.0019	0.0018	0.0017	0.0017	0.0016	0.0016	0.0015	0.0015	0.0014	0.0014
− 2.8	0.0026	0.0025	0.0024	0.0023	0.0023	0.0022	0.0021	0.0021	0.0020	0.0019
− 2.7	0.0035	0.0034	0.0033	0.0032	0.0031	0.0030	0.0029	0.0028	0.0027	0.0026
− 2.6	0.0047	0.0045	0.0044	0.0043	0.0041	0.0040	0.0039	0.0038	0.0037	0.0036
− 2.5	0.0062	0.0060	0.0059	0.0057	0.0055	0.0054	0.0052	0.0051	0.0049	0.0048
− 2.4	0.0082	0.0080	0.0078	0.0075	0.0073	0.0071	0.0069	0.0068	0.0066	0.0064
− 2.3	0.0107	0.0104	0.0102	0.0099	0.0096	0.0094	0.0091	0.0089	0.0087	0.0084
− 2.2	0.0139	0.0136	0.0132	0.0129	0.0125	0.0122	0.0119	0.0116	0.0113	0.0110
− 2.1	0.0179	0.0174	0.0170	0.0166	0.0162	0.0158	0.0154	0.0150	0.0146	0.0143
− 2.0	0.0228	0.0222	0.0217	0.0212	0.0207	0.0202	0.0197	0.0192	0.0188	0.0183
− 1.9	0.0287	0.0281	0.0274	0.0268	0.0262	0.0256	0.0250	0.0244	0.0239	0.0233
− 1.8	0.0359	0.0351	0.0344	0.0336	0.0329	0.0322	0.0314	0.0307	0.0301	0.0294
− 1.7	0.0446	0.0436	0.0427	0.0481	0.0409	0.0401	0.0392	0.0384	0.0375	0.0367
− 1.6	0.0548	0.0537	0.0526	0.0516	0.0505	0.0495	0.0485	0.0475	0.0465	0.0455
− 1.5	0.0668	0.0655	0.0643	0.0630	0.0618	0.0606	0.0594	0.0582	0.0571	0.0559
− 1.4	0.0808	0.0793	0.0778	0.0764	0.0749	0.0735	0.0721	0.0708	0.0694	0.0681
− 1.3	0.0968	0.0951	0.0934	0.0918	0.0901	0.0885	0.0869	0.0853	0.0838	0.0823
− 1.2	0.1151	0.1131	0.1112	0.1093	0.1075	0.1057	0.1038	0.1020	0.1003	0.0985
− 1.1	0.1357	0.1335	0.1314	0.1292	0.1271	0.1251	0.1230	0.1210	0.1190	0.1170
− 1.0	0.1587	0.1562	0.1539	0.1515	0.1492	0.1469	0.1446	0.1423	0.1401	0.1379

− 0.9	0.1841	0.1814	0.1788	0.1762	0.1736	0.1711	0.1685	0.1660	0.1635	0.1611
− 0.8	0.2119	0.2090	0.2061	0.2033	0.2005	0.1977	0.1949	0.1922	0.1894	0.1867
− 0.7	0.2420	0.2389	0.2358	0.2327	0.2297	0.2266	0.2236	0.2207	0.2177	0.2148
− 0.6	0.2743	0.2709	0.2676	0.2643	0.2611	0.2578	0.2546	0.2514	0.2483	0.2451
−0.5	0.3085	0.3050	0.3015	0.2981	0.2946	0.2912	0.2877	0.2843	0.2810	0.2776
−0.4	0.3446	0.3409	0.3372	0.3336	0.3300	0.3264	0.3228	0.3192	0.3156	0.3121
−0.3	0.3821	0.3783	0.3745	0.3707	0.3669	0.3632	0.3594	0.3557	0.3520	0.3483
−0.2	0.4207	0.4168	0.4129	0.4090	0.4052	0.4013	0.3974	0.3936	0.3897	0.3859
−0.1	0.4602	0.4562	0.4522	0.4483	0.4443	0.4404	0.4364	0.4325	0.4286	0.4247
−0.0	0.5000	0.4960	0.4920	0.4880	0.4840	0.4801	0.4761	0.4721	0.4681	0.4641
Z	+0.00	+0.01	+0.02	+0.03	+0.04	+0.05	+0.06	+0.07	+0.08	+0.09
+0.0	0.5000	0.5040	0.5080	0.5120	0.5160	0.5199	0.5239	0.5279	0.5319	0.5359
+0.1	0.5398	0.5438	0.5478	0.5517	0.5557	0.5596	0.5636	0.5675	0.5714	0.5753
+0.2	0.5793	0.5832	0.5871	0.5910	0.5948	0.5987	0.6026	0.6064	0.6103	0.6141
+0.3	0.6179	0.6217	0.6255	0.6293	0.6331	0.6368	0.6406	0.6443	0.6480	0.6517
+0.4	0.6554	0.6591	0.6628	0.6664	0.6700	0.6736	0.6772	0.6808	0.6844	0.6879
+0.5	0.6915	0.6950	0.6985	0.7019	0.7054	0.7088	0.7123	0.7157	0.7190	·0.7224
+0.6	0.7257	0.7291	0.7324	0.7357	0.7389	0.7422	0.7454	0.7486	0.7517	0.7549
+0.7	0.7580	0.7611	0.7642	0.7673	0.7704	0.7734	0.7764	0.7794	0.7823	0.7852
+0.8	0.7881	0.7910	0.7939	0.7967	0.7995	0.8023	0.8051	0.8079	0.8106	0.8133
+0.9	0.8159	0.8186	0.8212	0.8238	0.8264	0.8289	0.8315	0.8340	0.8365	0.8389
+1.0	0.8413	0.8438	0.8461	0.8485	0.8508	0.8531	0.8554	0.8577	0.8599	0.8621
+1.1	0.8643	0.8665	0.8686	0.8708	0.8729	0.8749	0.8770	0.8790	0.8810	0.8830
+1.2	0.8849	0.8869	0.8888	0.8907	0.8925	0.8944	0.8962	0.8980	0.8997	0.9015
+1.3	0.9032	0.9049	0.9066	0.9082	0.9099	0.9115	0.9131	0.9147	0.9162	0.9177
+1.4	0.9192	0.9207	0.9222	0.9236	0.9251	0.9265	0.9279	0.9292	0.9306	0.9319
+1.5	0.9332	0.9345	0.9357	0.9370	0.9382	0.9394	0.9406	0.9418	0.9429	0.9441
+1.6	0.9452	0.9463	0.9474	0.9484	0.9495	0.9505	0.9515	0.9525	0.9535	0.9545
+1.7	0.9554	0.9564	0.9573	0.9582	0.9591	0.9599	0.9608	0.9616	0.9625	0.9633

TABLE A.1 Areas under the Normal Curve* ***(Continued)***

Z	+0.00	+0.01	+0.02	+0.03	+0.04	+0.05	+0.06	+0.07	+0.08	+0.09
+1.8	0.9641	0.9649	0.9656	0.9664	0.9671	0.9678	0.9686	0.9693	0.9699	0.9706
+1.9	0.9713	0.9719	0.9726	0.9732	0.9738	0.9744	0.9750	0.9756	0.9761	0.9767
+2.0	0.9773	0.9778	0.9783	0.9788	0.9793	0.9798	0.9803	0.9808	0.9812	0.9817
+2.1	0.9821	0.9826	0.9830	0.9834	0.9838	0.9842	0.9846	0.9850	0.9854	0.9857
+2.2	0.9861	0.9864	0.9868	0.9871	0.9875	0.9878	0.9881	0.9884	0.9887	0.9890
+2.3	0.9893	0.9896	0.9898	0.9901	0.9904	0.9906	0.9909	0.9911	0.9913	0.9916
+2.4	0.9918	0.9920	0.9922	0.9925	0.9927	0.9929	0.9931	0.9932	0.9934	0.9936
+2.5	0.9938	0.9940	0.9941	0.9943	0.9945	0.9946	0.9948	0.9949	0.9951	0.9952
+2.6	0.9953	0.9955	0.9956	0.9957	0.9959	0.9960	0.9961	0.9962	0.9963	0.9964
+2.7	0.9965	0.9966	0.9967	0.9968	0.9969	0.9970	0.9971	0.9972	0.9973	0.9974
+2.8	0.9974	0.9975	0.9976	0.9977	0.9977	0.9978	0.9979	0.9979	0.9980	0.9981
+2.9	0.9981	0.9982	0.9983	0.9983	0.9984	0.9984	0.9985	0.9985	0.9986	0.9986
+3.0	0.99865	0.99869	0.99874	0.99878	0.99882	0.99886	0.99889	0.99893	0.99896	0.99900

SOURCE: This table is a modification of one which appears in Grant and Leavenworth, *Statistical Quality Control*, 4th ed., McGraw-Hill, New York, 1972, pp. 642–643.

*Following are specific areas under the normal curve.

Cumulative probability	Tail probability	Z	Cumulative probability	Z
0.5	0.5	0	0.99903	3.1
0.75	0.25	0.675	0.99931	3.2
0.80	0.20	0.842	0.99952	3.3
0.90	0.10	1.282	0.99966	3.4
0.95	0.05	1.645	0.99977	3.5
0.975	0.025	1.96	0.99984	3.6
0.98	0.02	2.055	0.99989	3.7
0.99	0.01	2.33	0.99993	3.8
0.995	0.005	2.575	0.99995	3.9
0.998	0.002	2.88	0.99997	4.0
0.999	0.001	3.09		

TABLE A.2 Critical Values of the Number of Runs N_R above and below the Median in $k = 2\ m$ Observations (One-Tail)

		Significantly small critical values of N_R		Significantly large critical values of N_R	
k	m	$\alpha = .01$	$\alpha = .05$	$\alpha = .05$	$\alpha = .01$
10	5	2	3	8	9
12	6	2	3	10	11
14	7	3	4	11	12
16	8	4	5	12	13
18	9	4	6	13	15
20	10	5	6	15	16
22	11	6	7	16	17
24	12	7	8	17	18
26	13	7	9	18	20
28	14	8	10	19	21
30	15	9	11	20	22
32	16	10	11	22	23
34	17	10	12	23	25
36	18	11	13	24	26
38	19	12	14	25	27
40	20	13	15	26	28
42	21	14	16	27	29
44	22	14	17	28	31
46	23	15	17	30	32
48	24	16	18	31	33
50	25	17	19	32	34
60	30	21	24	37	40
70	35	25	28	43	46
80	40	30	33	48	51
90	45	34	37	54	57
100	50	38	42	59	63
110	55	43	46	65	68
120	60	47	51	70	74

SOURCE: S. Swed and Churchill Eisenhart, Tables for Testing Randomness of Sampling in a Sequence of Alternatives, *Ann. Math. Stat.*, vol. 14, pp. 66–87, 1943. (Reproduced by permission of the editor.)

TABLE A.3 **Runs above and below the Median of Length s in $k = 2m$ Observations with k As Large As 16 or 20**

s	Expected number of length exactly s	Expected number of length greater than or equal to s
1	$\frac{k(k+2)}{2^2(k-1)} \cong k/2^2 = \frac{k}{4}$	$(k+2)/2 =$ (total number)
2	$\frac{k(k+2)}{2^3(k-1)} \cong k/2^3 = \frac{k}{8}$	$\frac{(k+2)(k-2)}{2^2(k-1)} \cong k/2^2 = \frac{k}{4}$
3	$\frac{k(k+2)(k-4)}{2^4(k-1)(k-3)} \cong k/2^4 = \frac{k}{16}$	$\frac{(k-4)(k+2)}{2^3(k-1)} \cong k/2^3 = \frac{k}{8}$
4	$\frac{k(k+2)(k-6)}{2^5(k-1)(k-3)} \cong k/2^5 = \frac{k}{32}$	$\frac{(k+2)(k-4)(k-6)}{2^4(k-1)(k-3)} \cong k/2^4 = \frac{k}{16}$
5	$\frac{k(k+2)(k-6)(k-8)}{2^6(k-1)(k-3)(k-5)} \cong k/2^6 = \frac{k}{64}$	$\frac{(k+2)(k-6)(k-8)}{2^5(k-1)(k-3)} \cong k/2^5 = \frac{k}{32}$
6	$\frac{k(k+2)(k-8)(k-10)}{2^7(k-1)(k-3)(k-6)} \cong k/2^7 = \frac{k}{128}$	$\frac{(k+2)(k-8)(k-10)}{2^6(k-1)(k-3)} \cong k/2^6 = \frac{k}{64}$

TABLE A.4 Control Chart Limits for Samples of n_g

Plot	Sample mean $\bar{X}$ against standard μ with σ known	Sample mean $\bar{X}$ against past data using $\bar{X}$ and $\bar{s}$ or $\bar{R}$	Sample standard deviation s against standard (known) σ	Sample range R against standard (known) σ	Sample standard deviation s or range R against past data using $\bar{s}$ or $\bar{R}$	Sample* proportion $\hat{p}$ or defects per unit $\hat{u}$ against standard p or u	Sample* proportion $\hat{p}$ or defects per unit $\hat{u}$ against past data using $\bar{p}$ or $\bar{u}$
Upper control limit	$\mu + 3\sigma/\sqrt{n} = \mu + A\sigma$	$\bar{\bar{X}} + A_3\bar{s}$	$B_6\sigma$	$D_2\sigma$	$B_4\bar{s}$	$p + 3\sqrt{\frac{p(1-p)}{n}}$	$\bar{p} + 3\sqrt{\frac{\bar{p}(1-\bar{p})}{n}}$
		$\bar{\bar{X}} + A_2\bar{R}$			$D_4\bar{R}$	$u + 3\sqrt{\frac{u}{n}}$	$\bar{u} + 3\sqrt{\frac{\bar{u}}{n}}$
Centerline	μ	$\bar{\bar{X}}$	$c_4\sigma$	$d_2\sigma$	$\bar{s}$ R	p u	$\bar{p}$ $\bar{u}$
Lower control limit	$\mu - 3\sigma/\sqrt{n} = \mu - A\sigma$	$\bar{\bar{X}} - A_3\bar{s}$	$B_5\sigma$	$D_1\sigma$	$B_3\bar{s}$	$p - 3\sqrt{\frac{p(1-p)}{n}}$	$\bar{p} - 3\sqrt{\frac{\bar{p}(1-\bar{p})}{n}}$
		$\bar{\bar{X}} - A_2\bar{R}$			$D_3\bar{R}$	$u - 3\sqrt{\frac{u}{n}}$	$\bar{u} - 3\sqrt{\frac{\bar{u}}{n}}$

*For defects chart use u with $n = 1$.

n	A	A_2	A_3	B_3	B_4	B_5	B_6	c_4	d_2	D_1	D_2	D_3	D_4
2	2.121	1.880	2.659	0	3.267	0	2.606	0.7979	1.128	0	3.686	0	3.267
3	1.732	1.023	1.954	0	2.568	0	2.276	0.8862	1.693	0	4.358	0	2.575
4	1.500	0.729	1.628	0	2.266	0	2.088	0.9213	2.059	0	4.698	0	2.282
5	1.342	0.577	1.427	0	2.089	0	1.964	0.9400	2.326	0	4.918	0	2.115
6	1.225	0.483	1.287	0.030	1.970	0.029	1.874	0.9515	2.534	0	5.078	0	2.004
7	1.134	0.419	1.182	0.118	1.882	0.113	1.806	0.9594	2.704	0.205	5.203	0.076	1.924
8	1.061	0.373	1.099	0.185	1.815	0.179	1.751	0.9650	2.847	0.387	5.307	0.136	1.864
9	1.000	0.337	1.032	0.239	1.761	0.232	1.707	0.9693	2.970	0.546	5.394	0.184	1.816
10	0.949	0.308	0.975	0.284	1.716	0.276	1.669	0.9727	3.078	0.687	5.469	0.223	1.777

SOURCE: E. G. Schilling, *Acceptance Sampling in Quality Control*, Marcel Dekker, Inc., New York, 1982, p. 594. (Reprinted by courtesy of Marcel Dekker, Inc.)

TABLE A.5 Binomial Probability Tables

The cumulative probabilities of $x \le c$ are given in the column headed by p for any sample size. Note c is the sum of the row heading I and the column heading J, so $c = I + J$. Each value shown is $P(x \le c)$. To find the probability of exactly x in a sample of n, take $P(X=x) = P(X \le x) - P(X \le x - 1)$. To find $P(X \le x)$ when $p > 0.5$, use $c = (n - x - 1)$ under $(1 - p)$ and take the complement of the answer, that is $P(X \le x/n,p) = 1 - P(X \le n - x - 1/n, 1 - p)$

p	0.01	0.02	0.03	0.04	0.05	0.06	0.07	0.08	0.09	0.10	0.15	0.20	0.25	0.30	0.40	0.50
I \ J	0	0	0	0	0	0	0	0	0	0	0	0	0	0	0	0
								$n = 1; c = I + J$								
0	0.990	0.980	0.970	0.960	0.950	0.940	0.930	0.920	0.910	0.900	0.850	0.800	0.750	0.700	0.600	0.500
1	1.000	1.000	1.000	1.000	1.000	1.000	1.000	1.000	1.000	1.000	1.000	1.000	1.000	1.000	1.000	1.000
								$n = 2; c = I + J$								
0	0.990	0.960	0.941	0.922	0.902	0.884	0.846	0.828	0.810	0.722	0.640	0.562	0.490	0.360	0.250	0.500
1	1.000	1.000	0.999	0.998	0.998	0.996	0.995	0.994	0.992	0.990	0.978	0.960	0.938	0.910	0.840	0.750
2	1.000	1.000	1.000	1.000	1.000	1.000	1.000	1.000	1.000	1.000	1.000	1.000	1.000	1.000	1.000	1.000
								$n = 3; c = I + J$								
0	0.970	0.940	0.913	0.885	0.857	0.831	0.804	0.779	0.754	0.729	0.614	0.512	0.422	0.343	0.246	0.125
1	1.000	0.999	0.997	0.995	0.993	0.990	0.986	0.982	0.977	0.972	0.939	0.896	0.844	0.784	0.648	0.500
2	1.000	1.000	1.000	1.000	1.000	1.000	1.000	0.999	0.999	0.999	0.997	0.992	0.984	0.973	0.936	0.875
3	1.000	1.000	1.000	1.000	1.000	1.000	1.000	1.000	1.000	1.000	1.000	1.000	1.000	1.000	1.000	1.000
								$n = 4; c = I + J$								
0	0.961	0.922	0.855	0.849	0.815	0.781	0.748	0.716	0.686	0.656	0.522	0.410	0.316	0.240	0.130	0.063
1	0.999	0.998	0.995	0.991	0.986	0.980	0.973	0.996	0.957	0.948	0.890	0.819	0.738	0.652	0.475	0.313
2	1.000	1.000	1.000	1.000	1.000	0.999	0.999	0.998	0.997	0.996	0.988	0.973	0.949	0.916	0.821	0.688
3						1.000	1.000	1.000	1.000	1.000	0.999	0.998	0.996	0.992	0.974	0.938
4											1.000	1.000	1.000	1.000	1.000	1.000

	$n = 5; c = I + J$															
0	0.951	0.904	0.859	0.815	0.774	0.734	0.696	0.659	0.624	0.590	0.444	0.328	0.237	0.168	0.078	0.031
1	0.999	0.996	0.992	0.985	0.977	0.968	0.958	0.946	0.933	0.919	0.835	0.737	0.633	0.528	0.337	0.188
2	1.000	1.000	1.000	0.999	0.999	0.998	0.997	0.995	0.994	0.991	0.973	0.942	0.896	0.837	0.683	0.500
3				1.000	1.000	1.000	1.000	1.000	1.000	1.000	0.998	0.993	0.984	0.969	0.913	0.813
4											1.000	1.000	0.999	0.998	0.990	0.969
5													1.000	1.000	1.000	1.000
	$n = 10; c = I + J$															
0	0.904	0.817	0.737	0.665	0.599	0.539	0.484	0.434	0.389	0.349	0.197	0.107	0.056	0.028	0.006	0.001
1	0.996	0.984	0.965	0.942	0.914	0.882	0.848	0.812	0.775	0.736	0.544	0.376	0.244	0.149	0.046	0.011
2	1.000	0.999	0.997	0.994	0.988	0.981	0.972	0.960	0.946	0.930	0.820	0.678	0.526	0.383	0.167	0.055
3		1.000	1.000	1.000	0.999	0.998	0.996	0.994	0.991	0.987	0.950	0.879	0.776	0.650	0.382	0.172
4					1.000	1.000	1.000	0.999	0.999	0.998	0.990	0.967	0.922	0.850	0.633	0.377
5								1.000	1.000	1.000	0.999	0.994	0.980	0.953	0.834	0.623
6											1.000	0.999	0.996	0.989	0.945	0.828
7												1.000	1.000	0.998	0.988	0.945
8														1.000	0.998	0.989
9															1.000	0.999
10																1.000
I \ J	0	0	0	0	0	0	0	0	0	0	0	0	0	0	1	2
	$n = 15; c = I + J$															
0	0.860	0.739	0.633	0.542	0.463	0.395	0.337	0.286	0.243	0.206	0.087	0.035	0.013	0.005	0.005	0.004
1	0.990	0.965	0.927	0.881	0.829	0.774	0.717	0.660	0.603	0.549	0.319	0.167	0.080	0.035	0.027	0.018
2	1.000	0.997	0.991	0.980	0.964	0.943	0.917	0.887	0.853	0.816	0.604	0.398	0.236	0.127	0.091	0.059
3		1.000	1.000	0.998	0.995	0.990	0.982	0.973	0.960	0.944	0.823	0.648	0.461	0.297	0.217	0.151
4				1.000	0.999	0.999	0.997	0.995	0.992	0.987	0.938	0.836	0.686	0.515	0.403	0.304
5					1.000	1.000	1.000	0.999	0.999	0.998	0.983	0.939	0.851	0.722	0.610	0.500
6								1.000	1.000	1.000	0.996	0.982	0.943	0.869	0.787	0.696
7											0.999	0.996	0.983	0.950	0.905	0.849
8											1.000	0.999	0.996	0.985	0.966	0.941
9												1.000	0.999	0.996	0.991	0.982
10													1.000	0.999	0.998	0.996
11														1.000	1.000	1.000

TABLE A.5 Binomial Probability Tables *(Continued)*

p	0.01	0.02	0.03	0.04	0.05	0.06	0.07	0.08	0.09	0.10	0.15	0.20	0.25	0.30	0.40	0.50
J / I	0	0	0	0	0	0	0	0	0	0	0	0	0	0	1	3
								$n = 20; c = I + J$								
0	0.818	0.668	0.544	0.442	0.358	0.290	0.234	0.189	0.152	0.122	0.039	0.012	0.003	0.001	0.001	0.001
1	0.983	0.940	0.880	0.810	0.736	0.660	0.587	0.517	0.452	0.392	0.176	0.069	0.024	0.008	0.004	0.006
2	0.999	0.993	0.979	0.956	0.925	0.885	0.839	0.788	0.733	0.677	0.405	0.206	0.091	0.035	0.016	0.021
3	1.000	0.999	0.997	0.993	0.984	0.971	0.953	0.929	0.901	0.867	0.648	0.411	0.225	0.107	0.051	0.058
4		1.000	1.000	0.999	0.997	0.994	0.989	0.982	0.971	0.957	0.830	0.630	0.415	0.238	0.126	0.131
5				1.000	1.000	0.999	0.998	0.996	0.993	0.989	0.933	0.804	0.617	0.416	0.250	0.252
6						1.000	1.000	0.999	0.999	0.998	0.978	0.913	0.786	0.608	0.416	0.412
7								1.000	1.000	1.000	0.994	0.968	0.898	0.772	0.596	0.588
8											0.999	0.990	0.959	0.887	0.755	0.748
9											1.000	0.997	0.986	0.952	0.872	0.868
10												0.999	0.996	0.983	0.943	0.942
11												1.000	0.999	0.995	0.979	0.979
12													1.000	0.999	0.994	0.994
13														1.000	0.998	0.999
14															1.000	1.000
J / I	0	0	0	0	0	0	0	0	0	0	0	0	1	1	3	5
								$n = 25; c = I + J$								
0	0.778	0.603	0.467	0.360	0.277	0.213	0.163	0.124	0.095	0.072	0.017	0.004	0.007	0.002	0.002	0.002
1	0.974	0.911	0.828	0.736	0.642	0.553	0.470	0.395	0.329	0.271	0.093	0.027	0.032	0.009	0.009	0.007
2	0.998	0.987	0.962	0.924	0.873	0.813	0.747	0.677	0.606	0.537	0.254	0.098	0.096	0.033	0.029	0.022
3	1.000	0.999	0.994	0.983	0.966	0.940	0.906	0.865	0.817	0.764	0.471	0.234	0.214	0.090	0.074	0.054
4		1.000	0.999	0.997	0.993	0.985	0.973	0.955	0.931	0.902	0.682	0.421	0.378	0.193	0.154	0.115
5			1.000	1.000	0.999	0.997	0.993	0.988	0.979	0.967	0.838	0.617	0.561	0.341	0.274	0.212
6					1.000	0.999	0.999	0.997	0.995	0.991	0.930	0.780	0.727	0.512	0.425	0.345

J / I	0	0	0	0	0	0	0	0	0	0	0	0	1	2	4	6
7						1.000	1.000	0.999	0.999	0.998	0.975	0.891	0.851	0.677	0.586	0.500
8								1.000	1.000	1.000	0.992	0.953	0.929	0.811	0.732	0.655
9											0.998	0.983	0.970	0.902	0.846	0.788
10											1.000	0.994	0.989	0.956	0.922	0.885
11												0.998	0.997	0.983	0.966	0.946
12												1.000	0.999	0.994	0.987	0.978
13													1.000	0.998	0.996	0.993
14														1.000	0.999	0.998
15															1.000	1.000
	$n = 30; c = I + J$															
0	0.740	0.545	0.401	0.294	0.215	0.156	0.113	0.082	0.059	0.042	0.008	0.001	0.002	0.002	0.002	0.001
1	0.964	0.879	0.773	0.661	0.554	0.455	0.369	0.296	0.234	0.184	0.084	0.011	0.011	0.009	0.006	0.003
2	0.997	0.978	0.940	0.883	0.812	0.732	0.649	0.565	0.486	0.411	0.151	0.044	0.037	0.030	0.017	0.008
3	1.000	0.997	0.988	0.969	0.939	0.897	0.845	0.784	0.717	0.647	0.322	0.123	0.098	0.077	0.044	0.021
4		1.000	0.998	0.994	0.984	0.968	0.945	0.913	0.872	0.825	0.524	0.255	0.203	0.160	0.094	0.049
5			1.000	0.999	0.997	0.992	0.984	0.971	0.952	0.927	0.711	0.428	0.348	0.281	0.176	0.100
6				1.000	0.999	0.998	0.996	0.992	0.985	0.974	0.847	0.607	0.514	0.432	0.291	0.181
7					1.000	1.000	0.999	0.998	0.996	0.992	0.930	0.761	0.674	0.589	0.431	0.292
8							1.000	1.000	0.999	0.998	0.972	0.871	0.803	0.730	0.578	0.428
9									1.000	1.000	0.990	0.939	0.894	0.841	0.714	0.572
10											0.997	0.974	0.949	0.916	0.825	0.708
11											0.999	0.991	0.978	0.960	0.903	0.819
12											1.000	0.997	0.992	0.983	0.952	0.900
13												0.999	0.997	0.994	0.979	0.951
14												1.000	0.999	0.998	0.992	0.979
15													1.000	0.999	0.997	0.992
16														1.000	0.999	0.997
17															1.000	0.999
18																1.000

TABLE A.5 Binomial Probability Tables *(Continued)*

p	0.01	0.02	0.03	0.04	0.05	0.06	0.07	0.08	0.09	0.10	0.15	0.20	0.25	0.30	0.40	0.50
I \ J	0	0	0	0	0	0	0	0	0	0	0	1	1	3	5	8
							$n = 35; c = I + J$									
0	0.703	0.493	0.344	0.240	0.166	0.115	0.079	0.054	0.039	0.025	0.003	0.004	0.001	0.002	0.001	0.001
1	0.952	0.845	0.717	0.589	0.472	0.371	0.287	0.218	0.164	0.122	0.024	0.019	0.003	0.009	0.003	0.003
2	0.995	0.967	0.913	0.837	0.746	0.649	0.552	0.461	0.379	0.306	0.087	0.061	0.014	0.027	0.010	0.008
3	1.000	0.995	0.980	0.950	0.904	0.844	0.773	0.694	0.612	0.531	0.209	0.143	0.041	0.065	0.026	0.020
4		0.999	0.986	0.988	0.971	0.944	0.905	0.856	0.797	0.731	0.381	0.272	0.098	0.133	0.058	0.045
5		1.000	0.999	0.998	0.993	0.983	0.967	0.943	0.910	0.868	0.569	0.433	0.192	0.234	0.112	0.088
6			1.000	1.000	0.998	0.996	0.990	0.981	0.966	0.945	0.735	0.599	0.322	0.365	0.195	0.155
7					1.000	0.999	0.998	0.994	0.989	0.980	0.856	0.745	0.474	0.510	0.306	0.250
8						1.000	0.999	0.999	0.997	0.994	0.931	0.854	0.626	0.652	0.436	0.368
9							1.000	1.000	0.999	0.998	0.971	0.925	0.758	0.773	0.573	0.500
10									1.000	1.000	0.989	0.966	0.858	0.865	0.700	0.632
11											0.996	0.986	0.924	0.927	0.807	0.750
12											0.999	0.995	0.964	0.964	0.886	0.845
13											1.000	0.998	0.984	0.984	0.938	0.912
14												0.999	0.994	0.994	0.970	0.955
15												1.000	0.998	0.998	0.987	0.980
16													0.999	0.999	0.995	0.992
17													1.000	1.000	0.998	0.997
18															0.999	0.999
19															1.000	1.000

I \ J	0	0	0	0	0	0	0	0	0	0	0	1	2	3	6	10
	$n = 40; c = I + J$															
0	0.669	0.446	0.296	0.195	0.129	0.084	0.055	0.036	0.023	0.015	0.002	0.001	0.001	0.001	0.001	0.001
1	0.939	0.810	0.662	0.521	0.399	0.299	0.220	0.159	0.114	0.080	0.012	0.008	0.005	0.003	0.002	0.003
2	0.993	0.954	0.882	0.786	0.677	0.567	0.463	0.369	0.289	0.223	0.049	0.028	0.016	0.009	0.006	0.008
3	0.999	0.992	0.969	0.925	0.862	0.783	0.684	0.601	0.509	0.423	0.130	0.076	0.043	0.024	0.016	0.019
4	1.000	0.999	0.993	0.979	0.952	0.910	0.855	0.787	0.710	0.629	0.263	0.161	0.096	0.055	0.035	0.040
5		1.000	0.999	0.995	0.986	0.969	0.942	0.903	0.853	0.794	0.433	0.286	0.182	0.111	0.071	0.077
6			1.000	0.999	0.997	0.991	0.980	0.962	0.936	0.900	0.607	0.437	0.330	0.196	0.129	0.134
7				1.000	0.999	0.998	0.994	0.987	0.976	0.958	0.756	0.593	0.440	0.309	0.211	0.215
8					1.000	0.999	0.998	0.996	0.992	0.985	0.865	0.732	0.584	0.441	0.317	0.318
9						1.000	1.000	0.999	0.998	0.995	0.933	0.839	0.715	0.577	0.440	0.437
10								1.000	0.999	0.999	0.970	0.912	0.821	0.703	0.568	0.563
11									1.000	1.000	0.988	0.957	0.897	0.807	0.689	0.682
12											0.996	0.981	0.946	0.885	0.791	0.785
13											0.999	0.992	0.974	0.937	0.870	0.866
14											1.000	0.997	0.988	0.968	0.926	0.923
15												0.999	0.995	0.985	0.961	0.960
16												1.000	0.998	0.994	0.981	0.981
17													0.999	0.998	0.992	0.992
18													1.000	0.999	0.997	0.997
19														1.000	0.999	0.999
20															1.000	1.000

TABLE A.5 Binomial Probability Tables *(Continued)*

p	0.01	0.02	0.03	0.04	0.05	0.06	0.07	0.08	0.09	0.10	0.15	0.20	0.25	0.30	0.40	0.50
I \ J	0	0	0	0	0	0	0	0	0	0	0	1	3	4	8	12
	$n = 45; c = I + J$															
0	0.636	0.403	0.254	0.159	0.099	0.062	0.038	0.023	0.014	0.009	0.001	0.001	0.002	0.001	0.001	0.001
1	0.925	0.773	0.607	0.458	0.335	0.239	0.167	0.115	0.078	0.052	0.006	0.003	0.006	0.003	0.004	0.003
2	0.990	0.939	0.848	0.732	0.608	0.488	0.382	0.291	0.217	0.159	0.027	0.013	0.018	0.008	0.009	0.008
3	0.999	0.988	0.954	0.895	0.813	0.716	0.613	0.510	0.414	0.329	0.078	0.038	0.045	0.021	0.022	0.018
4	1.000	0.998	0.989	0.967	0.927	0.869	0.795	0.710	0.619	0.527	0.175	0.090	0.094	0.047	0.045	0.036
5		1.000	0.998	0.991	0.976	0.949	0.908	0.852	0.785	0.708	0.314	0.177	0.173	0.093	0.084	0.068
6			1.000	0.998	0.993	0.983	0.964	0.935	0.894	0.841	0.478	0.297	0.280	0.165	0.143	0.116
7				1.000	0.998	0.995	0.988	0.975	0.954	0.924	0.639	0.441	0.409	0.262	0.225	0.186
8					1.000	0.999	0.996	0.992	0.983	0.968	0.775	0.588	0.546	0.380	0.327	0.276
9						1.000	0.999	0.997	0.994	0.988	0.873	0.720	0.675	0.509	0.444	0.383
10							1.000	0.999	0.998	0.996	0.935	0.826	0.784	0.635	0.564	0.500
11								1.000	1.000	0.999	0.970	0.901	0.867	0.746	0.679	0.617
12										1.000	0.987	0.948	0.925	0.836	0.778	0.724
13											0.995	0.975	0.961	0.901	0.856	0.814
14											0.998	0.989	0.981	0.945	0.914	0.884
15											0.999	0.996	0.992	0.972	0.952	0.932
16											1.000	0.998	0.997	0.986	0.975	0.964
17												0.999	0.999	0.994	0.988	0.982
18												1.000	1.000	0.998	0.995	0.992
19														0.999	0.998	0.997
20														1.000	0.999	0.999
21															1.000	1.000

I \ J	0	0	0	0	0	0	0	0	0	0	1	2	4	5	9	14
								$n = 50; c = I + J$								
0	0.605	0.364	0.218	0.129	0.077	0.045	0.027	0.015	0.009	0.005	0.003	0.001	0.002	0.001	0.001	0.001
1	0.911	0.736	0.555	0.400	0.279	0.190	0.126	0.083	0.053	0.034	0.014	0.005	0.007	0.002	0.002	0.003
2	0.986	0.922	0.811	0.677	0.541	0.416	0.311	0.226	0.161	0.112	0.046	0.018	0.019	0.007	0.006	0.008
3	0.998	0.982	0.937	0.861	0.760	0.647	0.533	0.425	0.330	0.250	0.112	0.048	0.045	0.018	0.013	0.016
4	0.999	0.997	0.983	0.951	0.896	0.821	0.729	0.629	0.528	0.431	0.219	0.103	0.092	0.040	0.028	0.032
5		0.999	0.996	0.986	0.962	0.922	0.865	0.792	0.707	0.616	0.361	0.190	0.164	0.079	0.054	0.059
6			0.999	0.996	0.988	0.971	0.942	0.898	0.840	0.770	0.518	0.307	0.262	0.139	0.096	0.101
7				0.999	0.997	0.990	0.978	0.956	0.923	0.878	0.668	0.443	0.382	0.223	0.156	0.161
8					0.999	0.997	0.993	0.983	0.967	0.942	0.791	0.584	0.511	0.329	0.237	0.239
9						0.999	0.998	0.994	0.987	0.975	0.880	0.711	0.637	0.447	0.335	0.336
10							0.999	0.998	0.996	0.991	0.937	0.814	0.748	0.569	0.446	0.444
11								0.999	0.999	0.997	0.969	0.889	0.837	0.684	0.561	0.556
12									0.999	0.999	0.987	0.939	0.902	0.782	0.670	0.604
13											0.994	0.969	0.945	0.859	0.766	0.760
14											0.998	0.986	0.971	0.915	0.844	0.839
15											0.999	0.993	0.986	0.952	0.902	0.899
16												0.997	0.994	0.975	0.943	0.941
17												0.999	0.997	0.988	0.969	0.968
18												0.999	0.999	0.994	0.984	0.984
19													0.999	0.997	0.992	0.992
20														0.999	0.997	0.997
21															0.998	0.999
22															0.999	0.999

TABLE A.5 **Binomial Probability Tables** *(Continued)*

p	0.01	0.02	0.03	0.04	0.05	0.06	0.07	0.08	0.09	0.10	0.15	0.20	0.25	0.30	0.40	0.50
I \ J	0	0	0	0	0	0	0	0	0	1	3	5	7	10	17	25
							$n = 75; c = I + J$									
0	0.471	0.219	0.101	0.047	0.021	0.009	0.004	0.002	0.001	0.003	0.002	0.001	0.001	0.000	0.001	0.001
1	0.827	0.556	0.338	0.193	0.105	0.056	0.029	0.014	0.007	0.016	0.008	0.004	0.002	0.002	0.003	0.005
2	0.960	0.810	0.608	0.419	0.269	0.165	0.096	0.055	0.030	0.050	0.023	0.010	0.004	0.004	0.006	0.010
3	0.993	0.936	0.812	0.647	0.479	0.334	0.211	0.140	0.085	0.119	0.054	0.024	0.010	0.009	0.011	0.018
4	0.999	0.982	0.925	0.819	0.679	0.529	0.390	0.274	0.184	0.227	0.108	0.050	0.022	0.019	0.021	0.032
5	0.999	0.996	0.975	0.920	0.828	0.706	0.571	0.439	0.322	0.367	0.189	0.093	0.043	0.035	0.037	0.053
6		0.999	0.992	0.969	0.919	0.837	0.729	0.606	0.482	0.521	0.295	0.156	0.077	0.062	0.061	0.083
7			0.998	0.989	0.966	0.919	0.847	0.749	0.638	0.666	0.418	0.239	0.127	0.102	0.096	0.124
8			0.999	0.997	0.988	0.965	0.922	0.856	0.769	0.786	0.547	0.341	0.195	0.157	0.144	0.178
9				0.999	0.996	0.986	0.964	0.925	0.865	0.874	0.668	0.454	0.279	0.227	0.205	0.244
10				0.999	0.999	0.995	0.985	0.964	0.928	0.931	0.772	0.569	0.377	0.312	0.279	0.322
11					0.999	0.998	0.994	0.984	0.965	0.966	0.853	0.676	0.482	0.407	0.365	0.409
12						0.999	0.998	0.994	0.984	0.984	0.911	0.769	0.588	0.507	0.456	0.500
13							0.999	0.998	0.993	0.993	0.949	0.844	0.686	0.605	0.549	0.591
14								0.999	0.997	0.997	0.973	0.900	0.771	0.697	0.641	0.678
15									0.999	0.999	0.987	0.939	0.842	0.777	0.724	0.756
16									0.999	0.999	0.993	0.965	0.895	0.843	0.796	0.822
17											0.997	0.981	0.934	0.895	0.855	0.876
18											0.999	0.990	0.961	0.932	0.902	0.917
19											0.999	0.995	0.978	0.959	0.936	0.947
20												0.998	0.988	0.976	0.960	0.968
21												0.999	0.994	0.987	0.977	0.982
22													0.997	0.993	0.987	0.990
23													0.999	0.996	0.993	0.995
24														0.998	0.996	0.997
25														0.999	0.998	0.999
26															0.999	0.999

I \ J	0	0	0	0	0	0	0	1	1	2	5	8	12	16	24	34
								$n = 100; c = I + J$								
0	0.366	0.133	0.048	0.017	0.006	0.002	0.001	0.002	0.001	0.002	0.002	0.001	0.001	0.001	0.001	0.001
1	0.736	0.403	0.195	0.087	0.037	0.015	0.006	0.011	0.005	0.008	0.005	0.002	0.002	0.002	0.001	0.002
2	0.921	0.677	0.420	0.232	0.118	0.057	0.026	0.037	0.017	0.024	0.012	0.006	0.005	0.005	0.002	0.003
3	0.982	0.859	0.647	0.429	0.258	0.143	0.074	0.090	0.047	0.058	0.027	0.013	0.011	0.009	0.005	0.006
4	0.997	0.949	0.818	0.629	0.436	0.277	0.163	0.180	0.105	0.117	0.055	0.025	0.021	0.016	0.008	0.010
5	0.999	0.985	0.919	0.788	0.616	0.441	0.291	0.303	0.194	0.206	0.099	0.047	0.038	0.029	0.015	0.018
6	1.000	0.996	0.969	0.894	0.766	0.607	0.444	0.447	0.313	0.321	0.163	0.080	0.063	0.048	0.025	0.028
7		0.999	0.989	0.952	0.872	0.748	0.599	0.593	0.449	0.451	0.247	0.129	0.100	0.076	0.040	0.044
8		1.000	0.997	0.981	0.937	0.854	0.734	0.722	0.588	0.583	0.347	0.192	0.149	0.114	0.062	0.067
9			0.999	0.993	0.972	0.922	0.838	0.824	0.712	0.703	0.457	0.271	0.211	0.163	0.091	0.097
10			1.000	0.998	0.989	0.962	0.909	0.897	0.812	0.802	0.568	0.362	0.286	0.224	0.130	0.136
11				0.999	0.996	0.983	0.953	0.944	0.886	0.876	0.672	0.460	0.371	0.296	0.179	0.184
12				1.000	0.999	0.993	0.978	0.972	0.936	0.927	0.763	0.559	0.462	0.377	0.239	0.242
13					1.000	0.997	0.990	0.987	0.966	0.960	0.837	0.654	0.553	0.462	0.307	0.309
14						0.999	0.996	0.994	0.983	0.979	0.893	0.739	0.642	0.549	0.382	0.382
15						1.000	0.998	0.998	0.992	0.990	0.934	0.811	0.722	0.633	0.462	0.460
16							0.999	0.999	0.996	0.995	0.961	0.869	0.792	0.711	0.543	0.540
17							1.000	1.000	0.999	0.998	0.978	0.913	0.850	0.779	0.622	0.618
18									0.999	0.999	0.988	0.944	0.896	0.837	0.697	0.691
19									1.000	1.000	0.994	0.966	0.931	0.884	0.763	0.758
20											0.997	0.980	0.956	0.920	0.821	0.816
21											0.999	0.989	0.972	0.947	0.869	0.864
22											0.999	0.994	0.984	0.966	0.907	0.903
23											1.000	0.997	0.991	0.979	0.936	0.933
24												0.998	0.995	0.987	0.958	0.956
25												0.999	0.997	0.993	0.973	0.972
26												1.000	0.999	0.996	0.983	0.982
27													0.999	0.998	0.990	0.990
28														0.999	0.994	0.994
29															0.997	0.997
30															0.998	0.998
31															0.999	0.999
32															1.000	1.000

TABLE A.6 Poisson Probability Curves*

Probability of occurrence of c less defects in a sample of n

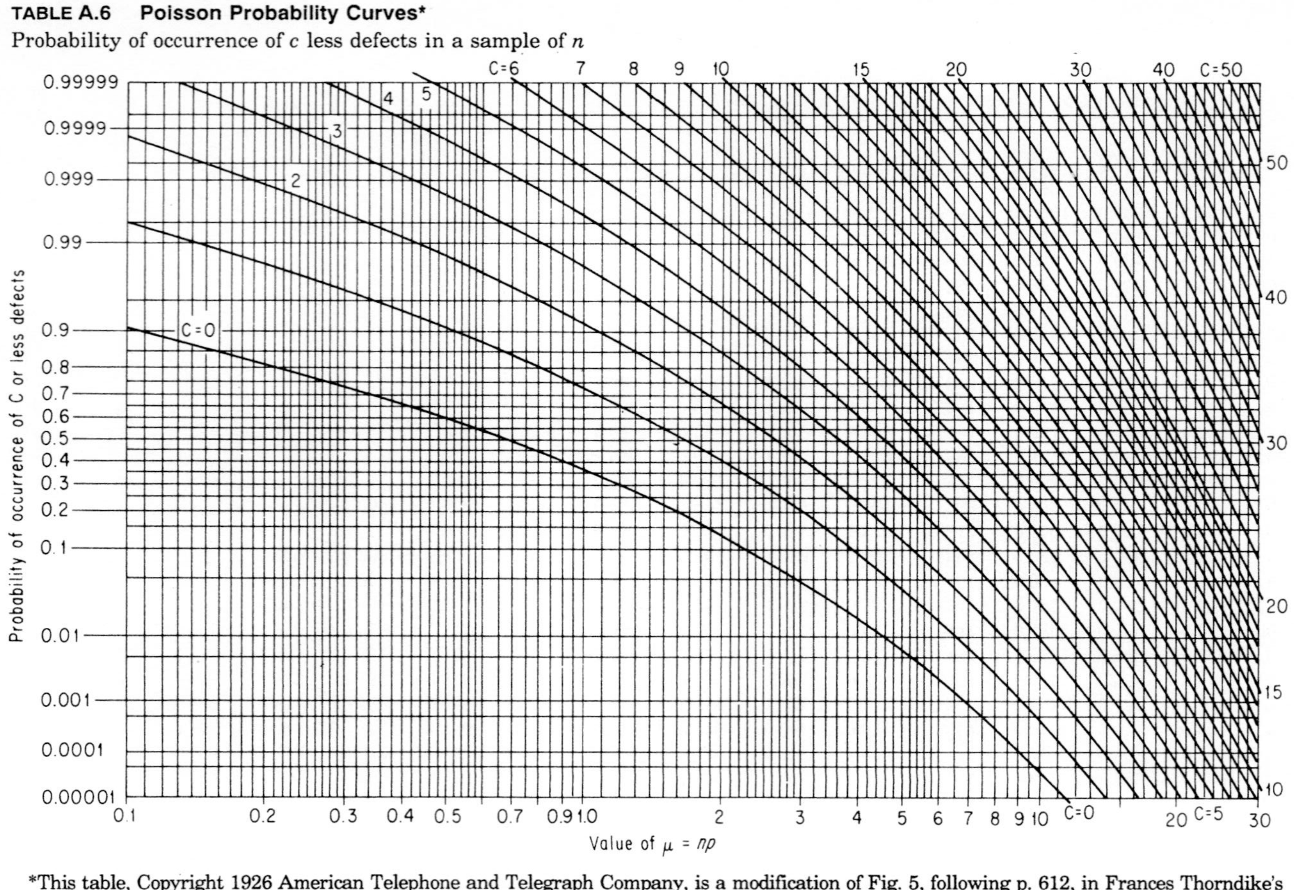

*This table, Copyright 1926 American Telephone and Telegraph Company, is a modification of Fig. 5, following p. 612, in Frances Thorndike's article, "Application of Poisson's Probability Summation," *The Bell System Technical Journal,* vol. 5, October 1926, and is reproduced by permission of the editor of *BSTJ*. It appears also as Fig. 2.6 on p. 35 of Harold F. Dodge and Harry G. Romig, *Sampling Inspection Tables,* 2d ed., John Wiley & Sons, Inc., New York, 1959, p. 35, copyright 1959 and has the permission of John Wiley & Sons, Inc., to be reproduced here.

TABLE A.7 Nonrandom Variability—Standard Given: df = ∞ (Two-Sided)

k	$Z_{.10}$	$Z_{.05}$	$Z_{.01}$
1	1.64	1.96	2.58
2	1.96	2.24	2.81
3	2.11	2.39	2.93
4	2.23	2.49	3.02
5	2.31	2.57	3.09
6	2.38	2.63	3.14
7	2.43	2.68	3.19
8	2.48	2.73	3.22
9	2.52	2.77	3.26
10	2.56	2.80	3.29
15	2.70	2.93	3.40
20	2.79	3.02	3.48
24	2.85	3.07	3.53
30	2.92	3.14	3.59
50	3.08	3.28	3.72
120	3.33	3.52	3.93

TABLE A.8 Exact Factors* for One-Way Analysis of Means, H_α (Two-Sided)

Significance Level = 0.10

	Number of Means, k																		
df	2	3	4	5	6	7	8	9	10	11	12	13	14	15	16	17	18	19	20
2	2.065																		
3	1.664	2.58																	
4	1.508	2.29	2.68																
5	1.425	2.15	2.49	2.73															
6	1.374	2.06	2.37	2.60	2.77														
7	1.340	1.99	2.29	2.51	2.67	2.80													
8	1.315	1.95	2.24	2.44	2.60	2.72	2.82												
9	1.296	1.91	2.20	2.40	2.55	2.67	2.76	2.84											
10	1.281	1.89	2.17	2.36	2.50	2.62	2.71	2.79	2.87										
11	1.270	1.87	2.14	2.33	2.46	2.58	2.68	2.75	2.82	2.88									
12	1.260	1.85	2.12	2.30	2.44	2.55	2.64	2.72	2.78	2.84	2.89								
13	1.252	1.84	2.10	2.28	2.42	2.53	2.61	2.69	2.75	2.81	2.86	2.91							
14	1.245	1.82	2.09	2.26	2.40	2.50	2.59	2.67	2.73	2.78	2.83	2.88	2.92						
15	1.240	1.81	2.07	2.25	2.38	2.48	2.57	2.64	2.70	2.77	2.81	2.85	2.90	2.94					
16	1.235	1.80	2.06	2.24	2.36	2.47	2.55	2.63	2.68	2.75	2.80	2.83	2.88	2.92	2.95				
17	1.230	1.80	2.05	2.23	2.36	2.45	2.54	2.61	2.68	2.73	2.78	2.82	2.86	2.90	2.93	2.96			
18	1.226	1.79	2.04	2.21	2.34	2.44	2.53	2.59	2.66	2.71	2.76	2.81	2.84	2.88	2.91	2.94	2.97		
19	1.223	1.78	2.04	2.20	2.33	2.43	2.52	2.58	2.65	2.70	2.75	2.79	2.83	2.86	2.90	2.93	2.95	2.98	
20	1.220	1.78	2.03	2.19	2.32	2.43	2.51	2.57	2.64	2.69	2.74	2.78	2.81	2.85	2.89	2.91	2.94	2.97	2.99
24	1.210	1.76	2.01	2.17	2.29	2.39	2.47	2.54	2.60	2.65	2.70	2.74	2.78	2.81	2.84	2.87	2.90	2.92	2.94
30	1.200	1.74	1.98	2.15	2.26	2.36	2.44	2.51	2.56	2.61	2.65	2.70	2.74	2.76	2.80	2.82	2.85	2.88	2.90
40	1.191	1.72	1.97	2.12	2.24	2.33	2.40	2.47	2.52	2.57	2.61	2.66	2.69	2.72	2.76	2.78	2.81	2.83	2.86
60	1.182	1.71	1.94	2.09	2.21	2.31	2.38	2.44	2.50	2.54	2.59	2.62	2.65	2.69	2.71	2.74	2.76	2.78	2.81
120	1.172	1.69	1.92	2.08	2.18	2.27	2.35	2.40	2.46	2.50	2.55	2.58	2.61	2.65	2.67	2.70	2.72	2.74	2.77
Inf	1.163	1.67	1.90	2.05	2.15	2.24	2.31	2.38	2.42	2.47	2.51	2.55	2.57	2.60	2.63	2.66	2.68	2.70	2.72
SG	1.949	2.11	2.23	2.31	2.38	2.43	2.48	2.52	2.56	2.59	2.62	2.65	2.67	2.70	2.72	2.74	2.76	2.77	2.79

	Significance Level = 0.05																		
	Number of Means, k																		
df	2	3	4	5	6	7	8	9	10	11	12	13	14	15	16	17	18	19	20
2	3.043																		
3	2.250	3.41																	
4	1.963	2.91	3.37																
5	1.818	2.65	3.06	3.33															
6	1.730	2.51	2.87	3.12	3.30														
7	1.672	2.40	2.75	2.98	3.15	3.30													
8	1.631	2.34	2.66	2.87	3.04	3.18	3.28												
9	1.599	2.28	2.59	2.80	2.96	3.08	3.19	3.28											
10	1.575	2.24	2.54	2.75	2.89	3.02	3.11	3.21	3.27										
11	1.556	2.20	2.49	2.69	2.85	2.96	3.06	3.14	3.22	3.28									
12	1.541	2.18	2.47	2.66	2.80	2.92	3.01	3.09	3.16	3.22	3.27								
13	1.527	2.16	2.43	2.63	2.77	2.88	2.97	3.05	3.12	3.18	3.24	3.29							
14	1.517	2.14	2.42	2.60	2.74	2.85	2.94	3.02	3.08	3.15	3.20	3.24	3.29						
15	1.507	2.12	2.39	2.58	2.71	2.82	2.91	2.99	3.05	3.11	3.16	3.21	3.25	3.28					
16	1.499	2.11	2.37	2.56	2.69	2.80	2.89	2.96	3.03	3.08	3.13	3.18	3.22	3.26	3.29				
17	1.492	2.10	2.36	2.54	2.67	2.78	2.86	2.94	3.00	3.06	3.11	3.15	3.19	3.23	3.26	3.30			
18	1.486	2.08	2.35	2.52	2.66	2.76	2.84	2.92	2.98	3.03	3.08	3.13	3.17	3.21	3.24	3.27	3.30		
19	1.480	2.07	2.34	2.51	2.64	2.74	2.82	2.90	2.96	3.01	3.06	3.11	3.15	3.19	3.21	3.25	3.28	3.31	
20	1.475	2.07	2.32	2.50	2.63	2.73	2.82	2.88	2.95	3.00	3.04	3.09	3.13	3.17	3.20	3.23	3.26	3.28	3.31
24	1.459	2.04	2.29	2.46	2.58	2.68	2.77	2.84	2.89	2.95	3.00	3.04	3.07	3.11	3.14	3.17	3.20	3.22	3.25
30	1.444	2.02	2.26	2.42	2.55	2.64	2.72	2.79	2.85	2.90	2.94	2.98	3.02	3.05	3.08	3.10	3.13	3.16	3.19
40	1.429	1.98	2.23	2.39	2.51	2.60	2.68	2.74	2.80	2.84	2.88	2.92	2.96	2.99	3.02	3.05	3.07	3.10	3.12
60	1.414	1.96	2.20	2.35	2.46	2.56	2.63	2.70	2.75	2.79	2.83	2.87	2.91	2.94	2.96	2.99	3.01	3.04	3.06
120	1.400	1.94	2.17	2.32	2.43	2.52	2.59	2.65	2.69	2.75	2.79	2.82	2.85	2.88	2.90	2.93	2.95	2.98	3.00
Inf	1.386	1.91	2.14	2.29	2.39	2.48	2.54	2.60	2.66	2.70	2.74	2.77	2.79	2.83	2.86	2.88	2.90	2.92	2.94
SG	2.236	2.39	2.49	2.57	2.63	2.68	2.73	2.77	2.80	2.83	2.86	2.88	2.91	2.93	2.95	2.97	2.98	3.00	3.02

TABLE A.8 Exact Factors* for One-Way Analysis of Means, H_α (Two-Sided) *(Continued)*

Significance Level = 0.01

	Number of Means, k																		
df	2	3	4	5	6	7	8	9	10	11	12	13	14	15	16	17	18	19	20
2	7.018																		
3	4.130	6.13																	
4	3.256	4.69	5.38																
5	2.851	4.03	4.58	4.96															
6	2.621	3.66	4.13	4.45	4.71														
7	2.474	3.41	3.85	4.14	4.36	4.54													
8	2.372	3.25	3.65	3.92	4.13	4.29	4.42												
9	2.298	3.14	3.51	3.76	3.95	4.10	4.22	4.33											
10	2.241	3.05	3.39	3.64	3.82	3.96	4.08	4.18	4.26										
11	2.196	2.97	3.31	3.54	3.72	3.85	3.96	4.05	4.14	4.20									
12	2.160	2.91	3.24	3.46	3.63	3.76	3.86	3.96	4.03	4.11	4.16								
13	2.130	2.87	3.19	3.40	3.56	3.68	3.79	3.87	3.96	4.02	4.08	4.13							
14	2.105	2.83	3.14	3.35	3.51	3.63	3.72	3.81	3.88	3.95	4.00	4.05	4.11						
15	2.084	2.79	3.10	3.30	3.46	3.57	3.67	3.75	3.82	3.89	3.94	4.00	4.04	4.08					
16	2.065	2.76	3.07	3.26	3.41	3.53	3.62	3.71	3.78	3.83	3.89	3.94	3.99	4.03	4.07				
17	2.049	2.74	3.03	3.23	3.38	3.49	3.58	3.67	3.73	3.79	3.85	3.89	3.94	3.98	4.01	4.05			
18	2.035	2.72	3.01	3.20	3.34	3.45	3.55	3.63	3.69	3.76	3.80	3.85	3.89	3.93	3.97	4.00	4.03		
19	2.023	2.69	2.99	3.18	3.31	3.43	3.52	3.59	3.66	3.72	3.77	3.81	3.85	3.89	3.93	3.96	3.99	4.02	
20	2.012	2.68	2.96	3.16	3.30	3.40	3.49	3.56	3.63	3.69	3.73	3.79	3.83	3.86	3.89	3.93	3.96	3.98	4.02
24	1.978	2.62	2.90	3.09	3.21	3.31	3.40	3.48	3.54	3.59	3.64	3.68	3.72	3.76	3.79	3.82	3.85	3.87	3.90
30	1.945	2.57	2.84	3.01	3.14	3.24	3.32	3.38	3.44	3.50	3.54	3.58	3.62	3.65	3.69	3.72	3.74	3.77	3.79
40	1.912	2.52	2.78	2.94	3.07	3.17	3.24	3.30	3.36	3.41	3.45	3.49	3.53	3.56	3.58	3.61	3.63	3.66	3.68
60	1.881	2.47	2.72	2.88	3.00	3.09	3.16	3.22	3.28	3.33	3.36	3.40	3.43	3.47	3.50	3.52	3.54	3.56	3.59
120	1.850	2.42	2.66	2.82	2.93	3.02	3.09	3.15	3.20	3.24	3.27	3.31	3.34	3.37	3.40	3.42	3.45	3.47	3.49
Inf	1.822	2.38	2.61	2.75	2.87	2.94	3.01	3.07	3.12	3.17	3.20	3.23	3.26	3.28	3.31	3.34	3.35	3.38	3.39
SG	2.806	2.93	3.02	3.09	3.14	3.19	3.23	3.26	3.29	3.32	3.34	3.36	3.38	3.40	3.42	3.44	3.45	3.47	3.48

Significance Level = 0.001

	Number of Means, k																		
df	2	3	4	5	6	7	8	9	10	11	12	13	14	15	16	17	18	19	20
2	22.343																		
3	9.139	13.39																	
4	6.088	8.65	9.87																
5	4.857	6.74	7.61	8.22															
6	4.214	5.75	6.45	6.94	7.30														
7	3.824	5.15	5.76	6.16	6.47	6.71													
8	3.565	4.76	5.30	5.65	5.92	6.14	6.31												
9	3.381	4.48	4.97	5.30	5.54	5.74	5.89	6.03											
10	3.243	4.28	4.73	5.04	5.26	5.43	5.58	5.70	5.82										
11	3.137	4.12	4.55	4.83	5.04	5.21	5.34	5.46	5.56	5.64									
12	3.053	3.99	4.40	4.67	4.87	5.03	5.15	5.26	5.36	5.44	5.51								
13	2.985	3.89	4.29	4.54	4.73	4.88	5.00	5.11	5.20	5.27	5.34	5.41							
14	2.927	3.80	4.18	4.44	4.62	4.76	4.87	4.98	5.06	5.13	5.20	5.27	5.31						
15	2.880	3.73	4.10	4.35	4.52	4.66	4.77	4.86	4.94	5.02	5.08	5.14	5.19	5.24					
16	2.839	3.67	4.04	4.27	4.44	4.57	4.68	4.77	4.85	4.92	4.98	5.03	5.09	5.13	5.17				
17	2.804	3.63	3.98	4.20	4.36	4.50	4.60	4.70	4.77	4.83	4.89	4.95	4.99	5.04	5.08	5.12			
18	2.773	3.58	3.92	4.14	4.31	4.43	4.54	4.62	4.70	4.76	4.82	4.87	4.91	4.97	5.00	5.04	5.07		
19	2.746	3.54	3.87	4.10	4.25	4.38	4.48	4.56	4.63	4.70	4.75	4.80	4.85	4.89	4.93	4.97	5.00	5.03	
20	2.722	3.50	3.83	4.05	4.21	4.32	4.42	4.51	4.58	4.64	4.69	4.75	4.79	4.83	4.87	4.90	4.94	4.96	4.99
24	2.648	3.40	3.71	3.91	4.06	4.18	4.27	4.35	4.41	4.47	4.52	4.56	4.61	4.65	4.68	4.71	4.74	4.77	4.80
30	2.578	3.29	3.59	3.78	3.93	4.03	4.12	4.19	4.25	4.30	4.35	4.39	4.43	4.46	4.49	4.53	4.56	4.58	4.60
40	2.511	3.19	3.47	3.66	3.79	3.89	3.98	4.04	4.10	4.15	4.19	4.23	4.27	4.30	4.33	4.36	4.37	4.40	4.43
60	2.447	3.10	3.37	3.54	3.67	3.76	3.84	3.90	3.96	4.00	4.04	4.07	4.11	4.14	4.16	4.19	4.21	4.23	4.26
120	2.385	3.01	3.26	3.43	3.55	3.64	3.70	3.77	3.82	3.86	3.90	3.93	3.96	3.99	4.02	4.04	4.05	4.08	4.10
Inf	2.327	2.92	3.17	3.33	3.43	3.52	3.59	3.65	3.69	3.73	3.76	3.80	3.83	3.85	3.87	3.90	3.92	3.93	3.96
SG	3.481	3.59	3.66	3.72	3.76	3.80	3.84	3.86	3.89	3.91	3.93	3.95	3.97	3.99	4.00	4.02	4.03	4.04	4.06

*The values for $k \geq 3$ in this table are exact values for the studentized maximum absolute deviate from the sample mean in normal samples, H_α, and represent modifications by E. G. Schilling and D. Smialek ("Simplified Analysis of Means for Crossed and Nested Experiments," *Proc. 43d Ann. Qual. Control Conf.*, Rochester Section, ASQC, March 10, 1987) of the exact values for the studentized maximum deviation from the population mean h_α calculated by L. S. Nelson ("Exact Critical Values for Use with the Analysis of Means," *J. Qual. Tech.*, vol. 15, no. 1, January 1983, pp. 40–44) using the relationship $H_\alpha = h_\alpha\sqrt{(k-1)/k}$. The values for $k = 2$ are from the Student's t distribution as calculated by Ott (E. R. Ott, "Analysis of Means," *Rutgers U. Stat. Cent. Tech. Rep. no. 1*, August 10, 1958).

TABLE A.9 Dixon Criteria for Testing Extreme Mean or Individual*

Statistic	No. of obs., k	$\alpha = 0.10$ P_{90}	$\alpha = 0.05$ P_{95}	$\alpha = 0.02$ P_{98}	$\alpha = 0.01$ P_{99}
$r_{10} = \frac{X_2 - X_1}{X_k - X_1}$	3	0.886	0.941	0.976	0.988
	4	0.679	0.765	0.846	0.889
	5	0.557	0.642	0.729	0.780
	6	0.482	0.560	0.644	0.698
	7	0.434	0.507	0.586	0.637
$r_{11} = \frac{X_2 - X_1}{X_{k-1} - X_1}$	8	0.479	0.554	0.631	0.683
	9	0.441	0.512	0.587	0.635
	10	0.409	0.477	0.551	0.597
$r_{21} = \frac{X_3 - X_1}{X_{k-1} - X_1}$	11	0.517	0.576	0.638	0.679
	12	0.490	0.546	0.605	0.642
	13	0.467	0.521	0.578	0.615
$r_{22} = \frac{X_3 - X_1}{X_{k-2} - X_1}$	14	0.492	0.546	0.602	0.641
	15	0.472	0.525	0.579	0.616
	16	0.454	0.507	0.559	0.595
	17	0.438	0.490	0.542	0.577
	18	0.424	0.475	0.527	0.561
	19	0.412	0.462	0.514	0.547
	20	0.401	0.450	0.502	0.535
	21	0.391	0.440	0.491	0.524
	22	0.382	0.430	0.481	0.514
	23	0.374	0.421	0.472	0.505
	24	0.367	0.413	0.464	0.497
	25	0.360	0.406	0.457	0.489

Note that:
X_1 = Suspected outlier
X_2 = Result next to suspect
X_k = Farthest from suspect
SOURCE: W. J. Dixon, "Processing Data for Outliers," *Biometrics,* vol. 9, no. 1, pp. 74-89. (Reprinted by permission of the editor of *Biometrics.*)

TABLE A.10 Grubbs Criteria for Simultaneously Testing the Two Largest or Two Smallest Observations

Compare computed values of $S^2_{n-1,n}/S^2$ or $S^2_{1,2}/S^2$ with the appropriate critical ratio in this table; smaller observed sample ratios call for rejection. $X_1 \le X_2 \le \ldots \le X_n$

Number of observations	10% level	5% level	1% level
4	.0031	.0008	.0000
5	.0376	.0183	.0035
6	.0921	.0565	.0186
7	.1479	.1020	.0440
8	.1994	.1478	.0750
9	.2454	.1909	.1082
10	.2853	.2305	.1415
11	.3226	.2666	.1736
12	.3552	.2996	.2044
13	.3843	.3295	.2333
14	.4106	.3568	.2605
15	.4345	.3818	.2859
16	.4562	.4048	.3098
17	.4761	.4259	.3321
18	.4944	.4455	.3530
19	.5113	.4636	.3725
20	.5269	.4804	.3909

$$S^2 = \sum_{i=1}^{n} (X_i - \bar{X})^2 \qquad \bar{\bar{X}} = \sum_{i=1}^{n} X_i/n$$

$$S^2_{1,2} = \sum_{i=3}^{n} (X_i - \bar{X}_{1,2})^2 \qquad \bar{X}_{1,2} = \sum_{i=3}^{n} X_i/(n-2)$$

$$S^2_{n-1,n} = \sum_{i=1}^{n-2} (X_i - \bar{X}_{n-1,n})^2 \qquad \bar{X}_{n-1,n} = \sum_{i=1}^{n-2} X_i/(n-2)$$

SOURCE: Frank E. Grubbs, Procedures for Detecting Outlying Observations in Samples, *Technometrics*, vol. 11, no. 1, pp. 1-21, February 1969. (Reproduced by permission of the editor.)

TABLE A.11 Values of Adjusted d_2 Factor (d_2*) and Degrees of Freedom (df)

To be used with estimates of σ based on k independent sample ranges of n_g each. (Unbiased estimate of σ^2 is $(\overline{R}/d_2^*)^2$; unbiased estimate of σ is $\overline{R}/d_2$.) Also, df $\cong (0.9)k(n_g - 1)$.

	Subgroup Size											
k	$n_g = 2$		$n_g = 3$		$n_g = 4$		$n_g = 5$		$n_g = 6$		$n_g = 7$	
Number of samples	df	d_2^*	df	d_2^*	df	d_2^*	df	d_2^*	df	d_2^*	df	d_2^*
1	1.0	1.41	2.0	1.91	2.9	2.24	3.8	2.48	4.7	2.67	5.5	2.83
2	1.9	1.28	3.8	1.81	5.7	2.15	7.5	2.40	9.2	2.60	10.8	2.77
3	2.8	1.23	5.7	1.77	8.4	2.12	11.1	2.38	13.6	2.58	16.0	2.75
4	3.7	1.21	7.5	1.75	11.2	2.11	14.7	2.37	18.1	2.57	21.3	2.74
5	4.6	1.19	9.3	1.74	13.9	2.10	18.4	2.36	22.6	2.56	26.6	2.73
6	5.5	1.18	11.1	1.73	16.6	2.09	22.9	2.35	27.1	2.56	31.8	2.73
7	6.4	1.17	12.9	1.73	19.4	2.09	25.6	2.35	31.5	2.55	37.1	2.72
8	7.2	1.17	14.8	1.72	22.1	2.08	29.3	2.35	36.0	2.55	42.4	2.72
9	8.1	1.16	16.6	1.72	24.8	2.08	32.9	2.34	40.5	2.55	47.7	2.72
10	9.0	1.16	18.4	1.72	27.6	2.08	36.5	2.34	44.9	2.55	52.9	2.72
11	9.9	1.16	20.2	1.71	30.3	2.08	40.1	2.34	49.4	2.55	58.2	2.72
12	10.8	1.15	22.0	1.71	33.0	2.07	43.7	2.34	53.9	2.55	63.5	2.72
13	11.6	1.15	23.9	1.71	35.7	2.07	47.4	2.34	58.4	2.55	68.8	2.71
14	12.5	1.15	25.7	1.71	38.5	2.07	51.0	2.34	62.8	2.54	74.0	2.71
15	13.4	1.15	27.5	1.71	41.2	2.07	54.6	2.34	67.3	2.54	79.3	2.71
16	14.3	1.15	29.3	1.71	43.9	2.07	58.2	2.34	71.8	2.54	84.6	2.71
17	15.2	1.15	31.1	1.71	46.7	2.07	61.8	2.34	76.2	2.54	89.8	2.71
18	16.0	1.15	33.0	1.71	49.4	2.07	65.5	2.33	80.7	2.54	95.1	2.71
19	16.9	1.14	34.8	1.70	52.2	2.07	69.1	2.33	85.2	2.54	100.4	2.71
20	17.8	1.14	36.6	1.70	54.9	2.07	72.7	2.33	89.7	2.54	105.7	2.71
25	22.2	1.14	45.7	1.70	68.5	2.07						
30	26.6	1.14	54.8	1.70								
50	44.2	1.13										
∞	d_2	1.128		1.6926		2.0588		2.3258		2.5344		2.7044

SOURCE: Acheson J. Duncan, "The Use of Ranges in Comparing Variabilities," *Ind. Qual. Control,* vol. 40, no. 5, February 1955; no. 8, May 1955, p. 70. (Reprinted by permission of the editor.)

TABLE A.12a ***F* Distribution, Upper 5 Percent Points ($F_{0.95}$) (One-Sided)**

ν_2 \ ν_1	1	2	3	4	5	6	7	8	9	10	12	15	20	24	30	40	60	120	∞
1	161.4	199.5	215.7	224.6	230.2	234.0	236.8	238.9	240.5	241.9	243.9	245.9	248.0	249.1	250.1	251.1	252.2	253.3	254.3
2	18.51	19.00	19.16	19.25	19.30	19.33	19.35	19.37	19.38	19.40	19.41	19.43	19.45	19.45	19.46	19.47	19.48	19.49	19.50
3	10.13	9.55	9.28	9.12	9.01	8.94	8.89	8.85	8.81	8.79	8.74	8.70	8.68	8.64	8.62	8.59	8.57	8.55	8.53
4	7.71	6.94	6.59	6.39	6.26	6.16	6.08	6.04	6.00	5.96	5.91	5.86	5.80	5.77	5.75	5.72	5.69	5.66	5.63
5	6.61	5.79	5.41	5.19	5.05	4.95	4.88	4.82	4.77	4.74	4.68	4.62	4.56	4.53	4.50	4.46	4.43	4.40	4.36
6	5.99	5.14	4.76	4.53	4.39	4.28	4.21	4.15	4.10	4.06	4.00	3.94	3.87	3.84	3.81	3.77	3.74	3.70	3.67
7	5.59	4.74	4.35	4.12	3.97	3.87	3.79	3.73	3.68	3.64	3.57	3.51	3.44	3.41	3.38	3.34	3.30	3.27	3.23
8	5.32	4.46	4.07	3.84	3.69	3.58	3.50	3.44	3.39	3.35	3.28	3.22	3.15	3.12	3.08	3.04	3.01	2.97	2.93
9	5.12	4.26	3.86	3.63	3.48	3.37	3.29	3.23	3.18	3.14	3.07	3.01	2.94	2.90	2.86	2.83	2.79	2.75	2.71
10	4.96	4.10	3.71	3.48	3.33	3.22	3.14	3.07	3.02	2.98	2.91	2.85	2.77	2.74	2.70	2.66	2.62	2.58	2.54
11	4.84	3.98	3.59	3.36	3.20	3.09	3.01	2.95	2.90	2.85	2.79	2.72	2.65	2.61	2.57	2.53	2.49	2.45	2.40
12	4.75	3.89	3.49	3.26	3.11	3.00	2.91	2.85	2.80	2.75	2.69	2.62	2.54	2.51	2.47	2.43	2.38	2.34	2.30
13	4.67	3.81	3.41	3.18	3.03	2.92	2.83	2.77	2.71	2.67	2.60	2.53	2.46	2.42	2.38	2.34	2.30	2.25	2.21
14	4.60	3.74	3.34	3.11	2.96	2.85	2.76	2.70	2.65	2.60	2.53	2.46	2.39	2.35	2.31	2.27	2.22	2.18	2.13
15	4.54	3.68	3.29	3.06	2.90	2.79	2.71	2.64	2.59	2.54	2.48	2.40	2.33	2.29	2.25	2.20	2.16	2.11	2.07
16	4.49	3.63	3.24	3.01	2.85	2.74	2.66	2.59	2.54	2.49	2.42	2.35	2.28	2.24	2.19	2.15	2.11	2.06	2.01
17	4.45	3.59	3.20	2.96	2.81	2.70	2.61	2.55	2.49	2.45	2.38	2.31	2.23	2.19	2.15	2.10	2.06	2.01	1.96
18	4.41	3.55	3.16	2.93	2.77	2.66	2.58	2.51	2.46	2.41	2.34	2.27	2.19	2.15	2.11	2.06	2.02	1.97	1.92
19	4.38	3.52	3.13	2.90	2.74	2.63	2.54	2.48	2.42	2.38	2.31	2.23	2.16	2.11	2.07	2.03	1.98	1.93	1.88
20	4.35	3.49	3.10	2.87	2.71	2.60	2.51	2.45	2.39	2.35	2.28	2.20	2.12	2.08	2.04	1.99	1.95	1.90	1.84
21	4.32	3.47	3.07	2.84	2.68	2.57	2.49	2.42	2.37	2.32	2.25	2.18	2.10	2.05	2.01	1.95	1.92	1.87	1.81
22	4.30	3.44	3.05	2.82	2.66	2.55	2.46	2.40	2.34	2.30	2.23	2.15	2.07	2.03	1.98	1.94	1.89	1.84	1.78
23	4.28	3.42	3.03	2.80	2.64	2.53	2.44	2.37	2.32	2.27	2.20	2.13	2.05	2.01	1.96	1.91	1.86	1.81	1.76
24	4.26	3.40	3.01	2.78	2.62	2.51	2.42	2.36	2.30	2.25	2.18	2.11	2.03	1.98	1.94	1.89	1.84	1.79	1.73

TABLE A.12a ***F* Distribution, Upper 5 Percent Points ($F_{0.95}$) (One-Sided) *(Continued)***

ν_2 \ ν_1	1	2	3	4	5	6	7	8	9	10	12	15	20	24	30	40	60	120	∞
25	4.24	3.39	2.99	2.76	2.60	2.49	2.40	2.34	2.28	2.24	2.16	2.09	2.01	1.96	1.92	1.87	1.82	1.77	1.71
26	4.23	3.37	2.98	2.74	2.59	2.47	2.39	2.32	2.27	2.22	2.15	2.07	1.99	1.95	1.90	1.85	1.80	1.75	1.69
27	4.21	3.35	2.96	2.73	2.57	2.46	2.37	2.31	2.25	2.20	2.13	2.06	1.97	1.93	1.88	1.84	1.79	1.73	1.67
28	4.20	3.34	2.95	2.71	2.56	2.45	2.36	2.29	2.24	2.19	2.12	2.04	1.96	1.91	1.87	1.82	1.77	1.71	1.65
29	4.18	3.33	2.93	2.70	2.55	2.43	2.35	2.28	2.22	2.18	2.10	2.03	1.94	1.90	1.85	1.81	1.75	1.70	1.64
30	4.17	3.32	2.92	2.69	2.53	2.42	2.33	2.27	2.21	2.16	2.09	2.01	1.93	1.89	1.84	1.79	1.74	1.68	1.62
40	4.08	3.23	2.84	2.61	2.45	2.34	2.25	2.18	2.12	2.08	2.00	1.92	1.84	1.79	1.74	1.69	1.64	1.58	1.51
60	4.00	3.15	2.76	2.53	2.37	2.25	2.17	2.10	2.04	1.99	1.92	1.84	1.75	1.70	1.65	1.59	1.53	1.47	1.39
120	3.92	3.07	2.68	2.45	2.29	2.17	2.09	2.02	1.96	1.91	1.83	1.75	1.66	1.61	1.55	1.50	1.43	1.35	1.25
∞	3.84	3.00	2.60	2.37	2.21	2.10	2.01	1.94	1.88	1.83	1.75	1.67	1.57	1.52	1.46	1.39	1.32	1.22	1.00

SOURCE: E. S. Pearson and H. O. Hartley, *Biometrika Tables for Statisticians*, 3d ed., University College, London, 1966. (Reproduced by permission of the Biometrica trustees.)

TABLE A.12b *F* **Distribution, Upper 2.5 Percent Points ($F_{0.975}$) (One-Sided)**

ν_2 \ ν_1	1	2	3	4	5	6	7	8	9	10	12	15	20	24	30	40	60	120	∞
1	647.8	799.5	864.2	899.6	921.8	937.1	948.2	956.7	963.3	968.6	976.7	984.9	993.1	997.2	1001	1006	1010	1014	1018
2	38.51	39.00	39.17	39.25	39.30	39.33	39.36	39.37	39.39	39.40	39.41	39.43	39.45	39.46	39.46	39.47	39.48	39.49	39.50
3	17.44	16.04	15.44	15.10	14.88	14.73	14.62	14.54	14.47	14.42	14.34	14.25	14.17	14.12	14.08	14.04	13.99	13.95	13.90
4	12.22	10.65	9.98	9.60	9.36	9.20	9.07	8.98	8.90	8.84	8.75	8.66	8.56	8.51	8.46	8.41	8.36	8.31	8.26
5	10.01	8.43	7.76	7.39	7.15	6.98	6.85	6.76	6.68	6.62	6.52	6.43	6.33	6.28	6.23	6.18	6.12	6.07	6.02
6	8.81	7.26	6.60	6.23	5.99	5.82	5.70	5.60	5.52	5.46	5.37	5.27	5.17	5.12	5.07	5.01	4.96	4.90	4.85
7	8.07	6.54	5.89	5.52	5.29	5.12	4.99	4.90	4.82	4.76	4.67	4.57	4.47	4.42	4.36	4.31	4.25	4.20	4.14
8	7.57	6.06	5.42	5.05	4.82	4.65	4.53	4.43	4.36	4.30	4.20	4.10	4.00	3.95	3.89	3.84	3.78	3.73	3.67
9	7.21	5.71	5.08	4.72	4.48	4.32	4.20	4.10	4.03	3.96	3.87	3.77	3.67	3.61	3.56	3.51	3.45	3.39	3.33
10	6.94	5.46	4.83	4.47	4.24	4.07	3.95	3.85	3.78	3.72	3.62	3.52	3.42	3.37	3.31	3.26	3.20	3.14	3.08
11	6.72	5.26	4.63	4.28	4.04	3.88	3.76	3.66	3.59	3.53	3.43	3.33	3.23	3.17	3.12	3.06	3.00	2.94	2.88
12	6.55	5.10	4.47	4.12	3.89	3.73	3.61	3.51	3.44	3.37	3.28	3.18	3.07	3.02	2.96	2.91	2.85	2.79	2.72
13	6.41	4.97	4.35	4.00	3.77	3.60	3.48	3.39	3.31	3.25	3.15	3.05	2.95	2.89	2.84	2.78	2.72	2.66	2.60
14	6.30	4.86	4.24	3.89	3.66	3.50	3.38	3.29	3.21	3.15	3.05	2.95	2.84	2.79	2.73	2.67	2.61	2.55	2.49
15	6.20	4.77	4.15	3.80	3.58	3.41	3.29	3.20	3.12	3.06	2.96	2.86	2.76	2.70	2.64	2.59	2.52	2.46	2.40
16	6.12	4.69	4.08	3.73	3.50	3.34	3.22	3.12	3.05	2.99	2.89	2.79	2.68	2.63	2.57	2.51	2.45	2.38	2.32
17	6.04	4.62	4.01	3.66	3.44	3.28	3.16	3.06	2.98	2.92	2.82	2.72	2.62	2.56	2.50	2.44	2.38	2.32	2.25
18	5.98	4.56	3.95	3.61	3.38	3.22	3.10	3.01	2.93	2.87	2.77	2.67	2.56	2.50	2.44	2.38	2.32	2.26	2.19
19	5.92	4.51	3.90	3.56	3.33	3.17	3.05	2.96	2.88	2.82	2.72	2.62	2.51	2.45	2.39	2.33	2.27	2.20	2.13
20	5.87	4.46	3.86	3.51	3.29	3.13	3.01	2.91	2.84	2.77	2.68	2.57	2.46	2.41	2.35	2.29	2.22	2.16	2.09
21	5.83	4.42	3.82	3.48	3.25	3.09	2.97	2.87	2.80	2.73	2.64	2.53	2.42	2.37	2.31	2.25	2.18	2.11	2.04
22	5.79	4.38	3.78	3.44	3.22	3.05	2.93	2.84	2.76	2.70	2.60	2.50	2.39	2.33	2.27	2.21	2.14	2.08	2.00
23	5.75	4.35	3.75	3.41	3.18	3.02	2.90	2.81	2.73	2.67	2.57	2.47	2.36	2.30	2.24	2.18	2.11	2.04	1.97
24	5.72	4.32	3.72	3.38	3.15	2.99	2.87	2.78	2.70	2.64	2.54	2.44	2.33	2.27	2.21	2.15	2.08	2.01	1.94

TABLE 12b ***F* Distribution, Upper 2.5 Percent Points ($F_{0.975}$) (One-Sided) *(Continued)***

ν_2 \ ν_1	1	2	3	4	5	6	7	8	9	10	12	15	20	24	30	40	60	120	∞
25	5.69	4.29	3.69	3.35	3.13	2.97	2.85	2.75	2.68	2.61	2.51	2.41	2.30	2.24	2.18	2.12	2.05	1.98	1.91
26	5.66	4.27	3.67	3.33	3.10	2.94	2.82	2.73	2.65	2.59	2.49	2.39	2.28	2.22	2.16	2.09	2.03	1.95	1.88
27	5.63	4.24	3.65	3.31	3.08	2.92	2.80	2.71	2.63	2.57	2.47	2.36	2.25	2.19	2.13	2.07	2.00	1.93	1.85
28	5.61	4.22	3.63	3.29	3.06	2.90	2.78	2.69	2.61	2.55	2.45	2.34	2.23	2.17	2.11	2.05	1.98	1.91	1.83
29	5.59	4.20	3.61	3.27	3.04	2.88	2.76	2.67	2.59	2.53	2.43	2.32	2.21	2.15	2.09	2.03	1.96	1.89	1.81
30	5.57	4.18	3.59	3.25	3.03	2.87	2.75	2.65	2.57	2.51	2.41	2.31	2.20	2.14	2.07	2.01	1.94	1.87	1.79
40	5.42	4.05	3.46	3.13	2.90	2.74	2.62	2.53	2.45	2.39	2.29	2.18	2.07	2.01	1.94	1.88	1.80	1.72	1.64
60	5.29	3.93	3.34	3.01	2.79	2.63	2.51	2.41	2.33	2.27	2.17	2.06	1.94	1.88	1.82	1.74	1.67	1.58	1.48
120	5.15	3.80	3.23	2.89	2.67	2.52	2.39	2.30	2.22	2.16	2.05	1.94	1.82	1.76	1.69	1.61	1.53	1.43	1.31
∞	5.02	3.69	3.12	2.79	2.57	2.41	2.29	2.19	2.11	2.05	1.94	1.83	1.71	1.64	1.57	1.48	1.39	1.27	1.00

SOURCE: E. S. Pearson and H. O. Hartley, *Biometrika Tables for Statisticians*, 3d ed., University College, London, 1966. (Reproduced by permission of the Biometrica trustees.)

TABLE A.12c ***F* Distribution, Upper 1 Percent Points ($F_{0.99}$) (One-Sided)**

ν_2 \ ν_1	1	2	3	4	5	6	7	8	9	10	12	15	20	24	30	40	60	120	∞
1	4052	4999.5	5403	5625	5764	5859	5928	5982	6022	6056	6106	6157	6209	6235	6261	6287	6313	6339	6366
2	98.50	99.00	99.17	99.25	99.30	99.33	99.36	99.37	99.39	99.40	99.42	99.43	99.45	99.46	99.47	99.47	99.48	99.49	99.50
3	34.12	30.82	29.46	28.71	28.24	27.91	27.67	27.49	27.35	27.23	27.05	26.87	26.69	26.60	26.50	26.41	26.32	26.22	26.13
4	21.20	18.00	16.69	15.98	15.52	15.21	14.98	14.80	14.66	14.55	14.37	14.20	14.02	13.93	13.84	13.75	13.65	13.56	13.46
5	16.26	13.27	12.06	11.39	10.97	10.67	10.46	10.29	10.16	10.05	9.89	9.72	9.55	9.47	9.38	9.29	9.20	9.11	9.02
6	13.75	10.92	9.78	9.15	8.75	8.47	8.26	8.10	7.98	7.87	7.72	7.53	7.40	7.31	7.23	7.14	7.06	6.97	6.88
7	12.25	9.55	8.45	7.85	7.46	7.19	6.99	6.84	6.72	6.62	6.47	6.31	6.16	6.07	5.99	5.91	5.82	5.74	5.65
8	11.26	8.65	7.59	7.01	6.63	6.37	6.18	6.03	5.91	5.81	5.67	5.52	5.35	5.28	5.20	5.12	5.03	4.95	4.86
9	10.56	8.02	6.99	6.42	6.06	5.80	5.61	5.47	5.35	5.26	5.11	4.96	4.81	4.73	4.65	4.57	4.48	4.40	4.31
10	10.04	7.56	6.55	5.99	5.64	5.39	5.20	5.06	4.94	4.85	4.71	4.56	4.41	4.33	4.25	4.17	4.08	4.00	3.91
11	9.65	7.21	6.22	5.67	5.32	5.07	4.89	4.74	4.63	4.54	4.40	4.25	4.10	4.02	3.94	3.86	3.78	3.69	3.60
12	9.33	6.93	5.95	5.41	5.06	4.82	4.64	4.50	4.39	4.30	4.16	4.01	3.83	3.78	3.70	3.62	3.54	3.45	3.36
13	9.07	6.70	5.74	5.21	4.86	4.62	4.44	4.30	4.19	4.10	3.96	3.82	3.66	3.59	3.51	3.43	3.34	3.25	3.17
14	8.86	6.51	5.56	5.04	4.69	4.46	4.28	4.14	4.03	3.94	3.80	3.66	3.51	3.43	3.35	3.27	3.18	3.09	3.00
15	8.68	6.36	5.42	4.89	4.56	4.32	4.14	4.00	3.89	3.80	3.67	3.52	3.37	3.29	3.21	3.13	3.05	2.96	2.87
16	8.53	6.23	5.29	4.77	4.44	4.20	4.03	3.89	3.78	3.69	3.55	3.41	3.26	3.18	3.10	3.02	2.93	2.84	2.75
17	8.40	6.11	5.18	4.67	4.34	4.10	3.93	3.79	3.68	3.59	3.46	3.31	3.16	3.08	3.00	2.92	2.83	2.75	2.65
18	8.29	6.01	5.09	4.58	4.25	4.01	3.84	3.71	3.60	3.51	3.37	3.23	3.03	3.00	2.92	2.84	2.75	2.66	2.57
19	8.18	5.93	5.01	4.50	4.17	3.94	3.77	3.63	3.52	3.43	3.30	3.15	3.00	2.92	2.84	2.70	2.67	2.58	2.49
20	8.10	5.85	4.94	4.43	4.10	3.87	3.70	3.56	3.46	3.37	3.23	3.09	2.94	2.86	2.78	2.69	2.61	2.52	2.42
21	8.02	5.78	4.87	4.37	4.04	3.81	3.64	3.51	3.40	3.31	3.17	3.03	2.88	2.80	2.72	2.64	2.55	2.46	2.36
22	7.95	5.72	4.82	4.31	3.99	3.76	3.59	3.45	3.35	3.26	3.12	2.98	2.83	2.75	2.67	2.58	2.50	2.40	2.31
23	7.88	5.66	4.76	4.26	3.94	3.71	3.54	3.41	3.30	3.21	3.07	2.93	2.78	2.70	2.62	2.54	2.45	2.35	2.26
24	7.82	5.61	4.72	4.22	3.90	3.67	3.50	3.36	3.26	3.17	3.03	2.89	2.74	2.66	2.58	2.49	2.40	2.31	2.21

TABLE 12c ***F* Distribution, Upper 1 Percent Points ($F_{0.975}$) (One-Sided)** ***(Continued)***

ν_2 \ ν_1	1	2	3	4	5	6	7	8	9	10	12	15	20	24	30	40	60	120	∞
25	7.77	5.57	4.68	4.18	3.85	3.63	3.46	3.32	3.22	3.13	2.99	2.85	2.70	2.62	2.54	2.45	2.36	2.27	2.17
26	7.72	5.53	4.64	4.14	3.82	3.59	3.42	3.29	3.18	3.09	2.96	2.81	2.66	2.58	2.50	2.42	2.33	2.23	2.13
27	7.68	5.49	4.60	4.11	3.78	3.56	3.39	3.26	3.15	3.06	2.93	2.78	2.63	2.55	2.47	2.38	2.29	2.20	2.10
28	7.64	5.45	4.57	4.07	3.75	3.53	3.36	3.23	3.12	3.03	2.90	2.75	2.60	2.52	2.44	2.35	2.26	2.17	2.06
29	7.60	5.42	4.54	4.04	3.73	3.50	3.33	3.20	3.09	3.00	2.87	2.73	2.57	2.49	2.41	2.33	2.23	2.14	2.03
30	7.56	5.39	4.51	4.02	3.70	3.47	3.30	3.17	3.07	2.93	2.84	2.70	2.55	2.47	2.39	2.30	2.21	2.11	2.01
40	7.31	5.18	4.31	3.83	3.51	3.29	3.12	2.99	2.89	2.80	2.63	2.52	2.37	2.29	2.20	2.11	2.02	1.92	1.80
60	7.08	4.98	4.13	3.65	3.34	3.12	2.95	2.82	2.72	2.63	2.50	2.35	2.20	2.12	2.03	1.94	1.84	1.73	1.60
120	6.85	4.79	3.95	3.48	3.17	2.96	2.79	2.66	2.56	2.47	2.34	2.19	2.03	1.95	1.86	1.76	1.66	1.53	1.38
∞	6.63	4.61	3.78	3.32	3.02	2.80	2.64	2.51	2.41	2.32	2.18	2.04	1.88	1.79	1.70	1.59	1.47	1.32	1.00

SOURCE: E. S. Pearson and H. O. Hartley, *Biometrika Tables for Statisticians,* 3d ed., University College, London, 1966.(Reproduced by permission of the Biometrika trustees.)

TABLE A.13 Critical Values of the Tukey-Duckworth Sum

Approximate risk	Two-sided critical values of the sum $a + b$	One-sided critical values of the sum
0.09	6	5
0.05	7	6
0.01	10	9
0.001	13	12

SOURCE: John W. Tukey, "A Quick, Compact, Two-sample Test to Duckworth's Specifications," *Technometrics,* vol. 1 no. 1, February 1959, pp. 21–48. (Reproduced by permission.) The critical 0.09 value was given by Peter C. Dickinson. One-sided values were contributed by Dr. Larry Rabinowitz.

TABLE A.14 Values of H_α, k = 2, ANOM (Two-Tailed Test)

df	$\alpha = 0.10$	0.05	0.01
2	2.06	3.04	7.02
3	1.66	2.25	4.13
4	1.51	1.96	3.26
5	1.42	1.82	2.85
6	1.37	1.73	2.62
7	1.34	1.67	2.47
8	1.32	1.63	2.37
9	1.30	1.60	2.30
10	1.28	1.58	2.24
12	1.26	1.54	2.16
15	1.24	1.51	2.08
18	1.23	1.49	2.04
20	1.22	1.48	2.01
25	1.21	1.46	1.97
30	1.20	1.44	1.94
40	1.19	1.43	1.91
60	1.18	1.41	1.88
∞	1.16	1.39	1.82

TABLE A.15 Distribution of Student's *t* (Two-Tail)

Values of t corresponding to selected probabilities. Each probability is the sum of two equal areas under the two tails of the t curve. For example, the probability is $0.05 = 2(0.025)$ that a difference with df = 20 would have $t \geq |2.09|$

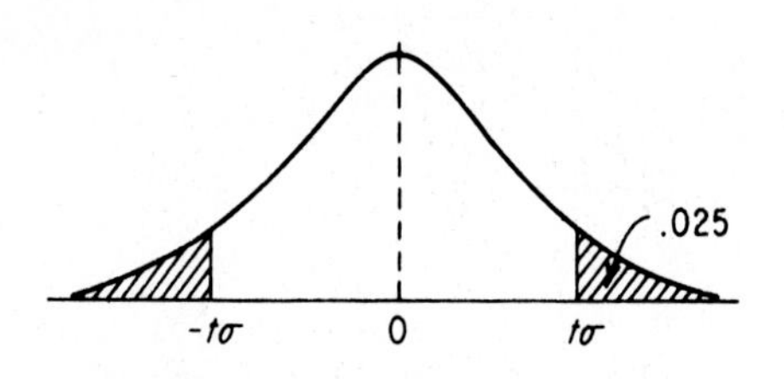

df	Probability .10	.05	.02	.01
6	1.94	2.45	3.14	3.71
7	1.90	2.37	3.00	3.50
8	1.86	2.31	2.90	3.35
9	1.83	2.26	2.82	3.25
10	1.81	2.23	2.76	3.17
11	1.80	2.20	2.72	3.11
12	1.78	2.18	2.68	3.06
13	1.77	2.16	2.65	3.01
14	1.76	2.15	2.62	2.98
15	1.75	2.13	2.60	2.95
20	1.73	2.09	2.52	2.85
25	1.70	2.06	2.49	2.79
30	1.70	2.04	2.46	2.75
50	1.68	2.01	2.40	2.68
∞	1.645	1.960	2.326	2.576

SOURCE: This table is a modification of the one by Enrico T. Federighi, "Extended Tables of the Percentage Points of Student's *t*-Distribution," *Jour. Am. Stat. Assoc.*, vol. 54, p. 684, 1959. (Reproduced by permission of ASA.)

TABLE A.16 Nonrandom Uniformity, N_α (No Standard Given)

	.05					.01			
k* \ df:	10	15	30	∞	k \ df:	10	15	30	∞
3	.20	.20	.20	.20	3	.09	.09	.09	.09
4	.35	.35	.35	.35	4	.19	.19	.20	.20
5	.46	.46	.46	.47	5	.29	.29	.29	.30
6	.55	.55	.56	.56	6	.37	.37	.38	.38
7	.62	.63	.64	.65	7	.43	.44	.45	.46
8	.69	.70	.70	.72	8	.49	.50	.51	.53
9	.74	.75	.77	.78	9	.54	.56	.57	.59

* k = number of means being compared

SOURCE: Nair, K. R., "The Distribution of the Extreme Deviate from the Sample Mean and its Studentized Form," *Biometrika,* vol. xxxv, 1948, pp. 118-144. (Reproduced by permission of the Biometrika trustees.)

TABLE A.17 Some Blocked Full Factorials

Design	0		1		2		2A		3	
Factors	2		3		4		4		5	
Blocks	2		2		2		4		2	
Runs	4		8		16		16		32	
	*B*1	*B*2	*B*1	*B*2	*B*1	*B*2	*B*1	*B*2	*B*1	*B*2
	1	b	1	*c*	1	*d*	1	*a*	1	*e*
	ab	*a*	*ac*	*a*	*ad*	*a*	*bc*	*abc*	*ae*	*a*
	AB		*bc*	*b*	*bd*	*b*	*abd*	*bd*	*be*	*b*
			ab	*abc*	*ab*	*abd*	*acd*	*cd*	*ab*	*abc*
			ABC		*cd*	*c*	*B*3	*B*4	*ce*	*c*
					ac	*acd*	*b*	*d*	*ac*	*ace*
					bc	*bcd*	*c*	*bcd*	*bc*	*bce*
					abcd	*abc*	*ad*	*ab*	*abce*	*abc*
					ABCD		*abcd*	*ac*	*de*	*d*
							AD, ABC, BCD		*ad*	*ade*
									bd	*bde*
									abde	*abd*
									cd	*cde*
									acde	*acd*
									bcde	*bcd*
									abcd	*abcde*
									ABCDE	

*B*1 = Block 1 *B*3 = Block 3
*B*2 = Block 2 *B*4 = Block 4
Interaction confounded with blocks shown at bottom of column

TABLE A.18 Some Fractional Factorials

Design	0		1		2		4		3	
Factors	2		3		4		5		5	
Fraction	½		½		½		¼		½	
Runs	2		4		8		8		16	
	TRT	EFF	TRT	EFF	TRT	EFF	TRT	EFF	TRT	EFF
	1	*T*	1	*T*	1	*T*	1	*T*	1	*T*
	a(b)	*A*=*B*	*a(c)*	*A*−*BC*	*a(d)*	*A*	*a(d)*	*A*−*DE*	*a(e)*	*A*
	I=*AB*		*b(c)*	*B*−*AC*	*b(d)*	*B*	*b(de)*	*B*−*CE*	*b(e)*	*B*
			ab	*AB*−*C*	*ab*	*AB*+*CD*	*ab(e)*	*AB*+*CD*	*ab*	*AB*
			I = −*ABC*		*c(d)*	*C*	*c(de)*	*C*−*BE*	*c(e)*	*C*
					ac	*AC*+*BD*	*ac(e)*	*AC*+*BD*	*ac*	*AC*
					bc	*BC*+*AD*	*bc*	−*E*+*BC*+*AD*	*bc*	*BC*
					abc(d)	*D*	*abc(d)*	*D*−*AE*	*abc(e)*	−*DE*
					I = *ABCD*		I = − *BCE* − *ADE* + *ABCD*		*d(e)*	*D*
									ad	*AD*
									bd	*BD*
									abd(e)	−*CE*
									cd	*CD*
									acd(e)	−*BE*
									bcd(e)	−*AE*
									abcd	−*E*
									I = −*ABCDE*	

TRT = Treatments in Yates order
EFF = Effect estimated for corresponding row in Yates (ignores higher than two factor interaction)

I = Defining contrast

TABLE A.19 Sidak Factors* for Analysis of Means for Treatment Effects, h_α^* (Two-Sided)

	Significance Level = 0.10																						
	Number of Means, k																						
df	2	3	4	5	6	7	8	9	10	11	12	13	14	15	16	17	18	19	20	24	30	40	60
2	2.920																						
3	2.353	3.690																					
4	2.132	3.150	3.452																				
5	2.015	2.882	3.129	3.327																			
6	1.943	2.723	2.939	3.110	3.253																		
7	1.895	2.618	2.814	2.969	3.097	3.206																	
8	1.860	2.544	2.726	2.869	2.987	3.088	3.176																
9	1.833	2.488	2.661	2.796	2.907	3.001	3.083	3.155															
10	1.812	2.446	2.611	2.739	2.845	2.934	3.012	3.080	3.142														
11	1.796	2.412	2.571	2.695	2.796	2.881	2.955	3.021	3.079	3.133													
12	1.782	2.384	2.539	2.658	2.756	2.838	2.910	2.973	3.029	3.080	3.127												
13	1.771	2.361	2.512	2.628	2.723	2.803	2.872	2.933	2.988	3.037	3.082	3.124											
14	1.761	2.342	2.489	2.603	2.696	2.774	2.841	2.900	2.953	3.001	3.045	3.085	3.122										
15	1.753	2.325	2.470	2.582	2.672	2.748	2.814	2.872	2.924	2.970	3.013	3.052	3.088	3.122									
16	1.746	2.311	2.453	2.563	2.652	2.726	2.791	2.848	2.898	2.944	2.985	3.024	3.059	3.092	3.122								
17	1.740	2.298	2.439	2.547	2.634	2.708	2.771	2.826	2.876	2.921	2.962	2.999	3.034	3.066	3.096	3.124							
18	1.734	2.287	2.426	2.532	2.619	2.691	2.753	2.808	2.857	2.901	2.941	2.977	3.011	3.043	3.072	3.100	3.126						
19	1.729	2.277	2.415	2.520	2.605	2.676	2.738	2.791	2.839	2.883	2.922	2.958	2.992	3.023	3.052	3.079	3.104	3.129					
20	1.725	2.269	2.405	2.508	2.593	2.663	2.724	2.777	2.824	2.867	2.906	2.941	2.974	3.005	3.033	3.060	3.085	3.109	3.132				
24	1.711	2.241	2.373	2.473	2.554	2.622	2.680	2.731	2.777	2.818	2.855	2.889	2.920	2.949	2.976	3.002	3.026	3.048	3.070	3.146			
30	1.697	2.215	2.342	2.439	2.517	2.582	2.638	2.687	2.731	2.770	2.805	2.838	2.868	2.895	2.921	2.946	2.968	2.990	3.010	3.082	3.169		
40	1.684	2.189	2.312	2.406	2.481	2.544	2.597	2.644	2.686	2.723	2.757	2.788	2.817	2.843	2.868	2.891	2.913	2.933	2.952	3.021	3.103	3.208	
60	1.671	2.163	2.283	2.373	2.446	2.506	2.558	2.603	2.643	2.678	2.711	2.740	2.768	2.793	2.816	2.838	2.859	2.878	2.897	2.962	3.040	3.139	3.276
120	1.658	2.138	2.254	2.342	2.411	2.469	2.519	2.562	2.600	2.635	2.666	2.694	2.720	2.744	2.766	2.787	2.807	2.826	2.843	2.904	2.978	3.072	3.201
Inf	1.645	2.114	2.226	2.311	2.378	2.434	2.481	2.523	2.560	2.592	2.622	2.649	2.674	2.697	2.718	2.738	2.757	2.774	2.791	2.849	2.920	3.008	3.129

TABLE A.19 Sidak Factors* for Analysis of Means for Treatment Effects, h_α^* (Two-Sided) *(Continued)*

	Significance Level = 0.05																						
	Number of Means, k																						
df	2	3	4	5	6	7	8	9	10	11	12	13	14	15	16	17	18	19	20	24	30	40	60
2	4.303																						
3	3.182	4.826																					
4	2.776	3.941	4.290																				
5	2.571	3.518	3.791	4.012																			
6	2.447	3.274	3.505	3.690	3.845																		
7	2.365	3.115	3.321	3.484	3.620	3.736																	
8	2.306	3.005	3.193	3.342	3.464	3.569	3.661																
9	2.262	2.923	3.099	3.237	3.351	3.448	3.532	3.607															
10	2.228	2.860	3.027	3.157	3.264	3.355	3.434	3.505	3.568														
11	2.201	2.811	2.970	3.094	3.196	3.283	3.358	3.424	3.484	3.538													
12	2.179	2.770	2.924	3.044	3.141	3.224	3.296	3.359	3.416	3.468	3.515												
13	2.160	2.737	2.886	3.002	3.095	3.176	3.245	3.306	3.360	3.410	3.455	3.497											
14	2.145	2.709	2.854	2.967	3.058	3.135	3.202	3.261	3.314	3.362	3.406	3.446	3.483										
15	2.131	2.685	2.827	2.937	3.026	3.101	3.166	3.224	3.275	3.321	3.364	3.402	3.439	3.472									
16	2.120	2.665	2.804	2.911	2.998	3.072	3.135	3.191	3.241	3.286	3.327	3.365	3.400	3.433	3.464								
17	2.110	2.647	2.783	2.889	2.974	3.046	3.108	3.163	3.212	3.256	3.296	3.333	3.367	3.399	3.429	3.457							
18	2.101	2.631	2.766	2.869	2.953	3.024	3.085	3.138	3.186	3.229	3.269	3.305	3.338	3.370	3.399	3.426	3.452						
19	2.093	2.617	2.750	2.852	2.934	3.004	3.064	3.116	3.163	3.206	3.245	3.280	3.313	3.343	3.372	3.399	3.424	3.448					
20	2.086	2.605	2.736	2.836	2.918	2.986	3.045	3.097	3.143	3.185	3.223	3.258	3.290	3.320	3.348	3.375	3.399	3.423	3.445				
24	2.064	2.566	2.692	2.788	2.866	2.931	2.988	3.037	3.081	3.121	3.157	3.190	3.220	3.249	3.275	3.300	3.323	3.345	3.366	3.440			
30	2.042	2.528	2.649	2.742	2.816	2.878	2.932	2.979	3.021	3.058	3.092	3.124	3.153	3.180	3.205	3.228	3.250	3.271	3.291	3.360	3.445		
40	2.021	2.492	2.608	2.696	2.768	2.827	2.878	2.923	2.953	2.998	3.031	3.060	3.088	3.113	3.137	3.159	3.180	3.199	3.218	3.283	3.363	3.464	
60	2.000	2.456	2.568	2.653	2.721	2.777	2.826	2.869	2.906	2.940	2.971	2.999	3.025	3.049	3.071	3.092	3.112	3.130	3.148	3.210	3.284	3.379	3.511
120	1.980	2.422	2.529	2.610	2.675	2.729	2.776	2.816	2.852	2.884	2.913	2.940	2.965	2.987	3.008	3.028	3.047	3.064	3.081	3.139	3.209	3.298	3.421
Inf	1.960	2.388	2.491	2.569	2.631	2.683	2.727	2.766	2.800	2.830	2.858	2.883	2.906	2.928	2.948	2.966	2.984	3.000	3.016	3.071	3.137	3.220	3.335

Significance Level = 0.01																							
	Number of Means, k																						
df	2	3	4	5	6	7	8	9	10	11	12	13	14	15	16	17	18	19	20	24	30	40	60
2	9.925																						
3	5.841	8.565																					
4	4.604	6.248	6.752																				
5	4.032	5.243	5.599	5.888																			
6	3.707	4.695	4.977	5.204	5.394																		
7	3.499	4.353	4.591	4.782	4.941	5.078																	
8	3.355	4.120	4.331	4.498	4.637	4.756	4.860																
9	3.250	3.952	4.143	4.294	4.419	4.526	4.619	4.703															
10	3.169	3.825	4.002	4.141	4.256	4.354	4.439	4.516	4.584														
11	3.106	3.726	3.892	4.022	4.129	4.221	4.300	4.371	4.434	4.492													
12	3.055	3.647	3.804	3.927	4.029	4.115	4.189	4.256	4.315	4.369	4.419												
13	3.012	3.582	3.733	3.850	3.946	4.028	4.099	4.162	4.218	4.270	4.317	4.360											
14	2.977	3.528	3.673	3.785	3.878	3.956	4.024	4.084	4.138	4.187	4.232	4.273	4.311										
15	2.947	3.482	3.622	3.731	3.820	3.895	3.961	4.019	4.071	4.117	4.160	4.200	4.237	4.271									
16	2.921	3.443	3.579	3.684	3.771	3.844	3.907	3.963	4.013	4.058	4.100	4.138	4.173	4.206	4.237								
17	2.898	3.409	3.541	3.644	3.728	3.799	3.860	3.914	3.963	4.007	4.047	4.084	4.118	4.150	4.180	4.208							
18	2.878	3.379	3.508	3.609	3.691	3.760	3.820	3.872	3.920	3.962	4.001	4.037	4.071	4.102	4.131	4.158	4.184						
19	2.861	3.353	3.480	3.578	3.658	3.725	3.784	3.835	3.881	3.923	3.961	3.996	4.029	4.059	4.087	4.114	4.139	4.162					
20	2.845	3.329	3.454	3.550	3.629	3.695	3.752	3.803	3.848	3.888	3.926	3.960	3.992	4.021	4.049	4.074	4.099	4.122	4.144				
24	2.797	3.257	3.375	3.465	3.539	3.601	3.654	3.702	3.744	3.782	3.816	3.848	3.878	3.905	3.931	3.955	3.977	3.999	4.019	4.091			
30	2.750	3.188	3.298	3.384	3.453	3.511	3.561	3.605	3.644	3.680	3.712	3.742	3.769	3.794	3.818	3.840	3.861	3.881	3.900	3.966	4.048		
40	2.704	3.121	3.225	3.305	3.370	3.425	3.472	3.513	3.549	3.582	3.612	3.640	3.665	3.689	3.711	3.732	3.751	3.769	3.787	3.848	3.923	4.019	
60	2.660	3.056	3.155	3.230	3.291	3.342	3.386	3.425	3.459	3.489	3.517	3.543	3.567	3.589	3.609	3.628	3.646	3.663	3.679	3.736	3.805	3.893	4.016
120	2.617	2.994	3.087	3.158	3.215	3.263	3.304	3.340	3.372	3.401	3.427	3.450	3.472	3.493	3.512	3.530	3.546	3.562	3.577	3.629	3.693	3.774	3.886
Inf	2.576	2.934	3.022	3.089	3.143	3.188	3.226	3.260	3.289	3.316	3.340	3.362	3.383	3.402	3.419	3.436	3.451	3.466	3.480	3.528	3.587	3.661	3.764

TABLE A.19 Sidak Factors* for Analysis of Means for Treatment Effects, h_α^* (Two-Sided) *(Continued)*

	Significance Level = 0.001																						
	Number of Means, k																						
df	2	3	4	5	6	7	8	9	10	11	12	13	14	15	16	17	18	19	20	24	30	40	60
2	31.599																						
3	12.924	18.706																					
4	8.610	11.438	12.311																				
5	6.869	8.692	9.234	9.677																			
6	5.959	7.314	7.708	8.025	8.292																		
7	5.408	6.502	6.813	7.063	7.272	7.452																	
8	5.041	5.973	6.234	6.442	6.616	6.765	6.896																
9	4.781	5.602	5.829	6.010	6.160	6.288	6.402	6.503															
10	4.587	5.329	5.533	5.694	5.827	5.941	6.042	6.131	6.211														
11	4.437	5.120	5.306	5.453	5.574	5.677	5.768	5.849	5.921	5.986													
12	4.318	4.955	5.128	5.263	5.375	5.470	5.554	5.628	5.694	5.755	5.811												
13	4.221	4.822	4.983	5.110	5.215	5.304	5.381	5.450	5.512	5.568	5.620	5.668											
14	4.140	4.712	4.865	4.985	5.084	5.167	5.240	5.305	5.363	5.416	5.465	5.510	5.551										
15	4.073	4.620	4.766	4.880	4.974	5.053	5.122	5.184	5.239	5.289	5.335	5.377	5.416	5.452									
16	4.015	4.542	4.682	4.791	4.881	4.956	5.023	5.081	5.134	5.182	5.226	5.266	5.303	5.338	5.371								
17	3.965	4.474	4.609	4.714	4.800	4.873	4.937	4.993	5.044	5.089	5.131	5.170	5.206	5.239	5.271	5.300							
18	3.922	4.416	4.546	4.648	4.731	4.802	4.863	4.918	4.966	5.010	5.050	5.088	5.122	5.154	5.184	5.212	5.239						
19	3.883	4.365	4.491	4.590	4.671	4.738	4.798	4.850	4.897	4.939	4.979	5.015	5.048	5.079	5.108	5.135	5.161	5.185					
20	3.850	4.319	4.442	4.539	4.617	4.683	4.741	4.792	4.838	4.878	4.916	4.951	4.983	5.014	5.042	5.068	5.094	5.117	5.139				
24	3.745	4.181	4.294	4.382	4.454	4.514	4.567	4.613	4.654	4.692	4.726	4.758	4.787	4.814	4.839	4.863	4.886	4.908	4.928	5.000			
30	3.646	4.049	4.153	4.234	4.300	4.355	4.402	4.445	4.482	4.517	4.547	4.576	4.603	4.627	4.650	4.672	4.692	4.711	4.730	4.795	4.874		
40	3.551	3.925	4.020	4.094	4.154	4.205	4.248	4.287	4.321	4.351	4.380	4.405	4.430	4.451	4.472	4.492	4.510	4.528	4.543	4.602	4.672	4.763	
60	3.460	3.806	3.894	3.962	4.017	4.063	4.102	4.138	4.169	4.197	4.222	4.246	4.267	4.287	4.306	4.325	4.340	4.356	4.371	4.423	4.486	4.569	4.683
120	3.374	3.694	3.775	3.837	3.888	3.929	3.965	3.997	4.025	4.051	4.074	4.095	4.114	4.133	4.150	4.165	4.179	4.195	4.208	4.255	4.312	4.384	4.488
Inf	3.291	3.588	3.662	3.719	3.765	3.803	3.836	3.865	3.891	3.914	3.935	3.954	3.972	3.988	4.004	4.018	4.031	4.044	4.056	4.099	4.150	4.215	4.306

*For unequal sample sizes plot individual limits with $h_\alpha^{**} = h_\alpha^* \sqrt{(n - n_g))/[n - (n/k)]}$ and $h_\alpha^{**} = h_\alpha^* \sqrt{(n - n_g)/n}$. The values for $k \geq 3$ in this table are upper bounds for the studentized imum absolute deviation from the population mean as calculated by E. G. Schilling and D. Smialek ("Simplified Analysis of Means for Crossed and Nested Experiments," *Proc. 43d Ann. Qual. Control Conf.*, Rochester Section, ASQC, March 10, 1987) from the inequality of Z. Sidak ("Rectangular Confidence Regions for the Means of Multivariate Normal Distributions," *J. Amer. Stat. Assoc.*, vol. 62, 1967, pp. 626–633) as suggested by L. S. Nelson ("Exact Critical Values for Use with the Analysis of Means," *J. Qual. Tech.*, vol. 15, no. 1, January 1983, pp. 40–44).

Index

ABOUT THE AUTHORS

The late **Ellis R. Ott** was Professor Emeritus of Experimental Statistics at Rutgers, The State University of New Jersey, and the Founding Director of the Rutgers Statistics Center. He received his Ph.D. from the University of Illinois. Dr. Ott was the recipient of numerous quality control awards, including honorary member of the American Society for Quality Control and its Brumbaugh Award, the Eugene L. Grant Award, and the Shewhart Medal. He was honored by an award established in his name by the Metropolitan Section of ASQC.

Dr. Edward G. Schilling is the Paul A. Miller Distinguished Professor and Chairman of the Graduate Statistics Department at Rochester Institute of Technology and the former Manager of the Lighting Quality Operation of the General Electric Company. He received his M.S. and Ph.D. degrees in statistics from Rutgers University, where he studied with Ellis Ott. Dr. Schilling is a fellow of the American Statistical Association and the American Society for Quality Control, and is the first person to win the American Society for Quality Control's Brumbaugh Award four times. He is also a recipient of the Shewhart Medal and the Ellis R. Ott award. He is the author of *Acceptance Sampling in Quality Control*.